# The Afar Volcanic Province within the East African Rift System

## Geological Society books refereeing procedures

The Society makes every effort to ensure that the scientific and production quality of its books matches that of its journals. Since 1997, all book proposals have been refereed by specialist reviewers as well as by the Society's Books Editorial Committee. If the referees identify weaknesses in the proposal, these must be addressed before the proposal is accepted.

Once the book is accepted, the Society Book Editors ensure that the volume editors follow strict guidelines on refereeing and quality control. We insist that individual papers can only be accepted after satisfactory review by two independent referees. The questions on the review forms are similar to those for *Journal of the Geological Society*. The referees' forms and comments must be available to the Society's Book Editors on request.

Although many of the books result from meetings, the editors are expected to commission papers that were not presented at the meeting to ensure that the book provides a balanced coverage of the subject. Being accepted for presentation at the meeting does not guarantee inclusion in the book.

More information about submitting a proposal and producing a book for the society can be found on its web site: www.geolsoc.org.uk.

It is recommended that reference to all or part of this book should be made in one of the following ways:

YIRGU, G., EBINGER, C.J. & MAGUIRE, P.K.H. (eds) 2006. *The Afar Volcanic Province within the East African Rift System*. Geological Society, London, Special Publications, **259**.

GARFUNKEL, Z. & BEYTH, M. 2006. Constraints on the structural development of Afar imposed by the kinematics of the major surrounding plates. *In*: YIRGU, G., EBINGER, C.J. & MAGUIRE, P.K.H. (eds) 2006. *The Afar Volcanic Province within the East African Rift System*. Geological Society, London, Special Publications, **259**, 23–42.

GEOLOGICAL SOCIETY SPECIAL PUBLICATION NO. 259

# The Afar Volcanic Province within the East African Rift System

EDITED BY

G. YIRGU
University of Addis Ababa, Ethiopia

C. J. EBINGER
Royal Holloway, University of London, UK

and

P. K. H. MAGUIRE
University of Leicester, UK

2006
Published by
The Geological Society
London

# THE GEOLOGICAL SOCIETY

The Geological Society of London (GSL) was founded in 1807. It is the oldest national geological society in the world and the largest in Europe. It was incorporated under Royal Charter in 1825 and is Registered Charity 210161.

The Society is the UK national learned and professional society for geology with a worldwide Fellowship (FGS) of 9000. The Society has the power to confer Chartered status on suitably qualified Fellows, and about 2000 of the Fellowship carry the title (CGeol). Chartered Geologists may also obtain the equivalent European title, European Geologist (EurGeol). One fifth of the Society's fellowship resides outside the UK. To find out more about the Society, log on to www.geolsoc.org.uk.

**The Geological Society Publishing House** (Bath, UK) produces the Society's international journals and books, and acts as European distributor for selected publications of the American Association of Petroleum Geologists (AAPG), the American Geological Institute (AGI), the Indonesian Petroleum Association (IPA), the Geological Society of America (GSA), the Society for Sedimentary Geology (SEPM) and the Geologists' Association (GA). Joint marketing agreements ensure that GSL Fellows may purchase these societies' publications at a discount. The Society's online bookshop (accessible from www.geolsoc.org.uk) offers secure book purchasing with your credit or debit card.

To find out about joining the Society and benefiting from substantial discounts on publications of GSL and other societies worldwide, consult www.geolsoc.org.uk, or contact the Fellowship Department at: The Geological Society, Burlington House, Piccadilly, London W1J 0BG: Tel. +44 (0)20 7434 9944; Fax +44 (0)20 7439 8975; E-mail: enquiries@geolsoc.org.uk.

For information about the Society's meetings, consult *Events* on www.geolsoc.org.uk. To find out more about the Society's Corporate Affiliates Scheme, write to enquiries@geolsoc.org.uk.

Published by The Geological Society from:
The Geological Society Publishing House, Unit 7, Brassmill Enterprise Centre, Brassmill Lane, Bath BA1 3JN, UK

(*Orders*: Tel. +44 (0)1225 445046, Fax +44 (0)1225 442836)
Online bookshop: www.geolsoc.org.uk/bookshop



**British Library Cataloguing in Publication Data**

A catalogue record for this book is available from the British Library.

ISBN-10 1-86239-196-3

ISBN-13 978-1-86239-196-3



Typeset by Techset Composition, Salisbury, UK

Printed by MPG Books Ltd, Bodmin, UK.

**Distributors**

***USA***
AAPG Bookstore, PO Box 979, Tulsa, OK 74101-0979, USA
*Orders*: Tel. +1 918 584-2555
Fax +1 918 560-2652
E-mail bookstore@aapg.org

***India***
Affiliated East-West Press Private Ltd, Marketing Division, G-1/16 Ansari Road, Darya Ganj, New Delhi 110 002, India
*Orders*: Tel. +91 11 2327-9113/2326-4180
Fax +91 11 2326-0538
E-mail affiliat@vsnl.com

***Japan***
Kanda Book Trading Company, Cityhouse Tama 204, Tsurumaki 1-3-10, Tama-shi, Tokyo 206-0034, Japan
*Orders*: Tel. +81 (0)423 57-7650
Fax +81 (0)423 57-7651
E-mail geokanda@ma.kcom.ne.jp

# Contents

Preface vii

YIRGU, G., EBINGER, C.J. & MAGUIRE, P.K.H. The Afar volcanic province within the East African Rift System: introduction 1

**Part 1: Plate kinematic and geodynamic framework of the Afar volcanic province**
Introduction 7

CALAIS, E., EBINGER, C.J., HARTNADY, C., & NOCQUET, J.M. Kinematics of the East African Rift from GPS and earthquake slip vector data 9

GARFUNKEL, Z. & BEYTH, M. Constraints on the structural development of Afar imposed by the kinematics of the major surrounding plates 23

BUCK, W.R. The role of magma in the development of the Afro-Arabian Rift System 43

KENDALL, J.-M., PILIDOU, S., KEIR, D., BASTOW, I.D., STUART, G.W. & AYELE, A. Mantle upwellings, melt migration and the rifting of Africa: insights from seismic anisotropy 55

**Part 2: Geochemical constraints on flood basalt and rift processes**
Introduction 73

ROGERS, N.W. Basaltic magmatism and the geodynamics of the East African Rift System 77

FURMAN, T., BRYCE, J., ROONEY, T., HANAN, B., YIRGU, G. & AYALEW, D. Heads and tails: 30 million years of the Afar plume 95

AYALEW, D., EBINGER, C., BOURDON, E., WOLFENDEN, E., YIRGU, G. & GRASSINEAU, N. Temporal compositional variation of syn-rift rhyolites along the western margin of the southern Red Sea and northern Main Ethiopian Rift 121

**Part 3: Rifting in the Afar volcanic province: Modelling and kinematics**
Introduction 131

AYELE, A. NYBLADE, A.A., LANGSTON, C.A., CARA, M. & LEVEQUE, J.-J. New evidence for Afro-Arabian plate separation in southern Afar 133

CASEY, M., EBINGER, C., KEIR, D., GLOAGUEN, R. & MOHAMED, F. Strain accommodation in transitional rifts: extension by magma intrusion and faulting in Ethiopian rift magmatic segments 143

KIDANE, T., PLATZMAN, E., EBINGER, C., ABEBE, B. & ROCHETTE, P. Palaeomagnetic constraints on continental break-up processes: observations from the Main Ethiopian Rift 165

ASFAW, L.M., BEYENE, H., MKONNEN, A. & OLI, T. Vertical deformation in the Main Ethiopian Rift: levelling results in its northern part, 1995–2004 185

PIZZI, A., COLTORTI, M., ABEBE, B., DISPERATI, L., SACCHI, G., & SALVINI, R. The Wonji fault belt (Main Ethiopian Rift): structural and geomorphological constraints and GPS monitoring 191

VETEL, W. & LE GALL, B. Dynamics of prolonged continental extension in magmatic rifts: the Turkana Rift case study (North Kenya) 209

**Part 4: Rifting in the Afar volcanic province: Geophysical studies of crustal structure and processes**
Introduction 235

DUGDA, M.T. & NYBLADE, A. New constraints on crustal structure in eastern Afar from the analysis of receiver functions and surface wave dispersion in Djibouti 239

STUART, G.W., BASTOW, I.D. & EBINGER, C.J. Crustal structure of the northern Main Ethiopian Rift from receiver function studies 253

MAGUIRE, P.K.H., KELLER, G.R., KLEMPERER, S.L., MACKENZIE, G.D., KERANEN, K., HARDER, S., O'REILLY, B., THYBO, H., ASFAW, L., KHAN, M.A. & AMHA, M. Crustal structure of the northern Main Ethiopian Rift from the EAGLE controlled-source survey; a snapshot of incipient lithospheric break-up 269

WHALER, K.A. & HAUTOT, S. The electrical resistivity structure of the crust beneath the northern Main Ethiopian Rift 293

CORNWELL, D.G., MACKENZIE, G.D., ENGLAND, R.W., MAGUIRE, P.K.H., ASFAW, L.M. & OLUMA, B. Northern Main Ethiopian Rift crustal structure from new high-precision gravity data 307

Index 323

# Preface

A full understanding of the structure and evolution of the Afar volcanic province requires a number of approaches to be applied, including geophysics, geochemistry, structure, geomorphology and other geoscience disciplines. Adopting this philosophy, we have assembled this collection of papers with the objective of providing an integrated study of the continental rupture processes above asthenospheric upwellings. This special publication of the Geological Society was inspired by an international conference entitled 'The East African Rift System: Geodynamics, Resources and Environment' held in Addis Ababa, Ethiopia, in June 2004. At this meeting, organized by the Ethiopian Geoscience and Mineral Engineering Association, more than 100 geoscientists were treated to 66 presentations on a broad range of topics, including rift geodynamics, geophysics, tectonics, magmatism, sedimentation, environment, geohazards and resources. A number of these papers are included in this volume, which covers various aspects of the deep structure, tectonic and magmatic evolution of the Afar volcanic province (see Introductions to Parts for summaries). The theme reflects a burgeoning interest in the international geoscience concerning the continental rifting and break-up processes associated with a mantle plume. We believe that the papers will help to unify some of the more fragmented aspects of previous research. Also, and in particular, the volume includes research outcomes from the recent Ethiopia Afar Geoscientific Lithospheric Experiment undertaken over the Northern Main Ethiopian Rift, which is believed to represent the transition between continental rifting and seafloor spreading. The results provide details about the structure and physical properties of the crust and upper mantle that have important implications concerning the geodynamics and magmatic evolution of the rift.

We wish to thank all the authors who have submitted manuscripts for their interest and patience in this volume's production. We would like to gratefully acknowledge reviews by the following colleagues: Massimo D'Antonio, Nicolas Bellahsen, Becky Bendick, Colin Brown, Jon Bull, Eric Calais, Lucy Flesch, Mike Fuller, Tanya Furman, Volker Haak, Zvi Garfunkel, Derek Keir, Randy Keller, Rainer Kind, Bernard LeGall, Sylvie Leroy, Dan Lizarralde, Ray MacDonald, Walter Mooney, Andy Nyblade, Angelo Peccerillo, Tim Redfield, Nick Rogers, Julie Rowland, Andy Saunders, Hans Schouten, Chris Swain, Christel Tiberi, Andrea Tommasi, Trond Torsvik, Christophe Vigny, Dave Wilson, Giday WoldeGabriel, anonymous reviewers, and our series editor, Bob Holdsworth.

The Ethiopian Geoscience and Mineral Engineering Association expresses its gratitude to the following institutions for their collaboration and support in organizing the conference: IGCP 482/489 (International Geological Correlation Program), UNESCO (United Nations Education, Scientific and Cultural Organisation), ILP (International Lithosphere Program), IASPEI (International Association of Seismology and Physics of the Earth's Interior), GSAf (Geological Society of Africa), Project EAGLE (Ethiopian Afar Geoscientific Lithospheric Experiment), The Ethiopian Science and Technology Commission, Addis Ababa University, Midroc Goldmine Plc, Petronas Plc, Si-Tech Plc, Southwest Development Plc, and K and S Plc.

G. Yirgu,
C.J. Ebinger
and P.K.H. Maguire

# The Afar volcanic province within the East African Rift System: introduction

G. YIRGU[1], C.J. EBINGER[2] & P.K.H. MAGUIRE[3]

[1]*Department of Geology and Geophysics, University of Addis Ababa, PO Box 176, Addis Ababa, Ethiopia (e-mail: yirgu.g@geol.aau.edu.et)*

[2]*Department of Geology, Royal Holloway, University of London, Egham, TW20 0EX, UK (e-mail: c.ebinger@gl.rhul.ac.uk)*

[3]*Department of Geology, University of Leicester, Leicester LE1 7RH, UK (e-mail: pkm@le.ac.uk)*

Continental rifting processes continually reshape the Earth's surface, producing sediment-filled rift basins, or rupturing the tectonic plates to form new ocean basins. Rift architecture and tectonics focus volcanic and seismic hazards, as well as geothermal energy resources, while rift systems in Africa have controlled faunal dispersal patterns and influenced human evolution in the past. The response of a plate to extension and heating provides fundamental clues into the plate rheology, and the underlying mantle convection patterns. A number of models have been proposed to explain the success and failure of continental rift zones, but there remains no consensus on how strain localizes to achieve rupture of initially 125–250 km-thick plates, or the interaction between the plates and asthenospheric processes.

The seismically and volcanically active East African rift system has long been a classic area for investigating rifting and break-up because its sectors encompass basins in all stages of rift to passive margin development. Its architecture is defined on the basis of both structural and magmatic components. It extends 3000 km from the Afar depression in the north to the Okavango Delta in the south, through Djibouti, Ethiopia, Kenya, Uganda, Burundi, Rwanda, Democratic Republic of Congo, Tanzania, Zambia, Malawi, Zimbabwe and Botswana. The East African rift system overlies one of the most extensive seismic velocity anomalies in the Earth's mantle, extending from the core–mantle boundary beneath the South Atlantic into the upper mantle beneath East Africa (Grand *et al.* 1997; van der Hilst & Karason 1999; Ritsema & van Heijst 2000). The rift system developed within Archaean–Proterozoic continental lithosphere, providing a unique opportunity to examine the mechanical response of strong, cold lithosphere to extension induced by asthenospheric upwelling and far-field forces.

Over 1 000 000 $km^3$ of basalts and more minor rhyolites cover the *c.* 1000 km-wide Ethiopia–Yemen plateau, which has experienced a maximum basement uplift of *c.* 1600 m above the mean altitude of the African plate to the west (Pik *et al.* 2003). The Ethiopian Rift, which transects the plateau, forms the third arm of the Red Sea–Gulf of Aden triple junction within the Eocene–Oligocene so-called Afar volcanic province (also known as the Ethiopia–Yemen flood basalt province) (Fig. 1). Thus, the region records both the evolution of Earth's youngest flood basalt province, its most youthful passive continental margins, as well as archetypal continental rift zones.

## The Afar volcanic province

The uplifted Ethiopia–Yemen plateau has experienced episodes of volcanism since *c.* 45 Ma, with eruptions continuing to the present day. Earliest volcanism in the East African rift system occurred in southwestern Ethiopia, southeastern Sudan, and northern Kenya at 40–45 Ma (Davidson & Rex 1980; Ebinger *et al.* 1993; George *et al.* 1998). Between *c.* 31 and 22 Ma, volcanism was widespread throughout central Ethiopia, Eritrea and Yemen where flood basalts and associated felsic pyroclastic rocks were erupted (Zumbo *et al.* 1995; Baker *et al.* 1996*b*; Hofmann *et al.* 1997; Kieffer *et al.* 2004). The thickest sequences exposed along the conjugate margins of the southern Red Sea were erupted between 31 and 29 Ma, prior to or concurrent with the onset of rifting (e.g. Hofmann *et al.* 1997; Ukstins *et al.* 2002; Wolfenden *et al.* 2005). Following the voluminous eruptions of flood basalts and associated felsic rocks, large shield volcanoes developed on the uplifted plateau hundreds of kilometres

*From*: YIRGU, G., EBINGER, C.J. & MAGUIRE, P.K.H. (eds) 2006. *The Afar Volcanic Province within the East African Rift System*. Geological Society, London, Special Publications, **259**, 1–6.
0305-8719/06/$15.00 

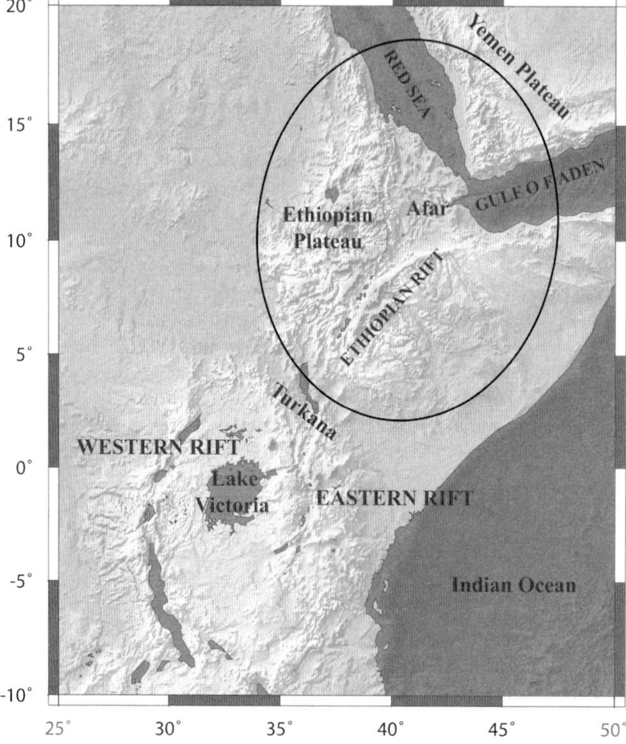


**Fig. 1.** East Africa and the main elements of the northern part of the East African rift system. The approximate extent of the Afar volcanic province is shown by the ellipse.

outside the faulted rift valleys (Kieffer *et al.* 2004). Red Sea and Aden rifting culminated in seafloor spreading at 4 and 16 Ma, respectively, effectively isolating part of the volcanic plateau in Yemen and Saudi Arabia (Chazot & Bertrand 1993; Baker *et al.* 1996*a*; Menzies *et al.* 2001). Recent volcanism is mainly confined to magmatic segments within the Ethiopian rift, and incipient spreading zones in Afar (e.g. Vidal *et al.* 1991; Schilling *et al.* 1992; Boccaletti *et al.* 1998). Thus, the Afar volcanic province has experienced a long, complex history of basaltic and explosive felsic volcanism since 45 Ma, with the volumetrically largest eruptions tied to the onset of rifting of Africa and Arabia at *c.* 30 Ma.

## Regional context of Cenozoic rifting

Although far from complete, systematic dating of volcanic and tectonic activity provides strong constraints on how the mantle source of the Afar volcanic province interacted with the lithosphere, and how this interaction was linked to rifting. Continental rifting in the Gulf of Aden initiated at *c.* 35 Ma, just prior to the eruption of the Ethiopia–Yemen flood basalts and associated felsic rocks at *c.* 30 Ma (e.g. Watchorn *et al.* 1998; d'Acremont *et al.* 2005). Rifting in the Red Sea began at 28 Ma, soon after the peak flood basalt volcanism (e.g. Wolfenden *et al.* 2005). The earliest extension documented in the East African rift system occurred in inactive basins west of present-day Lake Turkana at *c.* 25 Ma, within lithosphere previously stretched during a Mesozoic rifting episode (Morley *et al.* 1992; Hendrie *et al.* 1994). Extension between the Nubian and Somalian plates began in the southern and central sectors of the Main Ethiopian Rift by *c.* 18 Ma (WoldeGabriel *et al.* 1990) whereas in the northern sector it started at *c.* 11 Ma, more than 17 Ma after initial rifting in the southern Red Sea and Gulf of Aden rifts (Wolfenden *et al.* 2004).

The junction between the Red Sea and Gulf of Aden oceanic spreading centres is currently located within the complex Afar depression west of the Danakil microplate (Acton *et al.* 2000; Ebinger & Casey 2001). These data and observations have led to the proposal that the East African rift has propagated northward across older Red Sea and Aden structure, to form the Afar triple junction by *c.* 11 Ma (Tesfaye *et al.* 2003; Wolfenden *et al.* 2004). Global plate motions averaged over the last 3.2 Ma, and current geodetic and fault slip data predict an extensional velocity of around 4–6 mm $a^{-1}$ and an extension direction of *c.* 105–109° in the MER (Fig. 1, Calais *et al.* 2005; Bilham *et al.* 1999; Chu & Gordon 1999; Keir *et al.* 2005). Lemaux *et al.* (2002) used seafloor spreading anomaly patterns to deduce 23 km of opening along the Nubia and Somalia plates over the past 11 Ma with a change from NW–SE to approximately E–W extension at *c.* 3.2 Ma. Inversions of teleseismic events show present-day E–W to WNW–ESE extension directions (Ayele & Kulhanek 1997; Foster & Jackson 1999) consistent with the predictions of plate kinematic data. This realignment of the extensional stress direction is coincident with the development of the right-stepping en echelon magmatic centres along the axis of the Northern Main Ethiopian Rift and into southwestern Afar.

## Models of continental break-up

Observations within the active rift systems of the Afar volcanic province document the key processes of continental break-up: weakening of the lithosphere by mechanical stretching, intrusive heating and interactions with a dynamic asthenosphere. The relative importance of these processes remains debated, in large part owing to selective studies of non-volcanic passive margins and rift zones. Lister (1986) and Davis & Kusznir (2004) suggest that a lateral offset in the high strain zones within the upper and lower crust can be interpreted as evidence for lithospheric or crustal-scale detachment faults accommodating strain until break-up occurs. Buck (2004) uses numerical models to demonstrate the fundamental role of magma injection in strain accommodation, noting that lithospheric rupture can occur at a fraction of the tectonic driving force, where magma is available. Ebinger & Casey (2001) propose that magma injection localizes strain as rifting proceeds to break-up, leading to the abandonment of early syn-rift detachment faults. Dunbar & Sawyer (1989) use numerical models to demonstrate the role of zones of pre-existing lithospheric weakness in the localization of strain at break-up, with or without the development of lithospheric-scale detachments. Taylor *et al.* (1999) and Hebert *et al.* (2001) document the along-axis propagation of oceanic rifting into the continental lithosphere through the emplacement of discrete mantle upwellings and ridge jumps, while Kusznir & Tymms (2003) have recently introduced models of depth-dependent stretching in which upper crustal extension is much less than that of the lower crust and lithospheric mantle, to explain the formation of rifted margins. There are profound differences predicted by these various models for the 3-D geometry of the crust and upper mantle that develops during continental break-up and the initiation of oceanic rift segmentation.

The along-axis segmentation of oceanic rifts is produced by along-axis variations in magma supply (e.g. Batiza *et al.* 1996). In contrast the regular along-axis segmentation of continental rift basins is produced by the geometrical arrangement of large offset border fault systems whose length is dependent upon the integrated yield strength of the lithosphere (e.g. Hayward & Ebinger 1996; Jackson & Blenkinsop 1997). Thus, there is also a strong signal in the along-axis segmentation pattern of continental rifts. What is the genetic link between continental and oceanic rifting? It is apparent that mantle dynamics and magmatic processes become more important as the continental lithosphere is stretched to the point of rupture, but when and how does melt production facilitate continental break-up?

## The Afar mantle plume

Existing geophysical and geochemical data from the Afar volcanic province provide compelling evidence for rift initiation above a mantle plume, with controversy concerning the location, depth, extent and continuity of the hot asthenospheric material with the deep mantle anomaly beneath southern Africa. This volume addresses the fundamental questions of how mantle melt intrudes and is distributed through the plate, and how this magma intrusion process controls along-axis segmentation and facilitates continental break-up. Seismic data and tomographic models show that hot asthenosphere and associated crustal intrusions occur elsewhere throughout the length of the East African rift system (e.g. Nyblade *et al.* 2000; Ritter & Kaspar 1997; Achauer *et al.* 1994). Pre-rift flood basalts and Quaternary lavas in a large region surrounding the volcanically active Afar province have $^{3}He/^{4}He$ ratios as high as 20 Ra which are thought to characterize undegassed mantle material originating from below the 660 km discontinuity, consistent with a mantle plume (Marty *et al.* 1996; Pik *et al.* 2005). While volcanism in the Afar province and throughout East and South Africa has been attributed to the South African Superplume (Janney *et al.* 2002; Kieffer *et al.* 2004; Furman *et al.* 2004, 2006*a*), differences in the timing and composition of the thick flood basalt sequences erupted have led to a range of geodynamic models involving both single (Ebinger & Sleep 1998), modified single (Furman *et al.* 2004, 2006*b*) and double (George *et al.* 1998; Rogers 2006) plumes, the validity of which remain unresolved.

Regardless of plume head/stem location or number, magma should be generated by adiabatic melting beneath the thinning lithosphere underlying the Ethiopian, Red Sea and Aden rifts. Melt should rise to shallower levels with time, with this heat transfer process then weakening the lithosphere, enhancing strain localization (Buck 2004). The Afar volcanic province, including continental rifting in the Main Ethiopian Rift to nascent seafloor spreading in Afar, is a prime locale for examination of these processes, fundamental to the break-up of a continent.

## Motivation for future studies

The geophysical, geochemical and geomorphological studies presented in this volume resolve several issues. Yet these results provoke additional questions, and drive future research into mantle plumes, continental rifts and continental rupture processes. Specifically:

(a) Quaternary eruptive centres are found >100 km from the faulted rift valleys, and seismic and magneto-telluric data suggest pockets of melt persist beneath the uplifted plateau over 80 km from the rift valley. Where is this melt sourced, how does it move through the plate, and why is it restricted to the northwestern side of the rift valley? Are the seismic reflectors in the upper mantle correlated with melt extraction zones?

(b) Were there one or multiple plume events, and why has the widespread volcanism persisted long after the onset of rifting in the Red Sea, Gulf of Aden and Main Ethiopian Rifts?

(c) When was the thick underplate observed beneath the plateau accreted? During the period of rapid extrusion at *c.* 30 Ma, or throughout the past 40 Ma of extrusive volcanism? What are the thermal implications of this underplate?

(d) Are the magmatic segments mapped at the surface sourced by discrete upwellings at the lithosphere–asthenosphere boundary, or are they restricted to zones of dyke intrusion within the crust? Why are the largest thermal anomalies in the asthenosphere offset from the magmatic segments?

(e) Are magmatic segments the earliest stage in the formation of seaward-dipping reflector sequences imaged on magmatic margins worldwide?

(f) Are differences in crustal structure and thickness of underplate between the northwestern and southeastern sides of the rift relicts of Pan-African lithospheric structure?

The studies of continental break-up and flood basalt volcanism within the Afar volcanic province presented in this volume have implications for flood basalt provinces and magmatic passive continental margins worldwide. As well as addressing fundamental scientific problems, the results motivate the next stage of exploration.

## References

ACHAUER & THE KRISP TELESEISMIC WORKING GROUP 1994. New ideas on the Kenya rift based on the inversion of the combined dataset of the 1985 and 1989/90 seismic tomography experiments. *Tectonophysics*, **236**, 305–329.

D'ACREMONT, E., LEROY, S., BESLIER, M.-O., BELLAHSEN, N., FOURNIER, M., ROBIN, C., MAIA, M. & GENTE, P. 2005. Structure and evolution of the eastern Gulf of Aden conjugate margins from seismic reflection data. *Geophysical Journal International*, **160**, 869–890, doi:10.1111/j.1365-246X.2005.02524.x

ACTON, G.D., TESSEMA, A., JACKSON, M. & BILHAM, R. 2000. The tectonic and geomagnetic significance of paleomagnetic observations from volcanic rocks from central Afar, Africa. *Earth and Planetary Science Letters*, **180**, 225–241.

AYELE, A. & KULHANEK, O. 1997. Spatial and temporal variation of seismicity in the Horn of Africa from 1960 to 1993. *Geophysical Journal International*, **130**, 805–810.

BAKER, J., SNEE, L. & MENZIES, M. 1996*a*. A brief Oligocene period of flood volcanism in Yemen: implications for the duration and rate of continental flood volcanism at the Afro-Arabian triple junction. *Earth and Planetary Science Letters*, **138**, 39–55.

BAKER, J., THIRLWALL, M. & MENZIES, M.A. 1996*b*. Sr–Nd–Pb isotopic and trace element evidence for crustal contamination of plume-derived flood basalts: Oligocene flood volcanism in western Yemen. *Geochimica et Cosmochimica Acta*, **60**, 2559–2581.

BATIZA, R. 1996. Magmatic segmentation at mid-ocean ridges: A review. *In*: MACLEOD, C. ET AL. (eds) *Tectonic, Magmatic, Hydrothermal, and Biological Segmentation of Mid-Ocean Ridges*. Geological Society, London, Special Publications, **118**, 103–130.

BILHAM, R., BENDICK, R., LARSON, K., MOHR, P., BRAUN, J., TESFAYE, S. & ASFAW, L. 1999. Secular and tidal strain across the Main Ethiopian Rift. *Geophysical Research Letters*, **26**, 2789–2792.

BOCCALETTI, M., BONINI, M., MAZZUOLI, R., ABEBE, B., PICCARDI, L. & TORTORICI, L. 1998. Quaternary oblique extensional tectonics in the Ethiopian Rift (Horn of Africa). *Tectonophysics*, **287**, 97–116.

BUCK, W.R. 2004. Consequences of asthenospheric variability on continental rifting. *In*: KARNER, G.D., TAYLOR, B., DRISCOLL, N.W., & KOHLSTEDY, D.L. (eds) *Rheology and Deformation of the Lithosphere at Continental Margins*. Columbia University Press, New York, 1–30.

CHAZOT, G. & BERTRAND, H. 1993. Mantle sources and magma–continental crust interactions during early Red Sea–Aden rifting in Southern Yemen: elemental and Sr, Nd, Pb isotope evidence. *Journal of Geophysical Research*, **98**, 1818–1835.

CHU, D.H. & GORDON, R.G. 1999. Evidence for motion between Nubian and Somalia along the Southwest Indian Ridge. *Nature*, **298**, 64–66.

DAVIDSON, A. & REX, D.C. 1980. Age of volcanism and rifting in southwestern Ethiopia. *Nature*, **283**, 657–658.

DAVIS, M. & KUSZNIR, N.J. 2002. Are buoancy forces important during the formation of rifted margins? *Geophysical Journal International*, **149**, 524–533.

DUNBAR, J. & SAWYER, D. 1989. Three-dimensional dynamical model of continental rift propagation and margin formation. *Journal of Geophysical Research*, **101**, 27845–27863.

EBINGER, C.J., YEMANE, T., WOLDEGABRIEL, G., ARONSON, J.L. & WALTER, R.C. 1993. Late Eocene–Recent volcanism and rifting in the southern main Ethiopian rift. *Journal of the Geological Society, London*, **150**, 99–108.

EBINGER, C.J. & SLEEP, N.H. 1998. Cenozoic magmatism throughtout east Africa resulting from the impact of a single plume. *Nature*, **395**, 788–791.

EBINGER, C.J. & CASEY, M. 2001. Continental breakup in magmatic provinces: An Ethiopian example. *Geology*, **29**, 527–530.

FOSTER, A.N. & JACKSON, J.A. 1999. Source parameters of large African earthquakes; implications for crustal rheology and regional kinmeatics. *Geophysical Journal International*, **134**, 422–428.

FURMAN, T., BRYCE, J.G., KARSON, J. & IOTTI, A. 2004. East African Rift System plume structure: insights from Quaternary mafic lavas of Turkana, Kenya. *Journal of Petrology*, **45**, 1069–1088.

FURMAN, T., KALETA, K.M. & BRYCE, J.G. 2006*a*. Tertiary mafic lavas of Turkana, Kenya: constraints on temporal evolution of the EARS and the occurrence of HIMU volcanism in Africa. *Journal of Petrology* (in press).

FURMAN, T., BRYCE, J., ROONEY, T., HANAN, B., YIRGU, G. & AYALEW, D. 2006*b*. Heads and tails: 30 Ma of the Afar plume. *In*: YIRGU, G., EBINGER, C.J. & MAGUIRE, P.K.H. (eds) *The Afar Volcanic Province within the East African Rift System.* Geological Society, London Special Publications, **259**, 95–119.

GEORGE, R., ROGERS, N. & KELLY, S. 1998. Earliest magmatism in Ethiopia: evidence for two mantle plumes in one flood basalt province. *Geology*, **26**, 923–926.

GRAND, S.P., VAN DER HILST, R.D. & WIDIYANTORO, S. 1997. Global seismic tomography; a snapshot of convection in the Earth. *GSA Today*, **7**, 1–7.

HAYWARD, N. & EBINGER, C. 1996. Variations in the along-axis segmentation of the Afar rift system. *Tectonics*, **15**, 244–257.

HEBERT, H., DEPLUS, C., HUCHON, P., KHANBARI, K. & AUDIN, L. 2001. Lithospheric structure of a nascent spreading ridge inferred from gravity data: The western Gulf of Aden. *Journal of Geophysical Research*, **106**(B11), 26345–26363.

HENDRIE, D.B., KUSZNIR, N.J., MORLEY, C.K. & EBINGER, C.J. 1994. Cenozoic extension in northern Kenya: a quantitative model of rift basin development in the Turkana region. *Tectonophysics*, **236**, 409–438.

HOFMANN, C., COURTILLOT, V., FÉRAUD, G., ROCHETTE, P., YIRGU, G., KETEFO, E. & PIK, R. 1997. Timing of the Ethiopian flood basalt event and implications for plume birth and global change. *Nature*, **389**, 838–841.

JACKSON, J. & BLENKINSOP, T. 1993. The Malawi earthquake of March 10, 1989: Deep faulting within the East African rift system. *Tectonics*, **12**, 1131–1139.

JANNEY, P.E., LE ROEX, A.P., CARLSON, R.W. & VILJOEN, K.S. 2002. A chemical and multi-isotope study of the Western Cape melilite province, South Africa: implications for the sources of kimberlites and the origin of the HIMU signature in Africa. *Journal of Petrology*, **43**, 2339–2370.

KEIR, D., KENDALL, J.-M., EBINGER, C. & STUART, G. 2005. Variations in late syn-rift melt alignment inferred from shear-wave splitting in crustal earthquakes beneath the Ethiopian rift. *Geophysical Research Letters*, **32**, L23308, doi: 10.1029/2005 GL024150.

KIEFFER, B., ARNDT, N., *ET AL.* 2004. Flood and Shield basalts from Ethiopia: magmas from the African superswell. *Journal of Petrology*, **45**, 793–834.

KUSZNIR, N.J. & TYMMS, V. 2003. Modelling sea-floor spreading initiation and depth dependent stretching at rifted continental margins. *EOS, Transactions of the American Geophysical Union, 84(46) Fall Meeting Supplement*, Abstract T12A-0440.

LEMAUX, J., GORDON, R.G. & ROYER, J.-Y. 2002. The location of the Nubia–Somalia boundary along the Southwest Indian Ridge. *Geology*, **30**, 339–342.

LISTER, C.R.B. 1986 Differential thermal stresses in the Earth. *Geophysical Journal of the Royal Astronomical Society*, **86**, 319–330.

MAGUIRE, P., EBINGER, C., *ET AL.* 2003. Geophysics project in Ethiopia studies continental breakup. *EOS, Transactions of the American Geophysical Union*, **84**, 342–343.

MARTY, B., PIK, R. & YIRGU, G. 1996. Helium isotopic variations in Ethiopian plume lavas: nature of magmatic sources and limit on lower mantle contribution. *Earth and Planetary Science Letters*, **144**, 223–237.

MENZIES, M., BAKER, J. & CHAZOT, G. 2001. Cenozoic plume evolution and flood basalts in Yemen: a key to understanding older examples. *In*: ERNST, R.E. & BUCHAN, K.L. (eds) *Mantle plumes: their identification through time, Special Paper Geological Society of America*, **352**, pp. 23–36.

MORLEY, C.K., WESCOTT, W.A., STONE, D.M., HARPER, R.M., WIGGER, S.T. & KARANJA, F.M. 1992. Tectonic evolution of the northern Kenyan Rift, *Journal Geological Society of London*, **149**, 333–348.

NYBLADE, A., OWENS, T., GURROLA, H., RITSEMA, J. & LANGSTON, C.A. 2000. Seismic evidence for a deep upper mantle thermal anomaly beneath East Africa. *Geology*, **28**, 599–602.

PIK, R., MARTY, B., CARIGNAN, J. & LAVÉ, J. 2003. Stability of the Upper Nile drainage network (Ethiopia) deduced from (U–Th)/He thermochronometry: implications for uplift and erosion of the Afar plume dome. *Earth and Planetary Science Letters*, **215**, 73–88.

Pik, R., Marty, B. & Hilton, D.R. 2006. How many mantle plumes in Africa? The geochemical point of view. *Chemical Geology*, **226**, 100–114.

Ritsema, J., van Heijst, H. & Woodhouse, J. 1999. Complex shear wave velocity structure imaged beneath Africa and Iceland. *Science*, **286**, 1925–1928.

Ritter, J.R.R. & Kaspar, T. 1997. A tomography study of the Chyulu Hills, Kenya. *Tectonophysics*, **278**(1–4), 149–169.

Rogers, N. 2006. Basaltic magmatism and the geodynamics of the East African rift system. *In*: Yirgu, G., Ebinger, C.J. & Maguire, P.K.H. (eds) *The Afar Volcanic Province within the East African Rift System*. Geological Society, London, Special Publications, **259**, 77–93.

Schilling, J.-G., Kingsley, R., Hanan, B. & McCully, B. 1992. Nd–Sr–Pb isotopic variations along the Gulf of Aden: Evidence for the Afar mantle plume–lithosphere interaction. *Journal of Geophysical Research*, **97**, 10927–10966.

Taylor, B., Goodliffe, A. & Martinez, F. 1999. How continents break up: Insights from Papua New Guinea. *Journal of Geophysical Research*, **104**, 7497.

Tesfaye, S., Harding, D.J. & Kusky, T. 2003. Early continental breakup boundary and the migration of the Afar Triple Junction, Ethiopia. *Geological Society of America Bulletin*, **115**, 1053–1067.

Ukstins, I., Renne, P., Wolfenden, E., Baker, J., Ayalew, D. & Menzies, M. 2002. Matching conjugate volcanic rifted margins: $^{40}Ar/^{39}Ar$ chronostratigraphy of pre- and syn-rift bimodal flood volcanism in Ethiopia and Yemen. *Earth and Planetary Science Letters*, **198**, 289–306.

van der Hilst, R.D. & Karason, H. 1999. Composition heterogeneity in the bottom 1000 kilometres of Earth's mantle; toward a hybrid convection model. *Science*, **283**(5409), 1885–1888.

Vidal, P., Deniel, C., Vellutini, P.J., Piguet, P., Coulon, C., Vincent, J. & Audin, J. 1991. Changes of mantle source in the course of a rift evolution. *Geophysical Research Letters*, **18**, 1913–1916.

Watchorn, F., Nichols, G. & Bosence, D. 1998. Rift-related sedimentation and stratigraphy, southern Yemen (Gulf of Aden). *In*: Purser, B. & Bosence, D. (eds) *Sedimentary and Tectonic Evolution of Rift Basins*. Chapman & Hall, London, pp. 165–189.

WoldeGabriel, G., Aronson, J.L. & Walter, R.C. 1990. Geology, geochronology and rift basin development in the central sector of the main Ethiopian rift. *Geological Society of America Bulletin*, **102**, 439–485.

Wolfenden, E., Ebinger, C., Yirgu, G., Deino, A. & Ayalew, D. 2004. Evolution of the northern main Ethiopian rift: birth of a triple junction. *Earth and Planetary Science Letters*, **224**, 213–228.

Wolfenden, E., Ebinger, C., Yirgu, G., Renne, P. & Kelley, S.P. 2005. Evolution of the southern Red Sea rift: birth of a magmatic margin. *Geological Society of America Bulletin*, **117**, 846–864.

Zumbo, V., Feraud, G., Betrand, H. & Chazot, G. 1995. $^{40}Ar/^{39}Ar$ chronology of Tertiary magmatic activity in southern Yemen during the early Red Sea–Aden rifting. *Journal of Volcanology and Geothermal Research*, **65**(3–4), 265–279.

# Part 1: Plate kinematic and geodynamic framework of the Afar volcanic province

The four papers in this section establish the regional tectonic and geodynamic framework for subsequent papers in Parts 2–4. Each summarizes controversy regarding the mantle upwelling beneath Africa and its influence on rifting, as well as the role of magmatism in continental break-up.

**Calais *et al.*** evaluate geodetic data from Africa and adjoining plates to estimate a pole of rotation for the Nubia–Somalia plate system. Observations from a permanent GPS station on the Archaean Tanzania craton at Mbarara, Uganda (MBAR), cannot be fit by a simple two-plate model. These data, however, can be fit by a model that considers the strong, thick Archaean craton as a microplate that rotates relative to the Nubian and Somalian plates. Calais *et al.*'s new model is consistent with regional slip vectors derived from well-constrained earthquake focal mechanism solutions from East Africa, and explains the existence and persistence of the sub-parallel Western and Eastern rift systems. Although speculative, Calais *et al.* predict a second microplate, the Rovuma plate, lying between the Malawi rift and the Davie Ridge to the east. Calais *et al.* suggest that focused mantle flow may influence continental deformation in East Africa.

**Garfunkel & Beyth** constrain the Oligocene–Recent movements of the Arabia–Nubia–Somalia plate systems with regional geological and geophysical data. Rather than focus on the detailed movement of microplates within the diffuse rift–rift–rift triple junction zone, Garfunkel & Beyth estimate the size and location of new area (stretched continental and/or new igneous crust) to facilitate studies of the structural development of this evolving magmatic passive continental margin. Plate boundaries changed at several times in the past 30 Ma, but there are no resolvable changes in the motions of the plates themselves. Over the evolution of the Afar volcanic province, diffuse extension has given way to localized extension along a narrow plate boundary, and along-axis propagation has effectively shut off extension along now-inactive plate boundaries.

**Buck** notes the synchroneity of flood basalt magmatism and rifting in the Red Sea–Gulf of Aden–East African rift triple junction, indicating the vital role of magmatism in continental rifting and break-up processes. He then compares and contrasts forces required to rift continental lithosphere in the presence and absence of melt. These calculations demonstrate the role of melt in facilitating rifting and eventual break-up; thick, mechanically strong continental lithosphere requires magmatism to enable rifting. In his application to the Afar volcanic province, Buck notes the relative straightness of the Red Sea and East African rifts, and suggests that the shapes of the rift zones is in part controlled by the movement of melt-filled dykes through the plates.

**Kendall *et al.*** integrate seismic anisotropy measurements from local and teleseismic earthquakes to test models for rifting of East Africa. Although their paper focuses on patterns within the Afar volcanic province, they include observations from throughout East Africa to examine the role of melt in the rifting process. Surface wave studies show consistent NE-directed fast shear waves within the mantle at depths greater than 150 km, parallel to the trend of the African Superplume imaged in global seismic studies. At shallower depths corresponding to the crust and mantle lithosphere, the patterns are more variable. The fast directions are generally rift-parallel, and anisotropy is greatest beneath Quaternary eruptive centres within the highly evolved Main Ethiopian Rift. Kendall *et al.* conclude that the patterns of anisotropy within the lithosphere beneath the rift valleys argue for melt-assisted rifting models, rather than mechanical stretching models.

*From*: Yirgu, G., Ebinger, C.J. & Maguire, P.K.H. (eds) 2006. *The Afar Volcanic Province within the East African Rift System*. Geological Society, London, Special Publications, **259**, 7.
0305-8719/06/$15.00 

# Kinematics of the East African Rift from GPS and earthquake slip vector data

E. CALAIS[1], C. EBINGER[2], C. HARTNADY[3], & J.M. NOCQUET[4]

[1]*Purdue University, Department of Earth and Atmospheric Sciences, West Lafayette, Indiana, USA (e-mail: ecalais@purdue.edu)*

[2]*Department of Geology, Royal Holloway, University of London, Egham, UK*

[3]*Umvoto Africa (Pty) Ltd, PO Box 61, Muizenberg, South Africa 7950*

[4]*CNRS, UMR6526 Géosciences Azur, Valbonne, France*

**Abstract:** Although the East African Rift (EAR) System is often cited as the archetype for models of continental rifting and break-up, its present-day kinematics remains poorly constrained. We show that the currently available GPS and earthquake slip vector data are consistent with (1) a present-day Nubia–Somalia Euler pole located between the southern tip of Africa and the Southwest Indian ridge and (2) the existence of a distinct microplate (Victoria) between the Eastern and Western rifts, rotating counter-clockwise with respect to Nubia. Geodetic and geological data also suggest the existence of a (Rovuma) microplate between the Malawi rift and the Davie ridge, possibly rotating clockwise with respect to Nubia. The data indicate that the EAR comprises at least two rigid lithospheric blocks bounded by narrow belts of seismicity (<50 km wide) marking localized deformation rather than a wide zone of quasi-continuous, pervasive deformation. On the basis of this new kinematic model and mantle flow directions interpreted from seismic anisotropy measurements, we propose that regional asthenospheric upwelling and locally focused mantle flow may influence continental deformation in East Africa.

The East African Rift (EAR), a 5000 km-long series of fault-bounded depressions straddling east Africa in a roughly north–south direction, marks the divergent boundary between two major tectonic plates, Somalia and Nubia (Fig. 1). Although the EAR is often cited as a modern archetype for rifting and continental break-up and a Cenozoic continental flood basalt province, its current kinematics is still poorly understood and quantified, owing in part to its tremendous extent and inaccessibility.

Jestin *et al.* (1994) were among the first to quantify the kinematics of the EAR by estimating a 3 Ma average Nubia–Somalia angular velocity. Using Arabia–Nubia and Arabia–Somalia relative motions determined from marine geophysical data in the Red Sea and Gulf of Aden, they find a Nubia–Somalia Euler pole south of the Southwest Indian Ridge (Fig. 1). Their result differs significantly from that of Chu & Gordon (1999), who estimate a 3 Ma average Nubia–Somalia angular velocity from marine geophysical data along the Southwest Indian ridge and the Antarctica–Somalia and Antarctica–Nubia plate closure circuit (Fig. 1). More recently, direct estimates of the Nubia–Somalia plate motion have been made possible thanks to a limited number of permanent Global Positioning System (GPS) stations on both plates. For instance, Sella *et al.* (2002) used two GPS sites on the Somalian plate and four on the Nubian plate, while Fernandes *et al.* (2004) used three GPS sites on the Somalian plate and 11 on the Nubian plate and longer data time series. Again, these two GPS estimates of the Nubia–Somalia angular velocity differ significantly from each other, as well as from previous results derived from oceanic data (Fig. 1).

In addition to far-field plate motions, the kinematics of the EAR itself remains an open question. Some authors have proposed that the EAR consists of a mosaic of rigid lithospheric blocks bounded by localized deformation within narrow seismically and volcanically active rift valleys (e.g. Ebinger 1989; Hartnady 2002). Others favour a broad deformation zone (e.g. Gordon 1998) implicit in models that assume a weak mantle lithosphere beneath continents (e.g. Jackson 2002). But the distribution of strain across and along the EAR is currently unknown and no quantitative kinematic data are presently available for that plate boundary. In this paper, we use an updated GPS and earthquake slip vector data set to estimate the Somalia–Nubia angular velocity and propose a first-order kinematic model for present-day deformation in the EAR.

*From*: YIRGU, G., EBINGER, C.J. & MAGUIRE, P.K.H. (eds) 2006. *The Afar Volcanic Province within the East African Rift System*. Geological Society, London, Special Publications, **259**, 9–22.
0305-8719/06/$15.00 

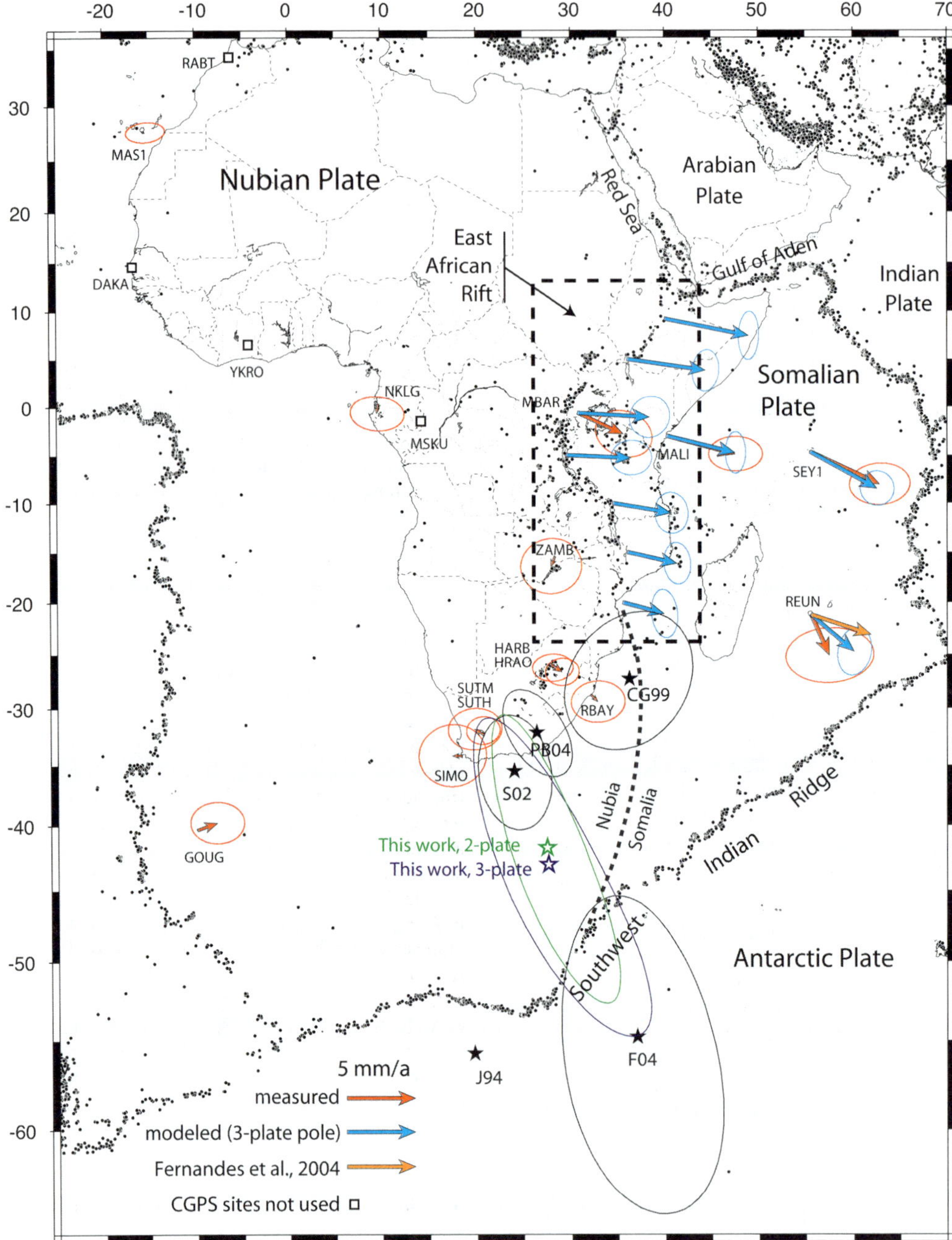


**Fig. 1.** GPS sites used in this study. Dots show seismicity (NEIC catalog). Stars are Euler poles for Somalia–Nubia relative motion with associated 1-sigma error ellipse. S02 = Sella *et al.* 2002; PB04 = Prawirodirdjo & Bock 2004; F04 = Fernandes *et al.* 2004; CG99 = Chu & Gordon 1999; J94 = Jestin *et al.* 1994. This work 2-plate = two-plate inversion using GPS velocities at MALI, REUN and SEY1 + nine earthquake slip vectors along the Main Ethiopian Rift. This work 3-plate = three-plate inversion using GPS velocities at MBAR, MALI, REUN, SEY1 and 47 earthquake slip vectors along the East African Rift (see Figure 2 and Table 1). Curved dashed line shows the Nubia–Somalia plate boundary in the SW Indian Ocean proposed by Lemaux *et al.* 2002.

## GPS data

We processed GPS data from 10 permanent GPS sites operating on the Nubian plate (MAS1, GOUG, NKLG, ZAMB, RBAY, SUTH, SUTM, HRAO, HARB, SIMO), the three sites available on the Somalian plate (MALI, SEY1, REUN), and one site in the EAR (MBAR). The results presented here include all the publicly available data from August 1998 to April 2005. We processed the GPS data using the GAMIT software version 10.2 (King & Bock 2005). We solved for station coordinates, satellite state vectors, one tropospheric delay every four hours at each site, and phase ambiguities using double-differenced GPS phase measurements, with International GPS Service (IGS) final orbits and International Earth Rotation Service (IERS) Earth orientation parameters relaxed. We then combined our regional daily solutions with global Solution Independent Exchange (SINEX) files from the IGS daily processing routinely done at Scripps Institution of Oceanography and imposed the reference frame by minimizing the position deviations of 38 globally distributed IGS core stations with respect to the International Terrestrial Reference Frame 2000 (ITRF2000; Altamimi *et al.* 2002), while estimating an orientation and translation transformation. Our primary result consists of precise positions and velocities at 14 continuous GPS sites on the Nubian and Somalian plates, expressed in ITRF2000 (Fig. 2; Table 1). In a second step, explained hereafter, we rotate these velocities in a Nubia-fixed frame.

GPS velocities, in a Cartesian geocentric frame, can be modelled as:

$$\vec{V} = \vec{\Omega} \times \vec{P}$$

where $P(x,y,z)$ is the unit vector defining the position of the GPS site, $V(vx, vy, vz)$ is the velocity vector at that site, and $\Omega(\omega_x,\omega_y,\omega_z)$ is the rotation vector defining the motion of the plate carrying the site. For a number of sites on a given plate, this cross product can be written in matrix form as:

$$\begin{pmatrix} vx_1 \\ vy_2 \\ vz_3 \\ \cdots \end{pmatrix} = \begin{pmatrix} 0 & Z_1 & -Y_1 \\ -Z_1 & 0 & X_1 \\ Y_1 & -X_1 & 0 \\ \cdots & & \end{pmatrix} \begin{pmatrix} \omega_x \\ \omega_y \\ \omega_z \end{pmatrix}$$

or

$$V = A\Omega(\Sigma)$$

where $V$ is the vector of observations with its associated covariance matrix $\Sigma$, $A$ the model matrix, and $\Omega$ the vector of unknowns. The least-squares solution is then given by:

$$\Omega = (A^T\Sigma^{-1}A)^{-1}A^T\Sigma^{-1}V$$

We estimated the Nubia-ITRF2000 angular velocity by inverting horizontal velocities at sites MAS1, NKLG, SUTH, SUTM and GOUG (Fig. 1). We formed the data covariance matrix using the 2-sigma velocity uncertainty (95% confidence) on GPS velocities and their site-by-site NS–EW correlation (i.e. we do not account for intersite correlations). We find a reduced $\chi^2$ ($\chi^2$ divided by the degrees of freedom) close to unity with a weighted RMS of 0.7 mm a$^{-1}$ for horizontal velocities (Table 2). We then use an $F$-ratio statistic to test whether the velocity at additional sites in Africa is consistent with the rigid rotation defined by the previous subset of sites. The $F$-ratio, defined by:

$$F = \frac{[\chi^2(p_1) - \chi^2(p_2)]/(p_1 - p_2)}{\chi^2(p_2)/p_2},$$

tests the significance of the decrease in $\chi^2$ from a model with $p_2$ versus $p_1$ degrees of freedom, with $p_1 > p_2$. This experimental $F$-ratio is compared to the expected value of a $F(p_1 - p_2, p_1)$ distribution for a given risk level $\alpha\%$ (equivalent to a $(100 - \alpha)\%$ confidence level) that the null hypothesis (the additional site is consistent with the rigid plate model) can be rejected. The degrees of freedom of a rigid rotation estimation is $p_1 = 2 \times N - 3$ for $N$ site velocities, it becomes $p_2 = 2 \times N - 3 - 2$ with an additional site. For site ZAMB, the $F$-test value is $F = [(8.52 - 7.53)/(9 - 7)]/[7.53/7] = 0.4$ (using values provided in Table 1), corresponding to $\alpha = 0.32$. The velocity at ZAMB is consistent with a Nubian plate model (defined by sites MAS1, NKLG, SUTH, SUTM and GOUG) at a 68% confidence level. This indicates that the relative motion between the Zambia craton and the stable Nubian plate is less than the residual velocity at ZAMB, or $\sim$1 mm a$^{-1}$. The same $F$-ratio test with sites HRAO, HARB and RBAY ($F = [(12.87 - 7.53)/(15 - 7)]/[7.53/7] = 0.62$, $\alpha = 0.25$) shows that these sites are consistent with the Nubian plate model defined above at a 75% confidence level. However, both HRAO and RBAY are separated from stable Nubia by the seismically active Okavango Rift in Botswana (Modisi *et al.* 2000) and the Senqu Seismic Belt in South Africa and Lesotho (Hartnady 1998) and may lie on a microplate separate from Nubia and Somalia (Transgariep block of Hartnady 2002). We therefore did not use them in our final Nubia-ITRF2000 angular velocity estimate.

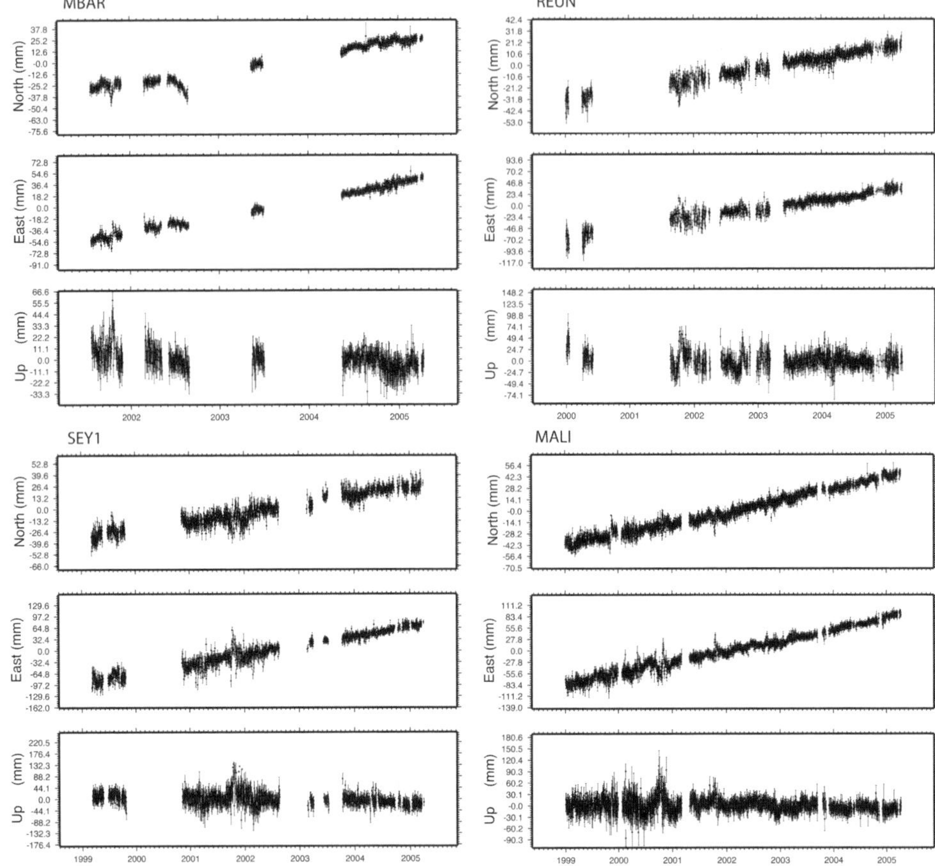


**Fig. 2.** Position time series for sites MBAR, REUN, MALI and SEY1, with one position per day and its associated one-sigma formal error, shown with respect to the weighted mean of the entire time series. Each component (north–south, east–west, vertical) is shown separately.

**Table 1.** *GPS velocities*

| Site | lon. | lat. | $V_e$ | $V_n$ | $\sigma_e$ | $\sigma_n$ | Corr. |
|---|---|---|---|---|---|---|---|
| GOUG | 350.12 | −40.35 | 22.3 | 17.9 | 0.8 | 0.6 | 0.033 |
| HARB | 27.71 | −25.89 | 17.7 | 17.1 | 0.6 | 0.4 | −0.014 |
| HRAO | 27.69 | −25.89 | 18.4 | 16.8 | 0.5 | 0.4 | −0.032 |
| MALI | 40.19 | −3.00 | 26.9 | 14.3 | 0.8 | 0.5 | 0.039 |
| MAS1 | 344.37 | 27.76 | 16.5 | 16.7 | 0.5 | 0.3 | −0.017 |
| MBAR | 30.74 | −0.60 | 26.1 | 16.6 | 1.0 | 0.7 | −0.006 |
| NKLG | 9.67 | 0.35 | 22.3 | 17.7 | 0.8 | 0.5 | −0.079 |
| RBAY | 32.08 | −28.80 | 16.6 | 16.3 | 0.8 | 0.6 | −0.002 |
| REUN | 55.57 | −21.21 | 17.3 | 9.6 | 1.4 | 0.8 | 0.064 |
| SEY1 | 55.48 | −4.67 | 26.2 | 10.3 | 1.0 | 0.6 | 0.087 |
| SIMO | 18.44 | −34.19 | 16.0 | 18.1 | 1.1 | 1.0 | 0.015 |
| SUTH | 20.81 | −32.38 | 16.7 | 18.2 | 0.5 | 0.4 | −0.016 |
| SUTM | 20.81 | −32.38 | 16.1 | 18.4 | 0.8 | 0.6 | 0.022 |
| ZAMB | 28.31 | −15.43 | 19.6 | 16.6 | 1.0 | 0.8 | 0.028 |

Velocities ($V$, mm $a^{-1}$) with respect to ITRF2000 and associated one standard deviation formal errors ($\sigma$, mm $a^{-1}$) and their correlation (Corr.).

## Somalia–Nubia plate motion

Finally, we rotate the ITRF2000 velocities into a Nubia-fixed frame using the Nubia-ITRF angular velocity estimated with MAS1, NKLG, SUTH, SUTM and GOUG. We use the Nubia-fixed velocities at sites MBAR, MALI, SEY1 and REUN in the following kinematic analysis. In addition to GPS velocities, we use slip vectors derived from the 53 focal mechanisms determined from body-waveform inversion by Foster & Jackson (1998), augmented by the Harvard Centroid Moment Tensor (CMT) database (Fig. 3a). We use the same slip vectors as Foster & Jackson (1998), who based their choice on structural framework of each epicentral region. We follow the same criteria for the additional CMT events. The joint inversion of geodetic and slip vector data is based on the maximum likelihood algorithm of Minster & Jordan (1978) and fitting functions of Chase (1978), modified to incorporate the ability to use geodetic vectors and their statistics.

First, we solved the Somalia–Nubia angular velocity in a two-plate inversion (Nubia–Somalia) using sites MALI, SEY1 and REUN plus nine earthquake slip vectors along the Main Ethiopian Rift, a fairly simple single structure that marks the boundary between Nubia and Somalia between latitudes 5° N and 10° N (Fig. 1; Table 2). We repeated the inversion for Somalia–Nubia angular velocity without using earthquake slip vectors along the Main Ethiopian Rift and found a similar result. The associated 1-sigma confidence ellipse includes, although barely, previous GPS-derived Somalia–Nubia estimates by Fernandes *et al.* (2004), Sella *et al.* (2002), and Prawirodirdjo & Bock (2004). Differences between GPS estimates of plate motions may stem from the length of the time series used in the analysis, the version of the global reference frame used and its implementation, and the choice of sites used to define stable plates. Here, it is likely that the major difference is due to site REUN on the Somalian plate. For instance, Sella *et al.* (2002) and Prawirodirdjo & Bock (2004) do not use REUN in their analysis. Sella *et al.* (2002) define Somalia using two sites only (SEY1 and MALI), to which Prawirodirdjo & Bock (2004) add RBAY, even though that site is located on the Nubian side of the Nubia–Somalia plate boundary of Lemaux *et al.*'s (2002; Fig. 1). Fernandes *et al.* (2004) define Somalia using the same sites as our study but obtain a significantly different velocity at REUN (Fig. 1), rotated counter-clockwise compared to our solution. This shifts their Somalia–Nubia Euler pole south compared to ours. The reason for the velocity difference at REUN between the two solutions is unclear, as

**Table 2.** *Plate angular velocities*

| | Plates | Angular velocity | | | Uncertainties | | | | $\chi^2$ [$\chi^2/dof$] | Data used |
|---|---|---|---|---|---|---|---|---|---|---|
| | | lat. | lon. | ang. | Major | minor | azim | $\sigma_{ang}$ | | |
| 1 | Nubia–ITRF | 50.60 | −80.97 | 0.261 | 1.30 | 0.68 | −74.37 | 0.003 | 7.53 [*1.08*] | MAS1, NKLG, SUTH, SUTM, GOUG |
| 2 | Nubia–ITRF | 50.66 | −81.30 | 0.261 | 1.26 | 0.66 | −75.84 | 0.003 | 8.52 [*0.95*] | idem + ZAMB |
| 3 | Nubia–ITRF | 51.24 | −81.92 | 0.261 | 1.04 | 0.52 | −79.80 | 0.002 | 12.87 [*0.86*] | idem + ZAMB, HRAO, HARB, RBAY |
| 4 | Victoria–Nubia | 16.61 | 35.99 | −0.075 | 16.55 | 1.85 | 17.86 | 0.040 | 45.34 [*0.92*] | 47 slip vectors + MBAR, MALI, SEY1, REUN |
| | Somalia–Nubia | −43.06 | 27.51 | 0.069 | 13.23 | 3.59 | −25.83 | 0.013 | | |
| 5 | Somalia–Nubia | −41.77 | 27.39 | 0.068 | 11.82 | 2.78 | −20.63 | 0.012 | 15.40 [*1.28*] | 9 slip vectors + MALI, SEY1, REUN |

Angular velocities, associated uncertainties, $\chi^2$, reduced $\chi^2$ ($dof$ = degrees of freedom), and data used in the estimation. Angular velocities are in degrees per million years, clockwise is positive. Azimuth is given clockwise from north. Note that row 4 is a three-plate inversion (Somalia–Nubia–Victoria).

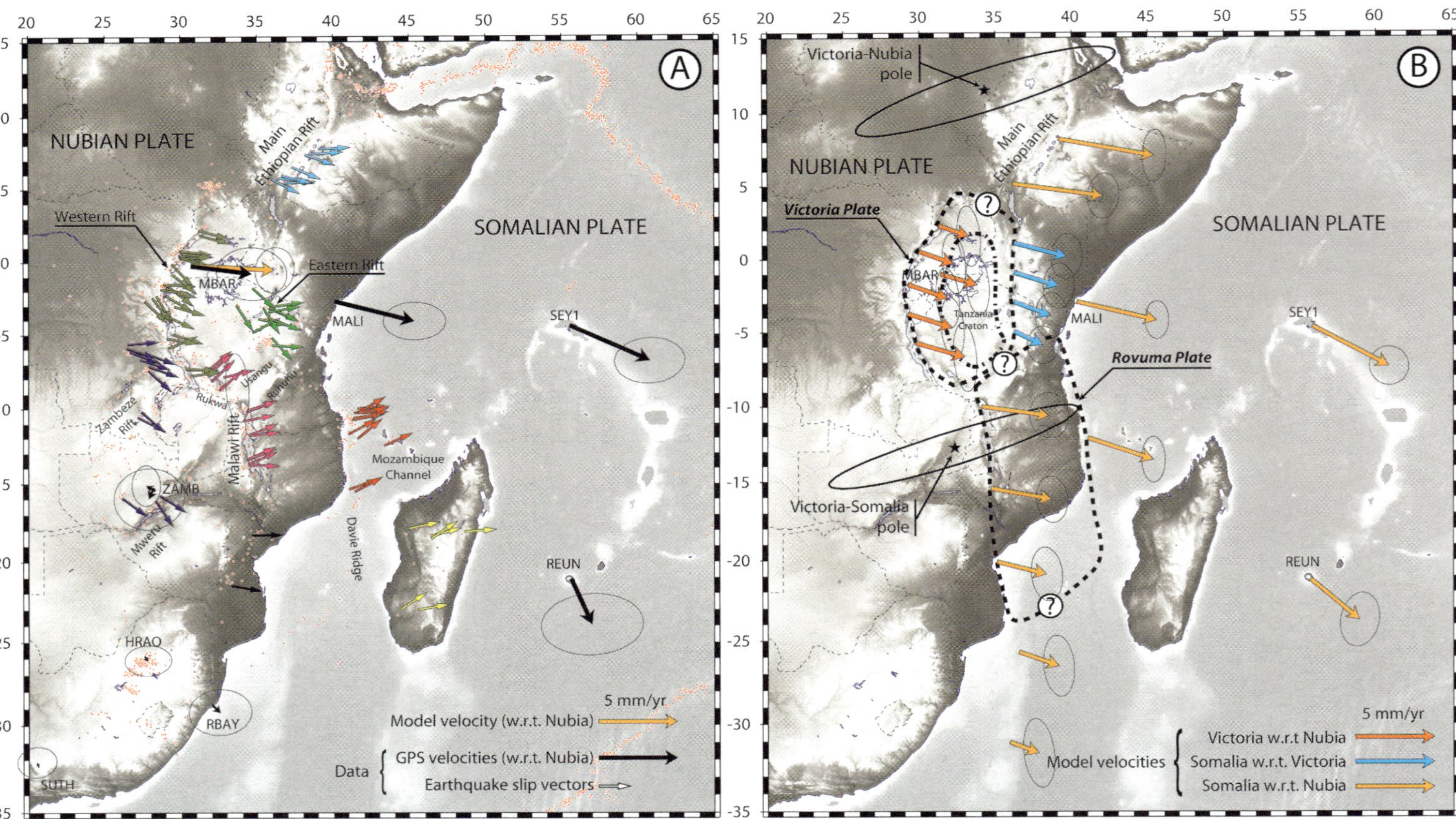


**Fig. 3.** (**a**) Data used to quantify the present-day kinematics of the EAR: GPS velocities (arrows, shown with respect to Nubia) and earthquake slip vector directions (colour bars – colour code indicates regional grouping). (**b**) Kinematic model of the EAR. Stars show the Euler pole of the Victoria plate and the associated 1-sigma error ellipse. Arrows show velocities predicted by our model along main branches of the EAR and at GPS sites.

they agree better for other sites. It may be due to the shorter time series used by Fernandes *et al.* (2004), which stops in mid-2003, or to possible volcanic deformation. However, the five-year-long position time series at REUN shows a linear displacement over that time period (in spite of a data gap between mid-2000 and mid-2001; Fig. 2), with no clear indication of transient deformation due to volcanic processes.

The reason for a difference between our Somalia–Nubia estimate and Chu & Gordon's (1999) is less clear, but could reflect changes in the regional plate kinematics over the past 3 Ma (Calais *et al.* 2003), a time of propagating plate boundaries in the Afar depression (e.g. Courtillot *et al.* 1999). Clearly, additional measurements at more sites on the Somalian plate (e.g. eastern Ethiopia, Somalia, Madagascar) are needed to firmly establish the present-day Somalia–Nubia motion.

We then compared the observed velocity at site MBAR with the one predicted by the Somalia–Nubia angular velocity defined above (Fig. 3a). Site MBAR shows a significantly smaller magnitude than predicted and is rotated clockwise, in a direction close to earthquake slip vectors along the Western rift. This is consistent with the fact that MBAR is located on cratonic lithosphere within the Nubia–Somalia plate boundary zone, east of the Western rift. In addition, Figure 3a shows that earthquake slip vector directions are constant or vary smoothly along each of the major tectonic segments of the EAR (thin coloured arrows on Fig. 3a), as also observed by Foster & Jackson (1998), but are inconsistent with the predicted Nubia–Somalia plate motion direction along most of the EAR (thick orange arrows on Fig. 3b). These misfits between GPS velocities or earthquake slip vector directions and Somalia–Nubia plate motion models indicate plate boundary zone deformation between Nubia and Somalia, in the form of continuously distributed strain or of rigid block motions.

Seismicity, active faulting and volcanism in the EAR are generally localized to the 50–80 km-wide Ethiopian rift, and the Western and Eastern rift valleys that bound the unfaulted and relatively aseismic Tanzania craton (Fig. 3). Seismic tomography models reveal a 200–250 km-deep keel beneath this small Archaean craton (Ritsema *et al.* 1998). We discuss hereafter a possible microplate geometry, on the basis of seismotectonic data, and test this hypothesis by jointly inverting GPS and earthquake slip vector data for a three-plate kinematic model. We do not attempt to account for the belts of seismicity and faulting west of Lakes Rukwa and Nyasa (Malawi) (Fig. 3a), where the lack of data would make the analysis too speculative.

## Kinematics of the EAR

The northernmost seismic belt of the EAR corresponds to the Main Ethiopian Rift, a single boundary between Nubia and Somalia (Fig. 3a). Near the Ethiopia–Sudan–Kenya border lies a complex zone of deformation where the EAR overprints Mesozoic to Palaeogene rifts (e.g. Hendrie *et al.* 1994). Seismicity is more diffuse and fault offsets are small in these poorly understood zones. South of about 3° N, teleseismic and tectonic activity splits into two branches, the Eastern and Western rifts (Fig. 3). Most of the teleseismic activity is concentrated in the Western rift, which contains a relatively small volume of volcanic material (e.g. Foster & Jackson 1998). On the contrary, the Eastern rift has eruptive centres along its length and moderate seismic activity, except near its southern termination at the edge of the Tanzania craton (e.g. Nyblade & Langston 1995; Foster & Jackson 1998). The Western rift wraps around the Archean Tanzania craton and connects southward with the Malawi rift via the reactivated Mesozoic Rukwa rift. Seismicity and field evidence for active deformation along the Malawi rift end between 20° S and 25° S (Fig. 3). Other seismic belts include a narrow, north–south trending zone of seismicity along the Davie Ridge in the Mozambique channel (Grimison & Chen 1988) and the NE–SW trending Mweru and Zambezi rifts on both sides of the Early Proterozoic Zambia (Bangweulu) block.

These seismic belts bound broad areas mostly devoid of seismic activity. A northern block, centred on the Tanzania Archaean craton and bounded by the western and eastern branches of the EAR, has been referred to as the Victoria (Kaz'min *et al.* 1987) or Ukerewe–Nyanza (Hartnady 2002; Hartnady & Mlisa 2004) block (Fig. 3b). A southern block, bounded to the west by the Malawi rift and to the east by the Davie Ridge, has been previously identified as the Rovuma block (Hartnady 2002). The eastern boundary of this block may encompass parts of Madagascar, where earthquake slip vectors have a similar direction as along the Davie Ridge (Fig. 3a). The boundary between the Victoria and Rovuma blocks is less clear as it is not well expressed in the seismicity or by recent faulting. Localized deformation occurs in the Usangu and Ruhuhu grabens (Fig. 3a), where seismic reflection, field, gravity and remote sensing data indicate recent extension along morphologically young normal faults that connect further north with the Eastern rift (Harper *et al.* 1999; Le Gall *et al.* 2004).

We estimate angular velocities for Somalia and Victoria with respect to Nubia by simultaneously inverting GPS velocities at sites MBAR (Victoria block) and REUN, SEY1 and MALI (Somalian plate) with earthquake slip vectors along the Main Ethiopian Rift (Somalia–Nubia boundary), the Western rift (Victoria–Nubia boundary) and the Eastern rift (Victoria–Somalia boundary). We use a 20° standard deviation for earthquake slip vector directions, the $2\sigma$ standard deviation for GPS velocities, and impose plate circuit closure in the inversion. Results are given in Table 2 and shown on Figures 1 and 3B. The reduced $\chi^2$ is close to unity, meaning that a three-plate model is consistent with the data within their uncertainties. We find a Somalia–Nubia velocity statistically similar to the two-plate inversion above, which is expected because no additional data are used to define the Somalian plate motion. We find that the kinematics of the Victoria block can be described by a counter-clockwise rotation with respect to Nubia about a pole located in NE Sudan (Fig. 3b). Equivalently, its kinematics with respect to Somalia can be described by a counter-clockwise rotation about a pole located in northern Zambia (Fig. 3b).

Figure 3b shows the extension direction and rate predicted by our model along the eastern and western branches of the EAR, which we assumed to represent the boundaries of the Victoria microplate. We find that 2 to 5 mm $a^{-1}$, or 40 to 100% of the total Somalia–Nubia plate motion, is accommodated by extension across the Western rift, with present-day rates increasing from north to south. Our model predicts oblique extension in the Albert rift at the northern end of the Western rift, consistent with seismic reflection data showing a significant strike-slip component (Abeinomugisha & Mugisha 2004). Conversely, 3.5 to 1 mm $a^{-1}$, or 60 to 20% of the total Somalia–Nubia plate motion, is accommodated by extension across the Eastern rift, with rates decreasing from north to south. This southward decrease of the extension rate is consistent with the decrease in seismicity and with the progressive disappearance of prominent active faults southward along the eastern branch, as it propagates into cold cratonic domain (Le Gall *et al.* 2004).

The kinematics of the Rovuma block cannot be fully quantified because there are no GPS data to estimate its rotation rate. However, earthquake slip vector directions and aligned chains of Quaternary eruptive centres along the Malawi rift (Rovuma–Nubia boundary) and the Davie ridge, the assumed Rovuma–Somalia boundary, provide some first-order constraints. ENE-directed slip vectors along the Malawi rift provide the direction of motion of the Rovuma block with respect to Nubia along the western border of the Rovuma block (Fig. 4b; Brazier *et al.*; Ebinger *et al.* 1989). A velocity triangle for a point on its eastern border (around 42° W/12° S, Fig. 4a) can be drawn using the Somalia–Nubia velocity, known from GPS data (see above), and the azimuth of the Somalia–Rovuma motion, given by the average slip vector direction on the Davie Ridge. Given that the magnitude of the latter vector is not known, the Rovuma–Nubia vector can be anywhere between N95° W if the Somalia–Rovuma motion is close to zero (i.e. 100% of the Somalia–Nubia motion is taken up in the Malawi rift), to N170° W if the Somalia–Rovuma motion is close to Somalia–Nubia in magnitude (i.e. 100% of the Somalia–Nubia motion is taken up on the eastern boundary of the Rovuma plate). According to earthquake magnitudes from the 30-year global NEIC catalogue, seismic strain release has been larger at the Malawi rift than along the Davie Ridge, suggesting larger displacement rates on the western boundary of the Rovuma plate than on its eastern boundary. If this is representative of longer-term strain, then the Rovuma–Nubia velocity along the Davie Ridge is closer to east–west than north–south. In any case, the Rovuma–Nubia plate motion vector at the eastern boundary of the Rovuma plate around 42° W/12° S must be pointing in a southeast quadrant (Fig. 4). This, together with the ENE-directed slip vectors along the western boundary of that plate (Malawi rift), implies a clockwise rotation of Rovuma with respect to Nubia.

Given the lack of significant active deformation features at the boundary between the Victoria and Rovuma plates, could the data used here be fit equally well with a single plate encompassing Victoria and Rovuma? The *c.* 45° systematic difference in earthquake slip vector directions along the Western and Malawi branches (Fig. 3a) argues for two distinct plates. This can be further quantified by testing whether slip vectors along the Malawi rift are consistent with a single plate model. Including the Malawi rift slip vectors in a Nubia–[Victoria+Rovuma]–Somalia inversion gives a $\chi^2$ of 63.9 for 14 additional degrees of freedom (14 additional slip vector data). Using values from Table 2 (row 4), this leads to an *F*-test value of 1.3, corresponding to $\alpha = 0.77$. Hence, the hypothesis that the data fit equally well a single Victoria + Rovuma plate can be rejected at a 77% confidence level. This confidence level increases as *a priori* uncertainties on slip vector directions are lowered (we used 20° here). The counter-clockwise rotation of Victoria and clockwise rotation of Rovuma (with respect to Nubia) found here may actually explain the lack of well-expressed active faults and seismicity along their common boundary, as schematically shown on Fig. 5.

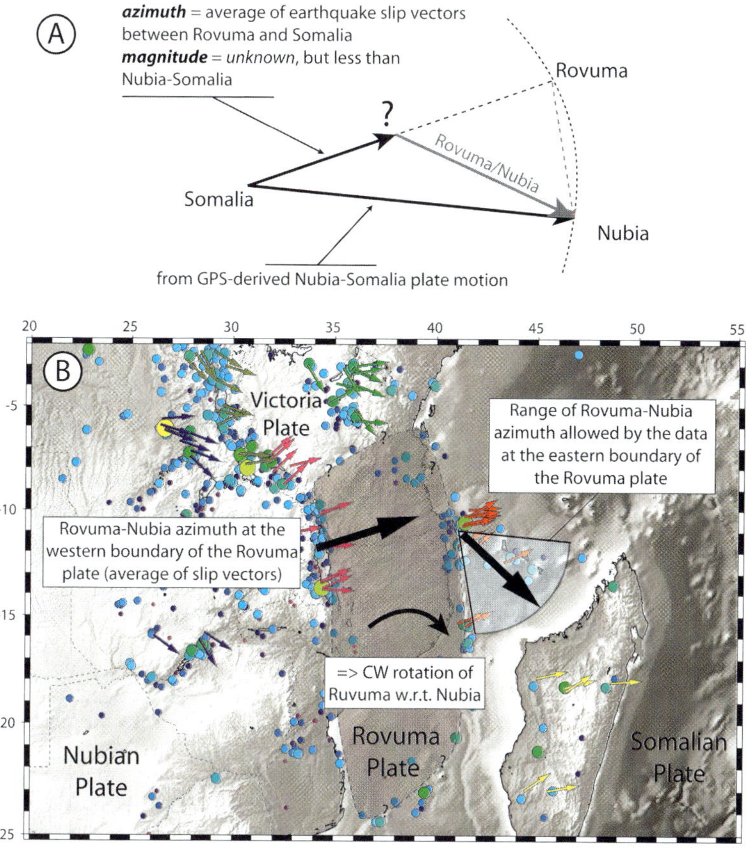


**Fig. 4.** Kinematics of the Rovuma block. (**a**) Velocity triangle for a location on the Rovuma–Somalia boundary around 42° N/12° S. (**b**) Zoom on the Rovuma plate. Large arrows show relative plate motions along its western and eastern boundaries (Malawi rift and Davie Ridge, respectively). They imply a CW rotation of Rovuma with respect to Nubia. Coloured arrows show earthquake slip vectors. Seismicity (NEIC catalog) is shown in background.

## Discussion and conclusion

The kinematic model proposed here, although consistent with the existing data, does not account for the widespread seismicity observed west of the Western rift (Fig. 1), where earthquake slip vectors show a consistent NW–SE trend. This seismicity has been interpreted as delineating an additional microplate (Transgariep block of Hartnady 2002). However, GPS station ZAMB, located on that potential microplate, does not show any significant relative motion with respect to Nubia. In addition, active tectonic features along the Zambia and Mweru rifts (Fig. 2a) have a limited length and do not delineate a continuous plate boundary. These observations do not preclude a relative motion of the Zambian craton with respect to Nubia, but it must happen at very slow rates, less than 1–2 mm $a^{-1}$.

Extrapolating the instantaneous kinematics found here to finite amounts of extension across the EAR is difficult given that extension rates may have varied in time. Assuming constant rates since the initiation of rifting (from 12–15 Ma to ~8 Ma from north to south along the Western rift and less than 5 Ma for the Eastern rift; Ebinger *et al.* 1997; Ebinger 1989; Abeinomugisha & Mugisha, 2004), the values found here lead to a maximum finite extension of about 30 km for the Western rift and 15 to 0 km (from north to south) for the Eastern rift. These values are at least twice the 15 km cumulative extension derived from reconstructions of surface fault geometries (Morley 1988; Ebinger 1989) and cantilever models of gravity and topography (Karner *et al.* 2000). The difference might be indicative of slower extension rates during the earlier phase of rifting.

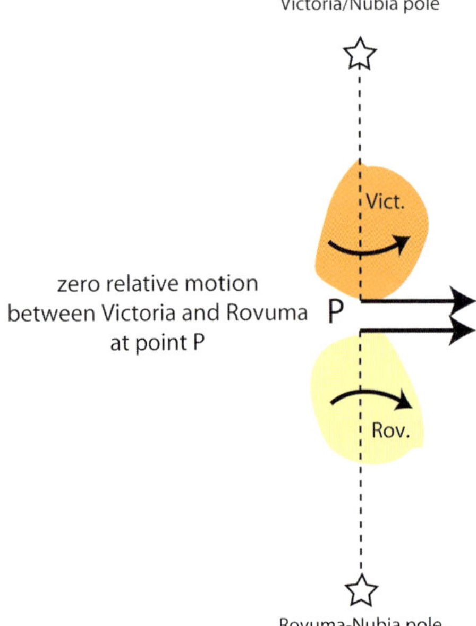


**Fig. 5.** Schematic kinematic diagram showing the counter-clockwise rotation of Victoria and clockwise rotation of Rovuma (with respect to Nubia). In the configuration shown here, the relative motion at point P along their common boundary is null.

The GPS dataset used here is clearly limited, but it correlates well with earthquake slip vectors. The combined interpretation of the two data sets is consistent with a plate boundary model where deformation of the African plate is restricted to the narrow rift valleys, and the surrounding lithospheric blocks remain undeformed. These kinematic data are consistent with tomographic models that also show lithospheric thinning and heating restricted to 100–150 km-wide zones beneath the faulted Ethiopian and Eastern rift valleys (e.g. Green *et al.* 1991; Ritsema *et al.* 1998; Bastow *et al.* 2005).

The identification of rigid lithospheric blocks that are kinematically independent from the Nubian or Somalian plates raises the question of the mechanism that drives their motion. The Ethiopian and Eastern rift systems form a diachronous but nearly contiguous belt that passes along the eastern side of the Tanzania craton. But what drives extension in the magma-poor Western rift system? A possible model stemming from studies of interactions between the cratonic keel and mantle flow provides insights into this problems. The core of the Victoria microplate is the 2.5–3 Ga Tanzania craton (Fig. 3b; Cahen *et al.* 1984), an assemblage of metamorphic and granitic terranes that has remained undisturbed tectonically since the Archaean, except for minor reheating attested by Tertiary kimberlites (e.g., Chesley *et al.* 1999). Seismic, xenolith and gravity data show that the lithosphere of the Tanzanian craton is colder and stronger than surrounding orogenic belts (e.g. Ebinger *et al.* 1997; Ritsema *et al.* 1998), and is underlain by a 170–250 km-thick lithospheric keel (Nyblade *et al.* 2000; Weeraratne *et al.* 2003; Debayle *et al.* 2005).

Viscous coupling between the convecting mantle and the lithosphere has been proposed as a significant driving force (either active or resistive) for plate motions, in particular in cratonic domains where a deep lithospheric keel is embedded in the convecting mantle (e.g. Ziegler 1992; Bokelmann 2002; Fouch *et al.* 2002; Sleep *et al.* 2002). Fouch *et al.*'s models (2002) predict focusing of flow around the boundaries of the keel which match anisotropy patterns determined in SKS-splitting observations from North America. Sleep *et al.* (2002) followed with the case of a mantle upwelling and a cratonic keel, demonstrating the strong focusing effect and directionality in both decompression melting and anisotropy. But without a consensus on the present location of plume stem(s) (e.g. Weeraratne *et al.* 2003; Furman *et al.* this volume Rogers, this volume), we cannot easily compare models and observations (e.g. Walker *et al.* 2004; Kendall *et al.* this volume).

Seismic anisotropy measurements and azimuthal variations in surface wave models provide independent constraints on upper mantle flow beneath East Africa and its possible interaction with the Tanzania craton (Fig. 6; Kendall *et al.* this volume; Gao *et al.* 1997; Walker *et al.* 2004). Local, surface wave, and SKS-splitting measurements along the Ethiopian and Eastern rifts show that strong, rift-parallel, anisotropy is primarily due to aligned melt zones in the mantle lithosphere (Kendall *et al.* 2005; Kendall *et al.* this volume). Azimuthal variations in surface wave models show sub-lithospheric fast shear waves coherently oriented in a NE direction from Tanzania to the Red Sea (Debayle *et al.* 2005) and near-radial orientations from a possible plume stem beneath Lake Eyasi (Weeraratne *et al.* 2003). Overall, seismic anisotropy fast directions in east Africa parallel the trend of the deep African superplume, but not the WNW motion of the African plate in a hot-spot frame (Fig. 6). At a larger scale, Behn *et al.* (2004) explain SKS splitting observations on islands surrounding Africa with density-driven upwelling flow from the African superplume and its interactions with a moving plate. Thus, existing data show both a strong signal from asthenosphere and mantle lithosphere, as well as regional variations in flow direction around the margins of the deep-rooted Tanzania craton.

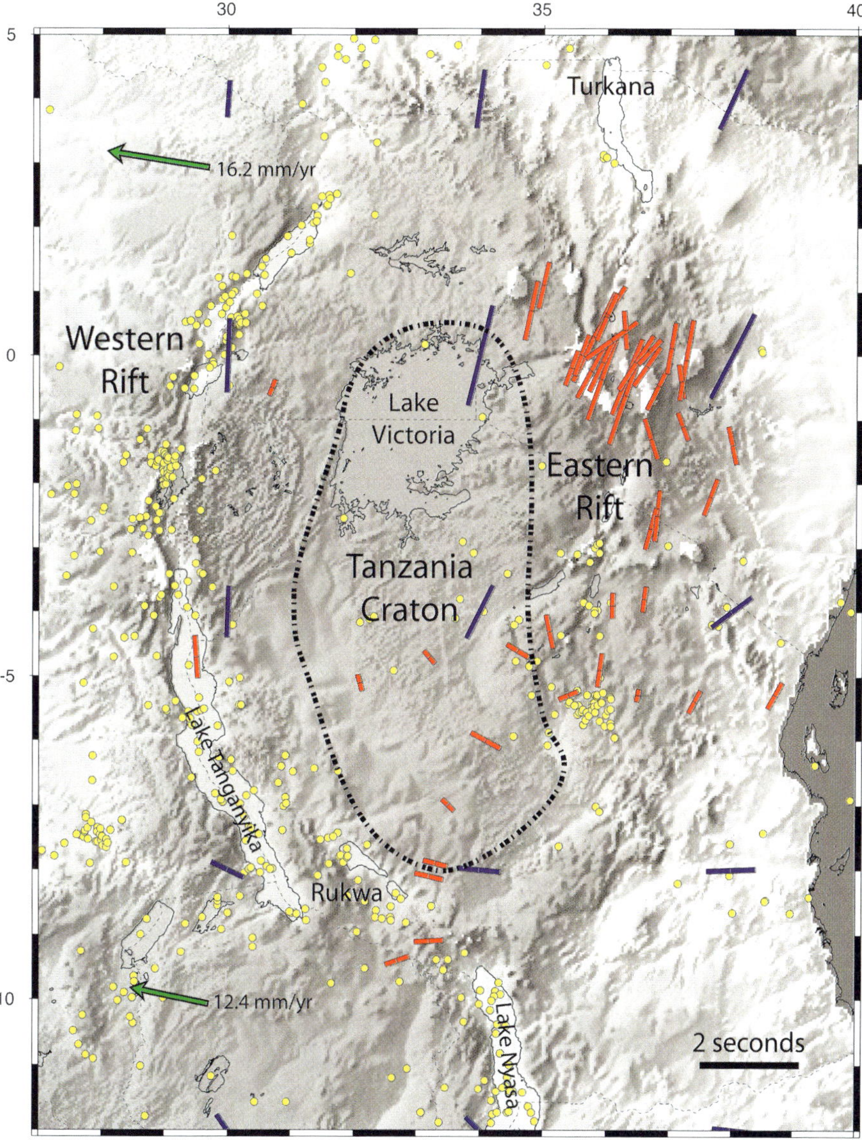


**Fig. 6.** Existing anisotropy data in the central part of the EAR. Red bars are from SKS shear-wave splitting data (Gao *et al.* 1997; Walker *et al.* 2004; Barruol & Ismail 2001), blue bars are from a surface wave study for a depth of 200 km (Debayle *et al.*, 2005). Green arrows show the current African plate motion in a hot-spot frame (Gripp & Gordon 2002).

On the basis of the asthenosphere–lithosphere flow models and anisotropy analyses, we propose that the independent motion of the Victoria plate shown in this study results from a combination of along-axis mantle flow and increased drag exerted on the Tanzania craton lithospheric keel by viscous coupling with normal or enhanced asthenospheric flow (e.g. Stoddard & Abbott 1996; Sleep *et al.* 2002; Fig. 7).

The NE-directed sublithospheric flow driven by the African Superplume upwelling acting on the Tanzanian craton keel, thickest in its southern part (Debayle *et al.* 2005), may force a counter-clockwise rotation of the craton, consistent with

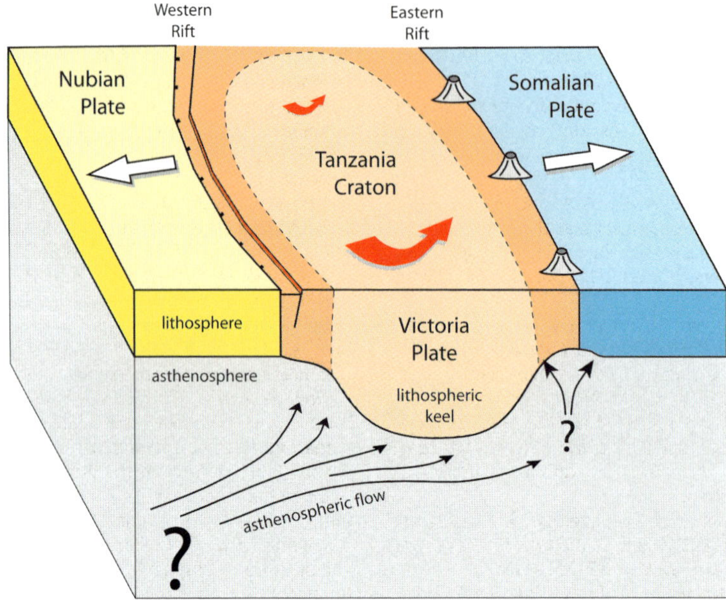


**Fig. 7.** Interpretative block diagram across the central part of the EAR. Viscous coupling between NE-directed asthenospheric flow driven by the African Superplume and the Tanzanian craton keel may force a counter-clockwise rotation of the craton, consistent with our preliminary kinematic data. The magma-poor Western rift accommodates differential movement of the deep-keeled Victoria plate, whereas the magma-rich Eastern rift formed in response to focused mantle flow with hot asthenosphere possibly ponding below thinned lithosphere.

our preliminary kinematic data. If confirmed, this differential rotation of the Victoria microplate can explain the origin and geometry of the enigmatic Western rift system without requiring a complex mechanism of stress transfer from the Eastern rift (Nyblade & Brazier 2002). In this model, the magma-poor Western rift accommodates differential movement of the deep-keeled Victoria plate, whereas the magma-rich Eastern rift formed in response to focused mantle flow with hot asthenosphere possibly ponding below thinned lithosphere. The interaction between southward-deepening asthenospheric upwelling may also explain the seismicity belts with minor or incipient normal faults that bound cratonic domains in southern Africa (Kaapvaal, Zimbabwe, Zambia cratons). With increased GPS and seismic coverage of the African continent, we should be able to map these variations in deformation and mantle flow and build a database to test current models for the fragmentation of continental plates in response to plate-driving forces.

This paper results in large part from discussions held during the US–Africa Workshop on Anatomy of Continental Rifts in Addis Ababa, Ethiopia, in June 2004. We are grateful to the local organizing committee and to M. Abdelselam, S. Klemperer and Gezahegn Yirgu for making this event possible. We thank M. Kendall, B. Le Gall, J. Rolet, J. Déverchère and W. Vétel for insightful discussions on the East African Rift, R. Gordon for reviewing an early version of this manuscript, and C. Vigny and Z. Garfunkel for their constructive comments.

## References

Abeinomugisha, D. & Mugisha, F. 2004. Structural analysis of the Albertine graben, Western Uganda. *Abstract*, East African Rift System Evolution, Resources, and Environment Conference, Addis Ababa, June.

Altamimi, Z., Sillard, P. & Boucher, C. 2002. ITRF2000: A new release of the International Terrestrial Reference Frame for earth science applications. *Journal of Geophysical Research*, 10.1029/2001JB000561.

Barruol, G. & Ben-Ismail, W. 2001. Upper mantle anisotropy beneath the African IRIS and GEOSCOPE stations. *Geophysical Journal International*, **146**, 549–561.

Bastow, I., Stuart, G.W., Kendall, J.M. & Ebinger, C. 2005. Upper mantle seismic structure in a region of incipient continental breakup: northern Ethiopian rift. *Geophysical Journal International*, **162**, 479–493 doi: 10.111/j.1365-246X.2005.02666.x.

BEHN, M.D., CONRAD, C.P. & SILVER, P.G. 2004. Detection of upper mantle flow associated with the African Superplume. *Earth and Planetary Science Letters*, **224**, 259–274.

BOKELMANN, G. 2002. Convection-driven motion of the North American craton: evidence from P-wave anisotropy. *Geophysical Journal International*, **148**, 278–287.

BRAZIER, R.A., NYBLADE, A.A. & FLORENTIN, J. 2005. Focal mechanisms and the stress regime in NE and SW Tanzania, East Africa. *Geophysical Research Letters*, **32**, L14315, doi: 10.1029/2005 GL023156.

CAHEN, L., SNELLING, N.J., DELHAL, J. & VAIL, J.R. 1984. *The Geochronology and Evolution of Africa.* 512 pp., Oxford University Press, New York.

CALAIS, E., DEMETS, C. & NOCQUET, J.M. 2003. Evidence for a post-3.16 Ma change in Nubia–Eurasia plate motion. *Earth and Planetary Science Letters*, doi:10.1016/S0012-821X(03)00482-5.

CHASE, C.G. 1978. Plate kinematics: the Americas, East Africa, and the rest of the world. *Earth and Planetary Science Letters* **37**, 353–368.

CHESLEY, J.T., RUDNICK, R.L. & LEE, C.T. 1999. Re–Os systematics of mantle xenoliths from the East African rift: Age, structure, and history of the Tanzanian craton. *Geochimica and Cosmochimica Acta*, **63**, 1203–127.

CHU, D. & GORDON, R. 1999. Evidence for motion between Nubia and Somalia along the Southwest Indian ridge. *Nature*, **398**, 64–66.

COURTILLOT, V., MANIGHETTI, I., TAPPONNIER, P. & BESSE, J. 1999. On causal links between flood basalts and continental breakup. *Earth and Planetary Science Letters*, **166**, 177–195.

DEBAYLE, E., KENNETT, B. & PRIESTLEY, K. 2005. Global azimuthal seismic anisotropy and the unique plate-motion deformation of Australia. *Nature*, **433**, 509–512, doi:10.1038/nature03247.

EBINGER, C. 1989. Tectonic development of the western branch of the East African Rift System. *Bulletin of the Geological Society of America*, **101**, 885–903.

EBINGER, C., POUDJOM-DJOMANI, Y., MBEDE, E. & FOSTER, A. 1997. Rifting the Archaean: Development of the Natron–Manyara–Eyasi basins. Tanzania, *Journal of the Geological Society London*, **154**, 947–960.

EBINGER, C., BECHTEL. T., FORSYTH, D. & BOWIN, C. 1989. Effective elastic plate thickness beneath the East African and Afar plateaux, and dynamic compensation of the uplifts. *Journal of Geophysical Research*, **94**, 2883–2901.

FERNANDES, R.M.S, AMBROSIUS, B.A.C., NOOMEN, R., BASTOS, L., COMBRINCK, L., MIRANDA, J.M. & SPAKMAN, W. 2004. Angular velocities of Nubia and Somalia from continuous GPS data: implications on present-day relative kinematics. *Earth and Planetary Science Letters*, **222**, 197–208.

FOSTER, A. & JACKSON, J.A. 1998. Source parameters of large African earthquakes: implications for crustal rheology and regional kinematics. *Geophysical Journal International*, **134**, 422–448.

FOUCH, M., FISCHER, K.M., PARMENTIER, E.M., WYSESSION, M.E. & CLARKE, T.J. 2000. Shear-wavesplitting, continental keels and patterns of mantle flow. *Journal of Geophysical Research*, **105**, 6255–6275.

FURMAN, T., BRYCE, J., ROONEY, T., HANAN, B., YIRGU, G. & AYALEW, D. 2006. Heads and Tails: 30 Million years of the Afar plume. *In*: YIRGU, G., EBINGER, C.J. & MAGUIRE, P.K.H. (eds) *The Afar Volcanic Province within the East African Rift System.* Geological Society, London, Special Publications, **259**, 95–119.

GAO, S., DAVIS, P.M., *ET AL.* 1997. SKS splitting beneath continental rift zones. *Journal of Geophysical Research*, **102**, 22781–22797.

GORDON, R. G. 1998. The plate tectonic approximation: Plate nonrigidity, diffuse plate boundaries, and global plate reconstructions. *Annual Reviews of Earth and Planetary Sciences*, **26**, 615–642.

GREEN, V., ACHAUER, U. & MEYER, R.P. 1991. A 3D seismic image of the crust and upper mantle beneath the Kenya rift. *Nature*, **354**, 199–203.

GRIMISON, N.L. & CHEN, W.P. 1988. Earthquakes in Davie Ridge–Madagascar region and the southern Nubian–Somalian plate boundary. *Journal of Geophysical Research*, **93**, 10,439–10,450.

GRIPP, A.E. & GORDON, R.G. 2002. Young tracks of hotspots and current plate velocities. *Geophysical Journal International*, **150**, 321–361.

HARPER, R., STONE, D. & MORLEY, C.K. 1999. Geophysics of the Usangu flats, Tanzania. *In*: MORLEY, C.K. (ed.) *Geoscience of Rift Systems – Evolution of East Africa, AAPG Studies in Geology*, No. 44, 111–114.

HARTNADY, C.J.H. 1998. Lesotho seismotectonics: SE African SCR in transition. *American Geophysical Union Chapman Conference on Stable Continental Region (SCR) Earthquakes*, National Geophysical Research Institute, Hyderabad, India (January 25–29), Abstract Volume, p.23.

HARTNADY, C.J.H. 2002. Earthquake hazard in Africa: perspectives on the Nubia–Somalia boundary. *South African Journal of Sciences*, **98**, 425–428.

HARTNADY, C.J.H. & MLISA, A. 2004. Boundaries and recent motions of the Ukerewe–Nyanza plate. *International Conference on the East African Rift System: Development, Evolution and Resources, Addis Ababa (June 20–24)*, Ethiopia, 81–84.

HENDRIE, D., KUSZNIR, N., MORLEY, C.K. & EBINGER, C.J. 1994. A quantitative model of rift basin development in the northern Kenya rift: Evidence for the Turkana region as an 'accommodation zone' during the Paleogene. *Tectonophysics*, **236**, 409–438.

JACKSON, J.A. 2002. Strength of the continental lithosphere: time to abandon the jelly sandwich? *GSA Today*, September issue.

JESTIN, F., HUCHON, P. & GAULIER, J.M. 1994. The Somali plate and the East African Rift system: present-day kinematics. *Geophysical Journal of International*, **116**, 637–654.

Karner, G., Byamungu, B., Ebinger, C., Kampunzu, A., Mukasa, R., Nyakaana, J., Rubondo, E. & Upcott, N. 2000. Distribution of crustal extension and regional basin architecture of the Albertine rift system, East Africa. *Marine and Petroleum Geology*, **17**, 1131–1150.

Kaz'min, V.G., Zonenshayn, L.P., Savostin, L.A. & Bershbitskaya, A.I. 1987. Kinematics of the Afro-Arabian rift system. *Geotectonics*, **21**, 452–460.

Kendall, J.-M., Stuart, G.W., Ebinger, C.J., Bastow, I.D. & Keir, D. 2005. Magma-assisted rifting in Ethiopia. *Nature*, **433**, 146–148.

Kendall, J-M., Pilidou, S., Keir, D., Bastow, I.D., Stuart, G.W. & Ayele, A. 2006. Mantle upwellings, melt migration and the rifting of Africa: Insights from seismic anisotropy. *In*: Yirgu, G., Ebinger, C.J. & Maguire, P.K.H. (eds) *The Afar Volcanic Province within the East African Rift System.* Geological Society, London, Special Publications, **259**, 55–72.

King, R.W. & Bock, Y. 2005. Documentation for the GAMIT GPS software analysis, release 10.2. unpublished.

Le Gall, B., Gernigon, L., *et al.* 2004. Neogene–Recent rift propagation in Central Tanzania: Morphostructural and aeromagnetic evidence from the Kilombero area. *Bulletin of the Geological Society of America*, **116**, 490–510.

Lemaux, J., Gordon, R.G. & Royer, J.-Y. 2002. The location of the Nubia–Somalia boundary along the Southwest Indian Ridge. *Geology*, **30**, 339–342.

McKenzie, D.P., Davies, D. & Molnar, P. 1970. Plate tectonics of the Red Sea and East Africa. *Nature*, **226**, 243–248.

Minster, J.B. & Jordan, T.H. 1978. Present-day plate motions. *Journal of Geophysical Research*, **83**, 5331–5354.

Modisi, M.P., Atekwana, E.A., Kampunzu, A.B. & Ngwisanyi, T.H. 2000. Rift kinematics during the incipient stages of continental extension: Evidence from the nascent Okavango rift basin, northwest Botswana. *Geology*, **28**, 939–942.

Morley, C. K. 1988. Variable extension in Lake Tanganyika. *Tectonics*, **7**, 785–801.

Nyblade, A.A. & Langston, C.A. 1995. East African earthquakes below 20 km depth and their implications for crustal structure. *Geophysical Journal of International*, **121**, 49–62.

Nyblade, A. & Brazier, R. 2002. Precambrian lithospheric controls on the development of the East African rift system. *Geology*, **30**, 755–758.

Nyblade, A.A., Owens, T.J., Gurrola, H., Ritsema, J. & Langston, C.A. 2000. Seismic evidence for a deep upper mantle thermal anomaly beneath East Africa. *Geology*, **7**, 599–602.

Prawirodirdjo, L. & Bock, Y. 2004. Instantaneous global plate motion model from 12 years of continuous GPS observations. *Journal of Geophysical Research*, **109**, B08405, doi:10.1029/2003JB002944.

Ritsema, J., Nyblade, A.A., Owens, T.J. & Langston, C.A. 1998. Upper mantle seismic velocity structure beneath Tanzania, east Africa: Implications for the stability of cratonic lithosphere. *Journal of Geophysical Research*, **103**, 21, 201–21,213.

Rogers, N.W. 2006. Basaltic magmatism and geodynamics of the East African Rift System. *In*: Yirgu, G., Ebinger, C.J. & Maguire, P.K.H. (eds) *The Afar Volcanic Province within the East African Rift System.* Geological Society, London, Special Publications, **259**, 77–93.

Sella, G. F., Dixon, T.H. & Mao, A. 2002. REVEL: A model for recent plate velocities from Space Geodesy. *Journal of Geophysical Research*, doi: 107, 10.1029/2000JB00033.

Sleep, N.H., Ebinger, C.J. & Kendall, J.M. 2002. Deflection of mantle plume by cratonic keels. *In*: Fowler, C.M.R., Ebinger, C.J. & Hawkesworth, C.J. (eds) *The Early Earth: Physical, Chemical, and Biological Development.* Geological Society, London, Special Publications, **199**, 135–150.

Stoddard, P.R. & Abbott, D. 1996. Influence of tectosphere upon plate motion. *Journal of Geophysical Research*, **101**, 5425–5433.

Walker, K.T., Nyblade, A.A., Klemperer, S.L., Bokelmann, G.H.R. & Owens, T.J. 2004. On the relationship between extension and anisotropy: Constraints from shear wave splitting across the East African Plateau. *Journal of Geophysical Research*, **109**, B08302, doi: 10.1029/2003JB002866.

Weereratne, D.S., Forsyth, D.W., Fischer, K.M. & Nyblade, A.A. 2003. Evidence for an upper mantle plume beneath the Tanzanian craton from Rayleigh wave tomography. *Journal of Geophysical Research*, **108**, doi:10,1029/2002JB002273.

Ziegler, P. 1992. Plate tectonics, plate moving mechanisms and rifting. *Tectonophysics*, **215**, 9–34.

# Constraints on the structural development of Afar imposed by the kinematics of the major surrounding plates

Z. GARFUNKEL[1] & M. BEYTH[2]

[1]*Institute of Earth Sciences, Hebrew University Jerusalem, 91904, Israel (e-mail: zvi.garfunkel@huji.ac.il)*

[2]*Geological Survey of Israel, 30 Malkhe Yisrael Str., Jerusalem, 95501, Israel (e-mail: mbeyth@gsi.gov.il)*

**Abstract:** The development of the Afar rift–rift–rift triple junction is analysed from the view-point of the Nubia, Arabia and Somali plate kinematics. A variety of constraints allow definition of a range of kinematic models that approximate well to the plate motions with a resolution of a couple to tens of kilometres. Rigid plate kinematics probably cannot resolve smaller motions. The size and location of the new area that opened between the major plates is inferred from plate kinematics and this provides a framework in which to assess the structural development that accommodates plate separation.

The development of the Afar region was complicated by the presence of microplates – the Danakil and Aisha blocks – which results in a complex plate boundary geometry. This led to local deformation that does not directly reflect the divergence of the major plates, e.g. rotations of microplates and of minor blocks about vertical axes and strike–slip faulting. The opening of new area was accommodated by various crustal growth and accretion mechanisms, e.g. building of thick new igneous crust, normal seafloor spreading, and/or crustal stretching. Thus, plate motions by themselves do not determine the development of the plate boundaries, as this is strongly influenced by other factors such as lateral variations of the rates of magma supply (e.g. away from, and over, the Afar plume).

The plate boundaries changed – e.g. the Gulf of Aden spreading centre propagated westward *c.* 2 Ma ago and normal seafloor spreading began along portions of the Red Sea axis since *c.* 5 Ma ago – while there were no resolvable changes in the plate motions. Such changes therefore signify a reorganization of the way in which plate divergence and addition of new area is accommodated: diffuse extension may give way to separation along a narrow spreading centre, or new plate boundaries may form at the expense of other boundaries that became inactive.

Plate tectonics is the framework for understanding geological processes, but the use of plate kinematics for tectonic interpretation is often difficult because a variety of processes accompany plate motions. One setting where it is advantageous to use plate kinematics is along divergent plate boundaries. There, the plate motions define the extent of new area that is opened, which provides a framework for evaluating the resulting geological features and growth of new crust. Here we apply this approach to the Afar depression and surrounding areas (Fig. 1).

The Afar depression is the site of active plate separation and intense magmatic activity at the triple junction between the Red Sea and Gulf of Aden nascent oceans and the Ethiopian rift (Fig. 1; Laughton 1966; Gass & Gibson 1969; Roberts 1969). McKenzie *et al.* (1970) treated it as a rift–rift–rift triple junction between the Nubia (Africa), Arabia and Somali plates, but the structure of Afar is much more complex. Here crustal divergence is accommodated by diffuse extension and by addition of igneous material in an up to 250 km-wide area (Mohr 1970, 1978; Merla *et al.* 1973; Barberi *et al.* 1975). At the same time, seafloor spreading along narrow spreading centres occurs in the adjacent Gulf of Aden and the southern Red Sea, but rather than meeting in a triple junction these spreading centres extend on the two sides of the Danakil block—a continental sliver in the midst of the triple junction area that acted as an independent microplate. As most of Afar is covered by Plio-Quaternary volcanics and sediments, attention was focused on the young tectonics (e.g. Barberi & Varet 1977; Courtillot *et al.* 1984; Tapponier *et al.* 1990; Souriot & Brun 1992; Acton *et al.* 1991; Manighetti *et al.* 1997, 2001; Eagles *et al.* 2002; Bayene & Abdelsalam 2005; and references therein) whereas the older structures masked by these rocks received less attention.

*From*: YIRGU, G., EBINGER, C.J. & MAGUIRE, P.K.H. (eds) 2006. *The Afar Volcanic Province within the East African Rift System*. Geological Society, London, Special Publications, **259**, 23–42.
0305-8719/06/$15.00 

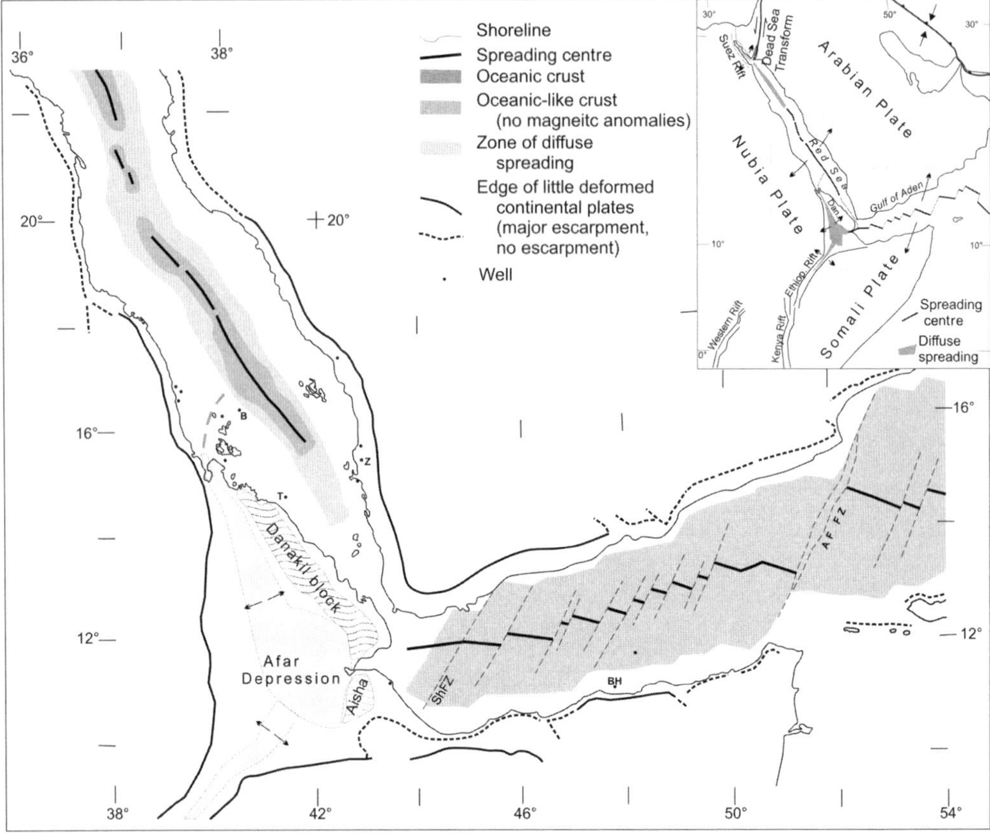


**Fig. 1.** Tectonic setting of the Afar depression.

One way of viewing the special features and early history of the Afar area, followed in this work, is to examine how its development was related to the Nubia–Arabia–Somali plate kinematics. This constrains the opening of new area between these plates and allows the study of how it was related to the structural development in time and space. In what follows, we first briefly review the geological setting, and then we re-examine the acceptable range of Nubia, Arabia, and Somali plate kinematic models; afterwards, we use these models to reconstruct the changes in the plate configuration around Afar and examine the large-scale structural development of the region within this framework. This discussion exemplifies the use of plate kinematics to help decipher the role of microplates (blocks) and of changes in the plate boundaries, and the operation of different crustal growth mechanisms. We also examine the limitations of this approach. Here, the time-scale of Gradstein *et al.* (2004) is used.

## Geological setting

### *Pre-break-up history*

The continental areas around the Afar depression were stabilized during the Late Proterozoic Pan-African (East African) orogeny. Subsequently, during most of the Phanerozoic, these areas were parts of a continuous stable platform, and though it was affected by intra-continental rifting in Mesozoic times, it retained its continuity until the Cenozoic continental break-up stage.

The Pan-African basement on the two sides of the Red Sea is built of juvenile terranes, while farther east in Arabia and on the two sides of the Gulf of Aden the basement consists of older continental units that were reworked during the Pan-African orogeny, and these terranes extend southward into Ethiopia (Stoeser & Camp 1985; Vail 1985; Brown *et al.* 1989; Patchett 1992; Abdelsalam &

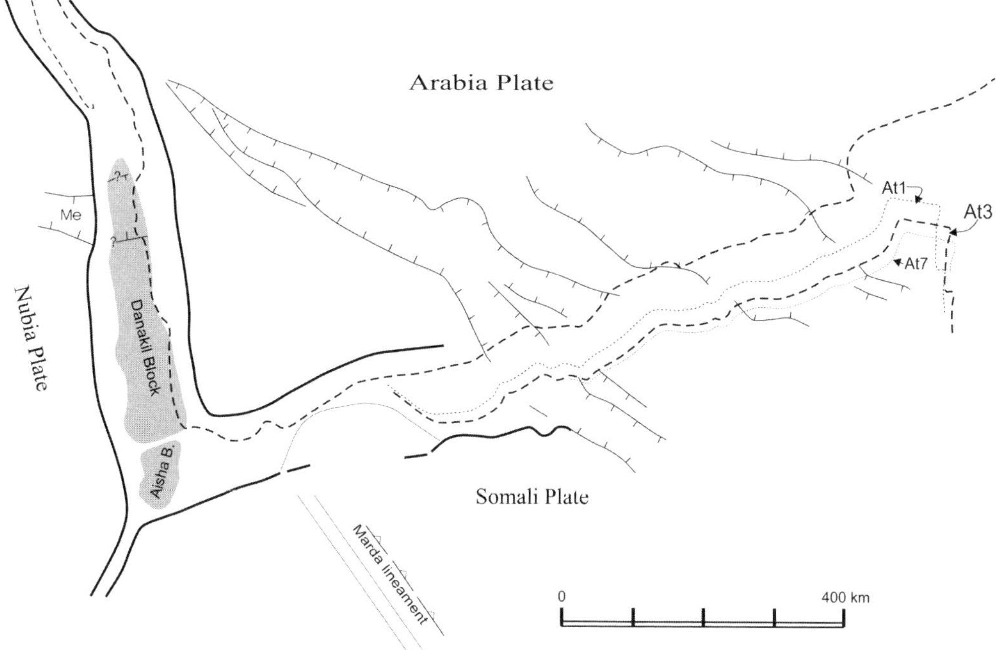


**Fig. 2.** Reconstruction of the Gulf of Aden, Afar and southern Red Sea at the end of the pre-rifting stage. Reconstruction corresponding to different models in Table 3 are shown.

Stern 1996; Kroner & Sassi 1994; Whitehouse *et al.* 1998; Asrat *et al.* 2001; Stern 2002).

After the Pan-African orogeny, the region became a stable platform (Beydoun 1970; Beyth 1973*a*, 1991; Merla *et al.* 1973; Bosellini 1989; Bott *et al.* 1992; Davidson *et al.* 1994; Tefera *et al.* 1996). On the flanks of the Cenozoic rifts, the sedimentary cover of the platform consists mainly of Jurassic, Cretaceous and Palaeogene series that were deposited when the region was repeatedly invaded by shallow seas which spread inland from the Indian Ocean. The Danakil block and the Aisha (also called Ali Sabieh) block farther south had the same history, which establishes that they are continental splinters stranded within the triple junction area (Merla *et al.* 1973; Clin 1991; Bunter *et al.* 1998; Sagri *et al.* 1998).

The Late Jurassic and Cretaceous were periods of significant intra-continental rifting in this region (Fig. 1; Beydoun 1970; Bott *et al.* 1992; Redfern & Jones 1996; Bosence 1997). Several WNW–SSE graben systems, mostly known from the subsurface, formed on the two sides of the Gulf of Aden. The Mekele 'outlier' on the west side of the Afar depression is another Late Jurassic rift structure (Beyth 1972; Martire *et al.* 1998). In the northern Danakil block, a particularly thick Jurassic section (Fig. 1; Bunter *et al.* 1998) may mark a similar *c.* 120 km-wide graben, but the structure is poorly understood. Another pre-rifting structure is the Marda 'fault' (Fig. 2)—a major NW trending faulted flexure, NE side down—that extends from the Afar depression across the Horn of Africa and determines the distribution of Late Cretaceous and Palaeogene sediments (Fig. 1; Merla *et al.* 1973; Boccaletti *et al.* 1991) but its nature is still not clear.

The pre-break-up period ended with a period of erosion that produced on the lands surrounding Afar an extensive flat erosion surface (peneplain) on which a cover of palaeosols (laterites) formed (Beyth 1973*b*; Brown *et al.* 1989; Drury *et al.* 1994; Davidson *et al.* 1994; Menzies *et al.* 1997).

## Continental break-up stage

The Cenozoic continental break-up marks a new stage in the tectonic history of East Africa and the Red Sea region. The most conspicuous early event in the Afar area was the outpouring of up to 1–2 km of mainly basaltic flood volcanics over a large area on the Ethiopian plateau and in Yemen between about 31 Ma to 29 or 28 Ma ago ($^{40}Ar/^{39}Ar$ ages; previous K–Ar ages are less reliable), i.e. in mid-Oligocene times, covering the flat peneplain and its laterite cover, but some volcanism persisted for several Ma and extended over large areas (Drury *et al.* 1994; Zumbo *et al.* 1995*a*; Hoffmann *et al.* 1997; Menzies *et al.* 1997; Ukstins *et al.* 2002; Coulie *et al.* 2003; Kieffer *et al.* 2004; Wolfenden *et al.* 2005). Less-voluminous volcanics of that age extend several hundred kilometres farther north (Sebai *et al.* 1991; Kenea *et al.* 2001). There is no evidence for earlier faulting. An earlier volcanic phase between 45 Ma and 33 Ma ago (Middle Eocene to basal Oligocene) occurred $\geq$500 km to the south in southern Ethiopia and in nearby Kenya, but this was not related to the rift system considered here (Behre *et al.* 1987; George *et al.* 1998; Ebinger *et al.* 2000).

Initiation of continental break-up is recorded by the formation of depocentres along the Gulf of Aden and the Red Sea and by faulting of their margins. The oldest rift-related depressions and structures documented along the margins of the eastern part of the Gulf of Aden are of Early Oligocene (pre-28 Ma) age, and they were accentuated in the later part of this period, but the scope of this initial faulting was limited (Platel & Roger 1989; Fantozzi 1996; Watchorn *et al.* 1998; Robertson & Bamakhalif 2001). Within the Gulf of Aden itself, the oldest-known fill, found in Bandar Harshau-1 well, consists of basinal sediments with middle Oligocene (26–27 Ma old) fossils that overlie pre-rift shallow water marine beds (Wyn-Hughes *et al.* 1991). Steckler & Gomm'a (1995) suggested that the Red Sea was initiated 35 Ma ago on the basis of fission track ages. However, such ages by themselves do not date faulting and may record various events. Since there is no evidence for faulting or major erosion of this age elsewhere along the Red Sea, the significance of this finding remains unclear. In the southern Red Sea, the oldest-known basinal deposits, in which the Thio-1 well bottomed, contain Late Oligocene (*c.* 25 Ma) fossils (Wyn-Hughes *et al.* 1991), but on its margin in Southern Sudan faulting began *c.* 28 Ma ago (Kenea *et al.* 2001). Another early rift structure is evidenced by nearly 1 km of lake sediments and volcanics along a *c.* 600 km long narrow belt inland of the coast of Saudi Arabia that were interpreted as the fill of an embryonic continental rift (Schmidt *et al.* 1982; Bohannon 1986: Brown *et al.* 1989). They were strongly faulted and then intruded by sills and dikes with $^{40}Ar/^{39}Ar$ ages of 24 Ma to *c.* 20 Ma (Sebai *et al.* 1991), so here too rifting was under way a few My after the end of the flood volcanism. Thus, shortly after the flood volcanism around Afar the rifts along the Gulf of Aden and the Red Sea subsided enough to be invaded by the sea, though minor initial faulting appears to have begun earlier.

Progressive plate separation led at first to stretching of the continental crust along the divergent plate boundaries, but in more advanced stages plate separation led to formation of new thin crust along the Red Sea and the Gulf of Aden (Cochran 1981, 1983, 2005; LePichon & Gaulier 1988; Mitchell et al. 1992; D'Acremont *et al.* 2005). The deep part of the Gulf of Aden east of the Shukra El-Sheikh fracture zone is underlain by oceanic crust which formed by normal seafloor spreading since 16–18 Ma ago along a mid-oceanic ridge offset by transform faults (Fig. 1; Sahota 1990; Prodehl & Mechie 1991; D'Acremont 2002; Leroy *et al.* 2004; D'Acremont *et al.* 2005). The structure of the Red Sea is less well defined. Except for a narrow axial trough that exposes basaltic crust, this basin is underlain by sediments, up to 4–6 km thick, much of which consist of *c.* 16 to 5 Ma old (Middle and Late Miocene) evaporites (Coleman 1984; Hughes & Beydoun 1992; Mitchell *et al.* 1992). The underlying crust is thin ($\leq$10 km), but it probably varies across the basin (Makris *et al.* 1991; Egloff *et al.* 1991; Prodehl & Mechie 1991; Cochran 2005). The central third of the basin, especially in its southern two-thirds, is characterized by an accentuated Bouguer gravity high, and may well be underlain by oceanic crust. However, magnetic anomalies (2 Ma to 6 Ma old) are developed discontinuously only over the axial trough, and transform faults cannot be recognized (Fig. 1; Roeser 1975; Cochran 1983; Izzeldin 1987; Chu & Gordon 1998; Garfunkel *et al.* 1987). In the areas flanking this zone, the gravity anomaly is less pronounced, so there the crust may be somewhat thicker and may consist, in part or entirely, of strongly attenuated continental (*c.* one-fifth of its original thickness).

In contrast, the Afar depression has a very different character, being mostly close to or above sea level. Though it is also characterized by a positive Bouguer gravity anomaly, seismic refraction studies show that its crust differs from normal continental crust and is mostly 25–20 km thick, and it thins to 16 km in the north, i.e. it is much thicker than the crust of the Red Sea and Gulf of Aden (Makris *et al.* 1991; Makris & Ginzburg 1987; Tiberi *et al.* 2005). Dugda *et al.* (2005) found a similar thickness from receiver function analysis obtained in a single station in Afar. In the north of the depression, $>$1 km of sediments are exposed (Sargi *et al.* 1998) and evaporites may occur in the subsurface, but further south the seismic data show that the crust consists largely of crystalline rocks. In particular, the high Poisson's ratio found by Dugda *et al.* (2005) indicates a mafic crust in central Afar. The Danakil block, which is underlain by Precambrian basement, has a weak gravity signature and the seismic refraction data show that its crust is much thinner than normal continental crust, but thicker than in the neighbouring Afar depression (Makris & Ginzburg 1987; Beyth

1991; Makris *et al.* 1991). In the Afar depression, igneous activity remained widespread since the beginning of rifting (Barberi *et al.* 1975).

## Plate kinematics

The plate kinematics of the Afar region was much studied. Gass & Gibson (1969) and Roberts (1969) recognized already that the tectonics of this region was dominated by the divergence of the Nubia (Nub), Arabia (Arb) and Somali (Som) (Fig. 1). Later, several quantitative models were proposed for the entire system (McKenzie *et al.* 1970; Joffe & Garfunkel 1987; Le Pichon & Gaulier 1988; Jestin *et al.* 1994) or for parts of it, stressing the last few million years (e.g. Chu & Gordon 1998, 1999; Eagles *et al.* 2003; Horner-Johnson *et al.* 2005), while GPS measurements constrain the current motions (McClusky *et al.* 2003; Calais *et al.* 2006, and references therein). For plate kinematics to be useful for interpretation of geological features, the plate motions should be determined to within 20–30 km or less. This is not easy to achieve because the available constraints, especially regarding the older motions, are not very precise. However, combining data from different places and using the plate circuits at the Afar and Sinai triple junctions allows to obtain a range of permissible models for the entire system (e.g. as discussed by Joffe & Garfunkel 1987). Though the Euler poles and rotation angles in various models may appear to differ significantly, it turns out that the actual motions implied by these models are quite similar, so they can be used to constrain the evolution of areas about which there is little direct information.

With this in mind, we re-examine the relative motions of the Africa, Arabia and Somali plates. We consider the entire plate system, rather than individual plate pairs, and deal with the entire history of the plate separation. In addition, we do not seek a 'best' model and the formal uncertainties of the Euler poles and the rotation angles. Rather we evaluate a range of models that are permitted by the known constraints and then focus on the actual motions implied by these models (e.g. for the Afar region). Since it is not possible to choose between the permissible models on the basis of the available geological information, we consider this to be the resolving power of plate kinematics when applied to geological structures. Therefore, we avoid its application to fine structural details and to structures that are only a few million years old: in that period, the plate motions were similar to the resolving power of the regional plate kinematics. The young structures are better studied by mapping, seismic focal studies and GPS data, which are outside the scope of this study. In what follows, we examine the plate motions since 5–10 Ma and since the beginning of plate separation (*c.* 28 Ma ago). These motions are considered separately because they are estimated on the basis of different data and are constrained quite well for the purpose of this work, whereas the available data do not allow a finer time resolution of the plate motions.

### *Plate kinematics in the last 10 Ma*

The Nub–Arb plate motion, i.e. the opening of the Red Sea, during the last 5 Ma is constrained by magnetic anomalies along the axis of the Red Sea and by the relations at the Sinai triple junction at its northern end (Joffe & Garfunkel 1987; Le Pichon & Gaulier 1988; Chu & Gordon 1998). The proposed models are quite similar, and the motions that they predict during the last 5 Ma differ by a few kilometres only (Table 1). GPS measurements (McClusky *et al.* 2003) show a similar motion, but with a somewhat slower velocity. The motion since 10 Ma (Table 2) is not directly constrained. It can either be approximated by extrapolating the 5 Ma motion, by analogy with the Gulf of Aden,

**Table 1.** *Young opening of the Red Sea (Arb–Nub)*

| | Euler pole | | Rotation rate | Position 5 Ma ago of point on Arb now at | |
|---|---|---|---|---|---|
| | lat (°N) | long (°E) | (°/Ma, ccw) | 17.0° N | 40.4° E |
| (1) | 32.2 | 24.0 | 0.402 | 17.49 | 40.97 |
| (2) | 32.75 | 22.64 | 0.402 | 17.52 | 40.99 |
| (3) | 32.59 | 23.70 | 0.418 | 17.51 | 41.00 |
| (4) | 31.5 ± 1.2 | 23.0 ± 2.7 | 0.403 ± 0.05 | 17.52 | 40.94 |
| (5) | 30.5 ± 1.0 | 25.70 ± 2.3 | 0.37 ± 0.04 | 17.41 | 40.86 |

(1) Joffe & Garfunkel (1987) adjusted to the time-scale of Cande & Kent (1995). (2) LePichon & Gaulier (1988) average since 4.7 Ma ago. (3) Jestin *et al.* (1994). (4) Chu & Gordon (1998), average since 3.2 Ma. (5) McClusky *et al.* (2003), GPS data. ccw, counterclockwise.

**Table 2.** *Models of the relative motion between Nub, Arab, Som in the last 10 Ma*

| | Euler pole | | Rotation | | |
|---|---|---|---|---|---|
| | Lat (°N) | Long (°E) | (°ccw) | | |
| **A. Arb–Nub**: opening of Red Sea, extrapolated to 10 Ma from Table 1 | | | | | |
| (R1)* | 32.2 | 24.0 | 4.02 | No. 1 in Table 1 | |
| (R2) | 31.5 | 23.0 | 4.03 | No. 4 in Table 1 | |
| (R3) | 32.0 | 24.0 | 3.80 | Smaller rotation | |
| **B. Som–Arb**: opening of Gulf of Aden, 0–10 Ma | | | | | |
| (A1) | 25.00 | 26.0 | −4.50 | Joffe & Garfunkel 1987 | |
| (A2) | 23.30 | 27.4 | −4.40 | D'Acremont 2002 | |
| (A3) | 25.67 | 25.5 | −4.09 | Present study | |
| (A4) | 25.29 | 26.5 | −4.23 | Present study | |
| (A5) | 24.10 | 28.5 | −4.56 | Present study | |
| **C. Som–Nub** calculated from different motions in A and B | | | | | |
| | | | | Opening in NMER** | |
| (E1) | −19.55 | 34.26 | −0.730 | 39.2 km | (R1) + (A1) |
| (E2) | −31.28 | 42.72 | −0.786 | 57.8 km | (R1) + (A2) |
| (E3) | −51.33 | 39.40 | −0.477 | 46.0 km | (R1) + (A3) |
| (E4) | −36.93 | 43.33 | −0.563 | 45.1 km | (R1) + (A4) |
| (E5) | −18.00 | 46.41 | −0.837 | 43.6 km | (R1) + (A5) |
| (E6) | −27.42 | 33.96 | −0.531 | 35.4 km | (R3) + (A3) |
| (E7) | −18.62 | 37.99 | −0.657 | 33.9 km | (R1) + (A5) |

*(R1) etc. – model designation, referred to in part C.
**Opening across Ethiopian rift at 9° N, 40° E that is predicted by the rotations given in first three columns of part C; these were calculated by combining the models given in the last column of part C. MER, Main Ethiopian Rift; ccw, counterclockwise.

but since the long-term average was probably slower (see below) it is more likely that the 10 Ma average opening rate was slower than the 5 Ma average (model R3 in Table 2).

The Arb–Som plate separation, i.e. the opening of the Gulf of Aden, is constrained by the strikes of the fracture zones and mainly by the magnetic anomalies in the Gulf. The fracture zones constrain the latitude of the Euler poles at any given longitude, but allow a wide range of longitudes (e.g. Jestin *et al.* 1994) because they have similar strikes and these are not very tightly defined since the fracture zones are not simple linear features (e.g. Tamsett & Searle 1988). Fitting the magnetic anomalies considerably narrows the range and also gives the opening rate which is found to have remained quite constant. Table 2 lists representative permissible published models of the plate motion in the last 10 Ma and some models that were derived in the present work (A3–A5) based on anomaly 5 (9.74–10.95 Ma) as mapped by Cochran (1981) Sahota (1990) and D'Acremont *et al.* (2005).

These models for the Nub–Arb and Arb–Som separation should not be viewed in isolation, because when combined they should yield acceptable Nub–Som motions, e.g. along the Ethiopian rift. The implications of some combinations are given in Table 2. It is seen that for a given Arb–Nub motion, the more westerly Som–Arb poles lead to a more southerly position of the Som–Nub pole (models E3–E5). Here, only positions south of the East African rifts are accepted, because more northern poles imply overall shortening across the more southern part of the East African rift system, or that parts of the Nubia and the Somali plates (west of the rifts and in the Indian Ocean, respectively) are not rigid. An inevitable consequence of the different positions of the Arb–Nub and Arb–Som Euler poles is the Nub–Som separation along the Main Ethiopian Rift (MER). Models assuming for the Red Sea the same average motions 0–5 Ma and 5–10 Ma ago predict an opening of *c.* 40 km or more across the northern MER (about half its width) in the last 10 Ma, i.e. at an average rate of *c.* 4 mm $a^{-1}$, while models allowing slower past opening of the Red Sea give smaller numbers, which are more reasonable. Small changes in the parameters yield even smaller numbers, but such models are not well constrained on the basis of available data, so we do not pursue this issue except for noting the general relations. However, even the smaller numbers are large compared with estimates based on local data, e.g. Wolfenden *et al.* (2004), and this will be further discussed when dealing with the total plate motions.

Data about short-term motions provide further insights. Chu & Gordon (1999) estimated the Nub–Som separation since 3.2 Ma ago, based on data from the SW Indian Ridge, to be >5 mm $a^{-1}$ across the northern MER. An updated study (Horner-Johnson *et al.* 2005) shows even faster average opening (>8 mm $a^{-1}$) in this period. Geodetic measurements revealed an opening rate of $4.5 \pm 1.0$ mm $a^{-1}$ (Bilham *et al.* 1999). GPS studies gave similar results. Sella *et al.* (2002) found 6.9 mm $a^{-1}$ (Som–Nub motion: 0.085°/Ma about 35.49° S, 24.02° E), Fernandes *et al.* (2004) found 6.9 mm $a^{-1}$ (0.069°/Ma about 43.06° S, 36.97° E), and Calais *et al.* (this volume) found *c.* 6 mm $a^{-1}$ (0.069°/Ma about 43.06° S, 27.51° E). Thus these works point at opening rates that cannot be extrapolated much backwards, because this would imply an opening close to the entire width of the MER during its history. These rates also exceed the average rates since 5 Ma obtained in this work, so young acceleration of the Nub–Som plate separation is indicated. However, if the northern MER opened at a rate of $\geq$5 mm $a^{-1}$ since 2–3 Ma ago, then in this period alone the opening amounted to $\geq$10–15 km, so opening of $\geq$20–25 km during the last 10 Ma is not unreasonable. The acceleration of the Nub–Som opening is compatible with young acceleration of the opening of the Red Sea (as the Arb–Som opening effectively did not change), but we do not have good constraints for modeling these changes. Because of these problems, we consider that our plate reconstructions for the last 10 Ma may involve uncertainties of up to *c.* 20 km.

### *The entire plate divergence*

Several lines of evidence constrain the total opening of the Red Sea. One constraint is provided by the relations at its northern end, i.e. at the Sinai triple junction, where the Arb–Nub plate motion is partitioned between the Dead Sea transform (DST) and the Suez rift (Joffe & Garfunkel 1987; Le Pichon & Gaulier 1988). Matching of rock units, major shear zones and ophiolite sutures in the Precambrian basement provides additional constraints. Sultan *et al.* (1993) assumed that these features were linear in map view and inferred that they are matched best by fitting the coastlines of the Red Sea. However, Freund (1970) and Mohr (1970) showed that such a fit is unacceptable because it greatly overestimates the opening of the Suez rift and does not leave room for the continental Danakil block. It also ignores the stretching of the continental margins which is evidenced by the fact that wells along the both margins of the Red Sea reached Precambrian basement (Mitchell *et al.* 1992). In fact, the basement features examined by Sultan *et al.* (1993) are not as regular in map view as they envisaged, and most of them trend close to SW–NE which is close to the direction of plate separation, so they do not constrain well the amount of opening. Only matching of shear zones on the two sides of the northern Red Sea that trend *c.* NW–SE—nearly perpendicular to the opening direction—gives useful constraints on the motion, as was first recognized by Abdel Gawad (1970).

The reconstruction of the Gulf of Aden is constrained by the fracture zones that extend across most of this basin width and by matching the continental margins, but the total opening is best inferred from matching the Late Jurassic–Cretaceous rifts across this basin, as was already recognized by Beydoun (1970), though the somewhat irregular pattern of these rifts does not allow a tightly constrained reconstruction. One possible reconstruction is shown in Fig. 2. Matching of the Precambrian basement could provide an additional constraint, but this is not possible with the available data.

Some acceptable models for the Arb–Nub and Arb–Som motions are listed in Table 3. The different positions of the Arb–Nub and Arb–Som Euler poles require separation between Som and Nub along the MER. Table 3 lists the implications for Som–Nub of different combinations of the Arb–Nub and Arb–Som models. It is seen that some predict quite large opening in the northern MER, close to its entire width, which is excessive. Other models, with the more northern Nub–Som Euler poles, imply a total opening of 30–40 km. Such models imply also a fast southward decrease of the total opening, so they are compatible with estimates from southern Ethiopia (Ebinger *et al.* 2000). In these models, the Euler poles of the total opening of the Red Sea are in a narrow range, while the models for the Gulf of Aden—all of which depict similar motions in its western part—have a wider range of Euler poles.

These predictions for the opening along the MER should be compared with other data. Dugda *et al.* (2005) reported crustal thinning and intrusion of basic material (characterized by a high Poisson's ratio) under the length of the MER. A seismic reflection/refraction profile across the northern MER (EAGLE project, Mackenzie *et al.* 2005) revealed a 20–30 km wide zone of high seismic velocity beneath the eastern part of the rift in the upper two-thirds of the crust, which was interpreted as signifying basic material, and also some crustal thinning under much of the rift. This can be taken as evidence for extension of 20–30 km to accommodate the igneous intrusions. To this, we suggest, should be added the effect of faulting along the rift margins, including inactive faults that may exist beneath the mid-Miocene and younger volcanics that cover the rift floor and its

**Table 3.** *Models of total relative motion between the Nubia, Arabia and Somali plates*

| | Euler pole | | | | |
|---|---|---|---|---|---|
| | Lat (°N) | Long (°E) | Rotation (°ccw) | | |
| **A. Arb–Nub:** total opening of Red Sea | | | | | |
| (Rt1) | 32.0 | 25.0 | 7.75 | Joffe & Garfunkel 1987 | |
| (Rt2) | 31.9 | 24.0 | 7.32 | Present study | |
| (Rt3) | 32.5 | 24.0 | 7.11 | Present study | |
| **B. Som–Arb:** total opening of Gulf of Aden | | | | | |
| (At1) | 24.5 | 27.5 | −9.4 | Joffe & Garfunkel 1987 | |
| (At2) | 27.02 | 25.50 | −8.77 | Same displacement as (At1) in W of Gulf | |
| (At3) | 26.55 | 27.5 | −8.93 | Arb–Som fit less tight | |
| (At4) | 27.83 | 25.3 | −8.11 | Same displacement as (At3) in W of Gulf | |
| (At5) | 28.28 | 24.5 | −7.91 | ditto | |
| (At6) | 29.36 | 22.50 | −7.37 | ditto | |
| (At7) | 28.0 | 25.0 | −8.05 | ditto | |
| (At8) | 28.82 | 23.5 | −7.63 | ditto | |
| **C. Som–Nub:** prediction for total opening | | | | Opening N. Ethiopian rift | |
| (Et1) | −6.44 | 32.29 | −2.018 | 63.8 km | (Rt1) + (At1) |
| (Et2) | −21.09 | 25.10 | −1.208 | 73.4 km | (Rt1) + (At2) |
| (Et3) | −5.24 | 36.25 | −1.455 | 40.5 km | (Rt1) + (At3) |
| (Et4) | −25.47 | 24.53 | −0.704 | 47.3 km | (Rt1) + (At4) |
| (Et5) | −4.38 | 25.45 | −0.973 | 35.5 km | (Rt2) + (At5) |
| (Et6) | −23.36 | 0.72 | −0.450 | 43.0 km | (Rt2) + (At6) |
| (Et7) | −5.98 | 29.66 | −0.905 | 30.7 km | (Rt3) + (At7) |
| (Et8) | −21.37 | 13.78 | −0.510 | 35.9 km | (Rt3) + (At8) |

shoulders. In addition, the presence of much acid volcanics along the MER (Tefera *et al.* 1996) raises the possibility that such material was also intruded at depth, but this would have seismic velocities similar to normal continental crust, so it would be difficult to detect seismically. Together, these considerations show that the above estimates of extension across the northern MER are not much larger than allowed by the available constraints. To resolve the discrepancy, more data on the subsurface structure and intrusions would be helpful, but a need for fine tuning of the plate kinematics is also indicated, though this will not affect the subsequent discussion of Afar.

### *History of motions*

The above estimates of the total (since *c.* 28 Ma ago) and young (0–10 Ma) plate motions can be used as anchors for deriving plate positions at intermediate times by interpolation, as there are no good constraints for intermediate times. Still, some refinement is possible. The best information comes from the Gulf of Aden. There, Cochran (1981) identified magnetic anomalies back to only *c.* 11 Ma (anomaly 5), but later Sahota (1990), D'Acremont (2002), Leroy *et al.* (2004) and D'Acremont *et al.* (2005) identified older magnetic anomalies in various parts of the Gulf. These anomalies show that between the Shukra el-Sheikh and Alula–Fartak fracture zones seafloor spreading began just before *c.* 16 Ma ago (anomaly 5c), while farther east it began 17.5–18 Ma ago (just before anomaly 5d) and that since then the rate and direction of the Arb–Som plate separation did not change significantly. Thus the motion since the beginning of seafloor spreading can be approximated by extrapolating the young motion at a constant rate (rotation rate of *c.* 0.43–0.45°/Ma). Comparison with the rotation angles of the acceptable models of Table 3 shows that in all cases the opening since *c.* 16 Ma ago amounted to 80% or more of the total opening, so during the *c.* 10 Ma that elapsed between the beginning of rifting and the beginning of seafloor spreading the average opening rate was less than half the opening rate since seafloor spreading began.

The opening history of the Red Sea is not well constrained, but since the Nub–Som separation was small, the Arb–Som separation could not have become faster without a coeval increase in the rate of the Arb–Nub separation. Furthermore, since the lateral slip along the Dead Sea Transform took up most of the opening of the Red Sea, most of the opening of the latter ($\geq$80%) could have taken place only since the transform was active. The transform motion began later than *c.* 20 Ma ago (Eyal *et al.* 1981), most likely 17–18 Ma ago (Garfunkel 1989; Garfunkel & Ben-Avraham 2001). These considerations show that the opening of the Red Sea accelerated at about the same time as that of the Gulf of Aden, which is earlier than the 13 Ma ago proposed by Le Pichon & Gaulier (1988). The amount of the Red Sea opening before 16–17 Ma is not known, but even if it was only 10–15% of the total, then the subsequent long-term average opening rate was smaller than the rate since 5 Ma. This fits the other considerations for preferring the model in which the Red Sea opening somewhat accelerated after 5 Ma, as discussed above.

### *Summary*

The foregoing discussion reveals that integration of a variety of unrelated constraints of different kinds coming from different places allows us to infer a range of kinematic models that approximate the divergence of the Nub, Arb and Som plates since the beginning of rifting and since about 10 Ma ago, and to infer a change from an early stage of slow plate divergence to faster plate separation 16–18 Ma ago in the Gulf of Aden and the Red Sea. The plate motions can be defined to within 20–30 km. The corresponding range of Euler poles and rotation angles in the different models cannot be narrowed on the basis of known geological relations. Therefore we focus on the actual motions predicted by the models, because they are the basic information needed for structural interpretation, rather than seeking a 'best' kinematic model, recognizing that this limits the application of plate kinematics for interpreting structural relations.

## *Afar region in the light of plate kinematics*

Following the above considerations, we apply the plate kinematic models to trace the major features of the development of Afar and its relations with the nearby areas. We focus on the formation of new area between the diverging Nub, Arb and Som plates and on the active plate boundaries, but in view of the limited resolving power of plate kinematics we do not discuss structural details, including those of the exposed Pliocene–Quaternary deformation. In much of Afar, this young deformation affects rocks and crust that did not exist or were strongly modified in earlier times, so it is doubtful that current deformation represents faithfully the older setting, especially in view of the structural changes in the last few million years (Manighetti *et al.* 1998, 2001).

### *Pre-rifting situation*

Figure 3 shows a reconstruction of the Afar region just prior to the onset of rifting, i.e. at the end of the flood volcanic activity (28–29 Ma ago). To facilitate comparison with the exposed geology, the figure shows the coastlines in their present positions relative to the major continents.

It is seen that the reconstruction aligns the main bodies of the flood volcanics in Yemen and Ethiopia. It also aligns the northern eroded edge of the Jurassic sediments on the two sides of the Red Sea. Taking into account the stretching of the edges of the Arabia and Nubia plates, enough room is left for the Danakil block, but it must be rotated from its present position, as recognized before (e.g. Sichler 1980; Eagles *et al.* 2002). The preservation of a thick Jurassic section in this block (Merla *et al.* 1973; Sagri *et al.* 1998) constrains its original position to south of the erosive edge of the Jurassic beds in the flanking areas. If the thick Jurassic section in the northern part of this block (Bunter *et al.* 1998) is indeed the continuation of the Mekele graben, then the Danakil block has to be shifted some 50 km southward with respect to Nubia. This position, based on geological criteria, is not very different from previous reconstructions (e.g. Eagles *et al.* 2002 and references therein) that were based on other considerations.

The Aisha (Ali Sabieh) block could not have been originally in its present position with respect to the Somali plate, because then the reconstruction would place it over the basement terrain of Yemen (Fig. 3). It is placed south of the Danakil block, because this is the only available position and this puts its Jurassic section in an area where it was not eroded before rifting, and also because this was their relative position before the opening of the Gulf of Tadjura a couple of million years ago. Such an original position raises the possibility that these two blocks, built of similar continental rocks, moved in unison during much of the history of Afar, though they need not have been rigidly attached to each other. To reach its present position, the Aisha block should have moved a couple of hundred kilometres right laterally with respect to the Somali plateau. In fact, these areas have different structures. In the Aisha block, the older faults strike close to north–south, whereas the adjacent part of the Somali plateau is dominated by WNW–ESE faults typical of the margins of the Gulf of Aden (Chessex *et al.* 1975; Merla *et al.* 1973), which suggests a major structural boundary between them. In the pre-rifting situation, the eastern

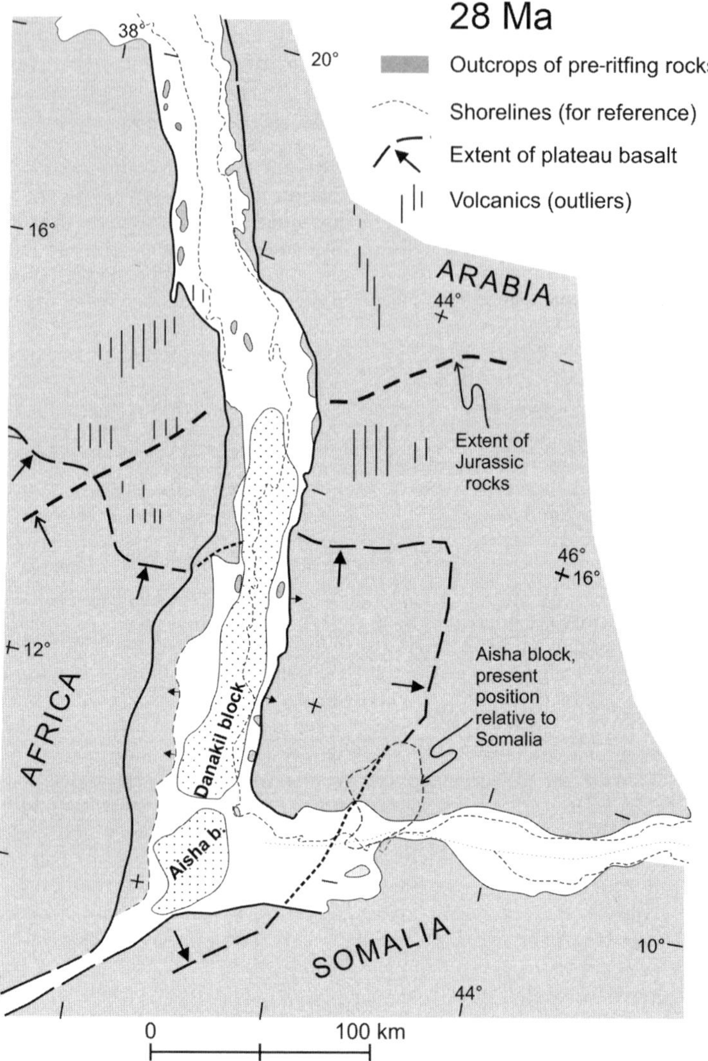


**Fig. 3.** Reconstruction of the Afar depression and the nearby areas at the onset of continental break-up. Note the original alignment of various features and the constraints on the original positions of the Danakil and Aisha blocks (see text for discussion).

boundary of the Danakil block was approximately aligned with the Marda fault. However, in the reconstruction, these structures have different strikes, so it is not clear whether the Madra lineament influenced the initial fracturing.

Though the Danakil block is placed between the thick flood volcanics of Ethiopia and Yemen, none of the dated volcanics from this block yielded ages older than 25 Ma, and thick old volcanics are absent (Barberi *et al.* 1975). This anomalous situation may indicate that an original cover of flood volcanics was eroded from the Danakil block, or that this block was already high standing during flood volcanism, so there the flood volcanic cover was reduced (or absent?). Both explanations raise the possibility of fracturing and uplifting of the Danakil block before or during the flood volcanism. On the Aisha block, the oldest volcanics are 25–27 Ma old (Chessex *et al.* 1975; Clin 1991; Zumbo *et al.* 1995; Audin *et al.* 2004), similar to the adjacent part of the Somali plateau (Behre *et al.* 1987).

Also noteworthy is that in the reconstruction the pre-rift rocks on the two sides of the westernmost Gulf of Aden are placed much closer to each other than farther east or along the Red Sea. This arises primarily because there the Somali plate forms a northward-bulging promontory, which raises the question as to whether this promontory is a pre-rift feature or whether it formed by younger deformation of the Somali margin. This requires further study.

### *Initial rifting stages (until about 15 Ma ago: Late Oligocene and Early Miocene)*

Since the early plate motions cannot be well resolved, as discussed above, we only present an approximate reconstruction at 15–16 Ma ago (Fig. 4) which summarizes the effects of the early stages of plate divergence. This reconstruction was obtained by extrapolating the plate motions backward from 10 Ma by increasing the rotation angles by 50% (models R3, A3, Table 2). The total new area formed is indicated. Since in part this was accommodated by stretching of the plate margins, the deformed (stretched) area is somewhat wider than the new area in the figure. At the latitude of Afar, the opening was 70–80 km, and farther east it reached 100 km in the Gulf of Aden.

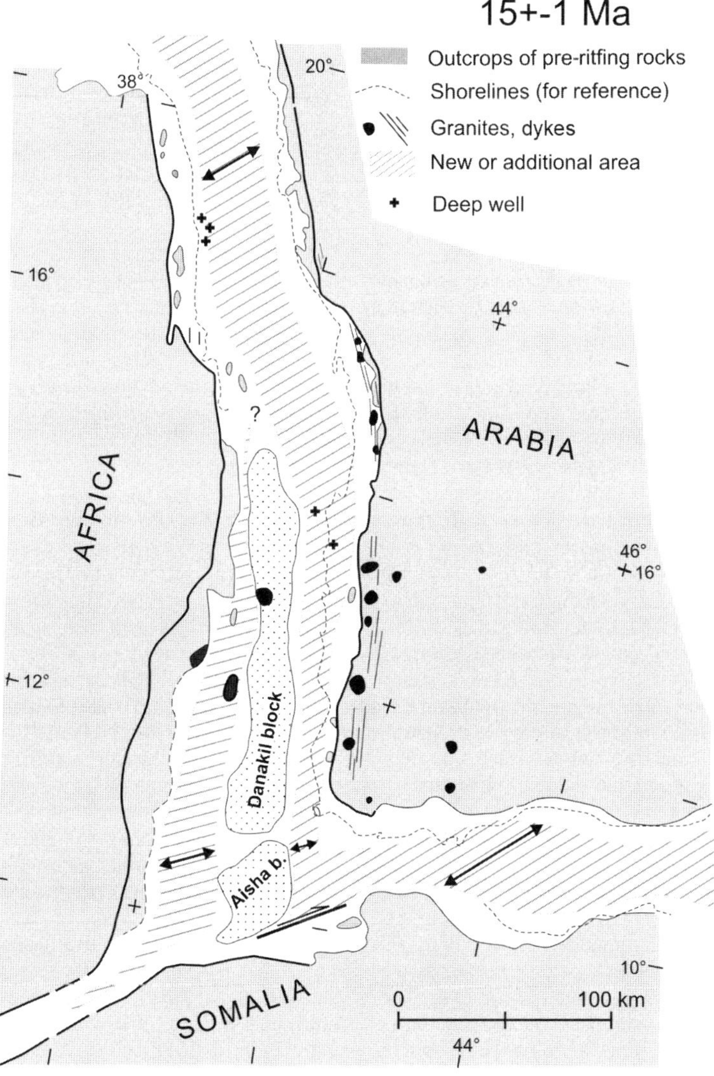


**Fig. 4.** Reconstruction of the Afar region and the nearby areas *c*. 15 Ma (±1 Ma) ago, showing the new area opened since the beginning of rifting.

In the Afar region two fault zones/rifts formed early in the continental break-up history. West of the Danakil block, faulting began *c*. 26 Ma ago and was accompanied by much volcanism between this block and the Ethiopian plateau (Wolfenden *et al.* 2005). Another subsiding rift formed east of the Danakil block, evidenced by the Thio-1 well which penetrated several hundred metres of Late Oligocene basinal marine sediments interbedded with some volcanics (base not reached) overlain by *c*. 1 km of Early Miocene (23–16 Ma) marine sediments (Mitchell *et al.* 1992; Wyn-Hughes *et al.* 1991). This depression extended northward to form the embryonic Red Sea where very early Miocene or somewhat older marine beds accumulated. This basin must have connected with the marine embayment along the Gulf of Aden, because this is the only way that the sea could have reached the southern Red Sea.

Thus the Danakil block (together with the Aisha block) acted as an independent microplate that was flanked by two rifts since the earliest stages of continental break-up. Formation of new area between the adjacent diverging Nubia and Arabia plates was partitioned between these rifts (Fig. 4). As noted above, the Aisha block moved in unison with the Danakil block, forming a single (but not necessarily rigid) microplate. As plate divergence continued, the northern and southern parts of this microplate remained close to the Eritrea–Ethiopia and Yemen plateaus, respectively (its northward motion was small, as discussed above). Otherwise, new area that no longer exists would have formed between them, which is unlikely. To maintain such a position while the Nubia and Arabia plates moved apart, the Danakil microplate must have rotated counter-clockwise with respect to these two major plates while remaining decoupled from them. Consequently, the rifts on is two sides were wedge-shaped. As the Aisha block did not move much with respect to Arabia, its motion relative to Somali should have resembled the Arb–Som motion, i.e. approximately parallel to the fracture zones in the Gulf of Aden. As there is no candidate for new area that could have opened between the Somali plate and the Aisha block, and given their relative motion, their boundary was most likely a right-lateral transform trending between east–west to NNE–SSE, so the Aisha block moved eastward relative to the part of the Somali plate on the southern side of the Afar depression (Fig. 4; Garfunkel & Beyth 2004). This boundary connected the site of crustal spreading in southern Afar with the spreading centre in the Gulf of Aden, and perhaps joined the ancestral Shukra el Sheikh fracture zone (Audin *et al.* 2005, but the early structure require more study).

The geological data from the rifts on the two sides of the Danakil block reveal that they developed in very different ways. In the faulted region west of this block—the precursor of the present-day Afar depression—much sub-aerial volcanism continued in the Early Miocene (Ukstins *et al.* 2002; Wolfenden *et al.* 2005). It seems that there, much igneous material was added to the crust and kept it above sea level, offsetting the stretching of the crust resulting from the plate separation. On the other hand, the thick marine Early Miocene section in Thio-1 (and in nearby wells) shows that the rift east of the Danakil block was a subsiding marine basin, where volcanic activity was restricted. Thus this block separated areas where the intensity of igneous activity was quite different. Farther east, however, considerable volcanism occurred along the coastal area of Yemen as evidenced by the >1.5 km of Early (and Middle?) Miocene volcanics penetrated in Zeidiya 1 well (Hughes & Beydoun 1992; Mitchell *et al.* 1992), and by 25–20 Ma-old basic and granitic intrusions along the western margin of Arabia (Fig. 4; Barberi *et al.* 1975; Coleman 1984; Brown *et al.* 1989; Sebai *et al.* 1991; Mohr 1991; Chazot & Bertrand 1995) that include the 'Red Sea dyke system' that extends to north of the Red Sea (Eyal *et al.* 1981; Garfunkel 1989). These relations show that while the plate motions allow to estimate the sizes and locations of the new areas that formed between the diverging plates, it does predict the processes occurring in the new areas. These are determined by additional factors, e.g. the supply of magma.

As discussed above, the plate motions accelerated some 17 Ma ago, and since then oceanic crust began to form in the Gulf of Aden and perhaps also in the Red Sea, though there stretching of continental crust could have been important (or even dominant at first). This terminated the initial phase of slow plate separation and diffuse crustal stretching. Wolfenden *et al.* (2004) raised the interesting possibility that in that early period the Nubia and Somali plates have not separated yet, implying a single Euler pole for the Arb–Nub and Arb–Som separation. This cannot be resolved with the available constraints on these motions. However, the similarity of the entire motions to the shorter-term motions shows that most (>>80%?) of the separation between these plate pairs took place with distinct Euler poles, which requires coeval Nub–Som separation. The history of motions inferred above shows that this situation should have existed since 16–18 Ma or probably earlier. On the other hand, shaping of the MER was documented only since some 11 Ma ago (WoldeGabriel *et al.* 1990; Chernet *et al.* 1998; Wolfenden *et al.* 2004). This discrepancy may arise from the fact that along the MER the base of the Cenozoic sequence is not exposed, being hidden by Middle Miocene and younger volcanics. In southern Ethiopia, where the basement is exposed, rifting took place in a wide zone since 18–20 Ma ago (Ebinger *et al.* 2000). This raises the possibility that additional old structures, connecting Afar with the south Ethiopian structures, are hidden beneath the MER or the plateau west of it. In fact, Kieffer *et al.* (2004) confirmed earlier reports of faulting and strong tilting of the flood volcanics on the plateau farther north, which shows that early faulting did take place in Ethiopia. Clearly, more needs to be known about the subsurface structure along the MER and its margins in order to establish their pre-Middle Miocene history.

### *Advanced plate divergence (15 Ma to 5 Ma ago, Middle and Late Miocene)*

After 15 Ma, the plate boundaries and the sites of creation of new area remained without notable changes for some time, so the plate arrangement can be approximated by extrapolating the motion of the last 10 Ma (Fig. 5). As before, opening of new area between the diverging Nub–Arb plates was partitioned on the two sides of the Danakil microplate. This block must have continued to rotate for its northern and southern ends to remain close to the neighbouring major plates. At that time, the Aisha block, still located next to the

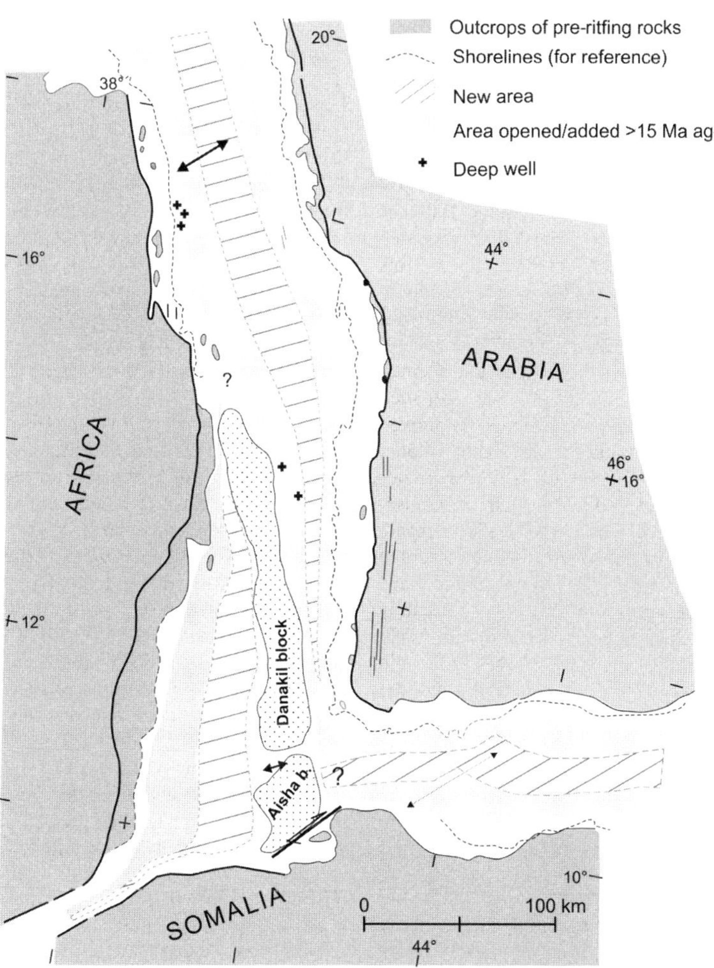


**Fig. 5.** Reconstruction of the Afar region and the nearby areas *c.* 10 Ma ago, showing the new areas added before and after 15 Ma.

southern end of the Danakil block, approached the promontory of the Somali plateau. As before, the junction between them was of the transform type, and was most likely marked by a strike–slip fault that Chessex *et al.* (1975) mapped along their junction. However, the clockwise rotation of the Aisha block since 8 Ma (Audin *et al.* 2004) proves deformation of a wide zone in this area, rather than motion entirely concentrated along a narrow transform.

The reconstruction places the Arabia–Somali–Danakil triple junction in the westernmost part of the Gulf of Aden (Fig. 5), where the deep structure is hidden under a cover of younger rocks. The Danakil–Nubia motion was small, but perhaps allowed some separation in the north of the Afar depression. The Arabia–Somali divergence was accommodated by a spreading centre east of the Shukra el-Sheikh fracture zone, but the absence of magnetic anomalies of that age west of this fracture zone in the westernmost Gulf of Aden (Cochran 1981; Sahota 1990) suggests that here a well-defined spreading centre did not form. As before, a triple junction (Arb–Som–Aisha) must have existed in this area. The Somali–Aisha (Danakil) boundary was still a complex transform that extended from this triple junction area to the southern part of the Afar depression. There, it joined the zone of plate separation within Afar and the Ethiopian rift. This junction was, most likely, a broad zone of faulting, similar to the present situation in Afar. Because of the continuing widening of the Afar depression, this junction migrated eastward (Tesfaye *et al.* 2003).

The western part of the Gulf of Aden narrows westward much faster than the more eastern parts of the Gulf, which cannot result from divergence of rigid plates, because this leads to a gradual change in the width of the new area. Thus the margin of at least one of the bordering plates must have been considerably deformed. The most likely possibility is that the promontory of the Somali plate, which has an anomalous position in the pre-rift reconstruction (Fig. 3; see above), was deformed and decoupled from the rest of the Somali plate and moved into the space opened between this and the Arabia plates, but the deformation mechanism still need clarification.

Like in earlier periods, the crust underlying the newly opened areas formed in different ways. In most of the Gulf of Aden, normal seafloor spreading continued, but west of the Shukra el-Sheikh fracture it was probably less orderly. In the Red Sea, thin igneous crust could have formed, but non-magmatic stretching of continental crust could also have been important (dominant?), especially close to the basin margins (cf. Cochran 2005). The relative importance of these processes is still not clear. The thin, probably oceanic, crust under the central gravity anomaly (Fig. 1) began to form *c.* 10 Ma ago or somewhat earlier (Izzeldin 1987). It probably formed by underplating of igneous material beneath the sediment fill of the basin, so magnetic anomalies were not produced. This is suggested by some volcanism in the southern Red Sea (e.g. well B-1: Mitchell *et al.* 1992). Most likely the situation was similar to the northern Red Sea where young diffuse deformation occurs, but the high heat flow anomaly raises the possibility of underplating by igneous material (Martinez & Cochran 1988; Cochran 2005). In the expanding Afar depression, abundant volcanic activity continued and extended also over the southern half of the Danakil microplate. Thus emplacement of igneous material that accompanied the extension probably contributed considerably to crustal growth.

### *Young evolution (5 Ma to present, Pliocene–Quaternary)*

This period was characterized by changes in the plate boundaries in the Afar depression and in the surrounding areas, which produced the present configuration (change from the situation of Fig. 5 to that of Fig. 1). The triple junction in the westernmost Gulf of Aden was reorganized and the spreading centre along the Arb–Som plate boundary grew westward since 2–3 Ma ago and propagated through the Gulf of Tadjura into the Afar depression, separating the Danakil and Aisha blocks (Manighetti *et al.* 1997, 2001). Still farther west, this boundary extends to a broad zone of faulting, block rotation and extension within the Afar depression (Fig. 1; Tapponier *et al.* 1990; Acton *et al.* 2000; Manighetti *et al.* 2001 and references therein).

From the point of view of plate kinematics, the important point is that while the new spreading centre developed, the Arb–Som and the Arb–Nub divergence and opening of new area between them continued without notable changes. Within this framework, the westward propagation of the Gulf of Aden spreading centre just changed the sites and manner of formation of new area, i.e. the new segment of the Gulf of Aden spreading centre must have formed at the expense of older plate boundaries that became inactive. Courtillot (1982) presented an example of how this might happen between plates that continue to separate. In the present case, a new oblique spreading centre replaced the pre-existing Danakil (Aisha)–Somali transform junction which became inactive. As a result, the Aisha block and the adjacent southeastern corner of the Afar depression became attached to the Somali plate and the complex triple junction within the Afar depression shifted northward, but the

Danakil block north of the Gulf of Tadjura still remained an independent microplate.

In the Red Sea, normal seafloor spreading that produces magnetic anomalies began along the basin axis. From lat. 15.5° N, east of the Danakil block, to about lat. 18° N, it began 4–6 Ma ago, but farther north it began 3–4 Ma ago and even later along several isolated segments of the basin axis (Fig. 1; Roeser 1975; Izzeldin 1987; Chu & Gordon 1998; Cochran 2005). The southern part of this new spreading centre and its southward continuation define the Danakil–Arabia boundary which extends to the triple junction area in the westernmost Gulf of Aden (Manighetti *et al.* 1998). The diachronous beginning of seafloor spreading along segments of the Red Sea axis is not related to a resolvable change in Arb–Nub plate separation and obviously does not reflect lateral local variations of the amount of plate separation. Rather, it expresses a change in the mechanism of basin widening and creation of new area (Garfunkel *et al.* 1987). The widening of the northern half of the basin is still largely accommodated by diffuse extension across much of the basin that is accompanied by magmatic activity (and underplating?), though it tends to concentrate along the basin axis (Martinez & Cochran 1988; Cochran 2005). This was probably the situation all along the Red Sea before the axial spreading centre formed.

In contrast to the narrow plate boundary east of the Danakil microplate, the plate junctions in the Afar depression remained a wide zone of diffuse extension, involving rotation of blocks on vertical axes and of crustal accretion by magma injection (Fig. 1; Barberi *et al.* 1975; Acton *et al.* 2000; Manighetti *et al.* 2001 and references therein). Thus the Danakil–Nubia plate junction is a *c.* 180 km-wide zone of faulting and volcanism at lat. 11° N to 12° N, but it narrows to *c.* 50 km at lat. 14° N. This zone of deformation continues northward into a NNE-trending seismically active zone in the Red Sea (Fig. 1; Chu & Gordon 1998; Hofstetter & Beyth 2003) but this zone is not marked by any recorded feature, and neither or its junction with the Red Sea spreading centre (Garfunkel *et al.* 1987). Any deformation in this zone appears to be much weaker than in Afar and is probably masked by young sediments, though in this general area the rotating Danakil block must have been decoupled from the adjacent Nubia plate since the beginning of rifting. Thus, the geometry of the plate boundaries shows that the NNW-trending zone of crustal spreading along the Afar depression is not the direct continuation of the Red Sea spreading centre, which extends east of the Danakil block in the centre of the southernmost Red Sea, but is a distinct plate boundary which links with the Red Sea axis through a diffuse deformation zone.

The position of most of Afar above sea level, in contrast with the Red Sea and the Gulf of Aden, is most likely due to addition of igneous material to the crust. Under a *c.* 200 km-wide area in central Afar (lat. 12° N), the crust is 20–25 km thick and consists of basic material (Makris & Ginzburg 1987; Makris *et al.* 1991; Dugda *et al.* 2005; Tiberi *et al.* 2005). If it consisted entirely of stretched continental crust, as proposed by Makris & Ginzburg (1987), this would imply less than 100 km extension, i.e. considerably less than the opening indicated by the plate kinematics, so the crust under Afar must be new igneous material, as recognized by Mohr (1989). This implies that the amount of igneous material supplied to this area far exceeded what is seen on the surface and it was delivered during most, if not the entire, history of the Afar depression. The thinning of the crust to *c.* 16 km towards northern Afar (Makris & Ginzburg 1987), where little new area formed between the Danakil block and Nubia (so there stretching was small), indicates that magma supply decreased in this direction. The crust of the southern Red Sea is about half as thick as in Afar for at least as far south as lat. 14° N (Egloff *et al.* 1991), so here too magma supply was less than in central Afar. Though several volcanoes exist in the southernmost Red Sea, magma delivery was not sufficient to bring the crust of this area to sea level.

## Discussion and conclusion

This work attempted to relate the evolution and structure of Afar and of the neighbouring areas to the Nubia, Arabia and Somali plate kinematics. Knowledge of these motions allows to determine the size and shape of new area that opened between these plates and allows to construct palinspastic maps. These provide a framework in which the evolution of the region can be viewed as accommodating the formation of new area. The Afar region demonstrates that the actual geological structures formed depend also on other factors independent of motions of major plates. The presence and behaviour of microplates—the Danakil and Aisha blocks in our case—is important because it determines their own boundaries and secondary triple junctions. The increase in surface area is accommodated by different crustal growth/accretion mechanisms along different plate boundaries or even along boundaries between the same plates. Abundant magma delivery led to growth of thick new igneous crust under Afar. There, deformation remained diffuse and complex (e.g. rotation of blocks on vertical axes: Acton *et al.* 1991; Manighetti *et al.* 2001) and the plate boundaries changed with time, which cannot be directly related to the simple and rather constant motions of the major surrounding major plates. In the Red

Sea and Gulf of Aden, where magma supply was much less, thinner crust and narrow spreading centres formed.

These features are considered to be the result of continuous activity of the Afar mantle plume that produced an abundant magma supply to a wide area. This kept the new crust hot and weak, promoting diffuse deformation. It probably also allowed, or even led to, the local deformation patterns that are independent of the motions of the major plates, e.g. involving rotations about vertical axes. In contrast, under the Gulf of Aden, decompression melting of the asthenosphere that rises into the gap between the diverging plates was the dominant process and controlled the development of narrow spreading centres. Thus here the evolution of the plate margins expresses directly the plate motions. This is also the situation in much of the Red Sea, though here magma supply is diffuse and in certain periods it was probably limited and did not suffice to build new crust, so non-magmatic was important or even dominant in some periods.

The plate kinematic models were inferred from different types of constraints from various places. This allows inferring the changes in surface area all along the boundaries of the plates under consideration, which provides a fremework for evaluating the development of the plate boundaries. The Afar region demonstrates that the resulting structure depends critically on additional factors such as the presence of microplates and magma supply.

The foregoing considerations also show that the nature of the available constraints allows only a range of permissible models to be inferred, and often makes it difficult to define formal uncertainties of the locations of Euler poles or of rotation angles. Despite these limitations, the important thing in our opinion is the possibility to constrain the motions along plate boundaries to within narrow limits, (20–30 km in our case), that are useful for geological interpretation. This appears to be the resolving power of plate kinematics. Rigid plate kinematics probably cannot be used to reliably interpret deformation that involves changes of distances of $\leq$10–15 km.

The Afar region illustrates these points. It history and relations to the plate motions show the complexities and uncertainties of the use of plate kinematics and demonstrate what can and cannot be done with this approach.

We are very grateful to an anonymous reviewer, S. Leroy, and especially to C. Ebinger for constructive and helpful reviews.

## References

ABDEL-GAWAD, M. 1970. Interpretation of satellite photographs of the Red Sea and Gulf of Aden. *Philosophical Transactions of the Royal Society of London*, **A267**, 23–40.

ABDELSALAM, M.G. & STERN, R.J. 1996. Sutures and shear zones in the Arabian–Nubian Shield. *Journal of African Earth Sciences*, **23**, 289–310.

ACTON, G.D., STEIN, S. & ENGLEN, J.F. 1991. Block rotation and continental extension in Afar: A comparison to oceanic microplate systems. *Tectonics*, **10**, 501–526.

ASRAT, A., BERBEY, P. & GLEIZES, G. 2001. The Precambrian geology of Ethiopia: a review. *Africa Geoscience Review*, **8**, 271–288.

AUDIN, L., QUIDELLEUR, X., COULIE, E., COURTILLOT, V., GILDER, S., MANIGHETTI, I., GILLOT, P.Y., TAPPONIER, P. & KIDANE, T. 2004. Palaeomagnetism and K–Ar and $^{40}Ar/^{39}Ar$ ages in the Ali Sabieh area (republic of Djibouti and Ethiopia): constraints on the mechanism of Aden ridge propagation into southeastern Afar during the last 10 Myr. *Geophysical Journal International*, **158**, 327–345.

BARBERI, F. & VARET, J. 1977. Volcanism of Afar: small-scale plate tectonic implications. *Geological Society of America Bulletin*, **88**, 1251–1266.

BARBERI, F., FERRARA, G., SANTACROCE, R. & VARET, J. 1975. Structural evolution of the Afar triple junction. *In*: PILGER, A. & ROSLER, A. (eds) *Afar Depression of Ethiopia*. Schweitzersche Verlag, Stuttgart, pp. 38–54.

BAYENE, A. & ABDELSALAM, M.G. 2005. Tectonics of the Afar depression: A review and synthesis. *Journal of African Earth Sciences*, **41**, 41–59.

BEHRE, S.M., DESTA, B., NICOLETTI, M. & TEFERRA, M. 1987. Geology, geochonology and geodynamic implications of the Cenozoic magmatic province in W and SE Ethiopia. *Journal of the Geological Society, London*, **144**, 213–226.

BEYDOUN, Z.R. 1970. Southern Arabia and northern Somalia: comparative geology. *Philosophical Transactions of the Royal Society of London*, **A267**, 267–292.

BEYTH, M. 1972. Paleozoic–Mesozoic sedimentary basin of Mekele outlier, Northern Ethiopia. *American Association of Petroleum Geologists Bulletin*, **56**, 2426–2439.

BEYTH, M. 1973*a*. Correlation of Palaeozoic–Mesozoic sediments in northern Yemen–Tigrai and northern Ethiopia. *American Association of Petroleum Geologists Bulletin*, **57**, 2440–2443.

BEYTH, M. 1973*b*. A suggestion for interpretation of the stratigraphy of northern Ethiopia according to the model of plate tectonics. *Geology*, **1**, 81–82.

BEYTH, M. 1991. 'Smooth' and 'rough' propagation of spreading, Southern Red Sea–Afar depression. *Journal of African Earth Sciences*, **13**, 157–171.

BILHAM, R., BENDICK, R., LARSON, K., MOHR, P., BRAUN J., TESFAYE, S. & ASFAW, S. 1999. Secular and tidal strain across the Ethiopian rift. *Geophysical Research Letters*, **26**, 2789–2792.

BOCCALETTI, M., GETANEH, A. & BONAVIA, F.F. 1991. The Marda Fault: a remnant of an incipient aborted rift in the paleo-African Arabian plate. *Journal of Petroleum Geology*, **14**, 79–92.

BOHANNON, R.G. 1986. Tectonic configuration of the western Arabian continental margin, southern Red Sea. *Tectonics*, **5**, 477–499.

BOSELLINI, A. 1989. The continental margins of Somalia: their structural evolution and sequence stratigraphy. *Memoire degli Instituti di geologia e mineralogia dell'Universita di Padua*, **41**, 373–458.

BOSENCE, D.W.J. 1997. Mesozoic rift basins in Yemen. *Marine Petroleum Geology*, **14**, 611–616.

BOTT, W.F., SMITH, B.A., OAKES, G.S., SIKANDER, A.H. & IBRAHAM, A.I. 1992. The tectonic framework and regional hydrocarbon prospectivity of the Gulf of Aden. *Journal of Petroleum Geology*, **15**, 211–243.

BROWN, G.F., SCHMIDT, D.L. & HUFFMAN, A.C. JR. 1989. *Geology of the Arabian Peninsula: Shield area of western Saudi Arabia*. US Geological Survey Professional Paper 560-A, pp. A1–A178.

BUNTER, M.A.G., DEBRETSION, T. & WOLDEGIORGIS, L. 1998. New developments in the pre-rift prospectivity of the Eritrean Red Sea. *Journal of Petroleum Geology*, **21**, 373–400.

CALAIS, E., EBINGER, C. & HARTNADY, C. & NOCQET, J.M. 2006. Kinematics of the East African Rift from GPS and earthquake slip vector data. *In:* YIRGU, G., EBINGER, C.J. & MAGUIRE, P.K.H. (eds) *The Afar Volcanic Province within the East African Rift System*. Geological Society, London, Special Publications, **259**, 9–22.

CANDE, S.C. & KENT, D.V. 1995. Revised calibration of the geomagnetic polarity timescale for the Late Cretaceous and Cenozoic. *Journal of Geophysical Research*, **100B**, 6093–6095.

CHAZOT, G. & BERTRAND, H. 1995. Genesis of silicic magmas during Tertiary continental rifting in Yemen. *Lithos*, **36**, 69–83.

CHERNET, T., HART, W., ARONSON, J. & WALTER, R.C. 1998. New age constraints on the timing of volcanism and tectonism in the northern main Ethiopian Rift–southern Afar transition zone (Ethiopia). *Journal of Volcanology and Geothermal Research*, **80**, 267–280.

CHESSEX, R., DELALOYE, M., MULLER, J. & WEIDMANN, M. 1975. Evolution of the volcanic region of Ali Sabieh (T.F.A.I.) in the light of K–Ar age determinations. *In*: PILGER, A. & ROSLER, A. (eds) *Afar Depression of Ethiopia*, Schweitzersche Verlag, Stuttgart, 221–227.

CHU, D. & GORDON, R.G. 1998. Current plate motions across the Red Sea. *Geophysical Journal International*, **135**, 313–328.

CHU, D. & GORDON, R.G. 1999. Evidence for motion between Nubia and Somalia along the southwest Indian ridge. *Nature*, **398**, 64–67.

CLIN, M. 1991. Evolution of Eastern afar and the Gulf of Tadjura. *Tectonophysics*, **198**, 355–368.

COCHRAN, J.R. 1981. The Gulf of Aden: structure and evolution of a young ocean basin and continental margin. *Journal of Geophysical Research*, **86B**, 263–288.

COCHRAN, J.R. 1983. A model for the development of the Red Sea. *American Association of Petroleum Geologists Bulletin*, **67**, 41–69.

COCHRAN, J.R. 2005. Northern Red Sea: Nucleation of an oceanic spreading center within a continental rift. *Geochemistry, Geophysics, Geosystems*, **6**, doi:10.1029/2004GC000826.

COLEMAN, R.G. 1984. The Red Sea: a small ocean basin formed by continental extension and seafloor spreading. International Geological Congress, 27th, **23** (*Origin and History of Continental Margins and Inland Seas*), 93–121.

COULIE, E., QUIDELLEUR, X., GILLOT, P.Y., COURTILLOT, V., LEFEVRE, J.C. & CHIESA, S. 2003. Comparative K–Ar and Ar/Ar dating of Ethiopian and Yemenite Oligocene volcanism: implications for timing and duration of the Ethiopian traps. *Earth and Planetary Science Letters*, **206**, 477–492.

COURTILLOT, V.E. 1982. Propagating rifts and continental breakup. *Tectonics*, **1**, 239–250.

COURTILLOT, V., ACHACHE, J., LNADRE, F., BONHOMMET, N., MONTIGNY, R. & FERAUD, G. 1984. Episodic spreading and rift propagation: new paleomagnetic and geochronologic data from the Afar nascent passive margin. *Journal of Geophysical Research*, **89**, 3315–3333.

D'ACREMONT, E. 2002. De la dechirure continental a l'accretion oceanqiue: ouverture du Golfe D'Aden. Ph.D. thesis, Université Pierre et Marie Curie, Paris, 323pp.

D'ACREMONT, E., LEROY, S., BESLIER, M.O., BELLAHSEN, N., FOURNIER, M., ROBIN, C., MAIA, M. & GENTE, P. 2005. Structure and evolution of the eastern Gulf of Aden conjugate margins from seismic reflection data. *Geophysical Journal International*, **160**, 869–890.

DAVIDSON, I., AL-KADASI, M., *ET AL.* 1994. Geological evolution of the southeastern Red Sea rift margin, republic of Yemen. *Geological Society of America Bulletin*, **106**, 1474–1493.

DRURY, S.A., KELLEY, S.P., BEHRE, S.M., COLLIER, R. & ARAHA, M. 1994. Structures related to Red Sea evolution in northern Eritrea. *Tectonics*, **13**, 1371–1380.

DUGDA, M.T., NYBLADE, A.A., JULIA, J., LANGSTON, C.A., AMMON, C.J. & SIMIYU, S. 2005. Crustal structure in Ethiopia and Kenya from receiver function analysis: Implications for rift development in eastern Africa. *Journal of Geophysical Research*, **110**, B01303, doi: 1029/2004JB003065.

EAGLES, G., GLOAGUEN, R. & EBINGER, C. 2002. Kinematics of the Danakil microplate. *Earth and Planetary Science Letters*, **203**, 607–620.

EBINGER, C.J. 1989. Tectonic development of the western branch of the East African Rift System. *Geological Society of America Bulletin*, **101**, 885–903.

EBINGER, C.J. & CASEY, M. 2001. Continental breakup in magmatic provinces: an Ethiopian example. *Geology*, **29**, 527–530.

EBINGER, C.J., YEMANE, T., HARDING, D.J., TESFAYE, S., KELLEY, S. & REX, D.C. 2000. Rift deflection, migration, and propagation: Linkage of Ethiopian and Eastern rifts, Africa. *Geological Society of America Bulletin*, **112**, 163–176.

EGLOFF, F., RIHM, J., MAKRIS, J., IZZELDIN, Y.A., BOBSIEN, M., MEIER, K., JUNGE, P.K., NOMAN, T. & WARSI, W. 1991. Contrasting structural styles of the eastern and western margins of the southern Red Sea: the 1988 SONNE experiment. *Tectonophysics*, **198**, 329–353.

EYAL, M., EYAL, Y., BARTOV, Y. & STEINITZ, G. 1981. Tectonic development of the western margin of the Gulf of Eilat (Aqaba) rift. *Tectonophysics*, **80**, 39–66.

FANTOZZI, P.L. 1996. Transition from continental to oceanic rifting in the Gulf of Aden: structural evidence from field mapping in Somalia and Yemen. *Tectonophysics*, **259**, 285–311.

FERNANDES, R.M.S., AMROSIUS, B.A.C., NOOMEN, R., BASTOS, L., COMBRINCK, L., MIRANDA, J.M. & SPAKMAN, W. 2004. Angular velocities of Nubia and Somalia from continuous GPS data: implications on present-day relative kinematics. *Earth and Planetary Science Letters*, **222**, 197–208.

FREUND, R. 1970. Plate tectonics of the Red Sea and East Africa. *Nature*, **220**, 453.

GARFUNKEL, Z. 1989. Tectonic setting of Phanerozoic volcanism in Israel. *Israel Journal of Earth Sciences*, **38**, 51–74.

GARFUNKEL, Z. & BEN-AVRAHAM, Z. 2001. *Basins along the Dead Sea transform. Peri-Tethys Memoir 6 (Memoires du Museum National d'Histoire Naturelle, Paris*, **186**), 607–627.

GARFUNKEL, Z. & BEYTH, M. 2004. Constraints on the structural development of Afar imposed by the kinematics of the major surrounding plates. *International Conference on the East African Rift System*, June 2004, Addis Ababa, p. 74.

GARFUNKEL, Z., GINZBURG, A. & SEARLE, R.C. 1987. Fault pattern and mechanism of crustal spreading along the axis of the Red Sea from side-scan sonar (GLORIA) data. *Annals Geophysicae*, **5B**, 187–200.

GASS, I.G. 1970. The evolution of volcanism in the junction area of the Red Sea, Gulf of Aden and Ethiopian rifts. *Philosophical Transactions of the Royal Society of London*, **A267**, 369–381.

GASS, I.G. & GIBSON, I.L. 1969. Structural evolution of the rift zones in the Middle East. *Nature*, **221**, 926–930.

GEORGE, R., ROGERS, N. & KELLEY, S. 1988. Earliest magmatism in Ethiopia: Evidence for two mantle plumes in one floor basalt province. *Geology*, **26**, 923–926.

GOMA'A, O. & STECKLER, M.S. 1995. Fission track evidence on the initial rifting of the Red Sea. Two pulses, no propagation. *Science*, **270**, 1341–1344.

GRADSTEIN, F. M., OGG, J.G., SMITH, A.G., BLEEKER, W. & LOURENS, L.J. 2004. A new geologic time scale with special reference to Precambrian and Neogene. *Episodes*, **27**, 83–100.

HOFMANN, C., COURTILLOT, V., FERAUD, G., ROCHETTE, P., YIRGU, G., KETEFO, E. & PIK, R. 1997. Timing of the Ethiopian flood basalt event and implications for plume birth and global change. *Nature*, **389**, 838–841.

HOFSTETTER, R. & BEYTH, M. 2003. The Afar depression: interpretation of the 1960–2000 earthquakes. *Geophysical Journal International*, **155**, 715–732.

HORNER-JOHNSON, B.C., GORDON, R.G., COWLES, S.M. & ARGUS, D.F. 2005. The angular velocity of Nubia relative to Somalia and the location of the Nubia–Somalia–Antarctica triple junction. *Geophysical Journal International*, **162**, 221–238.

HUGHES, G.W. & BEYDOUN, Z.R. 1992. The Red Sea–Gulf of Aden: Biostratigraphy, lithostratigraphy and palaeoenvironments. *Journal of Petroleum Geology*, **15**, 135–156.

IZZELDIN, A.Y. 1987. Seismic, gravity and magnetic surveys in the central part of the Red Sea: their interpretation and implications for the structure and evolution of the Red Sea. *Tectonophysics*, **143**, 269–306.

JESTIN, F., HUCHON, P. & GAULIER, J.M. 1994. The Somalia plate and the East African Rift System: present day kinematics. *Geophysical Journal International*, **116**, 637–654.

JOFFE, S. & GARFUNKEL, Z. 1987. Plate kinematics of the circum Red Sea—a re-evaluation. *Tectonophysics*, **141**, 5–22.

KENEA, N.H., EBINBER, C.J. & REX, D.C. 2001. Late Oligocene volcanism and extension in the southern Red Sea Hills, Sudan. *Journal of the Geological Society*, London, **158**, 285–294.

KIEFFER, B., ARNDT, N., *ET AL*. 2004. Flood and shield basalts from Ethiopia: Magmas from the African superswell. *Journal of Petrology*, **45**, 793–834.

KRONER, A. & SASSI, F.P. 1994. Evolution of the northern Somali basement: new constraints from zircon ages. *Journal of African Earth Sciences*, **22**, 1–15.

LAUGHTON, A.S. 1966. The Gulf of Aden in relation to the Red Sea and the Afar depression of Ethiopia. *In*: *The World Rift System. Geological Survey of Canada, Paper 66–14*, 78–97.

LE PICHON, X. & GAULIER, J.M. 1988. The rotation of Arabia and the Levant fault system. *Tectonophysics*, **153**, 271–294.

LEROY, S. *ET AL*. 2004. From rifting to spreading in the eastern Gulf of Aden: a geophysical survey of a young oceanic basin from margin to margin. *Terra Nova*, **16**, 185–192.

MACKENZIE, G.D., THYBO, H. & MAGUIRE, P.K.H. 2005. Crustal velocity structure across the main Ethiopian Rift: results from two-dimensional wide-angle seismic modeling. *Geophysical Journal International*, **162**, 994–1006.

MAKRIS, J. & GINZBURG, A. 1987. The Afar depression: transition between continental rifting and sea-floor spreading. *Tectonophysics*, **141**, 199–214.

MAKRIS, J., HENKE, H., EGLOFF, F. & AKAMELUK, T. 1991. The graivty field of the Red Sea and East Africa. *Tectonophysics*, **198**, 369–381.

MANIGHETTI, I., TAPPONIER, P., COURTILLOT, V., GRUSZOW, S. & GILLOT, P.Y. 1997. Propagation of rifting along the Arabia–Somaila plate boundary: the Gulf of Aden and Tadjoura. *Journal of Geophysical Research*, **102B**, 2681–2710.

MANIGHETTI, I., TAPPONIER, P., COURTILLOT, V., GALLET, Y., JACQUES, E. & GILLOT, P.Y. 2001. Strain transfer between disconnected, propagating rifts in Afar. *Journal of Geophysical Research*, **106B**, 13613–13665.

MARTINEZ, F. & COCHRAN, J.R. 1988. Structure and tectonics of the northern Red Sea: catching a continental margin between rifting and drifting. *Tectonophysics*, **150**, 1–32.

MARTIRE, L., CLARI, P. & PAVIA, G. 1998. Stratigraphic analysis of the Upper Jurassic (Oxfordian–Kimmeridgian) Antalo limestone in the Mekele outlier (Tigrai, Northern Ethiopia): preliminary data. *Peri–Tethys Memoire. 4 (Mémoires du Museum National d'Histoire Naturelle, Paris,* **179**), 131–144.

MCCLUSKY, S., REILINGER, R., MAHMOUD, S., BEN SARI, D. & TEALEB, A. 2003. GPS constraints on Africa (Nubia) and Arabia plate motions. *Geophysical Journal International*, **155**, 126–138.

MCKENZIE, D.P., DAVIES, D. & MOLNAR, P. 1970. Plate tectonics of the Red Sea and East Africa. *Nature*, **226**, 243–248.

MENZIES, M., BAKER, J., CHAZOT, G. & AL'KADASI, M. 1997. Evolution of the Red Sea volcanic margin, western Yemen. *AGU Geophysical Monograph*, **100**, 29–43.

MERLA, G., ABBATE, E., CANUTI, P., SAGRI, M. & TACCONI, P. 1973. *Geological map of Ethiopia and Somalia*. Scale 1: 2 000 000, CNR, Italy.

MITCHELL, D.J.W., ALLEN, R.B., SALAMA. W. & ABOUZAKM, A. 1992. Tectonostratigraphic framework and hydrocarbon potential of the Red Sea. *Journal of Petroleum Geology*, **15**, 187–210.

MOHR, P. 1970. Plate tectonics of the Red Sea and East Africa. *Nature*, **228**, 547–548.

MOHR, P. 1978. Afar. *Annual Review of Earth and Planetary Science*, **6**, 145–172.

MOHR, P. 1989. Nature of the crust under Afar: new igneous, not thinned continental. *Tectonophysics*, **167**, 1–11.

MOHR, P. 1991. Structure of Yemeni Miocene dyke swarms, and emplacement of coeval granite plutons. *Tectonophysics*, **198**, 203–221.

PATCHETT, J. 1992. Isotopic studies of Proterozoic crustal growth and evolution. *In*: CONDIE, K.C. (ed.) *Proterozoic Crustal Evolution. Developments in Precambrian Geology*, **10**, 481–508. Elsevier.

PLATEL, J. & ROGER, J. 1989. Evolution géodynamique du Dhofar (Sultanate d'Oman) pendant le Crétacé et le Tertiare en relation avec l'ouverture du golfe d'Aden. *Bulletin de la Société Géologique de France*, **8**(2), 252–263.

PRODEHL, C. & MECHIE, J. 1991. Crustal thinning in relationship to the evolution of the Afro-Arabian rift system: a review of seismic refraction data. *Tectonophysics*, **198**, 311–327.

REDFERN, P. & JONES, J.A. 1995. The interior rifts of the Yemen—analysis of basin structure and stratigraphy in a regional plate tectonic context. *Basin Research*, **7**, 337–356.

ROBERTS, D.G. 1969. Structural evolution of the rift zones in the Middle East. *Nature*, **223**, 55–57.

ROBERTSON, A.H.F. & BAMAKHALIF, K.A.S. 2001. Late Oligocene–early Miocene rifting of the northeastern Gulf of Aden: basin evolution in Dhofar (southern Oman). *Peri-Tethys Memoir 6 (Mémoires du Museum National d'Histoire Naturelle, Paris*, **186**), 641–670.

ROESER, H.A. 1975. A detailed magnetic survey of the southern Red Sea. *Geologisches Jahrbuch*, **D13**, 131–153.

SAGRI, M., ABBATE, E., ET AL. 1998. New data on the Jurassic and Neogene sedimentation in the Danakil Horst and Northern Afar Depression, Eritrea. *Peri-Tethys Memoir 3 (Mémoires du Museum National d'Histoire Naturelle, Paris*, **177**, 193–214.

SAHOTA, G. 1990. Geophysical investigations of the Gulf of Aden continental margins: geodynamic implications for the development of the Afro-Arabian rift system. Ph.D. Thesis, Swansea.

SAVOYAT, E., SHIFERAW, A. & BALCHA, T. 1989. Petroleum exploration in the Ethiopian Red Sea. *Journal of Petroleum Geology*, **12**, 187–24.

SCHMIDT, D.L., HADLEY, D.G. & BROWN, G.F. 1982. Middle Tertiary continental rift and evolution of the Red Sea in southwestern Saudi Arabia. *US Geological Survey Open File Report USGS-OF-03-6*, p. 1–52.

SEBAI, A., ZUMBO, V., FÉRAUD, G., BERTRAND, H., HUSSAIN, A.G., GIANNÉRINI, G. & CAMPERDON, R. 1991. $^{40}Ar/^{39}Ar$ dating of alkaline and tholeiitic magmatism of Saudi Arabia related to the early Red Sea rifting. *Earth and Planetary Science Letters*, **104**, 473–487.

SELLA, G.F., DIXON, T.H. & MAO, A. 2002. REVEL A model for recent plate velocities from space geodesy. *Journal of Geophysical Research*, **107**, doi: 10.1029/2002/JB00033.

SICHLER, B. 1980. La biellette danakile: un modèle pur l'évolution géodynamique de l'Afar. *Bulletin de la Société Géologique de France, ser.* **7**, (22), 925–933.

SOURIOT, T. & BRUN, J.P. 1992. Faulting and block rotation in the triangle, East Africa; the Danakil "crank-arm" model. *Geology*, **20**, 911–914.

STERN, R.J. 2002. Crustal evolution in the East African orogen: a neodymium isotopic perspective. *Journal of African Earth Sciences*, **34**, 109–117.

STOESER, D.D. & CAMP, V.E. 1985. Pan-African microplate accretion of the Arabian Shield. *Geological Society of America Bulletin*, **96**, 817–826.

SULTAN, M., BECKER, R., ARVIDSON, R.E., SHORE, P., STERN, R.J., EL ALFY, Z. & ATTIA, R.L. 1993. New constraints on Red Sea rifting from correlations of Arabian and Nubian Neoproterozoic outcrops. *Tectonics*, **12**, 1303–1319.

TAMSETT, D. & SEARLE, R.C. 1988. Structure and development of the midocean ridge plate boundary in the Gulf of Aden: Evidence from GLORIA sidescan sonar. *Journal of Geophysical Research*, 93B, 3157–3178.

TAPPONIER, P., ARMIJO, R., MANIGHETTI, V. & COURTILLOT, V. 1990. Bookshelf faulting and horizontal block rotation between overlapping rifts in southern Afar. *Geophysical Research Letters*, **17**, 1–4.

TEFERA, M., CHERNET, T. & HARO, W. 1996. Explanation of the Geological Map of Ethiopia, 2nd edition. *Ethiopian Institute of Geological Survey, Bulletin*. **3**, 79pp., Addis Ababa, Ethiopia.

TESFAYE, S., HARDING, D.J. & KUSKY, T. 2003. Early continental break-up boundary migration of the Afar triple junction, Ethiopia. *Geological Society of America Bulletin*, **115**, 1053–1067.

TIBERI, C., EBINGER, C., BALLU, V., STUART, G. & BEFEKADU, O. 2005. Inverse models of gravity data from the Red Se–Aden–East African rifts triple junction zone. *Geophysical Journal International*, **163**, 775–787.

UKSTINS, I.A., RENNE, P.R., WOLFENDEN, E., BAKER, J., AYALEW, D. & MENZIES, M. 2002. Matching conjugate volcanic rifted margins. $^{40}Ar/^{39}Ar$ chrnostratigraphy of pre- and syn-rift bimodal flood volcanism in Ethiopia and Yemen. *Earth and Planetary Science Letters*, **198**, 289–306.

VAIL, J.R. 1985. Pan-African (late Precambrian) tectonic terrains and the reconstruction of the Arabian–Nubian Shield. *Geology*, **13**, 839–842.

WATCHORN, F., NICHOLS, G.J. & BOSENCE, W.G. 1998. Rift related sedimentation and stratigraphy, southern Yemen (Gulf of Aden). *In*: PURSER, B.H. & BOSENCE, W.G. (eds) *Sedimentary and Tectonic Evolution of Rift Basins: Red Sea–Gulf of Aden*. Chapman & Hall, 165–192.

WHITEHOUSE, M.J., WINDLEY, B.F., BA-BTTAT, M.A.O., FANNING, C.M. & REX, D.C. 1998. Crustal evolution and terrane correlation in the eastern Arabian Shield, Yemen: geochronological constraints. *Journal of the Geological Society of London*, **155**, 281–295.

WOLDEGABRIEL, G., ARONSON, J.L., WALTER, R.C., 1990. Geology, geochronology and rift basin developments in the central sector of the Main Ethiopian Rift. *Geological Society of America Bulletin*, **102**, 439–485.

WOLFENDEN, E., EBINGER, C., YIRGU, G., DEINO, A. & AYELAW, D. 2004. Evolution of the northern Main Ethiopian Rift: birth of a triple junction. *Earth and Planetary Science Letters*, **224**, 213–228.

WOLFENDEN, E., EBINGER, C., YIRGU, G., RENNE, P.R. & KELLEY, S.P. 2005. Evolution of a volcanic rifted margin: southern Red Sea, Ethiopia. *Geological Society of America Bulletin*, **117**, 846–864.

WYN-HUGHES, G., VAROL, O. & BEYDOUN, Z.R. 1991. Evidence for Middle Oligocene rifting of the Gulf of Aden and for late Oligocene rifting of the southern Red Sea. *Marine Petroleum Geology*, **8**, 354–358.

ZUMBO, V., FÉRAUD, G., BERTRAND, H. & CHAZOT, G. 1995*a*. $^{40}Ar/^{39}Ar$ chronology of Tertiary magmatic activity in southern Yemen during the early Red Sea–Aden rifting. *Journal of Volcanology and Geothermal Research*, **65**, 265–279.

ZUMBO, V., FÉRAUD, G., VELLUTINI, P., PIGUET, P. & VINCENT, J. 1995*b*. First $^{40}Ar/^{39}Ar$ dating on Early Pliocene to Plio-Pleistocene magmatic events of the Afar–Republic of Djibouti. *Journal of Volcanology and Geothermal Research*, **65**, 281–295.

# The role of magma in the development of the Afro-Arabian Rift System

W. R. BUCK

*Lamont-Doherty Earth Observatory of Columbia University, Palisades, NY 10964, USA (e-mail: buck@ldeo.columbia.edu)*

**Abstract:** The initiation of the Afro-Arabian Rift System on three nearly straight segments occurred shortly after massive amounts of basalt poured out of the triple junction of those segments in Afar. The synchroneity of magmatism and rifting may reflect the fact that normal continental lithosphere is too strong to rift without magmatic dyke intrusions and the straightness of rifts reflects a localized source for the magma feeding those dykes. Simple relations are derived for the minimum extensional force needed for lithosphere cutting dyke intrusions as functions of the density structure and thickness of the lithosphere. As long as the density contrast between continental crust and magma is small compared to the density contrast between the mantle and magma, then the force needed to rift scales with the square of the thickness of the mantle lithosphere. Thus, continental regions with normal-thickness lithosphere may rift when reasonable levels of extensional force and sufficient magma are available. Very thick mantle lithosphere may not rift at levels of force that are likely to arise on Earth. Two main sections of the Afro-Arabian Rift System, the Red Sea and the Ethiopian Rift, appear to have developed as magma-assisted rifts in normal continental lithosphere. The northern and southern ends of the system are bounded by regions of very thick mantle lithosphere where dykes could not open. In the south, the Tanzanian Craton, with normal-thickness crust and a very deep lithospheric root, was also not split by the rift. In the north, the rift opened along a nearly straight line from the centre of the flood basalt province 2000 km to the edge of the Mediterranean Sea. The old oceanic lithosphere of this margin may be no thicker than the adjacent continental lithosphere of Egypt, but the thinner oceanic crust means that Mediterranean lithosphere may be too thick and dense to rift magmatically. The role of magma in the third branch, the Gulf of Aden, is not so clear given the lack of syn-rift dykes.

A map of the Afro-Arabian Rift System (Fig. 1) shows the straightness of the three segments that radiate from the high topography around the Afar triple junction. The Red Sea is nearly straight for 2000 km going north from the triple junction. North of this straight segment, the Red Sea rift merges with the abandoned Gulf of Suez rift and the Aqaba–Dead Sea transform rift. The southern branch of the rift is relatively straight through Ethiopia and into Kenya, but it dies out in the Tanzanian Craton (Venkataraman *et al.* 2004), and another branch opens to the west of the craton. The Gulf of Aden branch proximal to the triple junction is straight, although the present-day spreading is oblique to the coastlines (Audin *et al.* 2001; Leroy *et al.* 2004), and the distal end links with the mid-ocean ridge system.

Most large-scale rifts develop in association with large igneous provinces (e.g. Hinz 1981; Hill 1991). The Afro-Arabian Rift System is the youngest example of a such a rift system and its opening is associated with the igneous province that includes the plateau basalts of Yemen and Ethiopia which are up to 3 km thick (e.g. Menzies 1992, Pallister 1987). As with most large igneous provinces, the period of high magma flux lasted for as little as one million years and magma effusion rates were as great as 1 km $a^{-1}$ (e.g. Courtillot *et al.* 1999; Courtillot & Renne 2003 and references therein). Away from the Red Sea Rift margins, flood volcanism was less voluminous and longer lived (e.g. Kieffer *et al.* 2004).

Models of rift formation and evolution often ignore any effects of magmatism or dyke intrusion (e.g. McKenzie 1978; Royden & Keen 1980; Braun & Beaumont 1989; Buck 1991; Davis & Kusznir 2004) or assume that excess magmatism is largely a consequence of rifting. This is surprising as magmatism often precedes rifting (e.g. Courtillot & Renne 2003 and references therein) and a relatively small amount of hot basalt intruded into lithosphere could significantly heat (e.g. Royden *et al.* 1980) and weaken that lithosphere (e.g. Buck 2004).

The temporal and structural development of the Afro-Arabian Rift System imply that magma had a major role in the initiation of that system. The basalts of Yemen and Ethiopia began to pour out as early as 45 Ma (e.g. Kieffer *et al.* 2004) and were particularly effusive at about 30 Ma (Courtillot & Renne 2003 and references therein).

*From*: YIRGU, G., EBINGER, C.J. & MAGUIRE, P.K.H. (eds) 2006. *The Afar Volcanic Province within the East African Rift System*. Geological Society, London, Special Publications, **259**, 43–54.
0305-8719/06/$15.00 

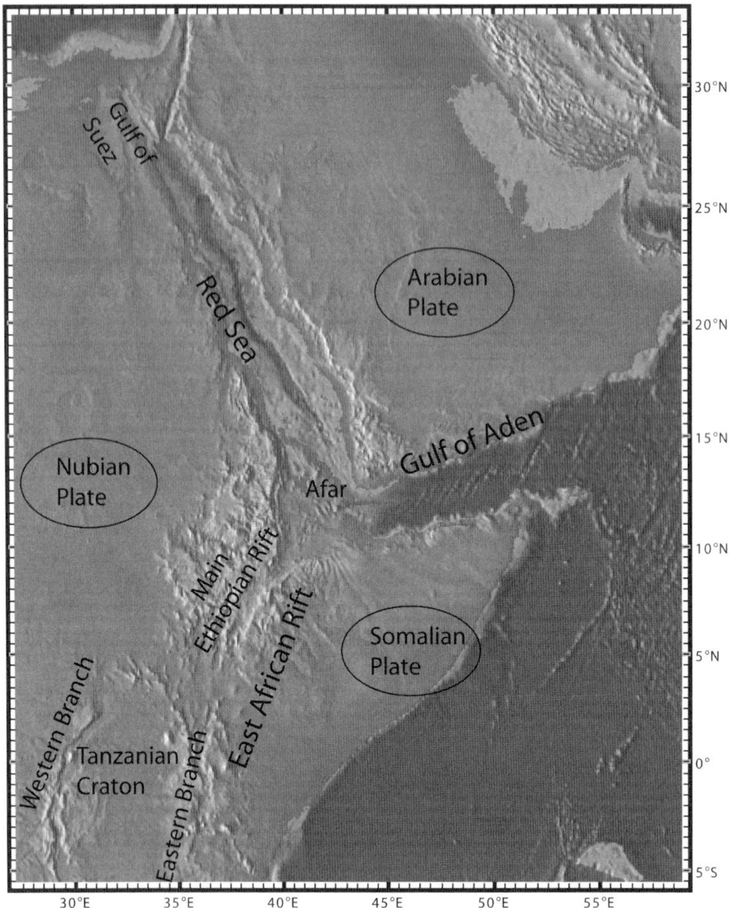


**Fig. 1.** Shaded relief map of the Afro-Arabian Rift System. Note the straightness of the rift arms proximal to the Afar Triple Junction and the complexity of the terminations of the Northern (Red Sea Rift) and Southern (East African Rift) segments.

It is difficult to determine the onset of rifting of the arms of the Afro-Arabian Rift, but Wolfenden *et al.* (2005) document aligned eruptive centres along border faults in the southern Red Sea Rift. On the northern end of that rift, 24 Ma volcanics fill palaeo-valleys that may have formed in association with rifting on the flanks of the Gulf of Suez, almost 2000 km from the centre of the igneous province (e.g. Patton *et al.* 1994). Dykes of similar age also cut the basement blocks exposed by normal faulting in the Suez region (e.g. Patton *et al.* 1994). Rift flank uplift has begun by about 20 Ma (Omar *et al.* 1989; Omar & Steckler 1995). The rift system is characterized by straight segments proximal to the triple junction with complex structures for the ends of the Red Sea and East African Rift branches. This paper examines how the mechanics of magmatic dyke intrusions may explain these large-scale structural features.

The first section of this paper describes simple scaling relations for the forces needed to extend lithosphere with and without dyke intrusions. Then arguments are advanced that straight rift segments imply that dykes fed from a localized source facilitated plate spreading during the initiation of rifting. Finally, possible causes for changes in the magma supply to parts of the Afro-Arabian Rift are briefly discussed.

## *Force available for driving rifting*

Many authors have discussed ways that plate tectonic and plume-related processes could produce relative tension at a rift (e.g. Forsyth & Uyeda 1975; Solomon *et al.* 1980), and these approaches give similar estimates of rift-driving forces. The simplest of these approaches relates to the uplift that may occur over a region of abnormally hot mantle. The radiation of rift branches away from the uplifted plateau areas of Ethiopia and Yemen are consistent with the extensional driving force being related to the uplift of that region (e.g. Sengor & Burke 1978; Ernst *et al.* 1995).

For uplift due to a low-density root, the extensional force scales with the magnitude of the uplift, $e$, and the depth of the low-density compensation layer, $d$ (Spohn & Schubert 1982; Bott 1991). The force is roughly $e\rho_m gd$, where $\rho_m$ is the mantle density and $g$ is the acceleration due to gravity. For a root of uniform-density hot mantle between 100 and 200 km, $d$ would be 150 km. Then ~1 km of uplift compensated at 150 km depth gives $\sim 5 \times 10^{12}$ Nt m$^{-1}$ of rift-driving force. Higher elevations and deeper compensation depths are possible, but the force may be spread over a longer rift than the width of the uplifted area. Thus, the average level of rift force is likely to be less than about $5 \times 10^{12}$ Nt m$^{-1}$.

## *Force needed for tectonic rifting*

Following standard practice (e.g. Brace & Kohlstedt 1980), we estimate the minimum tectonic force to rift by calculating the stress needed to allow normal fault slip on optimally oriented normal faults in an Andersonian stress field (Anderson 1951). For such a stress field, the vertical stress is taken to be the maximum principal stress and to be lithostatic:

$$\sigma_V(z) = \int_0^z g\rho_s(z)dz \qquad (1)$$

where $\rho_s(z)$ is the density of the solid lithosphere and $z$ is depth. For slip on a normal fault, the horizontal stress must be smaller than the vertical stress by a constant factor that depends on the rock friction coefficient, $f$. Therefore, the stress difference needed for tectonic extensional faulting at a given depth is:

$$\sigma_T(z) = \sigma_V(z) - \sigma_{TH}(z) = A[\sigma_V(z) - P_p(z)] \quad (2)$$

where $\sigma_{TH}$ is the horizontal stress for tectonic fault slip, $A = 2f/(f + (1+f^2)^{1/2})$ and $P_P$ is the pore pressure (see Turcotte & Schubert 2002). For $f = 0.6$ (Byerlee 1978), $A = 0.7$. The maximum pore pressure through the lithosphere is often assumed to equal hydrostatic pressure (i.e. $P_P = g\rho_w z$, where $\rho_w$ is the density of water).

The tectonic force needed for rifting is the integral of the tectonic stress difference over the thickness of the lithosphere, so:

$$F_T = \int_0^{h_l} \sigma_T(z)dz \qquad (3)$$

To the extent that we can take the density of the lithosphere and water to be constant with depth, we can get a simple estimate of this minimum tectonic force for rifting:

$$F_T = Bh_l^2 \qquad (4)$$

where $B = A(\rho_s - \rho_w)$. Taking $\rho_s - \rho_w = 2000$ kg m$^{-3}$ and $f = 0.6$ gives $B = 1.2 \times 10^4$ Pa m$^{-1}$. Figure 2 shows the quadratic relation between tectonic force and lithospheric thickness for this simplified estimate. Neglecting hydrostatic pore pressures increases the value of $B$, and thus the estimated force, by about 50%; and taking a mantle density for the lithospheric density increases it a further 10%. Assuming the creeping part of the lithosphere contributes to the tectonic force also increases this estimate, but for crust creeping more easily than the mantle the tectonic force is reduced by an amount that depends on the crustal thickness, rheological constants and thermal state (e.g. Brace & Kohlstedt 1980; Sawyer 1985; Kusznir & Park 1987). For the purpose of making simple comparisons with the force needed to open a magmatic rift in this paper, Equation 4 is sufficient.

## *Force needed for magmatic rifting*

To estimate the force required to open a dyke cutting the lithosphere, we need to estimate the vertical and horizontal lithospheric stresses (see Fig. 3). As assumed for Andersonian normal faulting, the vertical stress is taken to be the maximum principal stress and is lithostatic (as expressed in Equation 1). Neglecting dynamic (flow-related) stresses in a dyke, the stress at the walls of a dyke must equal the magma pressure. For a vertical dyke filled with magma up to the surface, the magma pressure is the static fluid pressure in the dyke, $P_f$. Thus:

$$\sigma_{MH}(z) = \int_0^z g\rho_f(z)dz = g\rho_f z \qquad (5)$$

where $\sigma_{MH}$ is the horizontal stress for magmatic dyke opening and $\rho_f$ is the magma density.

As long as the magma is less dense than the rock it intrudes, then some extensional stress has to be applied to keep the magma from rising to the

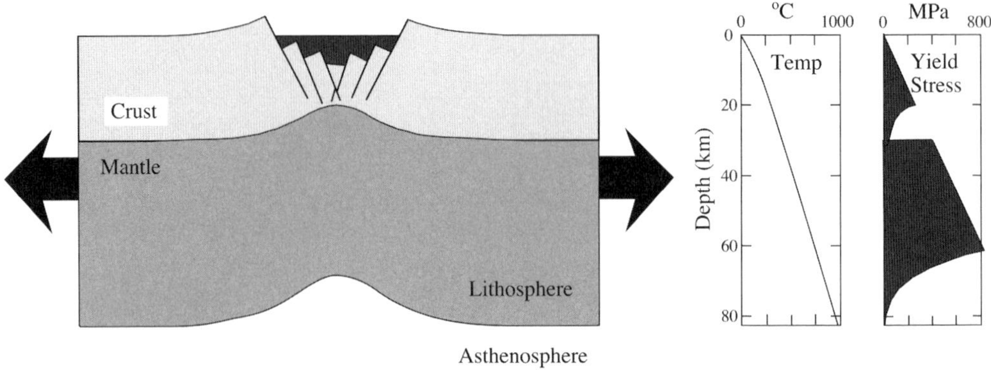


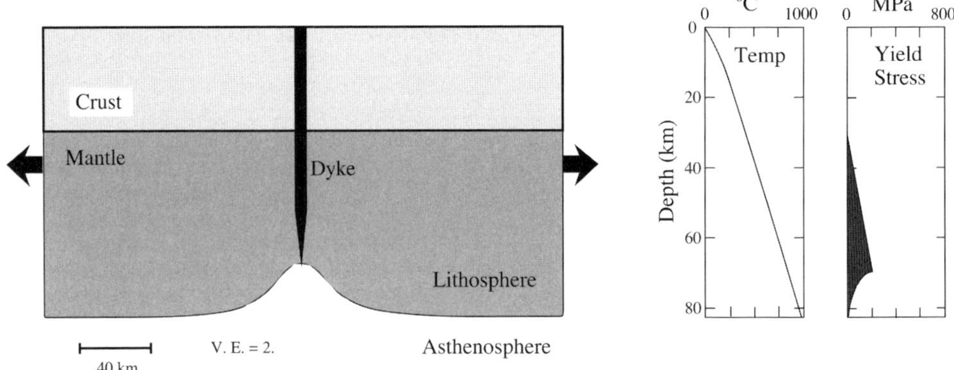


**Fig. 2.** Schematic of two possible ways to extend normal-thickness continental lithosphere. Note the large difference in the yield stress, the stress difference needed to get extensional separation of two lithospheric blocks (**a**) without and (**b**) with magmatic intrusion.

surface and extruding. Only when the magma is 'kept down' by this stress difference will the dyke open at depth and so allow plates to move apart. Neglecting any stress needed to open a crack, the maximum stress difference needed to open a magma-filled dyke is:

$$\sigma_M(z) = \sigma_V(z) - \sigma_{MH}(z) = \int_0^{h_l} g(\rho_s(z) - \rho_f)dz \quad (6)$$

Equation 6 is valid only when the fluid magma density is less than or equal to the average lithospheric density. For example, for constant lithospheric density that is less than the magma density, an extensional force would be required to pull dense magma up from the asthenosphere, magma could not rise to the surface and the limits of integration would have to be changed. Such cases are not likely to be important for lithosphere thicker than a few kilometres and for basaltic magma densities, so they are not discussed further.

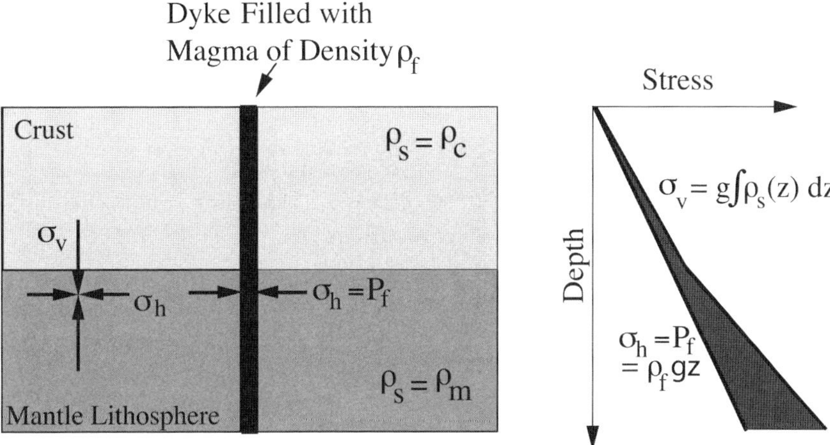


**Fig. 3.** The stress distribution for extensional separation of two lithospheric blocks by a vertical magmatic intrusion (a dyke). Here the solid lithosphere has a density, $\rho_s$, that is greater than the fluid magma density, $\rho_f$. The light grey region represents crust with a density, $\rho_c$, close to that of basaltic magma in dykes and the medium grey represents mantle lithosphere with a density, $\rho_m$, greater than that of the crust. The horizontal stress, $\sigma_h$, equals the pressure in the dyke, $P_f$, while the vertical stress, $\sigma_v$, equals the overburden pressure. The area between these two stresses (dark grey) equals the force that must be applied to keep the dyke open.

The extensional tectonic force to open a magma-filled dyke through denser lithosphere with a thickness $h_l$ is:

$$F_M = \int_0^{h_l} \sigma_M(z)\,dz \tag{7}$$

Consider the simple case that the densities of the crust and mantle, $\rho_c$ and $\rho_m$ respectively, are constant with depth. If the entire crust is brittle down to its base at $z = h_c$ and the thickness of mantle lithosphere is $h_m$ (so the total lithospheric thickness $h_l = h_c + h_m$), then:

$$F_M = g(\rho_c - \rho_f)\frac{h_c^2}{2} + g\left[(\rho_c - \rho_f)h_c + (\rho_m - \rho_f)\frac{h_m}{2}\right]h_m \tag{8}$$

The density of continental crust is not likely to be very different from the density of basaltic magma. If that is true, then it takes no force to open a dyke in the crust, but it still takes considerable force to open a dyke into the mantle lithosphere. For $\rho_c = \rho_f$, we get the further simplification that:

$$F_M = g(\rho_m - \rho_f)\frac{h_m^2}{2} \tag{9}$$

For reasonable density values, we plot this estimate of the force to dyke through the entire lithosphere versus the lithospheric thickness. For a density contrast between solid mantle and fluid magma of 500 kg m$^{-3}$ (based on mantle density of 3250 kg m$^{-3}$ and magma density of 2750 kg m$^{-3}$) it would take $4 \times 10^{12}$ Nt m$^{-1}$ to dyke through a 40 km-thick lithospheric mantle layer. To dyke through 100 km of mantle lithosphere would take $2.5 \times 10^{13}$ Nt m$^{-1}$ and this is considerably more force than is likely to be available to drive rifting. The term magmatic rifting will be used to describe lithospheric extension that is aided by dyke intrusion and the force for magmatic rifting is taken to be the force to open dykes through the lithosphere.

Dykes may not open through the entire thickness of the lithosphere if insufficient magma is available, and for such cases the force required to rift would be intermediate between the force for tectonic rifting and the force for magmatic rifting. Figure 4a shows how different these forces are likely to be as a function of lithospheric thickness. The tectonic force depends on the square of the whole-thickness

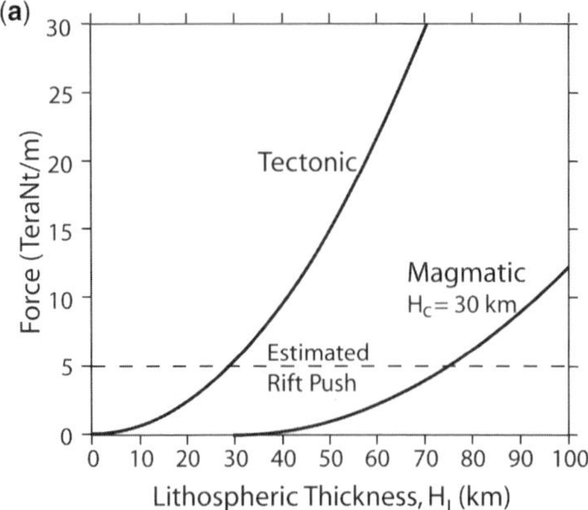


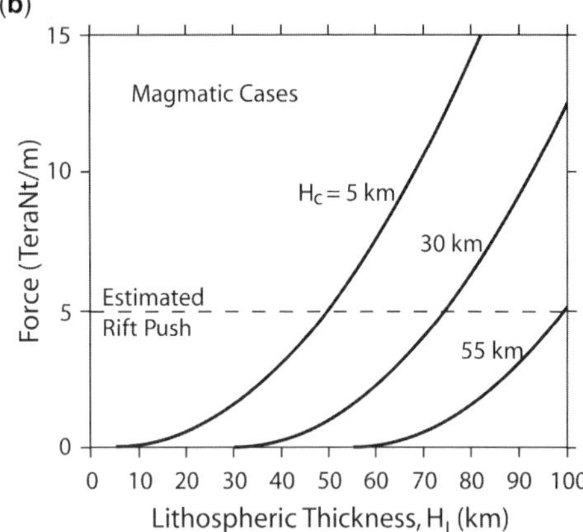


**Fig. 4.** Illustration of estimated force required for lithospheric extension for either tectonic or magmatic rifting as a function of lithospheric thickness lithosphere. Equation 4 was used to compute the tectonic force and Equation 9 was used to compute the magmatic force. (**a**) Shows the forces as a function of lithospheric thickness assuming a 30 km-thick crust. The horizontal dashed line is an estimate of the force available to drive lithospheric extension. (**b**) Shows how the crustal thickness affects the estimated magmatic force. It is the mantle lithospheric thickness, which equals the total lithospheric thickness minus the crustal thickness, that controls this force.

of the lithosphere ($\times \sim 6000$ Nt m$^{-3}$), while the magmatic force depends on the square of the mantle lithospheric thickness ($\times \sim 2500$ Nt m$^{-3}$). For reasonable driving force levels, only lithosphere thinner than $\sim$30 km thick should rift tectonically (i.e. in the absence of magmatic dyke intrusion). For a normal continental crustal thickness of 40 km, the base of the lithosphere could be as deep as *c.* 80 km and still allow magmatic rifting at reasonable force levels. Figure 4b shows that lithosphere with thicker continental crust should rift magmatically with much less force than for lithosphere with thin crust.

## *The meaning of rift straightness*

Lithospheric extension can be accomplished by either dyke opening or fault slip. Both dykes and faults form in response to stress difference in the lithosphere. Faults form and slip where the shear stress is great enough to overcome the strength of the material and allow brittle deformation. Faults change orientation where the stress orientations change or where there are strength variations. So, rifts where faults accommodate the extension do not have to be straight. They can curve where the stresses change or 'side-step' into weaker areas.

Dykes are narrow magma-filled tension cracks (e.g. Lister & Kerr 1991) and as such the plane of the dyke must be orthogonal to the least principal stress in a strong, brittle layer. When the least principal stress is horizontal, the dyke is vertical and these are the kinds of dykes discussed here. The key to the straightness of dykes may be that, unlike faults, they need a source of magma and a connection to that magma source. The open part of the dyke is the conduit connecting the source area to the tip of the dyke. If dykes are fed from a distributed magma source below a brittle layer, then the dykes do not have to be straight except on the scale of the layer thickness. One way to produce dykes that are straight over lateral distances longer than the brittle layer thickness is if the magma source is localized. The dykes can only remain connected to the source if they are straight.

If the magma does come from a central source, then the dyke cannot propagate if it loses connection to the magma supply because the connection is the open, unfrozen, straight dyke behind the propagating tip. If a dyke tip were to step laterally away from the plane of the open dyke by more than the width of the dyke, it would no longer be fed magma. Dykes exposed in ophiolites and bordering rifts are typically about 1 m wide (e.g. Varga 2003), so very small offsets are viable.

Straight dykes have been observed propagating from a central magma chamber along a sub-aerial segment of the Mid-Atlantic Ridge in Iceland. In 1975, an episode of approximately 15 dyke intrusion and magma extrusion events began, with the longest dyke propagating 70 km from the Krafla central volcano (Tryggvason 1980). Seismic and geodetic measurements unequivocally show that the dykes are sourced from a central magma chamber that subsided while the dyke was propagating (Einarsson 1991). Dykes on the flanks of active volcanoes, like those propagating down the east rift zone of Kilauea Volcano in Hawaii, are seen to be straight (e.g Segall *et al.* 2001) Ancient dykes in the MacKenzie Dyke Swarm of Canada are straight and traceable for thousands of kilometres (Fialko & Rubin 1999).

The fact that most mid-ocean ridge segments are straight and nearly orthogonal to the spreading direction may reflect dykes fed from central magma chambers. New data on ridge segments suggest that at the slowest spreading rates there are non-magmatic segments that are not straight. Dick *et al.* (2003) note that the slowest-spreading centres, such as the Gakkel Ridge in the Arctic and oblique sections of the Southwest Indian Ridge, show an alteration of volcanic segments and non-volcanic ones. Peridotite samples dredged from the surface of the non-volcanic segments indicate that the mantle there is stretched with little or no input of magma. When these non-volcanic segments are along oblique sections of the ridge, the usual pattern of ridge segments orthogonal to transform faults is not seen. Such segments are cut by numerous faults that trend oblique to the spreading direction.

Rifts and passive margins associated with large igneous provinces tend to be nearly straight. No rift is perfectly straight, but if a great circle can pass through part of every segment of a rift then that rift could be said to be straight. For many rifts, like the Aden rift, the small-scale segments do not parallel the coastline or the border faults of the rift (e.g. Leroy 2004). This is not surprising since the stress orientation and opening direction can change after the initial phase of opening.

The Red Sea is the clearest example of a straight rift with the coastlines showing only slight deviations from a great circle over 2000 km and the axis of spreading and rifting is even straighter than the coast. The South Atlantic Margins of South America and Africa, associated in space and time with the Parana Flood Basalts (Hinze 1981), are another very straight margin. The dyke-intruded and volcanic-covered North Atlantic Margins of Greenland (e.g. Holbrook *et al.* 2001) and the conjugate European margins are also remarkably straight on a scale of thousands of kilometres.

A magmatic rift does not have to be volcanic. The Northern Red Sea and the Gulf of Suez lack evidence of massive syn-rift volcanism (forming many kilometres thick seaward-dipping seismic reflector packages) that has have been documented along much of the South Atlantic and Greenland Margins (Hinz 1981; Mutter & Mutter 1988). However, syn-rift dykes striking parallel to the rift are seen as far north as the Gulf of Suez and valley-filling, syn-rift volcanics are common there as well (Patton *et al.* 1994). Buck (2004) argues that a relatively modest amount of basaltic dyke intrusion could have heated the lithosphere of the Northern Red Sea and Gulf of Suez sufficiently to allow continued rifting at moderate tectonic force levels with no further input of magma.

### *The distance of dyke/rift propagation*

Dyke propagation may be controlled by a dauntingly large number of thermal and mechanical processes. Among them are the pressure and flux of magma coming out of a source region, the viscous resistance to magma flow along the body of the dyke and into the tip region, elastic stresses in lithosphere and the freezing of magma (e.g. Lister & Kerr 1991; Rubin 1995). Although the details of dyke mechanics are controversial (e.g. Delaney & Pollard 1982; Fialko & Rubin 1999; Ida 1999), there is no doubt that the distance of dyke propagation should be related to the supply of magma. The volume of magma intruded into the thick lithosphere along a several-thousand-kilometre long rift has to be massive.

The size of a magma chamber should play a major role in determining how much magma can be supplied to dykes as they propagate. The magma pressure is likely to be reduced as volume is extracted from a magma chamber. Buck *et al.* (2004) argue that lithosphere-cutting dykes stop when the 'driving stress' (defined as the difference between magma pressure and tectonic stress orthogonal to the dyke) becomes too small, either because the dyke slows and the tip freezes, or because the driving stress is not enough to break open a new section of dyke. The larger and shallower the magma chamber, the smaller the pressure drop on extraction of a given volume of magma. Large magma chambers should supply large amounts of basalt to a propagating dyke and so could be necessary to the production of long dykes.

Simple thermal arguments would suggest that the size and depth of a magma chamber should correlate with the flux of magma coming into a region. The greatest-known fluxes of magma occur during the geologically short periods when large igneous provinces form. The volumes are up to several million cubic kilometres and the time interval of high rate magma output is one million years or less (Courtillot & Renne 2003 and references therein). Very large, near-surface magma chambers are likely to form during periods of high melt flux from localized mantle upwellings. Magma chambers the size of large gabbroic-layered intrusions found near the centres of some large igneous provinces could feed dykes propagating thousands of kilometres. For example, the Skaergaard layered intrusion of East Greenland is estimated to have a volume of *c.* 300 km$^3$ (Nielsen 2004), sufficient to fill a dyke 2000 km long, 50 km high and 3 m thick.

As noted in the previous section on forces for magmatic rifting, another necessary condition for dyking is extensional stress. The regions where the mantle lithosphere is very thick may not rift even if copious, high-pressure basaltic magma is present. Recall that extrusion of magma on the surface limits the maximum magma pressure. Cratonic regions, where the lithosphere may be well over 100 km thick (Jordan 1975; Venkataraman *et al.* 2004), and old oceanic lithosphere where the mantle lithosphere may be *c.* 60 km thick (e.g. Wiens & Stein 1983) should be too thick to rift. Figures 4 and 5 illustrate situations of crust and lithospheric thickness where rifting may not occur for reasonable extensional forces level.

Dykes and associated magmatic rift propagation can stop either because the magma pressure gets too low, or the stress is not sufficient to open a dyke through the lithosphere. For the Afro-Arabian Rift System, it may be the lack of extensional stress sufficient to open dykes through very thick mantle lithosphere that limits dyke propagation. This is suggested by the observation that the Northern and Southern Branches of the system end close to regions of thick mantle lithosphere (Fig. 6). The Northern Red Sea Branch ends close to the Mediterranean Sea where old oceanic lithosphere may be too thick to be cut by dykes (Fig. 5a).

Previous workers have noted that changes in lithospheric strength may have controlled the termination of the Red Sea Branch (e.g. Steckler & ten Brink 1986) or affected the structure of the East African Branch (e.g. Rosendahl 1987). Those workers were concerned with the difference in tectonic strength of the lithosphere. This paper contends that it is the large increase in force for magmatic rifting related to mantle lithosphere thickening that limits rift propagation.

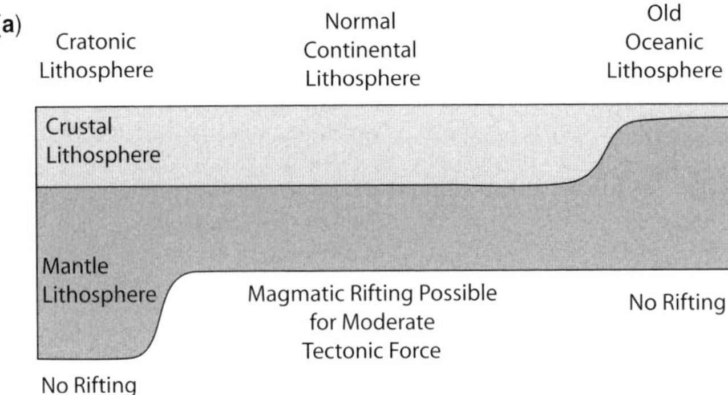


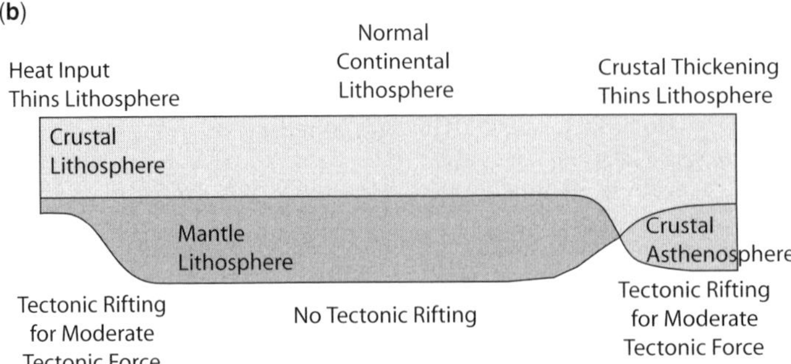


**Fig. 5.** Conditions under which rifting can or cannot occur with moderate levels of driving force. (**a**) The panel indicates that magmatic rifting can occur for normal thickness, or thinner, continental lithosphere (panel centre) while thick mantle lithosphere (panel ends) cannot rift. The ends show the two ways that magmatic rift propagation may be stopped: by running into either old, thick oceanic lithosphere or very thick cratonic lithosphere. (**b**) The panel indicates that tectonic rifting cannot happen for normal thickness, or thicker, continental lithosphere (panel centre) while tectonic rifting can occur in areas of thin lithosphere (panel ends).

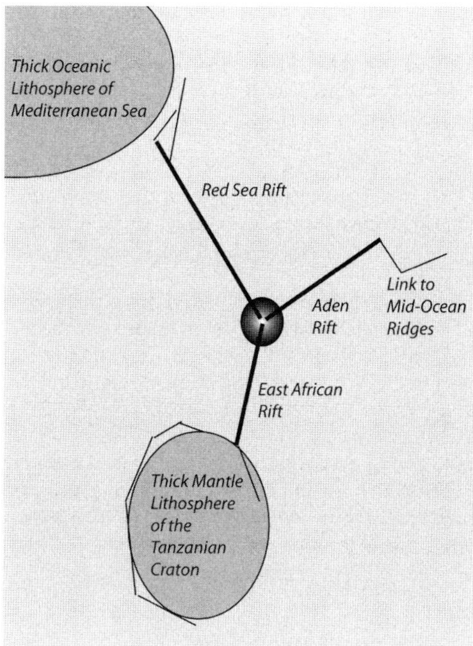


**Fig. 6.** Schematic of the structure of the Afro-Arabian Rift System initiation showing that the straight segments of the northern and southern parts (Red Sea and East African Rifts) may have stopped propagating when they encountered mantle lithosphere that was too thick to rift.

## *Stages of development of the Main Ethiopian Rift*

The branches of the Afro-Arabian Rift System, and other large rift systems, may go through several stages of development. These stages are likely to change depending on the flux of magma and the thickness of the lithosphere and crust. New data from the Main Ethiopian Rift may illustrate these stages particularly well (e.g. Kendall *et al.* 2005; Keranen *et al.* 2004).

The first stage of development of a magma-assisted rift is the active mantle melting phase when magma is produced in a localized area and supplied to distant parts of the rift. The earliest phase of opening of the Main Ethiopian Rift may have formed this way and, as argued above, this is consistent with the straightness of the rift. Dyke intrusion should produce thermal weakening of the lithosphere. The elevation of the rift would depend on the amount of magma intruded and might not involve significant subsidence (Buck 2004). The seismic anisotropy data of Kendall *et al.* (2005) shows a wide region of rift-parallel fast-shear wave directions which they interpret in terms of intruded rift-parallel dykes. Rift-parallel dykes are what would be expected for the first stage of magmatic rifting.

The second stage would involve an increasing fraction of rifting via tectonic stretching of the crust and lithosphere. This should occur with waning of hotspot magma production. During the earlier stage, magma injection into the lithosphere could lead to heating, and so lithospheric thinning sufficient to allow tectonic rifting at moderate levels of extensional force (see Fig. 4). During this tectonic stage, there would be advective thinning of the lithosphere and crust via stretching (e.g. McKenzie 1978) and production of block faults and border faults.

The third stage of rifting involves a return to more magmatic rifting, but now this involves thinner lithosphere and thus less magma is needed to cut through that lithosphere. Two sources are possible for the magma. One possibility is that passive mantle upwelling related to plate stretching brings hot mantle to shallow-enough depths that significant mantle melting occurs. This kind of initiation of seafloor spreading-type magmatism may be what is occurring along much of the Red Sea where the total amount of rift opening is over 100 km (e.g. Martinez & Cochran 1988). The second possibility is that continued low-level hotspot magma production is now sufficient to deliver enough magma to accommodate the plate separation. This is the more likely case for the Main Ethiopian Rift because it has undergone such a small amount of extension (Ebinger & Casey 2001) and so is not likely to have thin enough lithosphere and crust for significant advective melting. Another reason to suspect the source to be hotspot-related is due to the proximity of the Afar hotspot. The seismic anisotropy data of Kendall *et al.* (2005) show a narrow zone in the centre of the rift with fast shear wave directions parallel to the en echelon volcanic fissure system.

## Conclusions

Dyke intrusions radiating from a large igneous province may produce the pattern of straight segments proximal to a triple junction with complex structures for the distal segments, as for the Red Sea and East African Rift branches (Fig. 6). The arguments for relating the structural pattern of the Afro-Arabian Rift System to magmatic rifting are simple:

(1) Rifting of the normal continental lithosphere of the straight rift segments may not occur without magmatic intrusion to allow the plates to separate at moderate levels of tectonic force. The presence of large volumes of magma does not guarantee rifting.

Extensional force must be applied to allow magma-filled dykes to open if the (average) lithospheric density is different from the magma density.
(2) The complexity of the terminations of the straight segments indicates that dykes could not penetrate those regions because the mantle lithosphere there was too thick.
(3) Straight rift segments are most likely to occur when there is a localized magma source for the dykes allowing the rifting.

The flux of magma into large igneous provinces, combined with moderate tectonic extensional force, can lead to rifting of sections of normal continental lithosphere many thousands of kilometres long. The size of the Afro-Arabian Rifts appears to be set by the distance to thick cratons or old oceanic lithosphere that cannot rift at reasonable tectonic force levels even for unlimited magma supply. The fact that old oceanic lithosphere was rifted at the distal end of the Aden branch, while it was not rifted at the end of the Red Sea, may relate to the distance to the region of hotspot-related uplift that was a possible source of regional extensional tectonic stress. The Mediterranean is twice as far from the triple junction as is the Indian Ocean, so the tectonic force may have been large enough to split a section of the Indian Ocean lithosphere, allowing the Aden rift to link to the mid-ocean ridge system.

One corollary of this work is the suggestion that thick continental cratons may strongly resist rifting. Another is that rifts may be magmatic and not volcanic. In other words, dykes can intrude, allow lithospheric extension and weaken the lithosphere without being associated with copious volcanism. The accommodation of extension by injection reduces the amount of crustal thinning by stretching, even though the near-surface layer is stretching. This may explain why rifts like the Gulf of Suez and the Main Ethiopian Rift subside much less than would be predicted by lithospheric stretching models (Steckler 1985; Ebinger & Casey 2001).

Thanks to Gezahegan Yirgu, Cindy Ebinger and Genene Mulugeta for organizing the International Conference on the East African rift system which opened up the region for me. The suggestions of Sylvie Leroy and an anonymous reviewer greatly improved the manuscript. Lamont contribution number 6885.

## References

Anderson, E.M. 1951. *The Dynamics of Faulting*. Oliver and Boyd, Edinburgh.

Audin, L. 2001. Lithospheric structure of a nascent spreading ridge inferred from gravity data; the western Gulf of Aden. *Journal of Geophysical Research*, **106**, 26,345–26,363.

Bott, M.H.P. 1991. Ridge push and associated plate interior stress in normal and hot spot regions. *Tectonophysics*, **200**, 17–32.

Brace, W.F. & Kohlstedt, D.L. 1980. Limits on lithospheric stress imposed by laboratory experiment. *Journal of Geophysical Research*, **85**, 6248–6252.

Brandsdottir, B. & Einarsson, P. 1979. Seismic activity associated with the September 1977 deflation of Krafla Volcano in North-Eastern Iceland. *Journal of Volcanology and Geothermal Research*, **6**, 197–212.

Braun, J. & Beaumont, C. 1989. A physical explanation for the relation between rift flank uplift and breakup unconformity at rifted continental margins. *Geology*, **17**, 760–764.

Buck, W.R. 1991. Modes of continental lithospheric extension. *Journal of Geophysical Research*, **96**, 20,161–20,178.

Buck, W.R. 2004. Consequences of asthenospheric variability on continental rifting. *In:* Karner, G.D., Taylor, B., Driscoll, N.W. & Kohlstedt, D.L. (eds) *Rheology and deformation of the lithosphere at continental margins*. Columbia University Press, 1–31.

Byerlee, J.D. 1978. Friction of rocks. *Pure and Applied Geophysics*, **116**, 615–626.

Campbell, I.H. & Griffiths, R.W. 1990. Implications of mantle plume structure for the evolution of flood basalts. *Earth and Planetary Science Letters*, **99**, 79–93.

Courtillot, V.E. & Renne, P.R. 2003. On the ages of flood basalt events. *In*: Courtillot, V.E. (ed.) *The Earth's Dynamics*. Comptes Rendus—Academie des Sciences. *Geoscience*, **335**, 113–140.

Courtillot, V., Jaupart, C., Manighetti, I., Tapponnier, P. & Besse, J. 1999. On causal links between flood basalts and continental breakup. *Earth and Planetary Science Letters*, **166**, 177–195.

Davis, M. & Kusznir, N. 2004. Depth-dependent lithospheric stretching at rifted continental margins. *In*: Karner, G.D., Taylor, B., Driscoll, N.W. & Kohlstedt, D.L. (eds) *Rheology and Deformation of the Lithosphere at Continental Margins*. Columbia University Press, 92–137.

Delaney, P.T. & Pollard, D.D. 1982. Solidification of basaltic magma during flow in dike. *American Journal of Science*, **282**, 856–885.

Dick, H.J.B., Lin, J. & Schouten, H. 2003. An ultraslow-spreading class of ocean ridge. *Nature*, **426**, 405–412.

Ebinger, C.J. & Casey, M. 2001. Continental breakup in magmatic provinces; an Ethiopian example. *Geology*, **29**, 527–530.

Ebinger, C.J., Dieno, A.L., Drake, R.E. & Tesha, A.L. 1989. Chronology of volcanism and rift basin propagation: Rungwe volcanic province, East Africa. *Journal of Geophysical Research*, **94**, 15,585–15,803.

Einarsson, P. 1991. The Krafla rifting episode 1975–1989. *In*: Gardarsson, A. & Einarsson, Á. (eds)

*Náttúra M'yvatns, (The Nature of Lake M'yvatn).* Icelandic Nature Science Society, Reykjavik, 97–139.

EINARSSON, P. & BRANDSDOTTIR, B. 1980. Seismological evidence for lateral magma intrusion during the July 1978 deflation of the Krafla volcano in NE-Iceland. *Journal of Geophysics*, **47**, 160–165.

ERNST, R.E., HEAD, W.J., PARFITT, E., GROSFILS, E. & WILSON, L. 1995. Giant radiating dike swarms on Earth and Venus. *Earth Science Reviews*, **39**, 1–58.

FIALKO, Y.A. & RUBIN, A.M. 1999. Thermal and mechanical aspects of magma emplacement in giant dike swarms. *Journal of Geophysical Research*, **104**, 23,033–23,049.

FORSYTH, D.W. & UYEDA, S. 1975. On the relative importance of the driving forces of plate motion. *Geophysical Journal of the Royal Astronomical Society*, **43**, 163–200.

HILL, R.I. 1991. Starting plumes and continental break-up. *Earth and Planetary Science Letters*, **104**, 398–416.

HINZ, K. 1981. A hypothesis on terrestrial catastrophes—wedges of very thick oceanward dipping layers beneath passive margins. *Geologishes Jahrbuch Reihs E*, **22**, 3–28.

HOLBROOK, W.S., LARSEN, H.C., *ET AL.* 2001. Mantle thermal structure and active upwelling during continental breakup in the North Atlantic, *Earth and Planetary Science Letters*, **190**, 251–266.

IDA,Y. 1999. Effects of the crustal stress on the growth of dikes: conditions of intrusion and extrusion of magma. *Journal of Geophysical Research*, **104**, 17897–17910.

JORDAN, T.H. 1975. The continental lithosphere. *Reviews of Geophysics and Space Physics*, **13**, 1–12.

KENDALL, J.M., STUART, G.W., EBINGER, C.J., BASTOW, I.D. & KEIR, D. 2005. Magma-assisted rifting in Ethiopia. *Nature*, **433**, 146–148.

KERANEN, K., KLEMPERER, S.L., *ET AL.* 2004. Three-dimensional seismic imaging of a protoridge axis in the Main Ethiopian Rift. *Geology*, **32**, 949–952.

KIEFFER, B., ARNDT, N., *ET AL.* 2004. Flood and Shield Basalts from Ethiopia: Magmas from the African Superswell. *Journal of Petrology*, **45**, 793–834.

KUSZNIR, N.J. & PARK, R.G. 1987. The extensional strength of the continental lithosphere: its dependence on geothermal gradient, and crustal composition and thickness. *In*: COWARD, M.P., DEWEY, J.F. & HANCOCK, P.L. (eds) *Continental Extensional Tectonics*. Geological Society, London, Special Publications, **28**, 35–52.

LEROY, S., GENTE, P. *ET AL.* 2004. From rifting to spreading in the Eastern Gulf of Aden; a geophysical survey of young oceanic basin from margin to margin. *Terra Nova*, **16**, 185–192.

LISTER, J.R. 1994. The solidification of buoyancy-driven flow in a flexible-walled channel. Part 1, Constant-volume release. *Journal of Fluid Mechanics*, **272**, 21–44.

LISTER, J.R. & KERR, R.C. 1991. Fluid-mechanical models of crack propagation and their application to magma transport in dykes. *Journal of Geophysical Research*, **96**, 10,049–10,077.

MARTINEZ, F. & COCHRAN, J.R. 1988. Structure and tectonics of the northern Red Sea: Catching a continental margin between rifting and drifting. *Tectonophys.*, **150**, 1–32.

MCKENZIE, D.P. 1978. Some remarks on the development of sedimentary basins. *Earth and Planetary Science Letters*, **40**, 25–32.

MENZIES, M.A., BAKER, J., *ET AL.* 1992. The timing of magmatism, uplift and crustal extension: preliminary observations from Yemen. *In*: STOREY, B.C., ALABASTER, T. & PANKHURST, R.J. (eds) *Magmatism and the Causes of Continental Break-up*. Geological Society, London, Special Publications, **68**, 293–304.

MUTTER, J.C. & MUTTER, C.Z. 1988. Deep crustal structure and magmatic processes; the inception of seafloor spreading in the Norwegian-Greenland Sea. *In*: MORTON, A.C. & PARSON, L.M. (eds) *Early Tertiary Volcanism and the Opening of the NE Atlantic*. Geological Society, London, Special Publications, **39**, 35–48.

NIELSEN, T.F.D. 2004. The shape and volume of the Skaergaard Intrusion, Greenland; implications for mass balance and bulk composition. *Journal of Petrology*, **45**, 507–530.

OMAR, G.I. & STECKLER, M.S. 1995. Fission track evidence on the initial rifting of the Red Sea: two pulses, no propagation. *Science*, **270**, 1341–1344.

OMAR, G.I., STECKLER, M.S., BUCK, W.R. & KOHN, B.P. 1989. Fission-track analysis of basement apatites at the western margin of the Gulf of Suez rift, Egypt: evidence for sychroneity of uplift and subsidence. *Earth and Planetary Science Letters*, **94**, 316–328.

PALLISTER, J.S. 1987. Magmatic history of Red Sea rifting: perspective from the central Saudi Arabia coastal plain. *Bulletin of the Geological Society of America*, **98**, 400–417.

PATTON, T.L., MOUSTAFA, A.R., NELSON, R.A. & ABDINE, A.S. 1994. Tectonic evolution and structural setting of the Suez Rift. *In*: LANDON, S.M. (ed.) *Interior Rift Basins. AAPG Memoir*, **59**, 9–55.

RICHARDS, M.A., DUNCAN, R.A. & COURILLOT, V.E. 1989. Flood basalts and hot-spot tracks: Plume heads and tails. *Science*, **246**, 103–107.

ROSENDAHL, B.R. 1987. Architecture of continental rifts with special reference to East Africa. *Annual Review of Earth and Planetary Sciences*, **15**, 443–503.

ROYDEN, L. & KEEN, C.E. 1980. Rifting process and thermal evolution of the continental margin of eastern Canada determined from subsidence curves. *Earth and Planetary Science Letters*, **51**, 343–361.

ROYDEN, L., SCLATER, J.G. & VON HERZEN, R.P. 1980. Continental margin subsidence and heat flow: Important parameters in formation of petroleum hydrocarbons. *American Association of Petroleum Geologists Bulletin*, **64**, 173–187.

RUBIN, A.M. 1993. On the thermal viability of dikes leaving magma chambers. *Geophysics Research Letters*, **20**, 257–260.

RUBIN, A.M. 1995. Propagation of magma-filled cracks. *Annual Review of Earth and Planetary Sciences*, **23**, 287–336.

RUBIN, A.M. & POLLARD, D.D. 1987. Origins of blade-like dikes in volcanic rift zones. *US Geological Survey Professional Paper 1350*, 1449–1470.

SAWYER, D.S. 1985. Brittle failure in the upper mantle during extension of continental lithosphere. *Journal of Geophysical Research*, **90**, 3021–3025.

SEGALL, P. CERVELLI, P., OWEN, S., LISOWSKI, M. & MIKLUS, A. 2001. Constraints on dike propagation from continuous GPS measurements. *Journal of Geophysical Research*, **106**, 19 301–19 318.

SENGOR, A.M.C. & BURKE, K. 1978. Relative timing of rifting and volcanism on Earth and its tectonic implications. *Geophysical Research Letters*, **5**, 419–421.

SOLOMON, S.C., RICHARDSON, R.M. & BERGMAN, E.A. 1980. Tectonic stresses: models and magnitudes. *Journal of Geophysical Research*, **85**, 6086–6092.

SPOHN, T. & SCHUBERT, G. 1982. Convective thinning of the lithosphere; a mechanism for this initiation of continental rifting. *Journal of Geophysical Research*, **87**, 4669–4681.

STECKLER, M.S. 1985. Uplift and extension in the Gulf of Suez: indications of induced mantle convection. *Nature*, **317**, 135–139.

STECKLER, M.S. & TEN BRINK, U.S. 1986. Lithospheric strength variations as a control on new plate boundaries; examples from the northern Red Sea region. *Earth and Planetary Science Letters*, **79**, 120–132.

TRYGGVASON, E. 1980. Subsidence events in the Krafla area, north Iceland 1975–1979. *Journal of Geophysics*, **47**, 141–153.

TURCOTTE, D. & SCHUBERT, G. 2002. *Geodynamics*, Cambridge Univiversity Press, Cambridge, 456pp.

VARGA, R.J. 2003. The sheeted dike complex of the Troodos Ophiolite and its role in understanding mid-ocean ridge processes. *In*: DILEK-YILDIRIM, Y. & NEWCOMB, S. (eds) *Ophiolite concept and the evolution of geological thought*. Geological Society of America Special Paper, **373**, 323–336.

VENKATARAMAN, A., NYBLADE, A.A. & RITSEMA, J. 2004. Upper mantle Q and thermal structure beneath Tanzania, East Africa from teleseismic P-wave spectra. *Geophysical Research Letters*, **31**, doi:10.1029/2004GL020351.

WIENS, D.A. & STEIN, S. 1983. Age dependence of oceanic intraplate seismicity and implications for lithospheric evolution. *Journal of Geophysical Research*, **88**, 6455–6468.

WHITE, R.S. & MCKENZIE, D. 1989. Magmatism at rift zones: the generation of volcanic continental margins and flood basalts. *Journal of Geophysical Research*, **94**, 7685–7729.

WOLFENDEN, E., EBINGER, C. YIRGU, G., RENNE, P.R. & KELLEY, S.P. 2005. Evolution of a volcanic rifted margin: Southern Red Sea. Ethiopia. *Bulletin of the Geological Society of America*, **117**, 846–864.

# Mantle upwellings, melt migration and the rifting of Africa: insights from seismic anisotropy

J.-M. KENDALL[1], S. PILIDOU[2], D. KEIR[3], I.D. BASTOW[2], G.W. STUART[2] & A. AYELE[4]

[1]*Department of Earth Sciences, University of Bristol, Wills Memorial Building, Queen's Road, Bristol, BS8 1RJ, UK (e-mail: gljmk@bristol.ac.uk)*

[2]*School of Earth and Environment, Institute of Geophysics and Tectonics, University of Leeds, Leeds, LS2 9JT, UK*

[3]*Department of Geology, Royal Holloway, University of London, Egham, Surrey, TW20 0EX, UK*

[4]*Geophysical Observatory, Addis Ababa University, Addis Ababa, PO Box 1176, Ethiopia*

**Abstract:** The rifting of continents and eventual formation of ocean basins is a fundamental component of plate tectonics, yet the mechanism for break-up is poorly understood. The East African Rift System (EARS) is an ideal place to study this process as it captures the initiation of a rift in the south through to incipient oceanic spreading in north-eastern Ethiopia. Measurements of seismic anisotropy can be used to test models of rifting. Here we summarize observations of anisotropy beneath the EARS from local and teleseismic body-waves and azimuthal variations in surface-wave velocities. Special attention is given to the Ethiopian part of the rift where the recent EAGLE project has provided a detailed image of anisotropy in the portion of the Ethiopian Rift that spans the transition from continental rifting to incipient oceanic spreading. Analyses of regional surface-waves show sub-lithospheric fast shear-waves coherently oriented in a north-eastward direction from southern Kenya to the Red Sea. This parallels the trend of the deeper African superplume, which originates at the core–mantle boundary beneath southern Africa and rises towards the base of the lithosphere beneath Afar. The pattern of shear-wave anisotropy is more variable above depths of 150 km. Analyses of splitting in teleseismic phases (SKS) and local shear-waves within the rift valley consistently show rift-parallel orientations. The magnitude of the splitting correlates with the degree of magmatism and the polarizations of the shear-waves align with magmatic segmentation along the rift valley. Analysis of surface-wave propagation across the rift valley confirms that anisotropy in the uppermost 75 km is primarily due to melt alignment. Away from the rift valley, the anisotropy agrees reasonably well with the pre-existing Pan-African lithospheric fabric. An exception is the region beneath the Ethiopian plateau, where the anisotropy is variable and may correspond to pre-existing fabric and ongoing melt-migration processes. These observations support models of magma-assisted rifting, rather than those of simple mechanical stretching. Upwellings, which most probably originate from the larger superplume, thermally erode the lithosphere along sites of pre-existing weaknesses or topographic highs. Decompression leads to magmatism and dyke injection that weakens the lithosphere enough for rifting and the strain appears to be localized to plate boundaries, rather than wider zones of deformation.

The 3000 km-long East Africa Rift System (EARS) is a striking feature of the African continent (Fig. 1). The EARS marks the separation of the Nubian and Somalian plates and the possible isolation of smaller micropates (Calais *et al.*, this volume). In the north, it becomes an arm of a rift–rift–rift triple junction with the Gulf of Aden and Red Sea rifts. The formation of ocean basins through continental rifting is a fundamental feature of plate tectonics, yet the mechanism for break-up is poorly understood. The EARS is unique in that it captures each stage of the evolution of such a basin, from the birth of a rift in Tanzania, to the transition from continental rift to incipient oceanic spreading centre in northern Ethiopia (e.g. Ebinger & Casey 2001).

Another prominent feature of the African continent is its elevation and dynamic topography, which are due to deep-seated mantle upwelling beneath the continent (Davies 1998). The African superplume is one of two large-scale low-velocity anomalies that are clearly visible in the degree-two component of tomographic images of the mantle (the other lies beneath the southern Pacific Ocean). This low-velocity anomaly originates at the core–mantle boundary beneath southern

*From*: YIRGU, G., EBINGER, C.J. & MAGUIRE, P.K.H. (eds) 2006. *The Afar Volcanic Province within the East African Rift System*. Geological Society, London, Special Publications, **259**, 55–72.
0305-8719/06/$15.00 

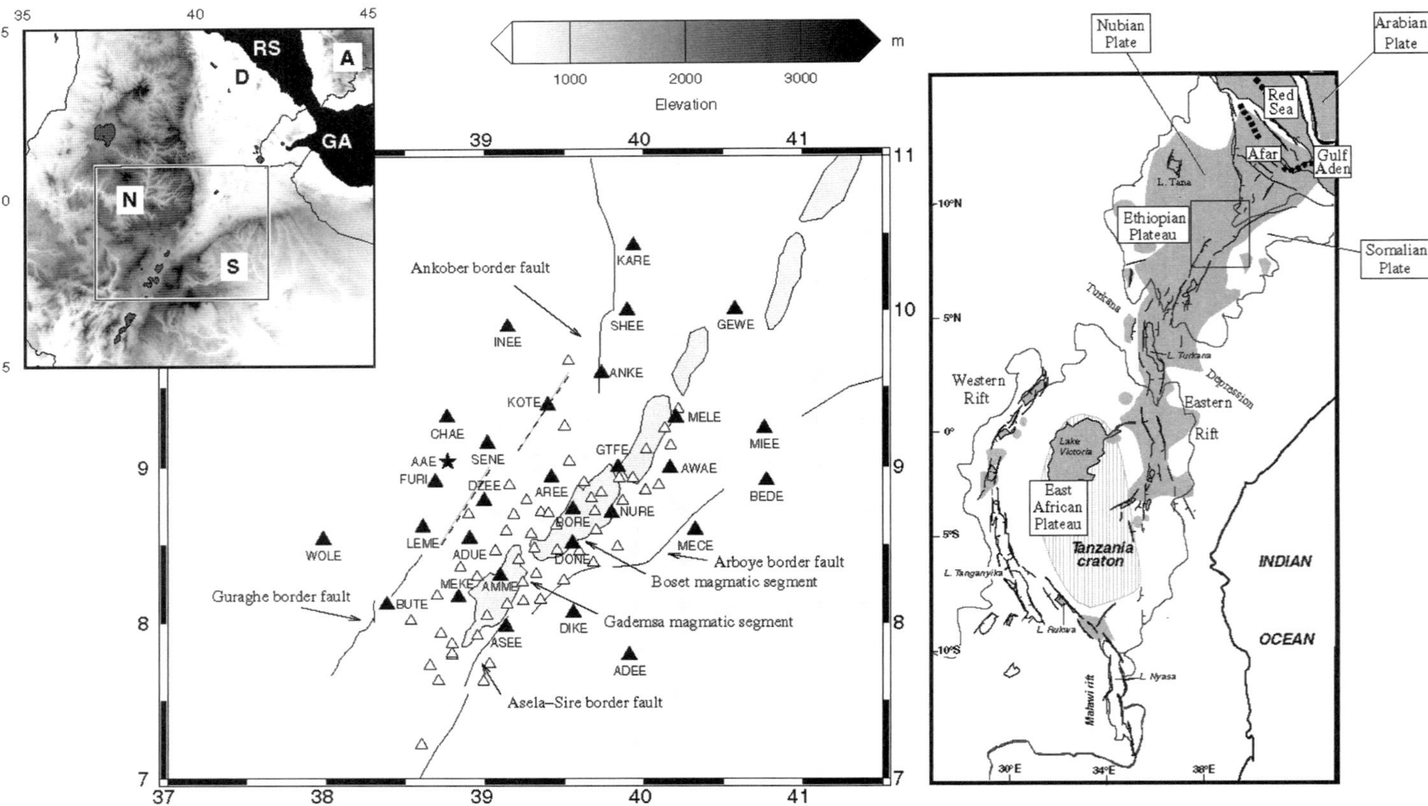


**Fig. 1.** (Right) Map of the East Africa Rift system (EARS). Box marks the EAGLE study area. Light lines enclose elevations >1000 m. (adapted from Ebinger, 2005). (Left) Location of the EAGLE passive seismic network. Filled triangles mark the phase I deployment (Oct. 2001–Feb. 2003) and the permanent IRIS station FURI. Open triangles mark stations of the phase II deployment (Nov. 2002–Feb. 2003). IRIS station AAE is marked by a star. Major mid-Miocene border faults and monoclines are marked with solid and dashed lines, respectively. Quaternary magmatic segments are outline by heavy black lines. Inset: regional tectonic setting and elevation. RS, Red Sea; GA, Gulf of Aden; A, Arabian plate; AD, Afar depression; D, Danakil microplate; N, Nubian plate; S, Somalian plate.

Africa, and rises in a north-northeastward direction (Ritsema *et al.* 1999), producing the African superswell (Nyblade & Robinson 1994). This buoys up Africa and the southwestern flank of Arabia (Lithgow-Bertelloni & Silver 1998; Gurnis *et al.* 2000; Daradich *et al.* 2003).

The connection between the African superplume/superswell and the EARS is unclear. Global tomography shows plume-like features extending into the lower mantle beneath the Afar region of Ethiopia, but its connection to the superplume is not well defined (compare, for example, models of Montelli *et al.* 2004 and Ritsema *et al.* 1999). Regional tomography shows a thin tabular low-velocity zone to depths of over 250 km beneath the continental part of the Ethiopian rift (Bastow *et al.* 2005). This anomaly broadens laterally in the more oceanic northeasterly region and appears to connect with the Afar plume (Benoit *et al.* 2003; Debayle *et al.* 2001). There are also numerous lines of evidence for a separate plume or upper-mantle instability under Tanzania or Uganda (Ebinger & Sleep 1998; Nyblade *et al.* 2000; Weeraratne *et al.* 2003), but its depth extent is unclear. The north–south propagation of volcanism and rifting, and general NNW movement of Nubia, suggest that the Tanzanian plume may have originally impacted beneath SW Ethiopia (George *et al.* 1998; Courtillot *et al.* 1999). However, this may be part of a larger single plume (Ebinger & Sleep 1998; Furman *et al.* this volume).

The driving forces for continental rifting come from distant plate forces (e.g. subduction), tractions imposed on the base of the plate due to underlying mantle convection, and forces associated with asthenospheric upwelling beneath the plate (e.g. small-scale convection or plumes) (Ebinger 2005). The perhaps simplest model of rifting involves mechanical stretching, where strain is accommodated by large offset border faults (McKenzie 1978). Such a mechanism requires large stresses in order to rupture cold, thick lithosphere. Alternatively, plume–lithosphere interactions may heat the lithosphere enough to promote magma production and more easily facilitate rifting of the consequently weakened lithosphere (Buck 2004). In the southern part of the EARS, where the rift is young, the strain is more predominantly accommodated by border faults, whilst in northeastern Ethiopia, where oceanic spreading is starting, the strain is largely accommodated by magma injection along segments in the rift valley (Ebinger & Casey 2001; Keranen *et al.* 2004). It is estimated that nearly 80% of the strain in the crust is localized to the magmatic segments (Bilham *et al.* 1999).

Seismic methods offer a means of testing such models of rifting. Tomography can be used to map the structure of low-velocity anomalies and hence the morphology of warm mantle upwellings beneath rifts (e.g. Evans & Achauer 1993). Additionally, measurements of seismic anisotropy can be used to infer patterns of strain and flow (Vauchez *et al.* 2000). The anisotropy can be due to the lattice-preferred orientation (LPO) of crystals or the preferred orientation of inclusions (e.g. oriented melt pockets (OMP)) or periodic thin layering (PTL) of contrasting materials (e.g. Kendall 2000). The resulting rock fabric produces a directional dependence in seismic velocities, or, in other words, seismic anisotropy. Here we use observations of anisotropy on a range of length scales to test models of rifting with the aim of achieving a better understanding of the connection between large-scale mantle flow and ocean basin formation.

There are a number of seismic methods for measuring anisotropy in the mantle (e.g. see Kendall 2000). The observation of two independent, orthogonally polarized shear-waves is perhaps the most unambiguous indicator of wave propagation through anisotropic media. A shear-wave in an isotropic medium will split into two shear-waves when it encounters an anisotropic medium. The orientation of the shear-waves and their difference in travel-times constrain the symmetry and magnitude of the anisotropy. Shear-wave splitting measured in teleseismic core phases such as SKS is routinely used to measure upper-mantle anisotropy (for review, see Savage 1999). Using networks or arrays of seismic stations, SKS splitting measurements offer good lateral resolution of anisotropy, but poor vertical resolution. Analyses of splitting in local events, if available, can be used to provide depth constraints on the extent and magnitude of the anisotropy.

Seismic anisotropy also affects surface-waves. It leads to azimuthal variations in surface-wave phase velocities, discrepancies between Love-wave and Rayleigh-wave-derived shear-wave velocity models, and particle motion anomalies (Kirkwood & Crampin 1981). The dispersive nature of surface-wave propagation leads to good resolution of anisotropy variation with depth, but long wavelengths mean poor horizontal resolution. The analysis of anisotropy using a combination of seismic body- and surface-waves provides good vertical and horizontal resolution and offers information about length scales of anisotropy.

Here we summarize observations of seismic anisotropy beneath the EARS. Special detail is given to the Ethiopian part of the rift, where we have detailed coverage from data acquired during the EAGLE (Ethiopia Afar Geoscientific Lithospheric Experiment) project (Maguire *et al.* 2003). After a review of anisotropy mechanisms, we start with the regional picture from teleseismic surface-wave analyses and progress to increasingly finer resolution beneath the Northern Main Ethiopian Rift (NMER).

## Mechanisms for seismic anisotropy (LPO vs. OMP)

The primary mechanism for seismic anisotropy in the upper mantle is the lattice-preferred orientation of olivine (and to a lesser extent enstatite) crystals in peridotites. Experimental measurements (Nicholas & Christensen 1987; Ben Ismaïl & Mainprice 1998) and numerical simulations (Blackman *et al.* 1996; Tommasi 1998) show that the *a*-axis [100] of olivine crystals will align in the flow direction. This is most effective when the rock is deforming by dislocation creep, which is controlled by strain rate and history, grain size, temperature and fluid content. There is some variability in the magnitude of anisotropy observed in xenoliths of peridotite, but it is on average 5% (Ben Ismaïl & Mainprice 1987). In general, such polycrystal rocks have triclinic elastic symmetry, but with peridotites this can be well-approximated with an orthorhombic symmetry (Blackman & Kendall 2002). Anisotropy is often observed to decrease beneath 220 km (e.g. Montagner 1994). This may be due to a transition from dislocation to diffusion creep, which is less effective in producing LPO (Karato 1992). More recently, it has been proposed that at high pressures the dominant slip direction in olivine changes to the *c*-axis [001], which produces little aggregate anisotropy (Mainprice *et al.* 2005).

Figure 2a shows P-wave and S-wave velocities as a function of propagation direction for a peridotite from Tanzanian upper-mantle xenoliths (Vauchez *et al.* 2005). P-wave velocities are fastest in the flow direction, whilst shear-wave splitting is minimal for S-wave propagation in the flow direction. With a horizontal flow direction and a vertical flow plane, a near-vertically propagating shear-wave (e.g. SKS) will show large amounts of shear-wave splitting and the fast shear-wave will be polarized in the direction of flow. The amount of SKS splitting will be much less for a horizontal flow plane. Assuming horizontal flow and horizontal wave propagation, the largest shear-wave splitting will occur in directions near normal to the flow direction and horizontally polarized shear-waves (Sh) will be faster than vertically polarized shear-waves (Sv); consequently, Love-wave analyses will yield faster shear-wave models than those derived from Rayleigh waves. Love–Rayleigh models will be much more similar for surface-waves propagating in the flow direction.

The preferred alignment of inclusions will also produce a long-wavelength anisotropy, provided the periodicity of the inclusions is much shorter than the seismic wavelength. In contrast, LPO is a more intrinsic cause of anisotropy and will not be seismic-wavelength dependent. Microcracks vertically oriented parallel to the direction of maximum horizontal stress are thought to be the main cause of anisotropy in the shallow crust (Crampin 1994). The resulting style of anisotropy is often referred to as azimuthal anisotropy as it produces azimuthal variations in velocities, and is often approximated by hexagonal symmetry with a horizontal symmetry axis or horizontal transverse isotropy (HTI). Alternatively, PTL of materials with contrasting velocities (e.g. volcanics and sediments) can be very effective in generating long-wavelength anisotropy (Backus 1962). If the layering is horizontal, the medium will act like a homogeneous hexagonally symmetric material with a vertical symmetry axis. This is commonly referred to as vertical transverse isotropy (VTI), but the terms polar or radial anisotropy are also often used to describe such symmetry. A characteristic of such media is a lack of azimuthal variation in velocities and a lack of shear-wave splitting in vertically travelling shear-waves.

The preferred alignment of melt inclusions is a very effective way to generate anisotropy (Kendall 1994). The magnitude of the anisotropy is not only very sensitive to the volume fraction of the melt, but also the shape of the melt inclusions. Thin disk-like inclusions are more effective at generating anisotropy than long tube-shaped inclusions (Kendall 2000). Spherical melt pockets are obviously the least-effective shapes for generating anisotropy. The shape and orientation of melt is controlled by wetting angles and strains in the medium (Schmeling 1985; Faul *et al.* 1994). As strain and melt fraction increases, the melt will start to align along grain edges and then crystal faces.

Figure 2b shows P- and S-wave velocities in a medium with vertically oriented melt pockets (OMP) that are tabular and disk-like in shape. P-wave velocities are highest for wave propagation parallel to the melt pockets, and shear-wave splitting is largest in these directions. For directions of horizontal wave propagation parallel to the OMP, vertically polarized shear-waves (Sv) are faster than horizontally polarized shear-waves (Sh). With surface waves, Love waves would produce a slower shear-wave model than that derived from Rayleigh waves. No shear-wave splitting occurs when waves propagate in a direction normal to the flat face of the inclusion. These predictions are sensitive to the inclusion aspect ratio, but they are most sensitive to the shape of the inclusion and the material in the inclusion. For example, flat oblate inclusions of melt are much more effective

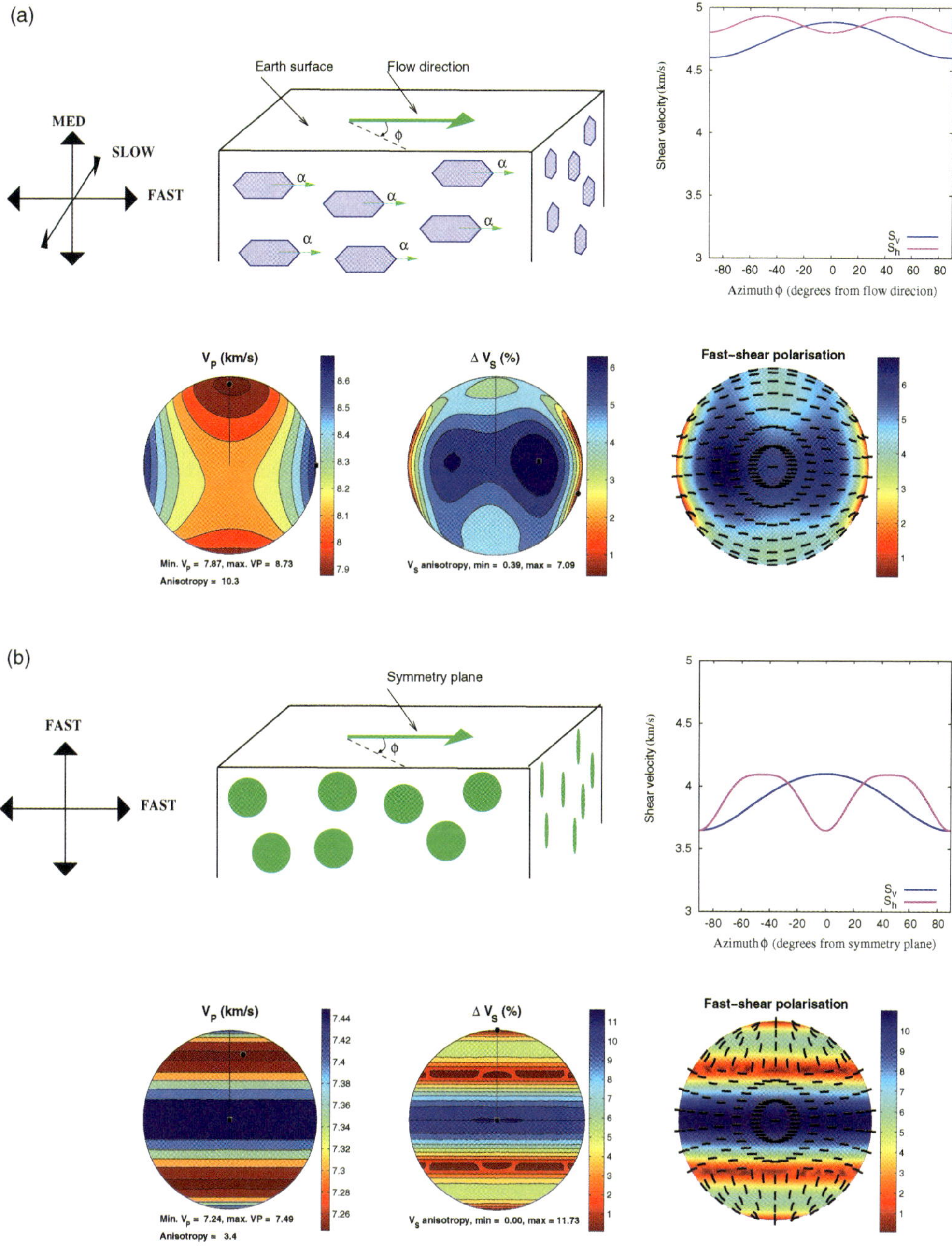


**Fig. 2.** Mechanisms for seismic anisotropy. (**a**) Anisotropy due LPO of upper-mantle minerals. Top left shows directions of fast, slow and medium P-wave velocities, assuming a horizontal flow direction and a vertical flow plane as shown in middle cartoon. Top right shows azimuthal variations in vertically and horizontally polarized shear-waves (Sv vs. Sh) Velocities are those determined for a peridotite from a Labait xenolith, Tanzania (Vauchez *et al.* 2005). Lower-hemisphere pole figures show seismic velocities: P-wave anisotropy is shown at left, % S-wave splitting is shown in middle, and the right shows polarization of fast shear-waves superimposed over S-wave splitting contours. Flow direction is to the right, the flow plane is vertical and the vertical direction is the centre of pole figure. (**b**) Anisotropy due to OMP. Melt-induced anisotropy is modelled using the theory of Tandon & Weng (1984) (e.g. see Kendall 2000). Melt volume fraction is assumed to be 0.1% and the melt lies in disk-like in pockets (oblate spheroids) with an aspect ratio of 0.02.

in generating anisotropy than prolate inclusions of high-velocity material. A further discussion of anisotropy sensitivity to inclusion shape and material can be found in Blackman & Kendall (1997) and Kendall (2000).

These diagnostics of wave propagation in various styles of anisotropy can be used to help guide interpretations of the cause of anisotropy.

## Regional patterns of surface-wave anisotropy

A large-scale picture of upper-mantle anisotropy can be derived from observations of azimuthal variations in surface-wave phase velocities. For example, Hadiouche *et al.* (1989) derived anisotropy models for Africa, noting a general north–south trend in the fastest direction for 40–100 sec Rayleigh waves in eastern Africa.

More recently, Debayle *et al.* (2005) have constructed a high-resolution, 3D tomographic model for the shear-wave velocity and azimuthal anisotropy of the African upper mantle. Their model resulted from the analysis of 9000 multi-mode (fundamental and first three higher modes) Rayleigh waveforms from 1022 regional earthquakes recorded at 250 stations. To construct the model, they first obtained path-averaged Sv-velocity models that best predicted the observations. Periods of 50–160 sec were used and path lengths ranged from 1000 km to 10 000 km. The path-average models were then inverted for horizontal variations in shear-velocity and azimuthal anisotropy at different depths. Tests showed that a few hundred kilometres of lateral resolution and 40–50 km vertical resolution is achieved in the top 400 km of the mantle.

Figure 3 shows depth slices from the Debayle *et al.* (2005) model centred on Ethiopia. Diffuse low-velocity zones lie beneath Afar and the Ethiopian plateau, and are in general agreement with recent body-wave travel-time tomography models (Montelli *et al.* 2004; Bastow *et al.* 2005; Benoit *et al.* 2006). These results are also similar to recent surface-wave results of Sebai *et al.* (2006).

Within the lithosphere or upper 150 km, the anisotropy shows clear differences between the Nubian plate and Somalian plate. The anisotropy orientation varies with depth between SE–NW and east–west beneath the Ethiopian plateau. The trend in the upper 75 km agrees with the orientation of major faults and Mesozoic rift structures (Moore & Davidson 1978; Berhe 1990). The orientation in the underlying lithosphere aligns with an east–west trending suture (Stern *et al.* 1990), the interpretation of which is somewhat controversial (Church *et al.* 1991). Both directions are different from Pan-African north-south structural trends (Berhe 1990; Abdelselam & Stern 1996). However, beneath the Somalian lithosphere, the surface-wave model shows a coherent north–south trend. The surface waves average horizontal structure over a few hundred kilometres in length scales. Within the Ethiopian Rift valley, the anisotropy appears weak, presumably because the surface-waves average a range of azimuthally anisotropic structures. Beneath Tanzania, the fast shear-wave direction is oriented roughly east–west and parallels the palaeo-fabric of the Tanzanian craton (e.g. Shackleton 1986).

At greater depths, beneath the EARS lithosphere, the anisotropy shows a consistent NE orientation, which parallels the rift from Tanzania to Arabia. This is roughly the orientation of the inferred large deep-seated African superplume as it rises towards Arabia (Ritsema & Allen 2003).

## SKS splitting along the EAR

SKS splitting analysis has been performed in many regions of continental rifting. In general, the orientation of the fast shear-wave is usually rift parallel, but there are some interesting variations. Fast shear-wave polarizations beneath the Rio Grande Rift are consistently rift parallel (Sandvol *et al.* 1992; Gök *et al.* 2003), whereas within the Baikal Rift the fast polarization directions are perpendicular to the rift (Gao *et al.* 1994). Gao *et al.* (1997) argue that fossil fabric beneath a rift should be weakened due to the enhanced mobility of olivine crystals at higher temperatures. However, recent studies of lithospheric peridotities affected by plumes do not seem to show this effect (Vauchez & Garrido 2001; Vauchez *et al.* 2005). Gao *et al.* (1997) also note that tomography results suggest that mantle beneath the Baikal Rift is cooler than beneath the Rio Grande Rift or EARS. Where temperatures are high enough to produce melt, rift-related anisotropy may be due to OMP (Kendall 1994; Gao *et al.* 1997).

There have been a number of studies of SKS splitting along the EARS (see compilation in Fig. 4). SKS splitting at permanent stations in Africa exhibit little variation in splitting parameters with source backazimuth, and it is therefore argued that the upper-mantle anisotropy is due to a simple single layer (Ayele *et al.* 2004; Barruol & Ben Ismail 2001; Walker *et al.* 2003). Ayele *et al.* (2004) interpret splitting in Kenya, Ethiopia and Djibouti in terms of melt, noting that the magnitude of the splitting correlates with amount of melt production.

In Tanzania and Kenya, data from temporary seismic deployments show fast shear-wave

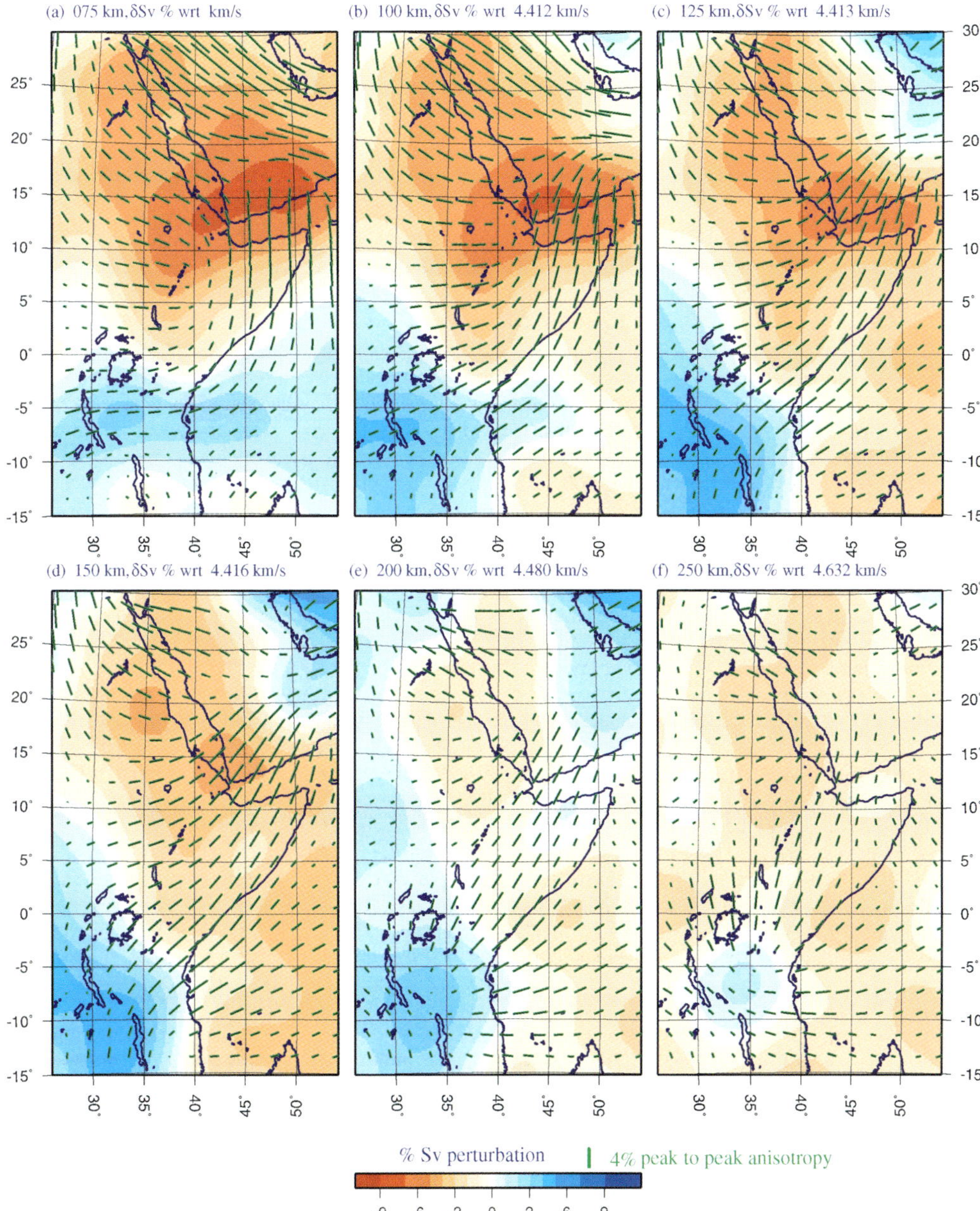


**Fig. 3.** Tomographic model of the regional, multimode surface-wave tomography study of Debayle *et al.* (2005). Maps show horizontal sections through the 3D model at depths of (**a**) 75 km; (**b**) 100 km; (**c**) 125 km; (**d**) 150 km; (**e**) 200 km; (**f**) 250 km. The background colours show Sv heterogeneity with respect to the starting background model (PREM); red and blue colours indicate areas of slower and faster wave propagation with respect to PREM and the reference velocity is shown at the top of each map. Green bars indicate fast directions for horizontally propagating Sv-waves and their length is proportional to the magnitude of the azimuthal anisotropy.

polarizations parallel to the edges of the eastern and western arms of the EARS that encircle the Tanzanian craton (Gao *et al.* 1997; Walker *et al.* 2003). Gao *et al.* (1997) interpreted rift-parallel anisotropy in terms of an OMP mechanism. The magnitude of splitting is much smaller in the south where the rift is youngest. Measurements away from the rift, from the craton and the SE part

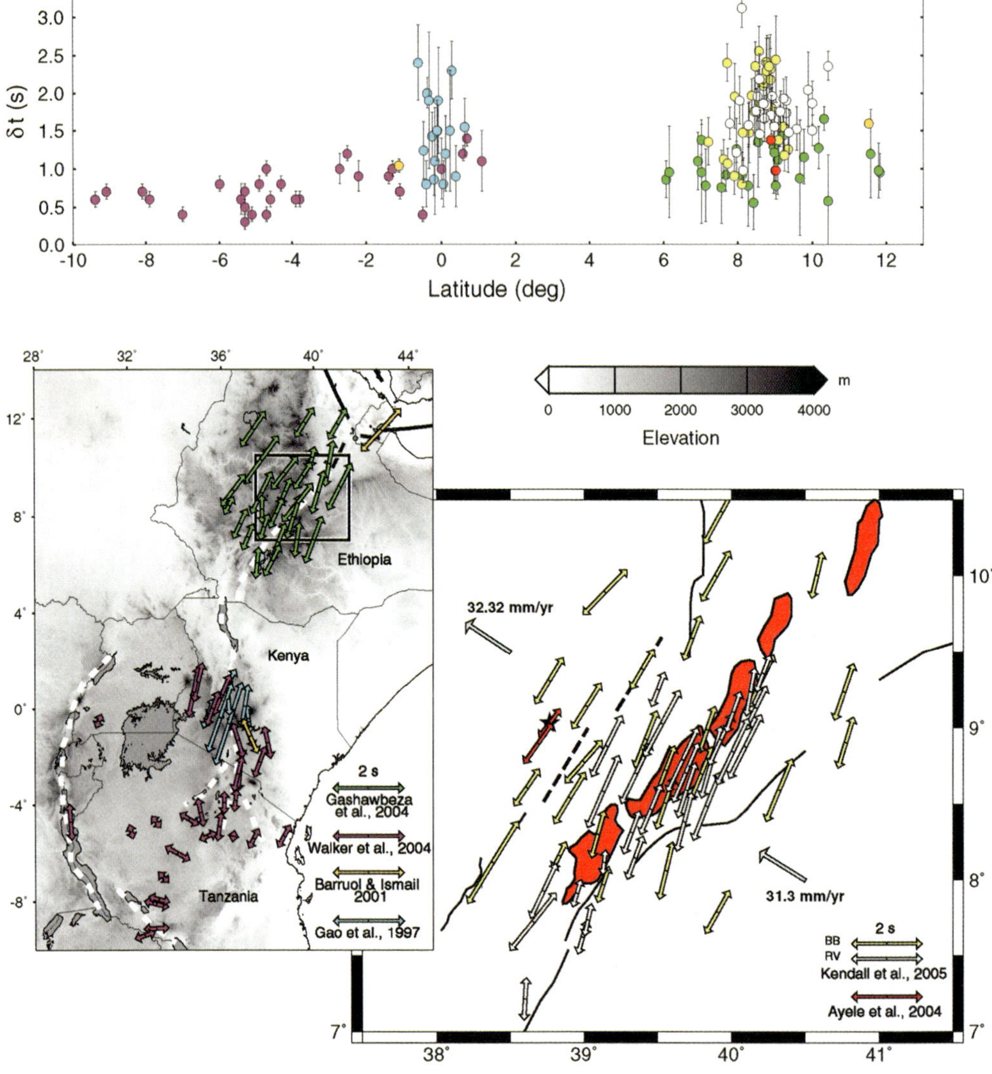


**Fig. 4.** SKS splitting measurements along the EARS; dashed white line on regional map (lower left) shows outline of rift. Lower right shows detail from Northern Ethiopian Rift. Major mid-Miocene border faults and monoclines are marked with solid and dashed lines, respectively. Larger grey arrows show direction of absolute plate motion. Quaternary magmatic segments are shaded in red. In both maps, arrows show orientation of fast shear-wave and length of arrow is proportional to magnitude of splitting. Magenta arrows show results of Walker *et al.* (2004); blue show results of Gao *et al.* (1997); green, Gashawbeza *et al.* (2004); orange, Barruol & Ben Ismail (2004); red, Ayele *et al.* (2004); yellow and white show results from NMER (see Kendall *et al.* 2005). Top figure shows SKS splitting (dt) as a function of latitude. Circle colours match those used for the maps.

of the rift extending into the Mozambique Belt, are somewhat different (Walker *et al.* 2003). Splitting beneath the Tanzanian craton is much weaker and oriented in a more east–west direction, in alignment with Precambrian structural trends (Shackleton 1986). Walker *et al.* (2003) interpret the splitting pattern in terms of a number of mechanisms including: asthenospheric flow around a cratonic keel, plume–lithosphere interactions, pre-existing lithospheric fabric and melt-induced anisotropy. Anisotropy due to asthenospheric flow around continental keels has been used to explain patterns of SKS splitting around cratonic North America (Fouch *et al.* 2000). SKS splitting due to plume–lithosphere interactions beneath Tanzania have been predicted by Sleep *et al.* (2002), but it is difficult to constrain such interpretations as the location or even present-day existence of the plume is unknown. Weeraratne *et al.* (2003) use Rayleigh-wave tomography to infer the location of a plume beneath the Tanzanian craton, but the predicted SKS splitting is inconsistent with the observations (Walker *et al.* 2003).

More recently, Kendall *et al.* (2005) made over 500 SKS splitting measurements at over 70 EAGLE temporary stations centred on the NMER (Fig. 4). The orientation of the fast shear-waves are roughly rift parallel. The magnitude of the splitting varies considerably over short distances (<40 km) and arguments based on the size of Fresnel zone (Rümpker & Ryberg, 2000) constrain the differences in anisotropy to the uppermost 100 km. There will also be a weaker, but more uniform, contribution to the SKS splitting from the underlying large-scale flow in the region, as revealed by surface waves. In general, the degree of splitting increases northward towards the more oceanic part of the rift, but is highest near the flanks of the rift. On the flanks of the rift the orientation of the fast shear-waves parallel the border faults and monoclines. Within the rift valley the orientation of the anisotropy rotates counterclockwise and follows the en-echelon orientation of the magmatic segments. The shallow source of anisotropy, increased splitting in more magmatic regions, and the anisotropy alignment with magmatic segments lead to the interpretation that the anisotropy is controlled by OMP.

In contrast, Gashawbeza *et al.* (2004) measure SKS splitting with a wider-aperture network, but again the fast shear-wave polarizations are parallel to the trend of the rift. They argue that the anisotropy beneath Ethiopia is Precambrian in origin with some Neogene reworking near the rift. They draw on interpreted suture orientations inferred from ophiolite belts (Berhe 1990), but as noted previously these interpretations are somewhat controversial (Church *et al.* 1991) (see also discussion).

In summary, observations of SKS splitting within EARS rift valleys are conformal to the trend of the rifts. Based on observations of SKS splitting alone, it is difficult to discriminate between anisotropy due to olivine LPO associated with asthenospheric flow, fossil anisotropy in the lithosphere surrounding the rift, and anisotropy due to OMP. However, the fast shear-wave orientations are clearly not aligned with directions of absolute plate motion (APM), thus ruling out anisotropy due to simple asthenospheric flow coupled to the base of the plates. The observations are also not in agreement with predictions of extension-induced olivine LPO (e.g. Blackman *et al.* 1996; Vauchez *et al.* 2000), as observed at the East Pacific Rise (Wolfe & Solomon 1998) and in the perhaps cooler Baikal Rift (Gao *et al.* 1997), where fast shear-wave polarizations are perpendicular to the spreading direction.

## Surface-wave anisotropy within the Northern Ethiopian Rift

A more-detailed picture of anisotropy beneath the Ethiopian Rift valley is obtained from the analysis of Rayleigh–Love wave propagation within the rift valley. Pilidou *et al.* (2004) analysed nearly 80 Rayleigh- and Love-wave seismograms from 23 local, regional and teleseismic events, recorded at 20 EAGLE broadband stations. Group velocities are measured from local rift earthquakes with propagation paths of length 400–800 km. Phase velocities are measured along interstation propagation paths of length 30–250 km. With both, the paths cross the rift valley with a wide range of azimuths. Such analyses allow us to test the LPO and OMP hypotheses. The predicted magnitude of Love–Rayleigh-wave model discrepancies can be similar for LPO and OMP mechanisms, but azimuthal variations predicted by these models are quite different (see Fig. 2).

Pilidou *et al.* (2004) follow a two-step procedure to determine path-averaged Earth models. In the first step, they calculate the group/phase velocity dispersion for the fundamental mode along a given path, and higher modes where the data allow it. In the second step, they match, through inversion, each dispersion curve with a theoretical dispersion curve for a suitably chosen layered-Earth model. The Rayleigh- and Love-waves are inverted separately. Data in the period range of 5–50 sec are analysed, which resolves Earth structure in the uppermost 75 km. Horizontal resolution at a depth of 50 km is roughly 50 km for the Sv-waves and 80 km for Sh-waves. Figure 5 summarizes the results.

In the upper 10 km, Sh-waves are faster than Sv-waves, independent of azimuth. Surface-wave anisotropy in the shallow crust beneath the NMER is therefore best explained with a VTI model, most probably caused by the layering of volcanics and sediments in the uppermost crust (e.g. see Mackenzie *et al.* 2005; Keller *et al.* 2003).

Beneath 20 km, there is a clear azimuthal variation in the shear-wave velocities. For wave propagation along the axis of the rift valley, the Sv model consistently shows faster velocities than the Sh model. At more oblique angles to the rift axis, the anisotropy is diminished and the Sh velocities

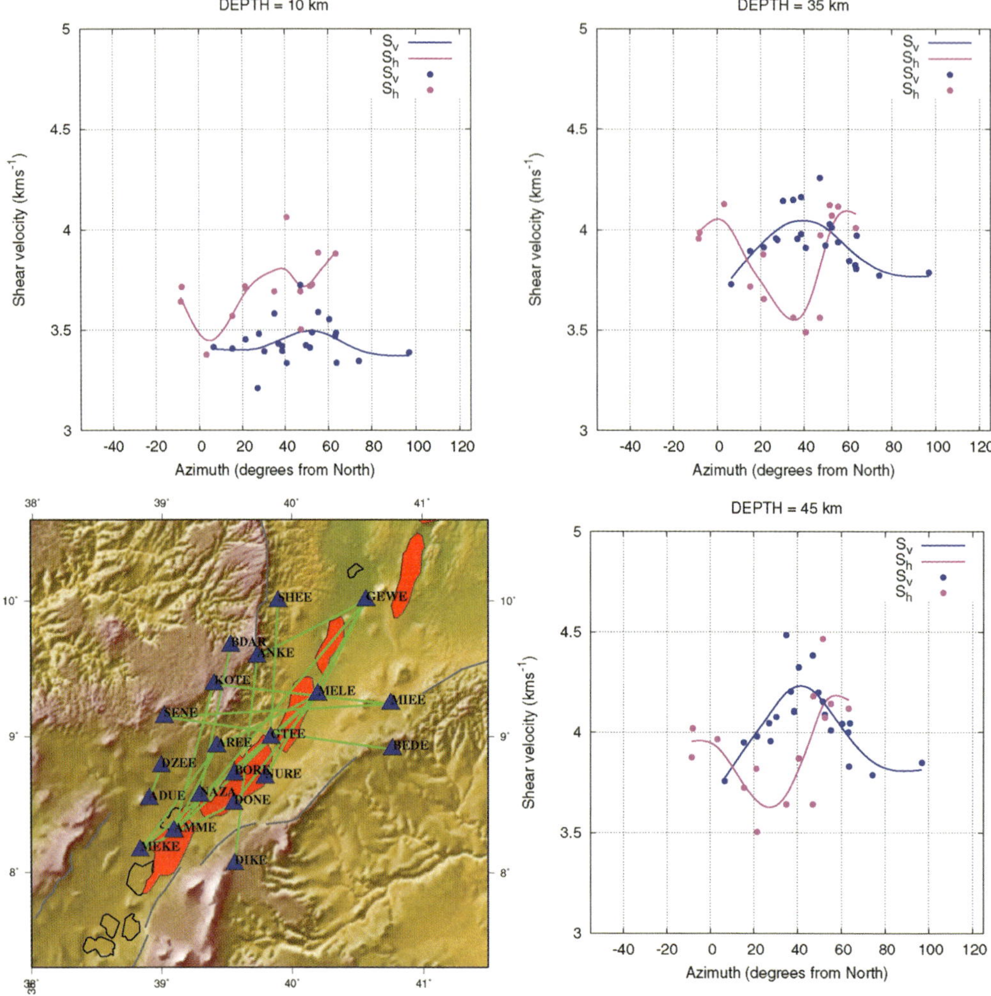


**Fig. 5.** Surface-wave anisotropy within the NMER. Azimuthal variation in Sv (blue) and Sh (pink) wavespeeds at depths of 10 km, 35 km and 45 km. Solid line is natural smooth spline fit to individual measurements and azimuth is measured clockwise from north. Estimates are made from group velocity measurements made on local events and interstation phase velocity measurements (see Pilidou *et al.* (2005) for more detail). Map shows the interstation paths for phase velocity measurements.

become faster than the Sv velocities. This pattern is similar to that predicted with the OMP model where the normal to the melt pockets is perpendicular to the rift axis (Fig. 2). This is in contrast to the velocities predicted from the peridotite anisotropy, which predict very little anisotropy for the limited range of directions where Sv velocities are faster than Sh velocities. These models therefore suggest OMP as the primary mechanism for anisotropy beneath the rift to the maximum depth of resolution (~70 km). The OMP anisotropy is over 12% and extends over at least 50 km of the upper mantle. This would produce over 1.5 seconds of SKS splitting, which is roughly the average amount of SKS splitting observed in the NMER valley (Kendall *et al.* 2005).

## Shear-wave splitting in crustal earthquakes

Shear-wave splitting in shallow earthquakes can be used to study the anisotropy within the crust if events lie within a near-vertical cone beneath the station (the so-called shear-wave window of Booth & Crampin 1985). Keir *et al.* (2005) analysed events at depths between 6 and 25 km beneath the EAGLE stations. The orientations of fast shear-waves at these stations parallel local surface features (Fig. 6). For example, the splitting at ANKE parallels the nearby mid-Miocene Ankober border fault. Within the rift valley, the anisotropy generally parallels Quaternary-age eruptive centres and active faults.

A VTI crust due to fine-scale layering, as suggested for the rift valley by the surface-wave analysis, will not generate any splitting in near-vertically propagating shear waves. Therefore, the anisotropy is most likely due to vertical cracks or inclusions aligned by regional and local stresses in the crust. The magnitude of the splitting varies dramatically across the EAGLE network, suggesting a heterogeneous stress field or variations in the underlying cause of anisotropy. The deepest events lie beneath the Ethiopia plateau and although they show small splitting values (up to 0.15 sec), they show clear depth-dependent variations in shear-wave splitting to a depth of 25 km. This suggests roughly 1% anisotropy beneath the plateau. In contrast, earthquakes 7 km beneath the Boset magmatic segment show 0.24 sec of splitting, which translates to over 6% anisotropy. Stations within the rift valley but located outside magmatic segments show less splitting (e.g. MELE or E79), but the average magnitude of splitting in the rift valley is nearly 3%, much larger than that beneath the plateau to the NW.

Where it is possible to compare, the orientation of the fast shear-wave agrees with the direction of maximum horizontal stress, as inferred from focal

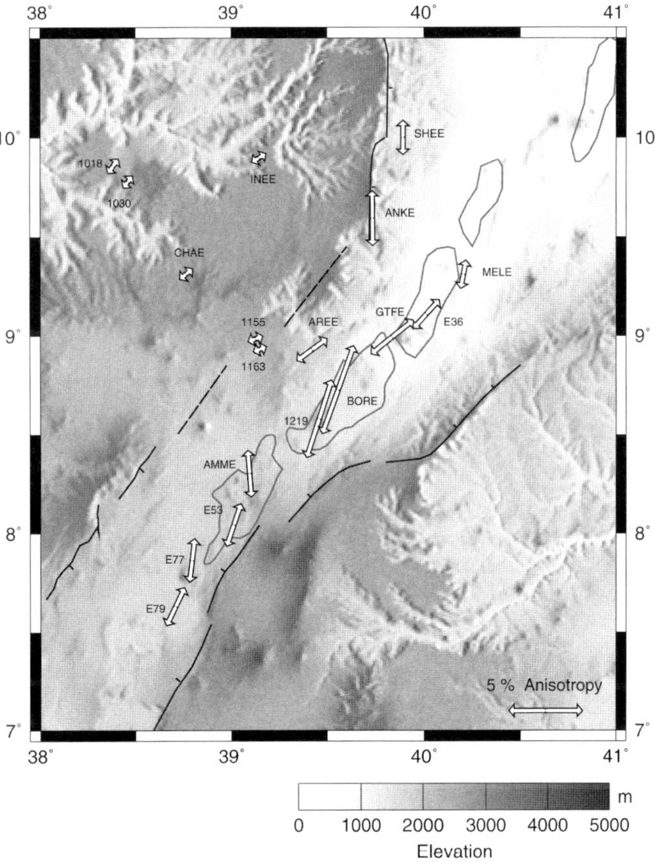


**Fig. 6.** Crustal anisotropy as determined from shear-wave splitting in crustal earthquakes. Calculation assumes anisotropy is uniformly distributed between the source and receiver. Orientation of arrows shows polarization of fast shear-wave, whilst the length of the arrow is proportional to the magnitude of the anisotropy. Heavy black lines show orientation of border faults, dashed lines show monoclines and thin lines show outline of magmatic segments in the rift valley.

mechanisms (Keir *et al.* 2006). Keir *et al.* (2005) interpret these observations in terms of the injection of vertically aligned magma intrusions or cracks beneath the Quaternary magmatic segments. Stations outside the rift show directions sub-parallel to metamorphic basement fabrics, consistent with melt decrease away from the rift. Support for this interpretation comes from other geophysical studies. Keranen *et al.* (2004) and Tiberi *et al.* (2005) interpret cooled mafic intrusions in the middle to lower crust beneath these magmatic segments using wide-angle refraction tomography and gravity models, respectively. The anomaly under Boset is especially pronounced, where melt-related anomalies have been interpreted in magnetotelluric (MT) (Whaler & Hautot 2006) and gravity (Cornwell *et al.* 2006) data. There is evidence for melt in the lower crust beneath the Ethiopian plateau, both in the past and at present, from wide-angle seismic refraction images of underplating (MacKenzie *et al.* 2005), mid-crustal conductive anomalies in MT data (Whaler & Hautot 2006), and Quaternary eruptive centres as far north as Lake Tana. Cumulatively, these results suggest the anisotropy is related to variable amounts of melt pocket alignment in the crust, with a higher degree of dyke intrusion in the crust beneath the rift.

## Synthesis of results

### *Large-scale flow*

Deep mantle flow associated with the African superswell is oriented in a rift-parallel direction beneath the EARS. The signature of this is perhaps evident in the pattern of surface-wave anisotropy at depths below 125 km (Debayle *et al.* 2005). It has been argued that this large-scale flow pattern is responsible for the dynamic topography of Africa (Lithgow-Bertelloni and Silver 1998; Gurnis *et al.* 2000) and Arabia (Daradich *et al.* 2003) and the EARS plate motions (Calais *et al.* 2006). The interaction of this density-driven flow with absolute plate motions has also been used to explain SKS splitting observations at permanent stations on islands surrounding Africa (Behn *et al.* 2004). In fact, the regional patterns of surface-wave anisotropy (Fig. 3) are reminiscent of the flow patterns predicted by Behn *et al.* (2004). It is therefore plausible that the sub-lithospheric anisotropy beneath the EARS is due to olivine LPO caused by viscous coupling between the overlying plates and a superplume that rises beneath Tanzania (e.g. Sleep *et al.* 2002).

The precise architecture of the upwelling upper-mantle beneath Africa is uncertain. The number of proposed plumes beneath Africa ranges from as few as one to as many as 40 (Davies 1998). Furthermore, based on laboratory experiments, the stability of a single large inclined superplume has been questioned (Davaille *et al.* 2005). Hence the pattern of sub-lithospheric anisotropy may be related to a more complicated pattern of upwellings from the deep mantle beneath Africa. Instead of a large single superplume beneath the upper mantle, there may be more than one plume head beneath our study region. For example, recent tomographic images suggest a plume beneath the Red Sea/Gulf of Aden that extends vertically into the lower mantle (Montelli *et al.* 2004). Future regional-scale seismic experiments will help address this issue by better delineating mantle structure beneath Africa (e.g. AfricaArray http://africaarray.psu.edu; Dalton 2005).

Anisotropy beneath Africa cannot be explained by the direction of absolute plate motion. Africa is nearly stationary in a hotspot reference frame and the Nubian and Somalian plates move slowly in a northwesterly direction (Fig. 4). However, recent analysis of GPS and earthquake slip vector data (Calais *et al.* 2006) show relative extension directions across the rift valleys, which are consistently perpendicular to the polarization of the fast shear-wave.

### *Magma assisted rifting*

It is difficult to reconcile the rift valley surface-wave anisotropy and SKS splitting directions with a model of olivine alignment beneath the Ethiopian rift. This precludes olivine alignment due to current processes (either mechanical extension or flow along the rift axis) or fossil fabric within pre-existing lithosphere. Instead, melt pockets oriented in a rift-parallel direction explain the observations remarkably well (see Fig. 2). The combined body-wave and surface-wave results provide robust evidence for aligned melt beneath the Northern Ethiopian Rift to depths of at least 75 km. This is further supported by structural (Ebinger & Casey 2001; Wolfenden *et al.* 2004), geochemical (Rooney *et al.* 2005), earthquake seismicity (Keir *et al.* 2006) and controlled-source seismic (Keranen *et al.* 2004) data, which indicate that the magmatic segments are zones of intense dyke injection and magmatic intrusion. Major element analyses of xenoliths from the Bishoftu area (near station DZEE, Fig. 1) suggest melting at depths up to 75 km (Rooney *et al.* 2005).

Regional tomography (Green *et al.* 1991; Ritsema *et al.* 1998; Bastow *et al.* 2005) show highly focused low-velocity anomalies beneath continental parts of the EARS. Vp/Vs ratios suggest that these anomalies may indicate partially molten mantle (Bastow *et al.* 2005). Anisotropy due to OMP explains how remarkably conformal the SKS splitting polarizations are to the rift valleys. This is further supported by the fact that the magnitude of the splitting (dt) increases moving northward in the direction of increased magma production (Fig. 4).

Shear-wave splitting in crustal earthquakes suggests that the melt-induced anisotropy persists well into the crust. There is an apparent contradiction between the rift-valley surface-wave results and the crustal splitting. This is resolved with a model of layered sediments and volcanics, punctuated by vertically aligned dykes and intrusions. The resulting anisotropy will be orthorhombic in symmetry. Vertically propagating shear-waves will not be sensitive to VTI anisotropy caused by the layering. In contrast, the surface-waves, with poorer lateral resolution, will not be able to resolve the localized and highly variable anisotropy beneath the magmatic segments and thus be most sensitive to the layering.

Although the anisotropy beneath the rifts appears to be dominated by melt, it is likely to be due to a range of mechanisms. Holtzman *et al.* (2003*a*) show strain partitioning and melt segregation in laboratory simulations of dunite deformation. These experiments suggest that anisotropy in deformed regions of partial melt may be due to OMP, LPO and the periodic layering of melt-rich and melt-poor bands. Interestingly, these experiments show olivine *a*-axis [100] alignment in the shear plane, but orthogonal to the shear direction. For vertical flow, the olivine *a*-axis alignment will be horizontal and rift-parallel, further accentuating the anisotropy due to aligned melt pockets and melt bands. Cumulatively, these mechanisms are highly effective in generating seismic anisotropy (Holtzman *et al.* 2003*b*).

Evidence of melt-related anisotropy is important as it supports ideas of magma-assisted rifting. An asthenospheric upwelling will heat the lithosphere and produce melt, thus reducing the strength of the lithosphere enough for it to rupture without large amounts of stretching (Buck 2004). Strain gradients will focus and align the melt in the mantle beneath rift shoulders, as predicted by Sleep (1997) and observed by Kendall *et al.* (2005). Melt is then focused up to the rift axis, eventually erupting at magmatic segments and perhaps following strain gradients (e.g. Phipps-Morgan 1987).

Regional tomography and geochemical studies indicate localized upwellings or small plumes that may originate from the African superplume. These will thermally thin the lithosphere and lead to decompression melting and magma injection into the lithosphere. The distant forces acting on the African plates are weak, but the extra forces due to small-scale convection beneath the rift may be sufficient to rift the weakened lithosphere (Burov & Guillou-Frottier 2005).

## *Pan-African fabric*

There is likely to be fossil anisotropy within the ancient lithosphere away from the rift valley. SKS splitting beneath the Tanzanian craton is weak, but parallels Precambrian Pan-African fabric (Walker *et al.* 2004). In general, cratonic fabric in Africa appears to generate small amounts of SKS splitting (e.g. Kaapvaal (Silver *et al.* 2001; Fouch *et al.* 2004); Congo (Ayele *et al.* 2004)).

The pre-existing lithospheric fabric in Ethiopia is less clear, owing to the cover of pervasive flood basalts. Mesozoic rift structures strike NW–SE, and Pan-African structural trends are oriented north–south (e.g. Moore & Davidson 1978). However, based on the orientation of ophiolite belts, Berhe (1990) inferred a NE–SW trending suture that connects the north–south Turkana trending suture and the north–south trending Nabitah suture in Arabia. In contrast, Stern *et al.* (1990) have used the same ophiolite belts to suggest a major east–west suture in this region that is in part marked by chains of eruptive volcanic centres (e.g. the Ambo lineament). Due to limited basement outcrop, the interpretation of such sutures is controversial (Church 1991). SKS splitting beneath the Ethiopian plateau roughly aligns (within 20–30 degrees) with the suture orientation proposed by Berhe (1991) (Gashawbeza 2004). Vauchez *et al.* (1997) have argued that there is correlation between mechanical anisotropy and pre-existing anisotropic fabric; in other words, rifting has a tendency to occur along previous sutures. In contrast, larger-scale surface-wave anisotropy shows a northwest orientation in the upper 75 km beneath the Ethiopian plateau and a north–south orientation in the deeper lithosphere. These surface waves will average structures over large horizontal distances and will see the cumulative effect of any pre-existing fabric and/or aligned melt. Any fossil anisotropy will be more dominant away from the rift valley. It would appear that the surface waves are not capable of resolving the near-surface near-rift melt-related anisotropy.

Beneath the Ethiopian plateau (NW of the rift), the fast shear-wave orientations remain rift-parallel over 200 km from the rift flanks (Gashawbeza *et al.* 2004; Kendall *et al.* 2005; Keir *et al.* 2005). As discussed, this may indicate pre-existing structural trends, but the signal may be still due to aligned melt in the lower crust and uppermost mantle, as it is beneath the rift valley. Indeed, there is further evidence for magmatism under the plateau. Whaler & Hautot (2006) see evidence from magneto-telluric surveying for large conductive bodies in the mid-crust. MacKenzie *et al.* (2005) interpret crustal underplating beneath this region in a wide-angle reflection profile (see also

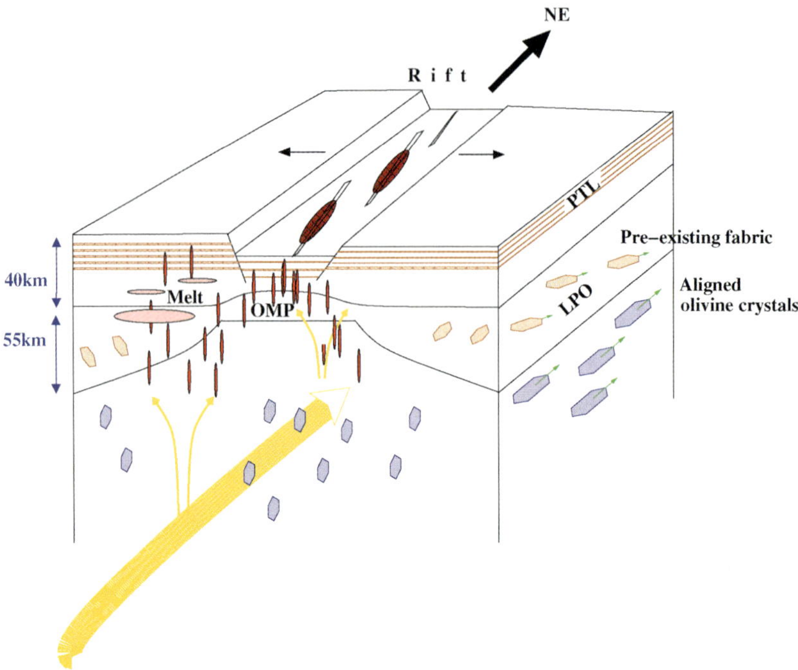


**Fig. 7.** Model of seismic anisotropy beneath the EARS. A range of mechanisms is responsible for anisotropy, including periodic thin layering (PTL) of contrasting materials, oriented pockets of melt (OMP), and the lattice-preferred orientation (LPO) of olivine crystals. OMP anisotropy is most effective if the anisotropy is aligned in thin disk-like pockets. Away from the rift, the lithospheric LPO is due to pre-existing fabric, whilst beneath the lithosphere the LPO is due to viscous coupling between the base of the lithosphere and large-scale mantle upwelling.

Maguire *et al.* 2006). Evidence from surface geology and magneto-telluric data of Quaternary volcanism and dyking is observed as far north as Lake Tana (Hautot *et al.* 2006). Finally, fine-scale tomographic images show a large low-velocity velocity anomaly beneath the Ethiopian plateau, offset northwestward from the rift valley. Analysis of Vp/Vs ratios derived from P-wave and S-wave models suggests this anomaly is most probably melt related and not purely thermal in origin (Bastow *et al.* 2005). Therefore, beneath the Ethiopian plateau the anisotropy may be due to a complicated combination of mechanisms associated with pre-existing fabric, fault trends and aligned melt. This may explain discrepancies in the surface-wave and body-wave results, as they are sensitive to different lateral and vertical length scales.

## Summary

Figure 7 summarizes our working model of anisotropy beneath the EARS. Shear tractions and buoyancy forces induced by deep mantle flow appear to drive rifting in Africa. In this model, the large-scale African superplume feeds smaller plumes or upwellings that impinge on the base of a heterogeneous lithosphere (e.g. Ebinger & Sleep 1998). The plume material flows around rigid keeled cratonic blocks, pooling beneath lithospheric thin spots (Sleep *et al.* 2002), which correlate with pre-existing rift zones and/or sutures (Dunbar & Sawyer 1989; Vauchez *et al.* 1997; Ebinger & Sleep 1998). The strain associated with rifting is then accommodated by magma injection along narrow plate boundaries, not diffuse regions. Extension is accommodated in focused zones of melt production where magma injection outpaces faulting. Anisotropy in the rift valley is dominated by the melt signature, whilst the pre-existing Pan-African anisotropic fabric contributes a weak signature away from the rift. The wide range of observations of seismic anisotropy beneath Africa hold a signature of the processes that lead to rifting and eventual formation of an ocean basin.

Cindy Ebinger is thanked for fruitful discussions and guidance. This work would not have been possible without cooperation and collaboration between members of the EAGLE working group (Maguire *et al.* 2003). Peter

Maguire, James Wookey, Kathy Whaler and Dave Cornwell are thanked for helpful comments. The manuscript was improved by insightful reviews from Andy Nyblade and two anonymous reviewers. Eric Debayle is thanked for making available his surface wave velocity model. SEIS-UK is acknowledged for equipment and technical support. This work was supported by NERC grant NER/A/S/2000/01003.

## References

ABDELSELAM, M. & STERN, R. 1996. Sutures and shear zones in the Arabia–Nubian shield. *Journal of African Earth Sciences*, **23**, 289–310.

AYELE, A., STUART, G.W. & KENDALL, J.-M. 2004. Insights into rifting from shear-wave splitting and receiver functions: an example from Ethiopia. *Geophysical Journal International*, **157**, 354–362.

BACKUS, G. 1962. Long-wave elastic anisotropy produced by horizontal layering. *Journal of Geophysical Research*, **67**, 4427–4440.

BARRUOL, G. & BEN-ISMAIL, W. 2001. Upper mantle anisotropy beneath the African IRIS and GEOSCOPE stations. *Geophysical Journal International*, **146**, 549–561.

BASTOW, I.D., STUART, G.W., KENDALL, J.M. & EBINGER, C.J. 2005. Upper mantle seismic structure in a region of incipient continental breakup: northern Ethiopian rift. *Geophysical Journal International*, **162**, 479–493.

BEHN, M.D., CONRAD, C.P. & SILVER, P.G. 2004. Detection of upper mantle flow associated with the African superplume. *Earth and Planetary Science Letters*, **224**, 259–274.

BEN ISMAÏL, W. & MAINPRICE, D. 1998. An olivine petrofabric database: An overview of upper mantle fabrics and seismic anisotropy. *Tectonophysics*, **296**, 145–157.

BENOIT, M., NYBLADE, A.A. VANDECAR, J.C. & GURROLA, H. 2003. Upper mantle P-wave velocity structure and transition zone thickness beneath the Arabian shield. *Geophysical Research Letters*, **30**, 1531, doi 10.1029/2002GLO16436.

BENOIT, M., NYBLADE, A.A. & VANDECAR, J.C. 2006. Upper mantle P-wave speed variations beneath Ethiopia and the origin of the Afar Hotspot. *Geology* (in press).

BERHE, S.M. 1990. Ophiolites in Northeast and East Africa: Implications for Proterozoic crustal growth. *Journal of the Geological Society, London*, **147**, 41–57.

BILHAM, R., BENDICK, R., LARSON, K., MOHR, P., BRAUN, J., TESFAYE, S. & ASFAW, L. 1999. Secular and tidal strain across the Main Ethiopian Rift. *Geophysics Research Letters*, **26**, 2789–2792.

BLACKMAN, D. & KENDALL, J.-M. 1997. Sensitivity of teleseismic body waves to mineral texture and melt in the mantle beneath a mid-ocean ridge. *Philosophical Transactions of the Royal Society of London*, **355**, 217–231.

BLACKMAN, D.K. & KENDALL, J.-M. 2002. Seismic anisotropy of the upper mantle: 2. Predictions for current plate boundary flow models. *Geochemistry, Geophysics, Geosystems*, **3**, 8602, doi:10.1029/2001GC000247.

BLACKMAN, D.K., KENDALL, J-M., DAWSON, P., WENK, H-R., BOYCE, D. & PHIPPS MORGAN, J. 1996. Teleseismic imaging of subaxial flow at mid-ocean ridges: travel-time effects of anisotropic mineral texture in the mantle. *Geophysical Journal International*, **127**, 415–426.

BOCCALETTI, M., BONINI, M., MAZZUOLI, R., ABEBE, B., PICCARDI, L. & TORTORICI, L. 1998. Quaternary oblique extensional tectonics in the Ethiopian Rift (Horn of Africa). *Tectonophysics*, **287**, 97–116.

BOOTH, D.C. & CRAMPIN, S. 1985. Shear-wave polarizations on a curved wavefront at an isotropic free-surface. *Geophysical Journal of the Royal Astronomical Society*, **83**, 31–45.

BUCK, W.R. 2004. Consequences of asthenospheric variability on continental rifting, *In*: KARNER, G.D. TAYLOR, B.N. DROSCOLL, W. & KOHLSTEDT, D.L. (eds) *Rheology and Deformation of the Lithosphere at Continental Margins*. pp. 1–30, Columbia University Press, New York.

BUROV, E. & GUILLOU-FROTTIER, L. 2005. The plume head-continental lithosphere interaction using a tectonically realistic formulation of the lithosphere. *Geophysical Journal International*, **161**, 469–490.

CALAIS, E., EBINGER, C., HARTNADY, C., NOCQUET, J.M. 2006. Kinematics of the East African Rift from GPS and earthquake slip vector data. *In*: YIRGU, G., EBINGER, C., MAGUIRE, P.K.H. (eds) *The Afar Volcanic Province within the East African Rift System*. Geological Society, London, Special Publications, **259**, 9–22.

CHURCH, W.R., BERHE, S.M., ABDELSALAM, M.G. & STERN, R.J. 1991. Discussion of ophiolites in Northesast and East Africa: Implications for Proterozoic crustal growth. *Journal of the Geological Society, London*, **148**, 600–606.

CORNWELL, D., MACKENZIE, G.D., MAGUIRE, P.K.H., ENGLAND, R.W., ASFAW, L.M. & OLUMA, B. 2006. Northern Main Ethiopian Rift crustal structure from new high-precision gravity data. *In*: YIRGU, G., EBINGER, C. & MAGUIRE, P.K.H. (eds) *The Afar Volcanic Province within the East African Rift System*. Geological Society Special Publications, **259**, 307–321.

COURTILLOT, V.C., MANIGHETTI, I., TAPPONIER, P. & BESSE, J. 1999. On causal links between flood basalts and continental breakup. *Earth and Planetary Science Letters*, **166**, 177–195.

CRAMPIN, S. 1994. The fracture criticality of crustal rocks. *Geophysical Journal International*, **118**, 428–438.

DALTON, R. 2005. African network set to boost Earth Sciences. *Nature*, **433**, 449.

DARADICH, A., MITROVICA, J.X., PYSKLYWEC, R.N., WILLET, S.D. & FORTRE, A.M. 2003. Mantle flow, dynamic topography, and rift-flank uplift of Arabia. *Geology*, **31**, 901–904.

DAVAILLE, A., STUTZMANN, E., SILVEIRA, G., BESSE, J. & COURTILLOT, V. 2005. Convective patterns under the Indo-Atlantic. *Earth and Planetary Science Letters*, **239**, 233–252.

DAVIES, G. 1998. A channelled plume under Africa. *Nature*, **395**, 743–744.

DEBAYLE, E., LÉVÊQUE, J.-J. & CARA, M. 2001. Seismic evidence for a deeply rooted low velocity anomaly in the upper mantle beneath the northeastern Afro/Arabian continent. *Earth and Planetary Science Letters*, **193**, 423–436.

DEBAYLE, E., KENNETT, B.L.N. & PRIESTLEY, K. 2005. Global azimuthal seismic anisotropy and the unique plate-motion deformation of Australia. *Nature*, **433**, 509–512, doi:10.1038/nature03247.

DUNBAR, J. & SAWYER, D. 1989. How pre-existing weaknesses control the style of continental breakup. *Journal of Geophysical Research*, **94**, 7278–7292.

EBINGER, C.J. 2005. Continental break-up: The East Africa perspective. *Astronomy and Geophysics*, **46**, 2.16–2.21.

EBINGER, C.J. & SLEEP, N.H. 1998. Cenozoic magmatism throughout East Africa resulting from impact of a single plume. *Nature*, **395**, 788–791.

EBINGER, C.J. & CASEY, M. 2001. Continental brekaup in magmatic provinces: an Ethiopian example. *Geology*, **29**, 527–530.

EVANS, J. & ACHAUER, U. 1993. Teleseismic tomography using the ACH method: Theory and application to continental scale studies. *In*: IYER, E. & HIRAHARA (eds) *Seismic Tomography: Theory and Practice*. Chapman and Hall, New York.

FAUL, U., TOOMEY, D.R. & WAFF, H.S. 1994. Intergranular basaltic melt is distributed in thin, elongated inclusions. *Geophysical Research Letters*, **21**, 29–32.

FOUCH, M., FISCHER, K.M. PARMENTIER, E.M. WYSESSION, M.E. & CLARKE, T.J. 2000. Shear-wave splitting, continental keels, and patterns of mantle flow. *Journal of Geophysical Resrarch*, **105**, 6255–6275.

FOUCH, M., SILVER, P.G. BELL, D.R. & LEE, J.N. 2004. Small-scale variations in seismic anisotropy near Kimberley, South Africa. *Geophysical Journal International*, **157**, 764–774, doi:10.1111/j.1365-246X.2004.02234.x.

GAO, S., DAVIS, P.M. LIU, H. SLACK, P.D. ZORIN, YU, A. MORDVINOVA, V.V. KOZHEVNIKOV, V.M. & MEYER, R.P. 1994. Seismic Anisotropy and Mantle Flow beneath the Baikal Rift Zone. *Nature*, **371**, 149–151.

GAO, S., DAVIS, P.M., *ET AL.* 1997. SKS splitting beneath continental rift zones. *Journal of Geophysical Research*, **102**, 22781–22797.

GASHAWBEZA, E.M., KLEMPERER, S.L., NYBLADE, A.A., WALKER, K.T. & KERANEN, K.M. 2004. Shear-wave splitting in Ethiopia: Precambrian mantle anisotropy locally modified by Neogene rifting. *Geophysical Research Letters*, **31**, doi:10.1029/2004GL020471.

GEORGE, R., ROGERS, N. & KELLEY, S. 1998. Earliest magmatism in Ethiopia: Evidence for two mantle plumes in one flood basalt province. *Geology*, **26**, 923–926.

GÖK., R., NI, J.F., WEST, M., SANDVOL, E., WILSON, D., Aster, R., BALDRIDGE, W.S., GRAND, S., GAO, W., TILLMAN, F. & SEMKEN, S. Shear wave splitting and mantle flow beneath LA RISTRA. *Geophysics Research Letters*, **30**(12), 1614, doi:10.1029/2002GL016616.

GREEN, W.V., ACHAUER, U. & MEYER, R.P. 1991. A three dimensional image of the crust and upper mantle beneath the Kenya Rift. *Nature*. **354**, 199–203.

GURNIS, M., MITROVICA, J.X., SU, W.J. & HEIJST, H. van 2000. Constraining mantle density structure using geological evidence of surface uplift rates: The case of the African plume. *Geochemistry, Geophysics, Geosystems*, **1**, 1999GC000035.

HADIOUCHE, O., JOBERT, N. & MONTAGNER, J.-P. 1989. Anisotropy of the African Continent inferred from surface waves. *Physics of the Earth Planetary Interiors*, **58**, 61–81.

HAUTOT, S., WHALER, K., GEBRU, W. & DESSISA, M. 2006. The structure of Mesozoic basin beneath the Lake Tana area, Ethiopia, revealed with magnetotellurics. *Journal of African Earth Sciences* (in press).

HOLTZMAN, B.K., KOHLSTEDT, D.L., ZIMMERMAN, M.E., HEIDELBACH, F., HIRAGA, T. & HUSTOFT, J. 2003*a*. Melt segregation and strain partitioning: Implications for seismic anisotropy and mantle flow. *Science*, **301**, 1227–1230.

HOLTZMAN, B.K., KOHLSTEDT, D.L., ZIMMERMAN, M.E. & KENDALL, J.-M. 2003*b*. Melt segregation, strain partitioning, olivine CPO, and the origin of seismic anisotropy in oceanic lithosphere. *Eos: Transactions of the American Geophysical Union*, **84**(46), Fall Meet. Suppl., Abstract T51H-04.

KARATO, S. 1992. On the Lehmann discontinuity. *Geophysics Research Letters*, **19**, 2255–2258.

KEIR, D., EBINGER, C.J., KENDALL, J.-M. & STUART, G.W. 2005. Variations in late syn-rift melt alignment inferred from shear-wave splitting in crustal earthquakes beneath the Ethiopian rift. *Geophysical Research Letters*, **32**, L23308, doi: 10.1029/2005GL024150.

KEIR, D., EBINGER, C.J., STUART, G.W., DALY, E. & AYELE, A. 2006. Strain accomodation by magmatism and faulting as rifting proceeds to breakup: Seismicity of the northern Ethiopian rift. Submitted to *Journal of Geophysical Research* (in press).

KELLER, G.R., HARDER, S.H., O'REILLY, B., MICKUS, K., TADESSE, K., MAGUIRE, P.K.H. & EAGLE WORKING GROUP 2004. A preliminary analysis of crustal structure variations along the Ethiopian Rift. *Proceedings of the International Conference on the East African Rift System*, June 20–24 2004, Addis Ababa, Ethiopia, 97–101.

KENDALL, J.-M. 1994. Teleseismic arrivals at a mid-ocean ridge: effects of mantle melt and anisotropy. *Geophysical Research Letters*, **21**, 301–304.

KENDALL, J.-M. 2000. Seismic anisotropy in the boundary layers of the mantle. *In*: KARATO, S., STIXRUDE, L., LIEBERMANN, R.C., MASTERS, T.G. & FORTE, A.M. (eds) *Earth's Deep Interior: Mineral Physics and Tomography from the Atomic to the Global Scale. Geophysical Monograph Series*, **117**, American Geophysical Union, 149–175.

KENDALL, J.-M., STUART, G.W., EBINGER, C.J., BASTOW, I.D. & KEIR, D. 2005. Magma assisted rifting in Ethiopia. *Nature*, **433**, 146–148.

KERANEN, K., KLEMPERER, S.L., GLOAGUEN, R. & EAGLE WORKING GROUP 2004. Three-dimensional seismic imaging of a protoridge axis in the Main Ethiopian Rift. *Geology*, **32**, 949–952.

KIRKWOOD, S.C. & CRAMPIN, S. 1981. Surface-wave propagation in an ocean basin with an anisotropic upper mantle: Numerical modelling. *Geophysical Journal of the Royal Astronomical Society*, **64**, 463–485.

LITHGOW-BERTELLONI, C. & SILVER, P.G. 1998. Dynamic topography, plate driving forces and the African superswell. *Nature*, **395**, 269–272.

MACKENZIE, G.D., THYBO, G.H. & MAGUIRE, P.K. H. 2005. Crustal velocity structure across the Main Ethiopian Rift: results from 2-dimenional wide-angle seismic modelling. *Geophysical Journal International*, **162**, 994, doi:10.1111/j.1365-246X.2005.02710.

MAGUIRE, P.K.H., EBINGER, C.J., *ET AL.* 2003. Geophysical project in Ethiopia studies continental breakup. *Eos: Transactions of the American Geophysical Union*, **84**, 337–343.

MAGUIRE, P.K.H., KELLER, C.R., *ET AL.* 2006. Crustal structure of the northern Main Ethiopian Rift from the EAGLE controlled source survey; a snapshot of incipient lithospheric break-up. *In*: YIRGU, G. EBINGER, C. & MAGUIRE, P.K.H. (eds) *The Afar Volcanic Province within the East African Rift System.* Geological Society, London, Special Publications, **259**, 269–291.

MAINPRICE, D., TOMMASI, A., COUVY, H., CORDIER, P. & FROST, D.J. 2005. Pressure sensitivity of olivine slip systems: Implications for the interpretation of seismic anisotropy of the Earth's upper mantle. *Nature*, **433**, 731–733.

MCKENZIE, D. 1978. Some remarks on the developement of sedimentary basins. *Earth and Planetary Science Letters*, **40**, 25–32, 1978.

MONTAGNER, J.-P. 1994. What can seismology tell us about mantle convection? *Reviews of Geophysics* **32**, 115–137.

MONTELLI, R., NOLET, G., DAHLEN, F.A., MASTERS, G., ENGDAHL, E.R. & HUNG, S.-H. 2004. Finite-frequency tomography reveals a variety of plumes in the mantle. *Science*, **303**, 338–343.

MOORE, H. & DAVIDSON, A. 1978. Rift structure in southern Ethiopia. *Tectonophysics*, **46**, 159–173.

NICHOLAS, A. & CHRISTENSEN, N.I. 1987. Formation of anisotropy in upper mantle peridotites – A review. *In*: FUCHS, K. & FROIDAVAUX, C. (eds) *Composition, Structure and Dynamics of the Lithosphere–Asthenosphere System, Geodynamics Series*, **16**, 111–123, American Geophysical Union, WASHINGTON, DC.

NYBLADE, A.A. & ROBINSON, S.W. 1994. The African superswell. *Geophysical Research Letters*, **21**, 765–768.

NYBLADE, A., KNOX, R. & GURROLA, H. 2000. Mantle transition zone thickness beneath Afar: Implications for the origin of the Afar hotspot. *Geophysical Journal International*, **142**, 615–619.

NYBLADE, A.A., OWENS, T.J., GURROLA, H., RITSEMA, J. & LANGSTON, C.A. 2000. Seismic evidence for a deep upper mantle thermal anomaly beneath East Africa. *Geology*, **28**, 599–602.

PHIPPS MORGAN, J. 1987. Melt migration beneath mid-ocean spreading centers. *Geophysical Research Letters*, **14**, 1238–1241.

PILIDOU, S., KENDALL, J.-M., STUART, G.W. & BASTOW, I.D. 2004. Evidence of melt-induced seismic anisotropy and magma assisted rifting in the Northern Ethiopian Rift. *EOS, Transactions of the American Geophysical Union, Fall Meeting Supplement*, T33A–1347.

RITSEMA, J. & ALLEN, R. 2003. The elusive mantle plume. *Earth and Planetary Science Letters*, **207**, 1–12.

RITSEMA, J., NYBLADE, A., OWENS, T. LANGSTON, C. & VANDECAR, J.C. 1998. Upper mantle seismic velocity structure beneath Tanzania, east Africa: Implications for the stability of cratonic lithosphere. *Journal of Geophysical Research*, **103**, 21201–21213.

RITSEMA, J., VAN HEIJST, H.-J. & WOODHOUSE, J.H. 1999. Complex shear wave velocity structure beneath Africa and Iceland. *Science*, **286**, 1925–1928.

ROONEY, T.O., FURMAN, T., YIRGU, G. & AYELEW, D. 2005. Structure of the Ethiopian lithosphere: Evidence from mantle xenoliths. *Geochimica and Cosmochimica Acta*, **69**(15), 3889–3910, doi:10.1016/j/gca.2005.03.043.

RUMPKER, G. & RYBERG, T. 2000. New 'Fresnel zone' estimates for shear-wave splitting observations from finite difference modelling. *Geophysical Research Letters*, **27**, 2005–2008.

SANDVOL, E., NI, J., OZALAYBEY, S. & SCHLUE, J. 1992. Shear-wave splitting in the Rio Grande Rift. *Geophysical Research Letters*, **19**, 2337–2340.

SAVAGE, M. 1999. Seismic anisotropy and mantle deformation: What have we learned from shear wave splitting. *Reviews of Geophysics*, **37**, 65–106.

SCHMELLING, H. 1985. Numerical models of the influence of partial melt on elastic, inelastic and electric properties of rocks, Part 1: elasticity and anelasticity. *Physics of the Earth Planetary Interiors*, **41**, 34–57.

SEBAI, A., STUTZMANN, E., MONTAGNER, J.-P., SICILIA, D. & BEUCLER, E. 2005. Anisotropic structure of the African upper mantle from Rayleigh and Love wave tomography. *Physics of the Earth Planetary Interiors* (in press).

SELLA, G.F., DIXON, T.H. & MAO, A. 2002. REVEL: A model for Recent plate velocities from space geodesy. *Journal of Geophysical Research*, **107**(B4), doi: 10.1029/2000JB000033.

SHACKLETON, R.M. 1986. Precambrian collision tectonics in Africa, *In*: COWARD, M.P. & RIES, A.C. (eds) *Collision Tectonics*. Geological Society, London, Special Publications, **19**, 329–349.

SILVER, P.G. 1996. Seismic anisotropy beneath the continents. *Annual Reviews of Earth and Planetary Sciences*, **24**, 385–432.

Silver, P.G., Gao, S.S., Liu, K.H. & the Kaapvaal working group 2001. Mantle Deformation beneath Southern Africa. *Geophysical Research Letters*, **28**, 2493–2496.

Sleep, N.H. 1997. Lateral flow and ponding of starting plume material. *Journal of Geophysical Research*, **102**, 10001–10012.

Sleep, N.H., Ebinger, C.J. & Kendall, J.-M. 2002. Deflection of mantle plume material by cratonic keels. *In*: Fowler, C.M.R., Ebinger, C.J., & Hawkesworth, C.J. (eds) *The Early Earth: Physical, Chemical and Biological Development.* Geological Society, London Special Publications, **199**, 135–150.

Stern, R.J., Nielsen, K.C., Best, E., Sultan, M., Arvidson, R.E. & Kroner, A. 1990. Orientation of the late Precambrian sutures in the Arabian–Nubian Shield. *Geology*, **18**, 1103–1106.

Tandon, G.P. & Weng, G.J. 1984. The effect of aspect ratio of inclusions on the elastic properties of unidirectionally aligned composites. *Polymer Composites*, **5**, 327–333.

Tiberi, C., Ebinger, C., Ballu, C., Stuart, G.W. & Oluma, B. 2005. Inverse models of gravity data from the Rea Sea – Gulf of Aden – Ethiopian rift triple junction zone. *Geophysical Journal International*, **163**, 775–787.

Tommasi, A. 1998. Forward modelling of the development of seismic anisotropy in the upper mantle. *Earth and Planetary Science Letters*, **160**, 1–13.

Vauchez, A., & Garrido, C. 2001. Seismic properties of an Asthenospherized lithospheric mantle: constraints from the lattice preferred orientation of peridotite minerals in the Ronda Massif. *Earth and Planetary Science Letters*, **192**, 235–249.

Vauchez, A., Barruol, G. & Tommasi, A. 1997. Why do continents break-up parallel to ancient orogenic belts? *Terra Nova*, **9**, 62–66.

Vauchez, A., Tommasi, A., Barruol, G. & Maumus, J. 2000. Upper mantle deformation and seismic anisotropy in continental rifts. *Physics and Chemistry of the Earth*, **25**, 111–117.

Vauchez, A., Dineur, F. & Rudnick, R. 2005. Microstructure, texture and seismic anisotropy of the lithospheric mantle: Insights from the Labait volcano xenoliths (Tanzania). *Earth and Planetary Science Letters*, **232**, 295–314.

Walker, K., Nyblade, A.A., Klemperer, S.L., Bokelmann, G.H. R. & Owens, T.J. 2003. On the relationship between extension and anisotropy: Constraints from shear wave splitting across the East Africa Plateau. *Journal of Geophysical Research*, **109**, doi:10.1029/2003JB002866.

Weereratne, D.S., Forsyth, D.W., Fischer, K.M. & Nyblade, A.A. 2003. Evidence for an upper mantle plume beneath the Tanzanian craton from Rayleigh wave tomography. *Journal of Geophysical Research*, **108**, doi:10.1029/2002JB002273.

Wessel, P. & Smith, W.H.F. 1998. New, improved version of generic mapping tools released. *Eos: Transactions of the American Geophysical Union*, **79**, 579.

Whaler, K. & Hautot, S. 2006. The electrical resistivity structure of the crust beneath the Northern Ethiopian Rift. *In*: Yirgu, G. Ebinger, C. & Maguire, P.K.H. (eds) *The Afar Volcanic Province within the East African Rift System.* Geological Society, London, Special Publications, **259**, 293–305.

Wolfe, C.J. & Solomon, S.C. 1998. Shear-wave splitting and implications for mantle flow beneath the MELT region of the East Pacific Rise. *Science*, **280**, 1230–1232.

Wolfenden, E., Ebinger, C., Yirgu, G., Deino, A. & Ayelew, D. 2004. Evolution of the northern Main Ethiopian Rift: birth of a triple junction. *Earth and Planetary Science Letters*, **224**, 213–228.

# Part 2: Geochemical constraints on flood basalt and rift processes

The Ethiopian volcanic province represents an ideal situation to study the dynamics of mantle sources involved in continental volcanism and the manner in which sources interacted with continental lithosphere. The first two papers in this section deal with the geochemistry of basalts in the province and their relevance to outstanding problems in basalt petrogenesis: conditions of melting, composition of source domains, and their longevity. The papers also address the controversy of whether one or more distinct mantle plumes are involved in the evolution of the present topography and tectonics of East Africa. The third paper addresses the earliest history and subsequent evolution of silicic magmatism at the junction between the southern Red Sea and northern Main Ethiopian Rifts.

The Ethiopian province, in contrast to other typical continental flood basalts such as those of the Deccan and Karoo provinces, is made up of a series of flood basalts overlain by large and conspicuous shield volcanoes. Distinctive petrological features of the Ethiopian plateau are the transitional tholeiitic to alkaline magmatic character of the mafic lavas, in contrast with the tholeiitic character of most continental flood basalts, and the high proportion of felsic pyroclastic rocks. The various lava types have similar Nd and Sr isotopic compositions but very different Pb isotopic compositions. Some have a weak 'HIMU' (high $^{206}Pb/^{204}Pb$ and low $^{87}Sr/^{86}Sr$) character. The modern Afar plume, as observed in Djibouti and Erta' Ale, is characterized by $^{206}Pb/^{204}Pb\sim19$, $^{207}Pb/^{204}Pb\sim15.6$; $^{208}Pb/^{204}Pb\sim38.8$ (Deniel *et al.* 1994; Barrat *et al.* 1998). Lavas in a large region surrounding the volcanically active Afar region have high $^{3}He/^{4}He$ ratios (Marty *et al.* 1996).

**Rogers** presents a comprehensive review of the major and trace element and radiogenic isotope data for basalts from throughout the East African rift system. Mafic volcanics of this system record a protracted history of continental extension that is linked to mantle plume activity. The author argues that only the MgO-rich picrites and ankaramites from the 29–31 Ma Ethiopian traps and the most recent basalts (<5 Ma) from Afar have an unambiguous Afar mantle plume signature He shows that Eocene Amaro basalts from southern Ethiopia also have a plume source but that their lower source temperatures and isotopic characteristics are distinct from those of Afar. The remaining basalts from the Ethiopian rift and throughout the Kenya and Western rifts are attributed to a lithospheric source region. The author interprets the regional geochemical variations in terms of a model involving the presence of two mantle plumes, the East African and Afar plumes, which dynamically support the East African and Ethiopian plateaus. It is suggested that the Amaro basalts represent the first manifestations of magmatism from the East African plume and as the African plate migrated north, subsequent magmatic activity is represented by progressively younger episodes further south through Turkana, Kenya and into northern Tanzania. This reinforces the two-plume model previously suggested by the same author and his co-workers (George *et al.* 1998; Rogers *et al.* 2000). This interpretation is in contrast to the views of **Furman *et al.*** and Keiffer *et al.* (2004). Keiffer *et al.* (2004) interpret the major, trace element and radiogenic isotope compositions of the Ethiopian flood and shield basalts in terms of variable degrees of melting of different components of a common, deep mantle source, heterogeneous both compositionally and thermally. **Rogers** further explains that lithospheric mantle heats up and becomes incorporated into the convecting mantle, causing greater degrees of lithospheric thinning than that shown by extension across individual rift basins.

**Furman *et al.*** present high-quality geochemical (major and trace element and Sr–Nd–Pb isotope) data on primitive recent mafic lavas from the northern sector of the Main Ethiopian Rift and integrate their results with data from previous geochemical studies in the entire Afar volcanic province. The authors show that modern rift basalts have been derived by moderate degrees of melting of fertile peridotite at depths corresponding to the base of the modern lithosphere (*c.* 100 km) through melt-induced binary mixing of melts from the Afar plume with melts from three mantle end-member compositions: namely, the convecting upper mantle and two enriched mantle sources. They argue that the Afar plume is a long-lived feature of the mantle and that its fundamental isotopic composition has effectively not changed during the

*From*: Yirgu, G., Ebinger, C.J. & Maguire, P.K.H. (eds) 2006. *The Afar Volcanic Province within the East African Rift System*. Geological Society, London, Special Publications, **259**, 73–75.
0305-8719/06/$15.00 

transition from Oligocene plume head volcanism to modern plume tail activity. They conclude, in contrast to the view of **Rogers**, that the overall geochemical data in the province (including the Eocene flood basalts of southern Ethiopia and northern Kenya) is consistent with a modified single plume model in which multiple plume stems rise from a common large plume originating at great depth in the mantle (i.e. the South African Superplume). This model is similar to that proposed by Keiffer *et al.* (2004) who interpret the geochemical variations of the magmas in terms of melting of different components of a common deep mantle source within the African 'superswell'. Recent geophysical studies provide evidence for possible plume stems in the upper mantle beneath both northern Ethiopia and the Tanzania craton (Debayle *et al.* 1985; Nyblade *et al.* 2000; Montelli *et al.* 2004; Bastow *et al.* 2005). It is important to note that the geochemical data alone cannot distinguish between models that call for a single homogeneous plume beneath East Africa (e.g. Ebinger & Sleep 1998), a complex heterogeneous plume (e.g. Furman *et al.* 2004, 2006), or two geochemically distinct mantle plumes (e.g. George *et al.* 1998; Rogers *et al.* 2000; Rogers, 2006). **Furman *et al.*** attribute temporal changes in Ethiopian volcanic geochemistry to the evolving tectonic environment and not to changes in sub-lithospheric source composition.

**Ayalew *et al.*** report geochemical results on a suite of silicic rocks spanning a time from 28 to 2.5 Ma from along the western margin of the southern Red Sea and northern Main Ethiopian Rifts, and attempt to constrain their origin and evolution through time. The study is based on recently published tectonic and geochronological work from this region (Wolfenden *et al.* 2004, 2005), which have shown a southward and riftward migration of the locus of magmatism with time. Here, rhyolites are preferentially localized on or near the border faults, indicating that rifting either preceded or was coeval with the felsic volcanism. The authors use the geochemical data (major and trace elements; Sr, Nd and O isotopic compositions) on well-dated silicic volcanic rocks, dominantly pyroclastic flows with some lavas, to explain the derivation of syn-rift rhyolites and constrain the degree of involvement of crustal material. According to them, the rhyolites cannot be derived from partial melting of the local crust or by a single stage partial melting of basalts or their derivatives. They concur that the geochemical data are consistent with derivation of the rhyolites from mantle-sourced basaltic magma through fractional crystallization accompanied by variable amounts of crustal contamination. The principal conclusion of the paper derives from observed temporal variations of Sr–Nd–O isotopes, which have been interpreted to reflect increasing participation of the basement with time and with the advancement of rifting in the genesis of silicic magma. This idea gets support from an earlier work by Trua *et al.* (1999) who have shown that more highly evolved rocks in the Ethiopian rift have clearly assimilated a crustal component. **Ayalew *et al.*** are of the opinion that the increasing crustal involvement with time may have been caused by progressive thermal and mechanical weakening of the crust within the faulted rift valley, in response to lithospheric thinning and magma injection during the past ~30 My. The interpretations agree with new geophysical observations for the northern MER and uplifted plateau region that demonstrate significant modification of the crust and mantle lithosphere by magmatic processes (e.g. Kendall *et al.* 2005; Keranen *et al.* 2004; Mackenzie *et al.* 2005). Derivation of silicic magmas from basaltic parents by fractional crystallization provides further supporting evidence for the underplate layer beneath the northwestern rift flank, imaged by the EAGLE-controlled source seismic data (Mackenzie *et al.* 2005; Maguire *et al.* 2006).

## References

BARRAT, J.A., FOURCADE, S., JAHN, B.M., CHEMINÉE, J.L. & CAPDEVILA, R. 1998. Isotope (Sr, Nd, Pb, O) and trace element geochemistry of volcanics from the Erta' Ale range (Ethiopia). *Journal of Volcanology and Geothermal Research*, **80**, 85–100.

BASTOW, I.D., STUART, G.W., KENDALL, J.-M., EBINGER, C.J., AYELE, A., CORNWELL, D.G. & MAGUIRE, P.K.H. 2005. Upper-mantle seismic structure of the northern Ethiopian Rift: a region of incipient continental breakup. *Geophysical Journal International*, **162**, 479–493.

DEBAYLE, E., LEVEQUE, J.-J. & CARA, M. 2001. Seismic evidence for a deeply rooted low-velocity anomaly in the upper mantle beneath the northeastern Afro/Arabian continent. *Earth and Planetary Science Letters*, **193**, 423–436.

DENIEL, C., VIDAL, P., COULON, C., VELLUTINI, P.-J. & PIGUET, P. 1994. Temporal evolution of mantle sources during continental rifting: the volcanism of Djibouti. *Journal of Geophysical Research*, **99**, 2853–2869.

EBINGER, C. & SLEEP, N. 1998. Cenozoic magmatism throughout E. Africa resulting from impact of a single plume. *Nature*, **395**, 788.

FURMAN, T., BRYCE, J.G., KARSON, J. & IOTTI, A. 2004. East African Rift System (EARS) plume structure: insights from Quaternary mafic lavas of Turkana, Kenya. *Journal of Petrology*, **45**, 1069–1088.

GEORGE, R., ROGERS, N. & KELLY, S. 1998. Earliest magmatism in Ethiopia: evidence for two mantle plumes in one flood basalt province. *Geology*, **26**, 923–926.

KENDALL, J.-M., STUART, G.W., EBINGER, C.J., BASTOW, I.D. & KEIR, D. 2005. Magma-assisted rifting in Ethiopia. *Nature*, **433**, 146–148.

KERANEN, K., KLEMPERER, S.L., GLOAGUEN, R. & EAGLE WORKING GROUP. 2004. Three-dimensional seismic imaging of a proto-ridge axis in the Main Ethiopian rift. *Geology*, **32**, 949–952.

KIEFFER, B., ARNDT, N., *et al.* 2004. Flood and Shield basalts from Ethiopia: magmas from the African superswell. *Journal of Petrology*, **45**, 793–834.

MACKENZIE, G., THYBO, H. & MAGUIRE, P.K.H. 2005. Crustal velocity structure across the Main Ethiopian Rift: results from 2-dimensional wide-angle seismic modeling. *Geophysical Journal International*, **162**, 994–1006.

MARTY, B., PIK, R. & YIRGU, G. 1996. Helium isotopic variations in Ethiopian plume lavas: nature of magmatic sources and limit on lower mantle contribution. *Earth and Planetary Science Letters*, **144**, 223–237.

MONTELLI, R., NOLET, G., DAHLEN, F.A., MASTERS, G., ENGDAHL, E.R. & HUNG, S.-H. (2004). Finite-frequency tomography reveals a variety of plumes in the mantle, *Science*, **303**, 338–343.

NYBLADE, A., OWENS, T., GURROLA, H., RITSEMA, J. & LANGSTON, C.A. 2000. Seismic evidence for a deep upper mantle thermal anomaly beneath East Africa. *Geology*, **28**, 599–602.

ROGERS, N., MACDONALD, R., FITTON, J.G., GEORGE, R., SMITH, M. & BARREIRO, B. 2000. Two mantle plumes beneath the East African Rift System: Sr, Nd and Pb isotope evidence from Kenya Rift basalts. *Earth and Planetary Science Letters*, **176**, 387–400.

TRUA, T., DENIEL, C. & MAZZUOLI, R. 1999. Crustal control in the genesis of Plio-Quaternary bimodal magmatism of the Main Ethiopian Rift (MER): geochemical and isotopic (Sr, Nd, Pb) evidence. *Chemical Geology*, **155**, 201–231.

WOLFENDEN, E., EBINGER, C., YIRGU, G., DEINO, A. & AYALEW, D. 2004. Evolution of the northern Main Ethiopian Rift: birth of a triple junction. *Earth and Planetary Science Letters*, **224**, 213–228.

WOLFENDEN, E., EBINGER, C., YIRGU, G., RENNE, P. & KELLEY, S.P. 2005. Evolution of the southern Red Sea rift: birth of a magmatic margin. *Geological Society of America Bulletin*, **117**, 846–864.

# Basaltic magmatism and the geodynamics of the East African Rift System

N.W. ROGERS

*Department of Earth Sciences, CEPSAR, The Open University, Walton Hall, Milton Keynes, MK7 6AA, UK (e-mail: n.w.rogers@open.ac.uk)*

**Abstract:** The major and trace element and radiogenic isotope compositions of basalts from throughout the East African rift system are reviewed in the context of constraints from previous geophysical studies. The data indicate the presence of two mantle plumes, the East African and Afar plumes, which dynamically support the East African and Ethiopian plateaus. Rifting across the plateaus is accompanied by the generation of large volumes of basaltic magma and associated evolved derivatives. Relatively few mafic magmas have an unambiguous Afar mantle plume signature, notably the MgO-rich picrites and ankaramites from the 29–31 Ma Ethiopian traps, and the most recent basalts (<5 Ma) from Afar. The Eocene Amaro basalts from southern Ethiopia also have a plume source but their lower source temperatures and isotopic characteristics are distinct from those of Afar. The remaining basalts from the Ethiopian rift, and throughout the Kenya and Western rifts, have a lithospheric source region as reflected in both radiogenic isotope and trace element characteristics. The Amaro basalts are suggested as the first manifestations of magmatism from the East African plume; subsequent magmatic activity being represented by progressively younger episodes further south through Turkana, Kenya and into Northern. Tanzania, as the African plate migrated north. Despite their clear lithospheric characteristics, U-series data on geologically recent basalts from the axis of the Kenya rift show that they were generated in a dynamic melting regime. Melting is effected when lithospheric mantle heats up and becomes incorporated into the convecting mantle, hence leading to greater degrees of lithospheric thinning than are indicated by extension across individual rift basins.

The frequent association in the geological record of continental break-up, large igneous provinces (LIP), and, by inference, mantle plumes is now well established (White & McKenzie 1989, 1995; Storey 1995). In most cases, all we can study is the end result of break-up, namely the rifted margins of the two continents and the developing ocean basin between, with an aseismic ridge marking the trace of the mantle plume. By contrast, the East African rift and the associated nascent oceans of the Red Sea and the Gulf of Aden, present examples of the various stages of continental rifting and break-up that are still in close proximity to the underlying mantle plumes. Within the region, there is a record of igneous activity stretching from the Eocene to the present day which offers an unrivalled opportunity to investigate the role of mantle plumes in this important geodynamic process.

Compared with other continents, Africa has a unique topographic character. Instead of linear mountain chains related to continental collision and tectonic processes at destructive plate margins, it is dominated by basins and plateaus which in some cases, most notably in East Africa, are associated with the development of rift basins (Fig. 1). This association is now established as causative, the topography at least in part driving extension (e.g. Coblenz & Sandiford 1994), and the topography has been further linked to upwelling in the mantle beneath the lithosphere (Ebinger *et al.* 1989; Nyblade & Brazier 2002). The relationship between topography and mantle convection, as reflected in tomographic images of the deep mantle, has been developed further to include the broad plateau of southern Africa, now sometimes referred to as the 'African superswell'. This feature is related to upwelling in the lower mantle, whereas the Ethiopian and East African plateaus, and possibly the smaller plateaus of North Africa (e.g. Tibesti, Hoggar and Darfur), are a product of upper mantle dynamics (Lithgow-Bertelloni & Silver 1998). In many respects, the topography of the African continent is mapping out major features of the convective regime in the underlying mantle.

The East and North African plateaus are also characterized by recent volcanic activity, much of which is basaltic and has a mantle origin (e.g. Ashwal & Burke 1989). Given the limited extension associated with the rifts across these plateaus, the presence of any magmatism implies elevated mantle potential temperatures (e.g. McKenzie & Bickle 1984), and the eruption of $10^5$–$10^6$ $km^3$ of mafic and differentiated magmas along the length of the Kenya rift (King 1978; Williams 1982;

*From*: YIRGU, G., EBINGER, C.J. & MAGUIRE, P.K.H. (eds) 2006. *The Afar Volcanic Province within the East African Rift System*. Geological Society, London, Special Publications, **259**, 77–93.
0305-8719/06/$15.00 

Baker 1987) throughout the Neogene, and similar volumes broadly associated with the Ethiopian rift, further implies that temperatures commensurate with those of mantle plumes have existed beneath this region for tens of millions of years. Although, in detail, most of the Ethiopian magmatism predates the onset of rifting in the Main Ethiopian Rift and is synchronous with or postdates extension across the Red Sea and Gulf of Aden, the eruption of such volumes of mafic lavas still requires the presence of elevated mantle potential temperatures. Aspects of the compositions of basalts from Ethiopia, notably their high $^{3}He$ contents (Marty *et al.* 1996), lend further strength to the argument for the presence of a mantle plume beneath the Ethiopian plateau.

The geology and composition of the basalts can therefore be used to infer properties of the underlying mantle. The volumes and rates of eruption combined with estimates of the depth, temperature and extent of melting can be exploited to develop models of the physical properties of the melting regime. The more detailed aspects of the basalt composition, notably trace elements and radiogenic isotopes, allow the distinction of different mantle source regions, such as asthenosphere versus mantle lithosphere, and possibly between different mantle plumes. Finally, analysis of U-series isotopes has the added potential to provide insights into the melting processes that have led to the production of the most recent basalts. All of these data can be used to test and complement physical models

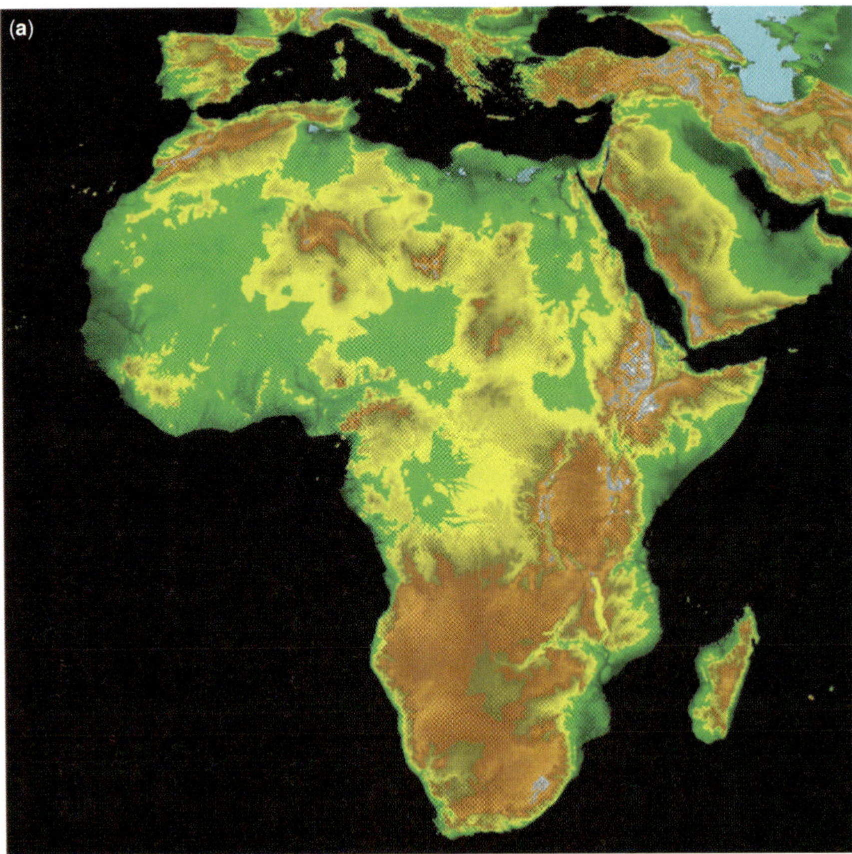


**Fig. 1.** **(a)** Digital elevation model of continental Africa and the Arabian peninsula with a larger image of the East African and Ethiopian plateaus. Note the broad plateau region of southern Africa (African superswell), and the roughly circular platforms of the East African and Ethiopian plateaus. Key: green $<1000$ m; yellow 1000–2000 m; brown 2000–3000 m; grey $>3000$ m. Image prepared from GTOPO30 digital elevation data, US National Intelligence Mapping Agency, by S. Drury.

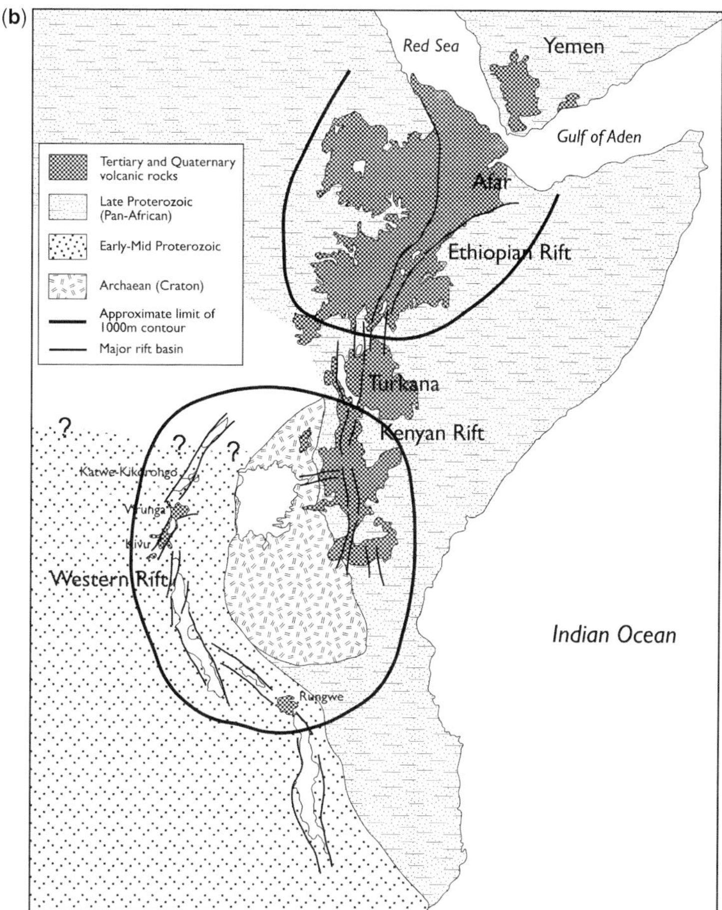


**Fig. 1.** (**b**) Sketch map of East Africa showing the location of the Main Ethiopian Rift and the Kenya and Western rifts, the distribution of Tertiary and Quaternary volcanic rocks and the disposition of basement of contrasting ages. The bold lines indicate the Ethiopian and East African plateaus with elevations above 1000 m (modified after Rogers *et al.* 1998).

of the mantle beneath the rift and associated plateaus.

This paper reviews critical aspects of the geochemical data from various studies of basaltic rocks from the Kenya, Western and Ethiopian rifts, the three major branches of the modern African rift system, and uses them to draw conclusions about the composition and structure of the mantle beneath this important example of continental rifting and break-up. It focuses on the interaction between mantle convection, lithosphere structure and tectonic processes along the length of the African rift, and in the process seeks to inform models of the rifting process. Specific questions that will be posed in this paper address the influence and involvement of the mantle lithosphere in the development of the rift, and the nature, number and longevity of underlying mantle plumes.

## Rifting and magmatism on the East African Plateau

### *Lithosphere control on the location of rifting*

One of the primary distinctions between the Ethiopian and East African plateaus is that whereas the first is cut by a single rift system, the latter is cut by two rifts, the Kenya or Gregory rift and the Western rift (Fig. 1). It is now well established that the development of the two rifts across the East African plateau is a consequence of lithospheric anisotropy (Nyblade & Brazier 2002). The presence of the old, cold and mechanically strong Tanzanian craton diverts rifting to the surrounding mobile belts, the Late Proterozoic Mozambique belt to the east and the Mid-Proterozoic Kibaran belt to the west. By contrast, the lack of a cratonic core underlying the Ethiopian plateau allows extension to propagate in a more uniform manner across the plateau (Fig. 1).

The Kenya and Western rifts have contrasting magmatic characteristics. The former rift is associated with large volumes (>$10^5$ $km^3$) of basaltic magma and evolved derivatives that have erupted throughout the Neogene (King 1978; Williams 1982; Baker 1987). As the rift has developed from wide half-graben basins into narrower full graben, so the magmatic activity has become increasingly focused on the active central rift valley. By contrast, volcanism along the length of the Western rift is sporadic and limited in extent and volume, and appears to be associated with accommodation zones between adjacent rift basins (Rosendahl 1987). The evolution of the Kenya rift from a broad depression, through half-graben to a narrow zone with the development of full-graben basins (Smith & Moseley 1993) is consistent with extension of thick lithosphere above hot mantle (Buck 1991). Extension facilitates magmatic intrusion, weakening the lithosphere which, when subject to further extensional stresses, continues to fail along the line of the original fracture; the rift thus acts as a focus for future extensional failure.

## *Compositional differences between the Western and Kenya rifts*

In addition to the differences in volumes of basaltic volcanism in the two rifts, there are clear compositional differences between the volcanic products of the Western and Kenya rifts. This is expressed in a plot of $K_2O$ versus $SiO_2$ (Fig. 2) in which mafic rocks from the Western rift at a given silica content have a higher potassium content than those from the Kenya rift. This difference is generally attributable to the presence of a potassic phase, either amphibole or phlogopite, in the mantle source region of at least the K-rich magmas (e.g. Edgar *et al.* 1976). The enrichment of potassium is accompanied by an enrichment of other incompatible elements, such as the LREE, Th, U and Ta, while the lack of fractionation between high-field-strength elements—HFSE such as Zr, Hf, Ti, Ta and Nb—and the large ion lithophile elements—LILE such as Rb, Cs, K, Ba and Sr—implies that the source underwent enrichment by addition of an alkaline mafic silicate melt (Rogers *et al.* 1992, 1998). Moreover, the radiogenic isotope characteristics of the Virunga potassic basalts (Fig. 3) further imply that trace element enrichment of the source region occurred about 1000 Ma ago (Davies & Lloyd 1989; Rogers *et al.* 1992, 1998; Furman 1995; Furman & Graham 1999), similar to the age of lithosphere stabilization after the Kibaran orogeny in this part of Africa. Thus, the majority of lavas from the Western rift are attributed to a lithospheric source region with probably no input from the underlying asthenosphere, although a sub-lithospheric component has been invoked to explain the Sr and Nd isotope characteristics of the Kivu lavas (Furman & Graham 1999).

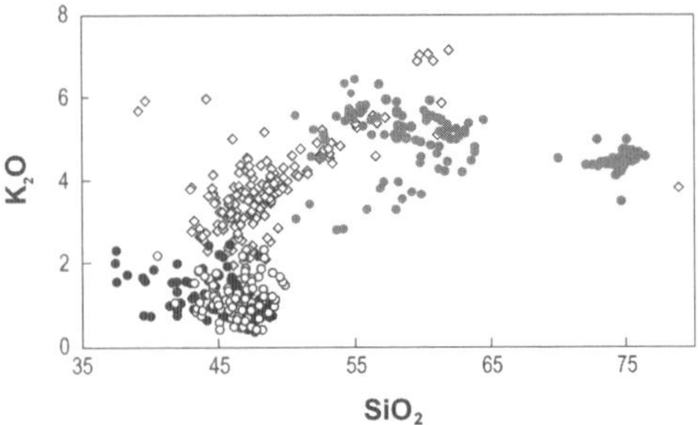


**Fig. 2.** Potassium–silica diagram for volcanic rocks from the Virunga province in the Western rift (open diamonds) and the Kenya rift (circles). Black circles: basalts from those parts of the Kenya rift underlain by the craton and its remobilized margins; white circles: basalts from those parts of the Kenya rift underlain by the late Proterozoic mobile belt. Grey circles represent analyses of intermediate and evolved rock types, including trachytes, comendites and phonolites. Data from Rogers *et al.* (1992, 1998) and Macdonald *et al.* (2001) and references therein.

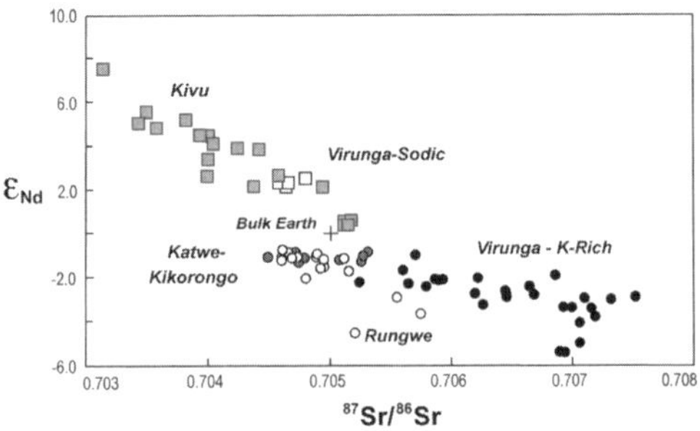


**Fig. 3.** Nd and Sr isotope analyses of mafic rocks from the Rungwe (open circles), Kivu (grey squares), Virunga potassic (black circles) and sodic (open squares), and Katwe–Kikorongo (grey circles) volcanic provinces in the Western rift. With the exception of the Kivu and Virunga-sodic data (from Nyiragongo), all analyses plot in the so-called enriched quadrant, signifying derivation from an old- trace element-enriched source region. (Data from Vollmer & Norry 1983; Davies & Lloyd 1989; Rogers *et al.* 1992, 1998; Furman 1995; and Furman & Graham 1999).

In contrast to the Western rift mafic lavas, basalts from the Kenya rift, while still alkaline, are not potassic and are superficially comparable with ocean island basalts (OIB) (Latin *et al.* 1993; Macdonald *et al.* 2001). They range from transitional tholeiites through alkali basalts to basanites and nephelinites, representing a spectrum of compositions that can be produced by decreasing melt fractions and increasing depths of melt segregation in a typical mantle melting regime (Macdonald *et al.* 2001). However, their radiogenic isotope ratios show a clear control exerted by the underlying lithosphere. The Kenya rift cuts across the major lithospheric boundary between the Tanzania Craton and the Mozambique Late Proterozoic mobile belt (Smith & Moseley 1993), and the basalts erupted through these contrasting basement types have distinct Sr and Nd isotope characteristics (Fig. 4). The major boundary appears to be between the mobile belt and the remobilized craton margin (Rogers *et al.* 2000).

Further differences in the trace elements of the Western and Kenya rift basalts are evident in the fractionation of the REE. All rift basalts are LREE-enriched, with LREE abundances ranging from a few tens times the chondritic abundance to many hundreds. However, significant differences are apparent in the fractionation of the HREE compared with overall LREE enrichment. These are summarized on plots of Tb/Yb against La/Yb which allow the effects of both melt fraction and residual mineralogy to be distinguished, most significantly whether or not a melt was generated in the presence of residual garnet.

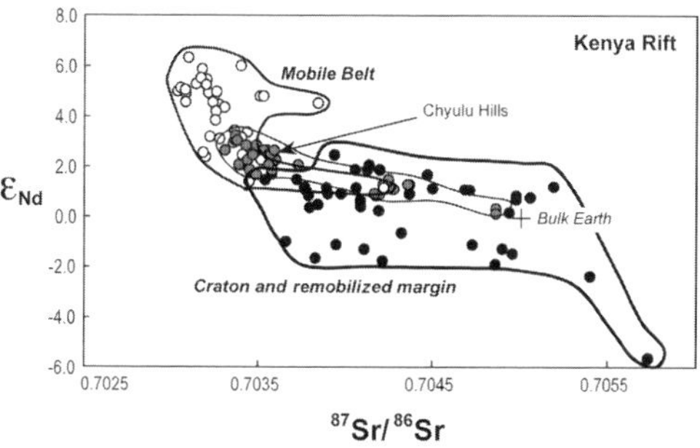


**Fig. 4.** Nd and Sr isotope analyses of basaltic rocks from the Kenya rift. Open symbols: basalts from the northern sector of the rift underlain by the late Proterozoic mobile belt. Black symbols: basalts from the southern sector of the rift underlain by the Tanzanian craton and its remobilized margins. Grey symbols: basalts from the Chyulu Hills (Spath *et al.* 2001), east of the main rift valley and straddling the craton–mobile belt boundary. All other data from Furman *et al.* (2004) and Rogers *et al.* (2000) and references therein.

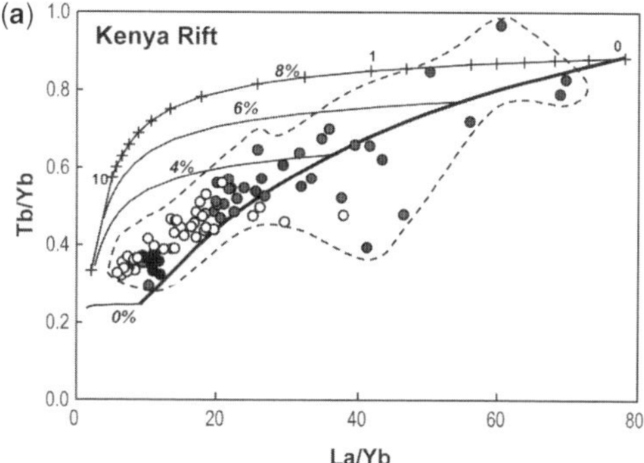


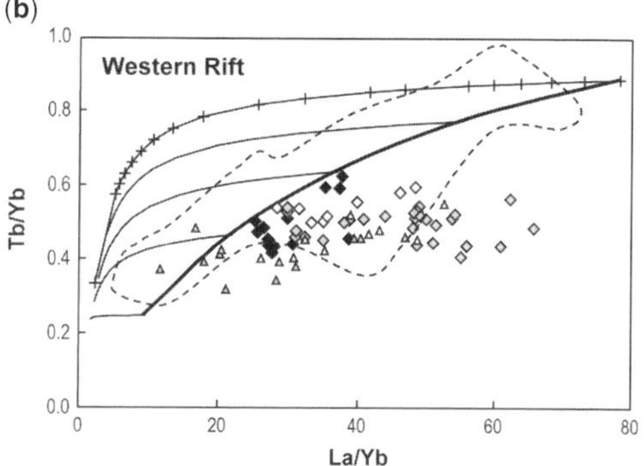


**Fig. 5.** Plots of La/Yb against Tb/Yb for basaltic rocks from (a) the Kenya rift and (b) the Western rift. The melting models in both diagrams are based on fractional melts derived from a fertile mantle source region with chondritic ratios of La/Yb and Tb/Yb. Partition coefficients are from Johnson (1998) and Blundy *et al.* (1998). Tick marks on uppermost curve illustrate melt fractions (labelled for 0, 1 and 10% melting). Lines labelled in italic numbers give the modal abundance of garnet in the source region (0, 2, 4, 6, 8%).
**(a)** Open circles represent analyses of basalts from the northern Kenya rift erupted through the Proterozoic mobile belt; and black circles, are basaltic rocks erupted through the craton and remobilized craton margin in the southern rift and northern Tanzania. Data from Macdonald *et al.* (2001) and Rogers (unpublished analyses).
**(b)** Grey triangles represent samples from the Kivu province (Furman & Graham 1999); white triangles, the Rungwe province (Furman 1995). Diamonds are samples from the Viunga province: black diamonds, K-basanites from Muhavura (Rogers *et al.* 1998); open and grey diamonds, K-basanites and more-evolved rocks from Karisimbi (Rogers *et al.* 1995).

The REE fractionation of basalts from the Kenya and Western rifts are summarized in Fig. 5. Superimposed on this diagram is a set of curves for melting a primitive mantle source region with 0–8% modal garnet. As the amount of garnet in the source region increases, so does the amount of variation in Tb/Yb, and so this diagram can be used to determine possible source mineralogy and depth of melting. The bold curved line at the high La/Yb extremity of the melting curves represents the locus of 0% melts, which is the theoretical maximum REE fractionation that can be produced in one melting stage from a fertile mantle source with unfractionated (i.e. chondritic) REE ratios. All samples that plot to the right of this bold line must have been derived from an enriched, probably lithospheric mantle source region whereas those to the left do not require an enriched source.

The significance of this line becomes apparent for the Western rift lavas (Fig. 5b), for which virtually all of the data plot to the right of the solid line, confirming the need for an enriched mantle source region as implied by the radiogenic isotope ratios. Note that this data field includes basalts from both the Rungwe and Kivu provinces, as well as the Virunga. In more detail, the trend within the data is shallow with limited variation in Tb/Yb for a large range of La/Yb, comparable with the shallow gradients of the melting curves with 2–4% garnet in the source. While some of this variation may be the result of clinopyroxene fractionation, overall it suggests that the melts were generated from a mantle source region with a low and constant garnet mode. Limited mineralogical variation in the mantle source is most consistent with a batch or equilibrium melting process, as is expected for melting the mantle lithosphere.

By contrast, the Kenya basalts define a much steeper trend in Fig. 5a that covers a greater range in source mineralogy, implying a source region with between 0 and 6% garnet, and melt fractions ranging between 2 and 10% (Macdonald *et al.* 2001). The most extreme REE fractionation is exhibited by those basalts erupted through the Tanzanian craton, although a few samples from all basement types plot beyond the limit of melting. However, most of the Kenya basalts plot to the left of the limiting line and so do not require an enriched source, although this does not preclude their origin from a source with fractionated REE. This behaviour is clearly quite distinct from that shown by the magmas of the Western rift and implies a marked difference in the depth and extent of the melting regime beneath Kenya, straddling the garnet–spinel lherzolite transition zone and extending possibly as shallow as the Moho (cf. Macdonald *et al.* 2001). Melting over such a depth range is only possible in a convecting system where the thermal gradient is adiabatic; in other words, where the asthenosphere extends to depths shallower than the normal base of the mantle lithosphere as has been revealed by the seismic profiles across the northern and central Kenya rift (Mechie *et al.* 1997).

Thus, the physical melting process is occurring in mantle that is currently imaged seismically as asthenosphere and so should have the compositional characteristics of ocean island basalts (OIB), yet the radiogenic isotope ratios suggest a lithospheric source region. Moreover, other aspects of the trace element compositions of Kenya basalts, in contrast to the REE, indicate a source region that is distinct from OIB. Macdonald *et al.* (2001) showed that they tend to be characterized by low Zr/Y ratios relative to Nb/Y compared with OIB. Moreover, recent studies based on more precise ICPMS analyses have shown that their Zr/Hf ratios are generally higher than many OIB with similar Zr contents (Le Roex *et al.* 2001; Spath *et al.* 2001; Furman *et al.* 2004). Such variations in the HFSE are unlikely to be a consequence of crustal contamination (cf. Macdonald *et al.* 2001), but are more easily attributed to a metasomatized mantle source probably located in the mantle lithosphere.

In summary, whereas plateau uplift is related to the dynamics of the underlying convecting mantle, the lithosphere exerts a controlling influence on the tectonics in both the Western and Kenya rifts. There is clear evidence for mantle with elevated temperatures, both from the presence of basaltic magma and from seismic attenuation (Weereratne *et al.* 2003). But the compositions of the basalts erupted within both rifts are strongly controlled by lithospheric structure and age. This is particularly apparent in the Western rift where all compositional parameters indicate a mantle lithosphere source region. In the Kenya rift, the involvement of the mantle lithosphere is most apparent in the Sr–Nd isotope systematics and particular aspects of the trace element contents. However, the REE reflect variations in the mineralogy of the mantle source that is most consistent with melting in the convecting asthenosphere, particularly in the central and northern sectors of the Kenya rift underlain by the Proterozoic mobile belt. This apparent contradiction will be discussed again below in relation to U-series isotopes.

Notwithstanding the nature of the melting process operating beneath the Kenya rift, the radiogenic isotopes reflect the age of the underlying mantle lithosphere and suggest that it represents the dominant source of basaltic magma. If the underlying mantle plume does contribute to Kenya basalts, then its isotopic composition will lie at the point common to basalts from both the mobile belt and cratonic regimes, as shown on Fig. 6. Alternatively, it may not contribute to surface magmatism in which case the basalts reveal little of the plume composition. However, a mantle plume remains a requirement as a source of heat to generate basaltic magma from the lithosphere (cf. Turner *et al.* 1996) and as a source of dynamic uplift to generate the plateau topography (Ebinger *et al.* 1989).

## Rifting and magmatism on the Ethiopian plateau

The Ethiopian rift and plateau appear superficially to be much simpler than East Africa, with one rift cutting across essentially isotropic lithosphere. However, beneath that deceptive simplicity lies a longer geological history related in part to the eruption of a thick sequence of flood basalts and subsequent continental break-up. Earliest volcanic activity is represented by a sequence of basalts in southern Ethiopia that erupted during the Eocene, between 35 and 45 Ma ago (Ebinger *et al.* 1993; George *et al.* 1998). These were followed by the eruption of the Ethiopian–Yemeni flood basalts, or traps, between 31 and 29 Ma ago (Baker *et al.* 1996; Hofmann *et al.* 1997; Rochette *et al.* 1998) that covered a large area of present-day Ethiopia, Eritrea and the southern Arabian peninsula with outliers occurring west and north into Sudan and east to the Somali border. Present-day volumes exceed 250 000 $km^3$ (Mohr 1983) and estimates of the original volume are as high as $10^6$ $km^3$ (Courtillot *et al.* 1999). Subsequent activity became more alkaline with the development of shield volcanoes overlying the flood basalt sequences (Kieffer *et al.* 2004) before becoming focused on the Afar depression. To the present-day volcanic and magmatic activity is largely confined to Afar, the tectonically active segments of the Ethiopian rift and the active spreading centres of the Red Sea and the Gulf of Aden (Wolfenden *et al.* 2004).

Compositionally, Ethiopian rift basalts are dominated by transitional tholeiites, falling close to the boundary between alkali basalts and olivine tholeiites (e.g. Hart *et al.* 1989; George & Rogers 2002). This characteristic extends to the Ethiopian flood basalts (Mohr 1983; Pik *et al.* 1998), making them unusual amongst flood basalts generally which are most frequently quartz-normative tholeiites (e.g. Turner & Hawkesworth 1995). Basalts from the Main Ethiopian Rift tend to be more alkalic and, as with the Kenya rift, retain a dominance of sodium over potassium (Hart *et al.* 1989; George & Rogers 2002).

### *Basalts from the Afar depression*

The Miocene to Recent basaltic activity associated with the development of the Afar depression shows a much simpler compositional pattern than that from the Kenya rift, reflecting both the greater homogeneity of the basement lithosphere and the developing tectonic regime (Vidal *et al.* 1991; Deniel *et al.* 1994). Figure 6 illustrates the Nd and Sr isotope composition of basalts from Afar and the Red Sea and Gulf of Aden spreading centres. The latter, as might be expected, are dominated by high $\varepsilon_{Nd}$ and low $^{87}Sr/^{86}Sr$, typical of MORB worldwide. Basalts from the Afar, by contrast, have lower $\varepsilon_{Nd}$ and higher $^{87}Sr/^{86}Sr$ ratios than MORB, and extend to values beyond that for the bulk Earth. Significantly, there is a secular change in the isotope characteristics, with the older Miocene lavas plotting at the most extreme isotope values, and the most recent trending towards a composition with $\varepsilon_{Nd}$ *c.* +6 and $^{87}Sr/^{86}Sr$ 0.7035 (Deniel *et al.* 1994). These most recent lavas are characterized by $^3He/^4He$ ratios up to 17 × atmospheric ($R/R_a = 17$) (Marty *et al.* 1996; Scarsi & Craig 1996) considerably greater than the value of 8 typical of most MORB, and in that respect have

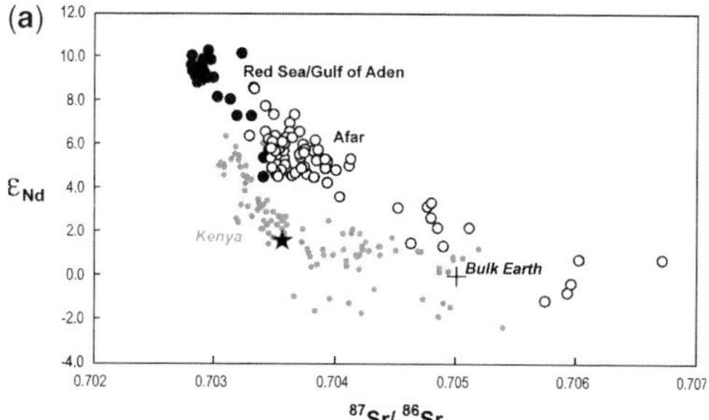


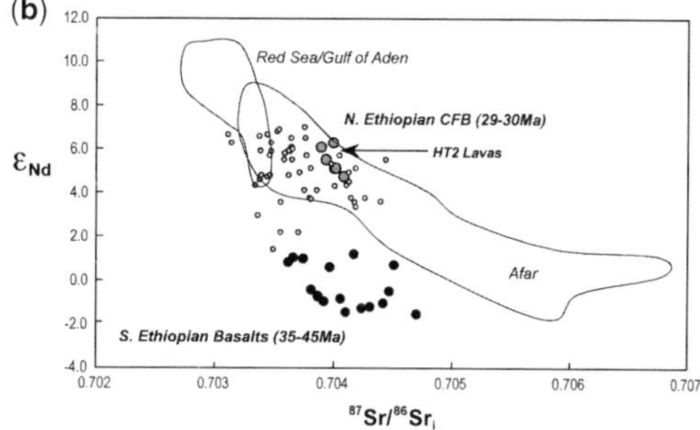


**Fig. 6.** **(a)** Sr and Nd isotope data from the Afar depression (Deniel *et al.* 1994; Vidal *et al.* 1991), (open circles), the Gulf of Aden (Schilling *et al.* 1992) and the Red Sea (Eissen *et al.* 1989) (black circles) compared with data from the Kenya rift (small grey circles). Note the difference between the probable composition of the Afar mantle plume and the whole array of data from Kenya. A possible composition of the East African mantle plume is marked at the common point between the 'mobile belt' and 'craton and margin' fields for the Kenya rift.
**(b)** Sr and Nd isotope data for northern Ethiopian flood basalts and shield volcanoes (Pik *et al.* 1998; Kieffer *et al.* 2004) compared with similar data for the southern Ethiopian basalts (George & Rogers 2002). Fields for Afar and Red Sea/Gulf of Aden based on (a). Note the similarity between the southern Ethiopian basalts and the data from Kenya.

much in common with plume-related OIB (e.g. Hannan & Graham 1996; Ellam & Stuart 2005). Hence the compositional variations within the Afar depression have been interpreted to reflect the waning influence of the lithosphere and the increasing influence of the underlying Afar mantle plume following extension as Arabia drifted slowly away from Africa (Vidal *et al.* 1991; Deniel *et al.* 1994).

The secular trends identified by Vidal *et al.* (1991) and Deniel *et al.* (1994) also showed an increase in the $^{206}Pb/^{204}Pb$ isotope ratio, which lead them to conclude that the Afar plume has a so-called HIMU (or high-$\mu$) signature. However, more recent analysis of basalts from across the Arabian plate has shown that the HIMU signature extends beyond the topographic influence of the plume and beyond regions characterized by elevated $^{3}He/^{4}He$ ratios (Bertrand *et al.* 2003; Pik *et al.* in press). Moreover, one of the characteristics of HIMU mantle plumes worldwide is that they have low $^{3}He/^{4}He$ ratios, not high. The alternative suggestion is that the HIMU characteristics are associated with the mantle lithosphere and that the plume is rarely sampled uncontaminated with lithospheric mantle by modern basaltic activity. The preferred composition for the Afar plume is that given by Pik *et al.* (1999) that relates to the composition of magnesian high-titanium basalts from the Ethiopian flood basalts.

## *The Ethiopian flood basalts (traps)*

The 29–30 Ma Ethiopian traps have been subdivided into three distinct magma types, the so-called low titanium (LT) and two high titanium groups, designated HT1 & HT2 (Pik *et al.* 1998). These magma groups are defined on the basis of their variation of $TiO_2$ with MgO and systematic variations in their immobile trace element contents. While variations within magma groups can be related to magmatic differentiation, both the LT and HT2 magma types include lavas with relatively primitive compositions such that variations between magma groups reflect either differences in source composition, mineralogy or both. In the case of the LT magmas, these may represent the mantle lithosphere or crustal contamination (Pik *et al.* 1999), although a more recent study invokes heterogeneity in the underlying mantle plume or 'Ethiopian superswell' (Kieffer *et al.* 2004).

Notwithstanding the different interpretations of the geochemical signals in the LT magmas, the HT2 magma group includes picritic and ankaramitic compositions, in addition to basalts, that are isotopically similar to modern-day Afar basalts with $\varepsilon_{Nd}$ *c.* +6 and elevated $^{3}He/^{4}He$ ratios (18 $R/R_a$) (Marty *et al.* 1996; Pik *et al.* 1999). Their REE are strongly fractionated and they have unusually low $Al_2O_3$ contents, reflecting the influence of garnet in their source region and implying melting at pressures >3 GPa and possibly as high as 4–5 GPa (120–150 km) (Pik *et al.* 1999).

The REE fractionation within the Ethiopian flood basalts is summarized in Fig. 7a. The HT2 magmas plot at high Tb/Yb ratios even though their La/Yb ratios are not as extreme as those of the Kenya basalts. The LT basalts by contrast plot at much lower Tb/Yb ratios and concomitantly low La/Yb, emphasizing the lower content of garnet in their source region and hence a lower pressure of melting. As with the Kenya basalts, the broad linear trend defined by the Ethiopian flood basalts in Fig. 6a reflects the development of a melting regime that ranges from depths within the garnet stability field to possibly as shallow as the Moho. The data plot well within the field of melts that can be derived from a fertile mantle source region and so none of the magmas require an enriched source. This supports the contention that all are plume-derived but that the LT magmas owe their characteristic incompatible element characteristics to crustal contamination (Pik *et al.* 1999).

Melting at the depths indicated by the HT2 magmas requires elevated mantle potential temperatures, probably in excess of 1580 °C, but more significantly, the restricted melting depths also imply melting prior to lithospheric thinning. This latter observation is consistent with the tectonic record of thinning along the flanks of the Gulf of Aden and the Red Sea where fission track evidence points to a major phase of extension associated with magmatic margins between 25 and 20 Ma, post-dating CFB eruption by at least 5 Ma (Omar & Steckler 1995; Menzies *et al.* 1997). The high eruption and source temperatures and melting depths of the HT2 magmas suggest that these are the earliest unambiguous manifestations of the Afar mantle plume (e.g. Pik *et al.* 1999). Indeed, the HT2 magmas may even result from melting of the plume head during the initial stages of plume development and probably represent the true isotopic composition of the Afar plume, relatively unaffected by lithosphere interaction.

## *Southern Ethiopian basalts*

Despite the high source temperatures and great melting depths of the HT2 magmas, they are not the earliest volcanic deposits in the Ethiopian province. These are represented by the Eocene basalts and evolved derivatives in southern Ethiopia (Ebinger *et al.* 1993), with ages between 45 and 35 Ma, and so predating the Ethiopian traps by as much as 15 Ma. They have been divided into two magma groups, the older Amaro basalts being overlain by the Gamo basalts. The latter have relatively

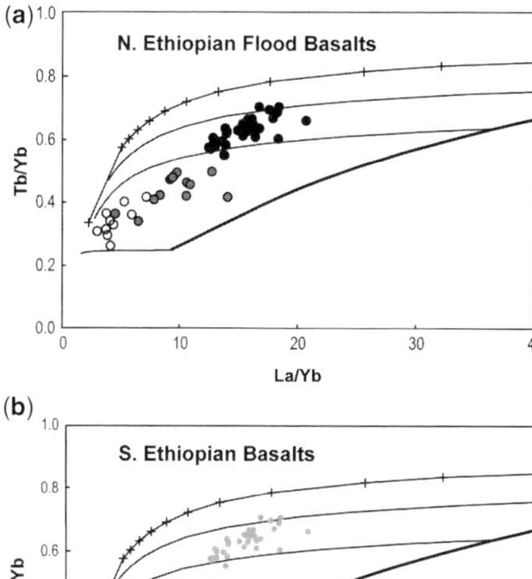


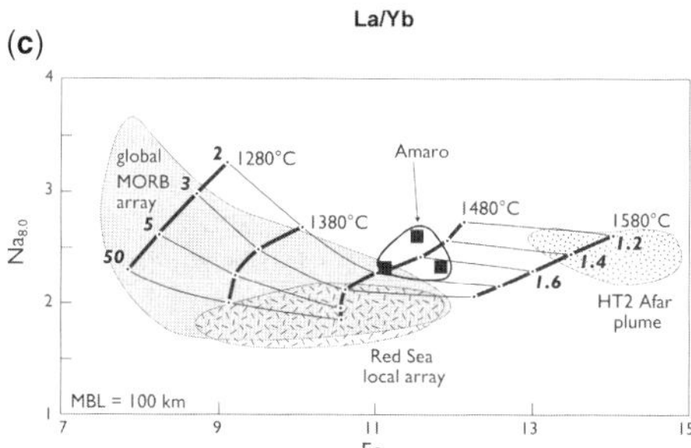


**Fig. 7.** (**a**) REE fractionation in the Oligocene (29–30 Ma) northern Ethiopian flood basalts. Model curves as in Figure 5 but note expanded La/Yb scale. Black circles = HT2 lavas; grey circles = HT1; open circles = LT lavas. Data from Pik *et al.* (1998) and Rogers (unpub. analyses). (**b**) REE fractionation in basalts from southern Ethiopia. Black circles = Amaro basalts; grey circles = Gamo basalts (both Eocen–Oligocene, 45–35 Ma); open circles = Miocene (19–11 Ma) Getra-Kele and Plio-Pleistocene (<2 Ma) Tosa-Sucha basalts. The latter two are associated with the development of the Main Ethiopian Rift, whereas the Amaro and Gamo basalts predate rifting in Ethiopia. (**c**) Comparison of the fractionation-corrected compositions of the Amaro basalts with the HT2 basalts from northern Ethiopia, MORB and basalts from the Red Sea. Superimposed are point- and depth-averaged melt compositions (McKenzie & Bickle 1988) for mantle of varying potential temperature and degrees of extension. Note that the HT2 basalts require derivation from elevated $T_p$ (~1580 °C) at modest amounts of extension ($\beta$ = 1.2–1.4) whereas the Amaro basalts were derived from a cooler mantle source ($T_p$ ~ 1480 °C). (After George & Rogers 2002.)

evolved compositions and appear to have developed their transitional tholeiitic characteristics from the more primitive Amaro basalts as a result of pyroxene fractionation in the mid-crust (George & Rogers 2002). The Amaro basalts are more primitive with a more pronounced tholeiitic affinity, and their major and trace element compositions reflect derivation from a mantle source region with elevated potential temperatures but lower than those typical of the HT2 from the Ethiopian traps. The REE are also less fractionated than those typical of the HT2, suggesting shallower melting depths as summarized in Fig. 7b. Lower mantle potential temperatures are also indicated by the major element compositions of the most primitive Amaro basalts, compared with the HT2 as shown in Fig. 7c. Superimposed on this diagram are point- and depth-averaged melts calculated using the parameterization of Watson & McKenzie (1991). These data show that while the Amaro basalts can be derived from a mantle source with a potential temperature of 1480 °C with moderate extension ($\beta = 1.8 - 1.6$), the HT2 magmas require considerably higher potential temperatures but with less extension ($\beta = 1.2 - 1.4$) (George & Rogers 2002).

The shallower melting regime of the Amaro basalts suggests derivation from beneath thinned continental lithosphere, yet extension in the southern Ethiopian rift began during the Miocene, at about 18 Ma (WoldeGabriel *et al.* 1990), 15–25 Ma after the their eruption. Extension contemporaneous with the Amaro–Gamo basalts occurs to the west and was responsible for the development of, amongst others, the Kaisut–Lokichar basins in NW Kenya and southern Sudan (Bosworth & Morley 1994; Hendrie *et al.* 1994), and the reactivation of the Pibor rift in southern Sudan (Ebinger & Ibrahim 1994). Moreover, the stratigraphic equivalents of the Amaro and Gamo basalts are thicker in the Omo region in SW Ethiopia (Davidson 1983), further suggesting that these Palaeogene rift systems were the main locus for Eocene mafic magmatism.

Geochemically, the Amaro and Gamo basalts were originally attributed to an early phase of Afar-related volcanism (Stewart & Rogers 1996) and their early eruption has provided a key element of the single plume model for the evolution of East African magmatism (Ebinger & Sleep 1998). However, the more detailed assessment of their compositional features summarized above (George & Rogers 2002) has shown that they have characteristics that are distinct from both recent Afar basalts and the HT2 magmas, and more similar to basalts from Kenya. Both the major and the trace element contents of the Amaro basalts imply marked petrological differences compared with the HT2 magmas from the northern Ethiopian CFB. Differences are also apparent in the Sr and Nd isotope systematics: the southern Ethiopian basalts have lower $\varepsilon_{Nd}$ values (<2) despite similar $^{87}Sr/^{86}Sr$ ratios to the northern Ethiopian traps (0.7035), compositions which overlap those of modern-day Kenya basalts (Fig. 6b). Together, these contrasts strongly imply that the Eocene basalts from southern Ethiopia were derived from a source that was distinct from the HT2 source region, and if the Amaro source is sub-lithospheric this further implies a different mantle plume. The isotopic affinity with the Kenya basalts also fits well with their geochronology in relation to the evolution of the Kenya rift (see below).

### *Basalts from the Ethiopian rift*

The magmatic evolution of the Ethiopian Rift is less well defined than that of the Kenya rift. In the south, after the Amaro and Gamo basalts, there is a hiatus until alkaline magmatism associated with extension across the Main Ethiopian Rift (MER) begins at 19 Ma, continuing to 11 Ma (Ebinger *et al.* 1993; George *et al.* 1998). These so-called Getra–Kele basalts were derived from an incompatible element-enriched garnet-free source region (George & Rogers 2002). The enriched nature of the source is illustrated by the REE fractionation (Fig. 7b). Both the Getra–Kele and the Plio-Pleistocene basalts from Tosa–Sucha (George & Rogers 1999) plot close to or outside the limit of melts that can be derived from primitive mantle, and their low Tb/Yb ratios indicate a garnet-poor source region. Once again, they appear to have been derived from the mantle lithosphere although they have an ambiguous isotopic signal with $\varepsilon_{Nd}$ values intermediate between the Afar plume and that proposed for the Kenya plume. These isotopic characteristics are also apparent in the basalts from other sectors of the rift (e.g. Hart *et al.* 1989) although their REE fractionation is less extreme (Rooney *et al.* 2005), and it is clear that further investigations into the igneous geochemistry of Ethiopian rift basalts are required before their significance can be deduced. The most recent volcanic activity is located in a series of en echelon volcanic segments (sometimes referred to as the Wonji fault belt) comprising extensive mafic dyke injection (Wolfenden *et al.* 2004) and central volcanoes characterized by evolved alkaline magmas (e.g. Gasparon *et al.* 1993; Peccerillo *et al.* 2003).

## U-series isotope analyses of basalts from the Kenya rift and Afar

Critical aspects of basalts from the African rift discussed above are contradictory. For example, in the Kenya rift and possibly in the Ethiopian rift, selected trace element (e.g. Zr/Hf ratios) and isotope characteristics indicate a lithospheric source region. By contrast, variations in the REE indicate melting at a range of depths, consistent with geophysical studies that show the presence of asthenospheric partial molten mantle as shallow as the base of the crust in parts of northern Kenya. Analyses of U-series isotopes in recent mantle-derived magmas allows the investigation of different aspects of the melting regime and comparison with physical models of melt generation and segregation (e.g. Bourdon & Sims 2003; Lundstrom 2003). Such studies are as yet in their infancy in the African rift but of particular relevance here is their use in relating geochemical variations to the physical nature of the melting regime beneath the rift axis.

Early studies using $\alpha$-counting techniques (Black *et al.* 1997, 1998) revealed that Kenya basalts are characterized by an excess of $^{230}Th$ over $^{238}U$, and this has been verified using mass spectrometric analyses. The results of more recent studies are summarized in Fig. 8 where they are compared with similar data from ocean island basalts which are generally considered to be plume-related. The available data from Afar (Vigier *et al.* 1999; Rogers *et al.* unpublished data) plot in a similar position to many OIB. In particular, they define a near-vertical trend within which the U/Th elemental ratio (expressed as the ($^{238}U/^{232}Th$) activity ratio) varies little, whereas the ($^{230}Th/^{232}Th$) ratio deviates from the equiline, resulting in a $^{230}Th$ excess of up to 40%. Young (<10 ka) basaltic rocks from the Kenya rift

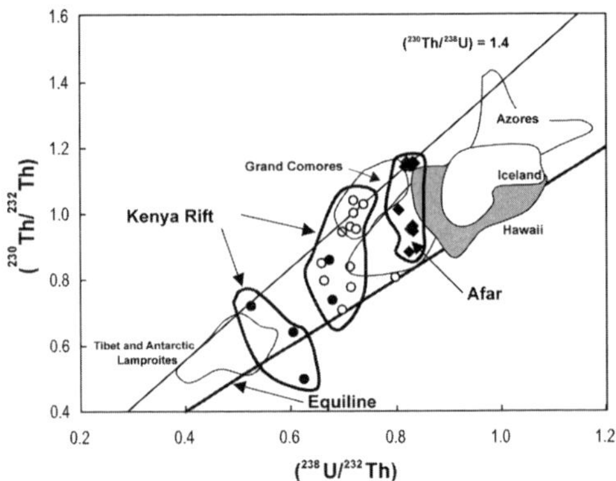


**Fig. 8.** U-series equiline diagram comparing analyses from the Kenya rift (Rogers *et al.* unpub. data) and Afar (Vigier *et al.* 1999; Rogers *et al.* unpub. data). Note that the Kenya data plot at lower ($^{238}U/^{232}Th$) ratios than Afar and the majority of OIB. The near-vertical arrays in both the Afar and Kenya data imply dynamic melting in the convecting mantle beneath both regions.

also plot above or close to the equiline but define two fields, one at low ($^{238}U/^{232}Th$) ~0.55 – 0.6, which exhibits limited $^{230}Th$ excess and a second with higher ($^{238}U/^{232}Th$) ~0.7 – 0.75 and more variable $^{230}Th$ excess, ranging from 0–46%.

The results emphasize the difference in U/Th ratios between Kenya basalts and OIB, in that U/Th ratios are generally lower than most OIB. Th is more incompatible in the mantle than U and so a lower U/Th ratio in the source region of the Kenya basalts is consistent with a trace element-enriched source region. Note also that the three samples that plot at the lowest U/Th ratios are comparable with lithosphere-derived lamproites from Tibet and Antarctica, further emphasizing the link with the mantle lithosphere. Finally, comparison with results from Ardoukoba in the Afar depression (Vigier *et al.* 1999) reveals the distinctive higher U/Th ratios of the latter relative to Kenya and more comparable with OIB. This is further evidence in favour of the distinctiveness of the mantle source regions tapped by basalts from the Kenya rift and those from at least the Afar segment of the Ethiopian rift.

The most significant feature of these data is the variable ($^{230}Th/^{232}Th$) at almost constant ($^{238}U/^{232}Th$) in both the Kenya basalts with the higher ($^{238}U/^{232}Th$) ratios (~0.7) and the Afar basalts. Given that the samples are all historic or recent and that crustal residence times have been negligible, such vertical arrays on the equiline diagram are indicative of a dynamic melting regime in convecting mantle. The 40% $^{230}Th$ excess in the basalts from both regions is the maximum that can be produced during melting of a garnet-bearing mantle source region (Bourdon & Sims 2003) and this is consistent with the variations in the REE described above. However, it is in conflict with the conclusion from radiogenic isotopes and aspects of the trace element characteristics of the Kenya basalts that they have compositions largely controlled by the mantle lithosphere, because the lithosphere is rigid and does not convect, and hence cannot easily develop a dynamic melting regime.

One possible mechanism that may reconcile this apparent contradiction involves the thermal erosion of the lithosphere, as has been suggested to account for the isotope characteristics of the Kivu volcanic rocks of the Western rift and basanites from the Huri Hills in Kenya (Furman & Graham 1999). Beneath the Kenya rift, the total lithospheric thinning revealed by seismic investigations is greater than that determined from upper crustal extension alone. The lithosphere underlying the rift flanks has a thickness of >120 km but beneath the axis the seismic lithosphere has been almost completely replaced by mantle with low seismic velocities. Part of this reduction in seismic velocity is related to the local presence of up to 6% melt, but significantly at least half of the seismic attenuation is due to temperature increases in the solid mantle (Green *et al.* 1991). As a consequence of this heating, the lithosphere has become more ductile and involved in convective movement, thus contributing to lithospheric thinning and allowing melts with lithospheric characteristics to be generated in a dynamic melting regime.

## Discussion

### *One plume or more beneath East Africa?*

The involvement of mantle plumes in the evolution of the present topography and tectonics of East Africa has led to competing models as to the number of distinct plumes that may lie beneath. At one extreme, Ebinger & Sleep (1999) suggest that magmatism across the whole of East and North Africa can be explained as a consequence of the impact of a single plume beneath southern Ethiopia 45 Ma ago. After impact, mobile plume mantle spread beneath the lithosphere, channelled by the inverse topography at its base, generated during Mesozoic and Tertiary extensional events. By contrast, George *et al.* (1998) and Rogers *et al.* (2000) have presented geochronological and geochemical evidence that suggests the migration and composition of basalts from southern Ethiopia and Kenya represent the products of one mantle plume whereas the basalts from the northern Ethiopian plateau and the Afar depression represent the products of another.

A single plume model is difficult to reconcile with geochemical data, especially the He, Sr and Nd isotopes. Comparing the Nd and Sr isotopic compilations from the Kenya rift and Afar for the most recent basaltic lavas (<12 Ma) emphasizes the distinct features of the two systems (Fig. 6). Significantly, there is little compositional overlap between the two, Kenya basalts being displaced to lower values of $\varepsilon_{Nd}$ for a given value of $^{87}Sr/^{86}Sr$. The Red Sea and Afar trends focus on isotopic values of $\varepsilon_{Nd} > +6$ and $^{87}Sr/^{86}Sr \approx 0.7035$, those for Kenya have a common composition at $\varepsilon_{Nd} \approx 1 - 2$ and $^{87}Sr/^{86}Sr < 0.7035$ (Fig. 6). While the latter may or may not represent the composition of the sub-lithospheric mantle beneath the East African plateau, there is clearly no contribution to the Kenya basalts from material with an Afar-like composition.

A recent review of the distribution of elevated $^{3}He/^{4}He$ ratios in both ground waters and basaltic rocks (Pik *et al.* 2006) further confirms that this signal is present in the East African rift only in samples derived from within the geographical confines of the Ethiopian plateau. Elevated $^3He/^4He$ ratios have yet to be confirmed beyond the limits of the plateau, suggesting that the compositional effects of the Afar plume do not extend beyond the limit of its physical influence. When high $^3He/^4He$ is present in basaltic rocks, it is a particularly unambiguous indicator of a mantle plume source of probable deep mantle provenance (e.g. Ellam & Stuart 2005; Hannan & Graham 1996). The absence of high $^3He/^4He$ ratios away from the topographic influence of the Afar plume (i.e. the Ethiopian plateau) is strong evidence against the spread of Afar plume material beyond these geographical confines (Pik *et al.* 2006). These observations are strong evidence that the two plateaus are supported by upwelling mantle with distinct compositions, even though the composition of that beneath East Africa cannot be defined with certainty.

Encouraged by the single plume model, Furman *et al.* (2004) related recent basalts from the Turkana region of northern Kenya to a mantle plume source region similar to Afar, chiefly on account of their 'high-$\mu$' Pb isotope ratios. However, as discussed above, the so-called 'high-$\mu$' Pb isotope characteristics are found throughout the region and are not confined to the Ethiopian plateau, and an alternative view places this material within the mantle lithosphere (Bertrand *et al.* 2003).

Moreover, other compositional characteristics of the Turkana basalts are inconsistent with plume derivation. For example, they have high $SiO_2$ and low $FeO_t$ contents, and lack a garnet signature in their REE fractionation, implying melt generation at pressures of 1.5–2.0 GPa (Furman *et al.* 2004), well within the depth limit of the extended mantle lithosphere. The Turkana depression shows the least topographic elevation of any magmatic region within the African rift (*s.l.*) south of Afar, coupled with the least negative regional Bouguer gravity anomaly (Tessema & Antoine 2004), and therefore shows the least dynamic uplift of any part of the East African rift system. In addition, the depth to the top of the melting column estimated from seismic velocity profiles is also deeper beneath Turkana and the northern Kenya rift than it is further south, which suggests that mantle potential temperatures are lower. Thus, it is unlikely that Turkana is underlain by a physical mantle plume, although this does not preclude the possible incorporation of plume-derived materials in the mantle beneath Turkana (cf. Furman *et al.* 2004).

## *Magma migration and plate motions*

Tracing back through the development of the two systems, it is clear that the most voluminous magmas generated at the greatest depth and the highest temperatures are found in the northern Ethiopian flood basalts, implying that these represent the earliest manifestation of the Afar mantle plume. The Eocene Amaro basalts of southern Ethiopia, by contrast, were produced at shallower depths and lower temperatures and have isotopic characteristics that are more comparable with those of more recent basalts erupted in the Kenya rift. Moreover, they define the oldest end of a time sequence related to the onset of magmatism that migrates from southern Ethiopia in the Eocene, through Turkana in the Oligocene and southwards across the Kenya dome and into northern Tanzania during the Miocene (Fig. 9) (George *et al.* 1998). Note that this age progression refers to the onset of magmatism, much of the length of this trail being characterized by magmatism and geothermal activity to the present day.

The southward migration of magmatism corresponds to ~10° latitude in 40 Ma and is equivalent to an average velocity of about 25 mm $a^{-1}$. This rate compares favourably with current and previous African plate vectors (O'Connor *et al.* 1999) which suggest counterclockwise rotation about a pole close to the Canary Islands with a decrease in velocity from 30 mm $a^{-1}$ to 20 mm $a^{-1}$ occurring 19 Ma ago. These plate motions result in a northward vector for the African plate in the vicinity of the African rift, consistent with the apparent southward migration of magmatism since 45 Ma.

The alternative one- and two-plume models for the African rift have recently been investigated numerically by Lin *et al.* (2005). Their results reveal that the distribution and age progression of magmatism in both the Kenya and Ethiopian rifts are reproduced most closely in a model with two mantle plumes. The distribution of magma predominantly along the Kenya rift is consistent with a plume currently beneath the Tanzanian craton, in

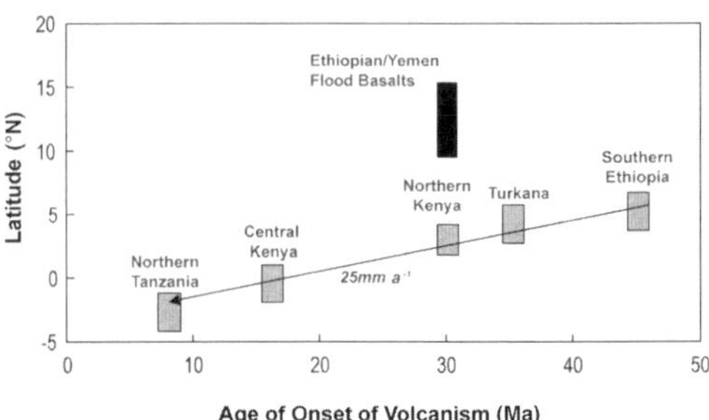


**Fig. 9.** Age progression in the onset of magmatism from southern Ethiopia south through Kenya into northern Tanzania at a rate of *c.* 25 mm $a^{-1}$ (after George *et al.* 1998). The onset of magmatism in the Ethiopian/Yemen flood basalts does not lie on this trend.

a position similar to that imaged seismically (Weeraratne *et al.* 2003), and requires the craton to approach the plume from the SW, consistent with plate motions. They also show that a stagnation streamline develops between the two plumes, allowing for the generation of basalts with distinct compositions and a limited zone between where hybrid magmas may occur (Lin *et al.* 2005). This may explain the ambiguous characteristics of the Turkana basalts, erupted in the extended region between the two plumes. While these models do not represent an independent test of the 'two-plumes' concept, they serve to demonstrate that a scenario involving two mantle plumes in positions that explain the present-day dynamic topography is consistent with our knowledge of the evolution of the system as a whole and our current understanding of mantle dynamics.

The change in plate velocity vector at 19 Ma reported by O'Connor *et al.* (1999) occurred shortly after major rifting along the flanks of the present-day Red Sea as reflected in fission track dating and other geological evidence. Prior to this time, the African and Arabian plates were moving as a single unit, largely driven by slab pull as the remnants of the Tethys Ocean were subducted beneath the Eurasian plate. Extension across the Red Sea, possibly aided by the earlier emplacement of the Afar mantle plume (Bellahsen *et al.* 2003), cut this driving force from the African plate, which consequently slowed and changed direction.

A major consequence of this important change in the geodynamics of the African plate was the change in the orientation of the major extensional stress. The current stress map of the African plate (Coblenz & Sandiford 1994) shows that the continent is everywhere under extension and that stress is at a maximum over the Ethiopian and East African plateaus, where the topography attains its maximum altitude. The present-day stress orientation across the Ethiopian and East African plateaus is roughly east–west, hence the orientation of the Neogene rift systems, albeit partly controlled by basement fabrics (Smith & Moseley 1993). Given the plate tectonic history of the African plate, it is likely that this regime has dominated since Arabia finally split away from Africa at *c.* 20 Ma (Omar & Steckler 1995; Menzies *et al.* 1997).

Prior to this event, the stress regime is more speculative, but considering that plate motions would then have been dominated by slab pull to the NE, it is likely that the main extensional stress would be orientated in a NE–SW direction resulting in NW–SE-orientated extensional basins. Indeed, this is the general observation and basins ranging in age from the Cretaceous (Anza graben) through the early Tertiary (Bosworth 1992; Bosworth & Morley 1994) and including the Red Sea itself, are oriented roughly in a NW–SE direction, but that during the late Palaeogene and throughout the Neogene, basins become oriented roughly north–south (Kenya rift) or NE–SW (MER) (Ebinger & Ibrahim 1994), corresponding to the modern-day stress orientation.

Thus a pattern emerges of extensional basins controlled in part by basement structure, but developing in response to the stress regime across the whole plate which is in turn a product of far-field, plate-boundary forces and topography. Moreover, given that the major topographic features of continental Africa are controlled by mantle convection and, in particular, mantle plumes, it is inevitable that many of these extensional basins will be accompanied by basaltic magmatism. In any continental environment, extension is an essential element for the development of basaltic volcanism which will consequently be focused on contemporary rift zones. The southward migration of magmatism from southern Ethiopia through Kenya to Tanzania followed the southward propagation of the rift system which was in turn caused by the southward migration of dynamic uplift caused by the underlying mantle plume. The main extensional driving forces in this part of the African plate are derived from variations in the gravitational potential energy across the plate. This contrasts with the situation during the break-up of the Red Sea and Gulf of Aden where extension once initiated, continued to develop into first broad rifted basins and then the young oceanic basins we now see. These well-developed extensional basins subsequently acted and continue to act as a focus for magmatic activity which consequently does not migrate in response to the movement of the overriding plate but exploits the lithospheric weaknesses generated at the extensional plate boundaries.

## Conclusions

Combining the evidence from basalt petrology and geochemistry with the large-scale topographic and geophysical features of the East African rift system leads to a system that is driven by plume-related dynamic uplift, the intraplate stress field and plate motions. The dynamic topography contributes significantly to the extensional force that drives rifting across both the East African and Ethiopian rifts, but that force also includes a contribution from far-field forces generated within the African plate. The association of magmatism with rifting in Africa is a direct consequence of the link between topography and extensional stress, simply because the greatest elevations are the result of dynamic support by the underlying mantle plume. In other words, extension only occurs where the lithosphere is underlain by a plume and that is also the region where magmatic activity will occur.

Magma migration from southern Ethiopia to northern Tanzania is a consequence of plate motion over the East African plume while magmatism in northern Ethiopia resulted initially from the impact of the Afar plume *c.* 30 Ma ago. Subsequently, as Arabia split away from the African plate, a rift formed along the line of the Red Sea and Gulf of Aden, prior to full separation and the onset of sea-floor spreading. Somewhat more speculatively, the change in orientation of extensional basins from NW–SE throughout the Cretaceous and Palaeogene, to broadly north–south during the latest Palaeogene throughout the Neogene and into Recent times, may be due to the isolation of the African plate from slab pull once Arabia separated fully and became an independent plate.

Mafic magma throughout the Kenya rift, Western rift and probably in large parts of the Ethiopian rift, were and continue to be derived from shallow and compositionally distinctive source regions within the mantle lithosphere. Melting in the mantle beneath the Kenya rift is achieved by thermal erosion of solid lithospheric material into the convecting asthenosphere, with melt generated as the material heats up and becomes incorporated in convective movements beneath the rift axis.

Notwithstanding the above description, not all of the evidence is consistent with the 'two-plume' model and not the least of these problems concerns the migration of magmatism and extension northwards from southern Ethiopia towards Afar in the development of the Main Ethiopian Rift (Wolfenden *et al.* 2004). This is in the opposite direction from that predicted by plate motions over a stationary hot spot. Moreover, present-day Nd and Sr isotope characteristics of Ethiopian rift basalts are distinct from those considered diagnostic of the Afar plume, yet they are characterized by Afar-like $^3He/^4He$ isotope ratios. Clearly, the details of the evolution of the Ethiopian rift and its magmatic history require further investigation and will no doubt provide further tests of the current model. While such topics represent targets for future research, the 'two-plume' model accounts for many of the large-scale geological, geophysical and petrological observations throughout the East African rift system and remains a viable explanation for the evolution and development of the system throughout the Palaeogene and Neogene.

I would like to thank numerous colleagues and friends over the years for their discussions of various ideas concerning the evolution of the African rift, in particular Cindy Ebinger, Rhiannon George, David Graham, Gezahegn Yirgu and Ray Macdonald. We may not agree all the time but the discussions are always fun. I also acknowledge and thank Andy Saunders, Raphael Pik and Tanya Furman for their constructively critical reviews that greatly helped to improve a hastily drafted initial manuscript. Finally, none of this would have been possible without the help of colleagues and friends in Ethiopia who have facilitated research in the field and who arranged the 2004 conference so successfully.

## References

Ashwal, L.D. & Burke, K. 1989. African lithospheric structure, volcanism and topography. *Earth and Planetary Science Letters*, **96**, 8–14.

Baker, B.H. 1987. Outline of the petrology of the Kenya rift alkaline province. *In*: Fitton, J.G. & Upton, B.G.J. (eds) *Alkaline Igneous Rocks*. Geological Society, London, Special Publications, **30**, 293–311.

Baker, J., Snee, L. & Menzies, M.A. 1996. A brief Oligocene period of flood basalt volcanism in Yemen; implications for the duration and rate of continental flood volcanism in the Afro-Arabian triple junction. *Earth and Planetary Science Letters*, **138**, 39–55.

Bellahsen, N., Faccenna, C., Funiciello, F., Daniel, J.M. & Jolivet, L. 2003. Why did Arabia separate from Africa? Insights from 3-D laboratory experiments. *Earth and Planetary Science Letters*, **216**, 365–381.

Bertrand, H., Chazot, G., Blichert-Toft, J. & Thoral, S. 2003. Implications of widespread high-$\mu$ volcanism on the Arabian Plate for Afar mantle plume and lithosphere composition. *Chemical Geology*, **198**, 47–61.

Black, S., Macdonald, R. & Kelly, M.R. 1997. Crustal origin for peralkaline rhyolites from Kenya: evidence from U-series disequilibria and Th isotopes. *Journal of Petrology*, **38**, 277–297.

Black, S., Macdonald, R., Barreiro, B., Dunkley, P. & Smith, M. 1998. Open system alkaline magmatism in northern Kenya: evidence from U-series disequilibria and radiogenic isotopes. *Contributions to Mineralogy and Petrology*, **131**, 365–378.

Blundy, J.D., Robinson, J.A.C. & Wood, B.J. 1998. Heavy REE are compatible in clinopyroxene on the spinel lherzolite solidus. *Earth and Planetary Science Letters*, **160**, 493–504.

Bosworth, W. 1992. Mesozoic and early Tertiary rift tectonics in East Africa. *Tectonophysics*, **209**, 115–137.

Bosworth, W. & Morley, C.K. 1994. Structural and stratigraphic evolution of the Anza rift, Kenya. *Tectonophysics*, **263**, 93–115.

Bourdon B. & Sims, K.W.W. 2003. U-series constraints on intraplate basaltic magmatism. *In*: Bourdon, B., Henderson, G.M., Lundstrom, C.C. & Turner, C.P. (eds) Uranium-series geochemistry. *Reviews in Mineralogy and Geochemistry*, **52**, 215–254.

BUCK, W.R. 1991. Modes of lithospheric extension. *Journal of Geophysical Research (Solid Earth)*, **96**, 20161–20178.

COBLENZ, D.D. & SANDIFORD, M. 1994. Tectonic stresses in the African plate: constraints on the ambient lithospheric stress state. *Geology*, **22**, 831–834.

COURTILLOT, V., JAUPART, C., MANIGHETTI, I., TAPPONIER, P. & BESSE, J. 1999. On causal links between flood basalts and continental break-up. *Earth and Planetary Science Letters*, **166**, 177–195.

DAVIDSON, A. 1983. The Omo River Project: reconnaissance geology and geochemistry of parts of Illubabor, Kefa, Gemu and Sidamo, Ethiopia. *Bulletin Ethiopian Institute of Geological Surveys, Ministry of Mines and Energy*, **2**.

DAVIES, G.R. & LLOYD, F.E. 1989. Pb–Sr–Nd isotope and trace element data bearing on the origin of the potassic subcontinental lithosphere beneath south-west Uganda. *Geological Society of Australia Special Publications*, **14**, 784–794.

DENIEL, C., VIDAL, PH., COULON, C., VELLUTINI, P.-J. & PIGUET, P. 1994. Temporal evolution of mantle sources during continental rifting: the volcanism of Djibouti (Afar). *Journal of Geophysical Research*, **99**, 2853–2869.

EBINGER, C.J. & IBRAHIM, A. 1994. Multiple episodes of rifting in Central and East Africa; a re-evaluation of gravity data. *Geologische Rundschau*, **83**, 689–702.

EBINGER, C.J. & SLEEP, N.H. 1998. Cenozoic magmatism throughout east Africa resulting from impact of a single plume. *Nature*, **395**, 788–791.

EBINGER, C., BECHTEL, T., FORSYTH, D. & BOWIN, C. 1989. Effective elastic thickness beneath the east African and Afar plateaus and dynamic compensation of the uplifts. *Journal of Geophysical Research*, **94**, 2883–2901.

EBINGER, C.J., YEMANE, T., WOLDEGABRIEL, G., ARONSON, J.L. & WALTER, R.C. 1993. Late Eocene to Recent volcanism and faulting in the southern Main Ethiopian Rift. *Journal of the Geological Society, London*, **150**, 99–108.

EDGAR, A.D., GREEN, D.H. & HIBBERSON, W.O. 1976. Experimental petrology of a highly potassic magma. *Journal of Petrology*, **17**, 339–356.

EISSEN, J.-P., JUTEAU, T., JORON, J.-L., DUPRE, B., HUMLER, E. & AL'MKHAMEDOV, A. 1989. Petrology and geochemistry of basalts from the Red Sea axial rift at 18° N. *Journal of Petrology*, **30**, 791–739.

ELLAM, R.M. & STUART, F.M. 2005. Coherent He–Nd–Sr isotope trends in high $^3He/^4He$ basalts: implications for a common reservoir, mantle heterogeneity and convection. *Earth and Planetary Science Letters*, **228**, 511–523.

FURMAN, T. 1995. Melting of metasomatised subcontinental lithosphere: undersaturated mafic lavas from Rungwe, Tanzania. *Contributions to Mineralogy and Petrology*, **122**, 97–115.

FURMAN, T. & GRAHAM, D. 1999. Erosion of lithospheric mantle beneath the East African Rift system: geochemical evidence from the Kivu volcanic province. *Lithos*, **48**, 237–262.

FURMAN, T., BRYCE, J.G., KARSON, J. & IOTTI, A. 2004. East African Rift System (EARS) plume structure: insights from Quaternary mafic lavas of Turkana, Kenya. *Journal of Petrology*, **45**, 1069–1088.

GASPARON, M., INNOCENTI, F., MANETTI, P., PECCERILLO, A. & TSEGAYE, A. 1993. Genesis of the Pliocene to recent bimodal mafic–felsic volcanism in the Debre Zeyt area, central Ethiopia: volcanological and geochemical constraints. *Journal of African Earth Sciences*, **17**, 145–165.

GEORGE, R.M. & ROGERS, N.W. 1999. The petrogenesis of Plio-Pleistocene alkaline volcanic rocks from the Tosa Sucha region, Arba Minch, southern Main Ethiopian Rift. *Acta Volcanologica*, **11**, 121–130.

GEORGE, R.M. & ROGERS, N.W. 2002. Plume dynamics beneath the African plate inferred from the geochemistry of the Tertiary basalts of southern Ethiopia. *Contributions to Mineralogy and Petrology*, **144**, 286–304.

GEORGE, R.M., ROGERS, N.W. & KELLEY, S. 1998. Earliest magmatism in Ethiopia: evidence for two mantle plumes in one flood basalt province. *Geology*, **26**, 923–926.

GREEN, W.V., ACHAUER, U. & MEYER, R.P. 1991. A three dimensional seismic image of the crust and upper mantle beneath the Kenya rift. *Nature*, **354**, 199–203.

HANNAN, B.B. & GRAHAM, D.W. 1996. Lead and helium isotope evidence from oceanic basalts for a common deep source of mantle plumes. *Science*, **272**, 991–995.

HART, W.K., WOLDEGABRIEL, G., WALTER, R.C. & MERTZMANN, S.A. 1989. Basaltic volcanism in Ethiopia: constraints on continental rifting and mantle interactions. *Journal of Geophysical Research*, **94**, 731–748.

HENDRIE, D.B., KUZNIR, N.J., MORLEY, C.K. & EBINGER, C.J. 1994. Cenozoic extension in northern Kenya: a quantitative model of rift basin development in the Turkana region. *Tectonophysics*, **236**, 409–438.

HOFMANN, C., COURTILLOT, V., FERAUD, G., ROCHETTE, P., YIRGU, G., KETEFO, E. & PIK, R. 1997. Timing of the Ethiopian flood basalt event and implications for plume birth and environmental change. *Nature*, **389**, 838–841.

JOHNSON, K.T.M. 1998. Experimental determination of partition coefficients for rare earth and high-field-strength elements between clin–pyroxene, garnet and basaltic melt at high pressures. *Contributions to Mineralogy and Petrology*, **133**, 60–68.

KIEFFER, B., ARNDT, N. *ET AL.* 2004. Flood and shield basalts from Ethiopia: magmas from the African superswell. *Journal of Petrology*, **45**, 793–834.

KING, B.C. 1978. Structural and volcanic evolution of the Gregory Rift Valley. *In*: BISHOP, W.W. (ed.) *Geological Background to Fossil Man*. Scottish Academic Press, Edinburgh.

LATIN, D., NORRY, M.J. & TARZEY, R.J.E. 1993. Magmatism in the Gregory Rift, East Africa: evidence for melt generation by a plume. *Journal of Petrology*, **34**, 1007–1027.

LE ROEX, A.P., SPÄTH, A. & ZARTMAN, R.E. 2001. Lithospheric thickness beneath the southern Kenya Rift: implications from basalt geochemistry. *Contributions to Mineralogy and Petrology*, **142**, 89–106.

LIN, S.-C., KUO, B-Y., CHIAO, L-Y. & VAN KEKEN, P. 2005. Numerical investigation of a two plume system and the effects of craton and lithosphere structure on melt generation in east Africa. *Earth and Planetary Science Letters*, **237**, 175–192.

LITHGOW-BERTELLONI, C. & SILVER, P.G. 1998. Dynamic topography, plate driving forces and the African superswell. *Nature*, **395**, 269–272.

LUNDSTROM, C.C. 2003. Uranium series disequilibria in mid-ocean ridge basalts: observations and models of basalt genesis. *In*: BOURDON, B., HENDERSON, G.M., LUNDSTROM, C.C. & TURNER, S.P. (eds) Uranium-series Geochemistry. *Reviews in Mineralogy and Geochemistry*, **52**, 175–214.

MACDONALD, R., ROGERS, N.W., FITTON, J.G., BLACK, S. & SMITH, M. 2001. Plume–lithosphere interactions in the generation of the basalts of the Kenya rift, East Africa. *Journal of Petrology*, **42**, 877–900.

MARTY, B., PIK, R. & YIRGU, G. 1996. Helium isotopic variations in Ethiopian plume lavas: nature of magmatic sources and limit on lower mantle contribution. *Earth and Planetary Science Letters*, **144**, 223–237.

MCKENZIE, D.P. & BICKLE, M.J. 1988. The volume and composition of melt generated by extension of the lithosphere. *Journal of Petrology*, **29**, 625–679.

MECHIE, J., KELLER, G.R., PRODEHL, C., KHAN, M.A. & GACIRI, S.J. 1997. A model for the structure, composition and evolution of the Kenya rift. *Tectonophysics*, **278**, 95–119.

MENZIES, M.A., GALLAGHER, K., YELLAND, A. & HURFORD, A.J. 1997. Volcanic and non-volcanic rifted margins of the Red Sea and Gulf of Aden: Crustal cooling and margin evolution in Yemen. *Geochimica et Cosmochimica Acta*, **61**, 2511–2527.

MOHR, P. 1983. Ethiopian flood basalt province. *Nature*, **303**, 577–584.

NYBLADE, A.A. & BRAZIER, R.A. 2002. Precambrian lithosphere controls on the development of the east African rift system. *Geology*, **30**, 755–758.

O'CONNOR, J.M., STOFFERS, P., VAN DEN BOGAARD, P. & MCWILLIAMS, M. 1999. First seamount age evidence for significantly slower African plate motion since 19–30 Ma. *Earth and Planetary Science Letters*, **171**, 575–589.

OMAR, G.I. & STECKLER, M.S. 1995. Fission track evidence on the initial rifting of the Red Sea: two pulses, no propagation. *Science*, **270**, 1341–1344.

PECCERILLO, A., BARBERIO, M.R., YIRGU, G., AYALEW, D., BARBIERI, M. & WU, T.W. 2003. Relationships between Mafic and Peralkaline Silicic Magmatism in Continental Rift Settings: A Petrological, Geochemical and Isotopic Study of the Gedemsa Volcano, Central Ethiopian Rift. *Journal of Petrology*, **44**, 2003–2032.

PIK, R., DENIEL, C., COULON, C., YIRGU, G., HOFMANN C. & AYELEW, D. 1998. The northwestern Ethiopian Plateau flood basalts: Classification and spatial distribution of magma types. *Journal Volcanology and Geothermal Research*, **81**, 91–111.

PIK, R., DENIEL, C., COULON, C., YIRGU, G. & MARTY, B. 1999. Isotopic and trace element signatures of Ethiopian flood basalts: evidence for plume–lithosphere interactions. *Geochimica et Cosmochimica Acta*, **63**, 2263–2279.

PIK, R., MARTY, B. & HILTON, D.R. 2006. How many mantle plumes in Africa? The geochemical point of view. *Chemical Geology*, **226**, 100–114.

ROCHETTE, P., TAMRAT, E., FERAUD, G., PIK, R., COURTILLOT, V., KETEFO, E., COULON, C., HOFFMANN, C., VANDAMME, D. & YIRGU, G. 1998. Magnetostratigraphy and timing of the Oligocene Ethiopian traps. *Earth and Planetary Science Letters*, **164**, 497–510.

ROGERS, N.W., DE MULDER, M. & HAWKESWORTH, C.J. 1992. An enriched mantle source for potassic basanites: evidence from Karisimbi volcano, Virunga volcanic province, Rwanda. *Contributions to Mineralogy and Petrology*, **111**, 543–546.

ROGERS, N.W., JAMES, D., KELLEY, S.P. & DE MULDER, M. 1998. The generation of potassic lavas from the eastern Virunga province, Rwanda. *Journal of Petrology*, **39**, 1223–1247.

ROGERS, N.W., MACDONALD, R., FITTON, J.G., GEORGE, R., SMITH, M. & BARREIRO, B. 2000. Two mantle plumes beneath the East African Rift system: Sr, Nd and Pb isotope evidence from Kenya rift basalts. *Earth and Planetary Science Letters*, **176**, 387–400.

ROONEY, T.O., FURMAN, T., YIRGU, G. & AYALEW, D. 2005. Structure of the Ethiopian lithosphere: xenolith evidence in the Main Ethiopian Rift. *Geochimica et Cosmochimica Acta*, **69**, 3889–3910.

ROSENDAHL, B. 1987. Architecture of continental rifts with special reference to east Africa. *Annual Reviews Earth and Planetary Science*, **15**, 445–503.

SCARSI, P. & CRAIG, H. 1996. Helium isotope ratios in Ethiopian Rift basalts. *Earth and Planetary Science Letters*, **144**, 505–516.

SCHILLING, J.-G., KINGSLEY, R.H., HANNAN, B.B. & MCCULLY, B.L. 1992. Nd–Sr–Pb isotopic variations along the Gulf of Aden: evidence for the Afar mantle plume–lithosphere interaction. *Journal of Geophysical Research*, **97**, 10927–10966.

SMITH, M. & MOSELEY, P. 1993. Crustal heterogeneity and basement influence on the development of the Kenya rift, East Africa. *Tectonics*, **236**, 591–606.

SPATH, A., LE ROEX, A.P. & OPIYO-AKECH, N. 2001. Plume–lithosphere interaction and the origin of continental rift-related alkaline volcanism—the Chyulu Hills volcanic province, southern Kenya. *Journal of Petrology*, **42**, 765–787.

STEWART, K. & ROGERS, N.W. 1996. Mantle plume and lithosphere contributions to basalts from southern Ethiopia. *Earth and Planetary Science Letters*, **139**, 195–211.

STOREY, B.C. 1995. The role of mantle plumes in continental break-up: case histories from Gondwanaland. *Nature*, **377**, 301–308.

TESSEMA, A. & ANTOINE, L.A.G. 2004. Processing and interpretation of the gravity field of the East African Rift: implication for crustal extension. *Tectonophysics*, **394**, 87–110.

TURNER, S. & HAWKESWORTH, C. 1995. The nature of the sub-continental mantle: constraints from the major element composition of continental flood basalts. *Chemical Geology*, **120**, 295–314.

TURNER, S., HAWKESWORTH, C., GALLAGHER, K., STEWART, K., PEATE, D. & MANTOVANI, M. 1996. Mantle plumes, flood basalts and thermal models for melt generation beneath continents: assessment of a conductive heating model and application to the Parana. *Journal of Geophysical Research*, **101**, 11503–11518.

VIDAL, P., DENIEL, C., VELLUTINI, P., PIGUET, P., COULON, C., VINCENT, J. & AUDIN, J. 1991. Changes in the mantle sources in the course of a rift evolution: the Afar case. *Geophysical Research Letters*, **18**, 1913–1916.

VIGIER, N., BOURDON, B., JORON, J.L. & ALLÈGRE, C.J. 1999. U-decay series and trace element systematics in the 1978 eruption of Ardoukoba, Asal rift: timescale of magma crystallization. *Earth and Planetary Science Letters*, **174**, 81–97.

VOLLMER, R. & NORRY, M.J. 1983. Unusual isotopic variations in Nyiragongo nephelinites. *Nature*, **301**, 141–143.

WATSON, S. & MCKENZIE, D.P. 1991. Melt generation by plumes: a study of Hawaiian volcanism. *Journal of Petrology*, **32**, 501–537.

WEERARATNE, D.S., FORSYTH, D.W., FISCHER, K.M. & NYBLADE, A.A. 2003. Evidence for an upper mantle plume beneath the Tanzanian Craton from Rayleigh wave tomography. *Journal of Geophysical Research, B, Solid Earth and Planets*, **108**, doi:10.1029/2002JB002273.

WHITE, R.S. & MCKENZIE, D.P. 1989. Magmatism at rift zones: the generation of volcanic continental margins and flood basalts. *Journal of Geophysical Research*, **94**, 253–267.

WHITE, R.S. & MCKENZIE, D.P. 1995. Mantle plumes and flood basalts. *Journal of Geophysical Research*, **100**, 17543–17585.

WILLIAMS, L.A.J. 1982. The volcanological development of the Kenya rift. *In*: NEUMANN, E.R. & RAMBERG, I.B. (eds) *Petrology and Geochemistry of Continental Rifts*. Reidel, Dordrecht, 101–121.

WOLDEGABRIEL, G., ARONSON, J. & WALTER, R.C. 1990. Geology, geochronology, and rift basin development in the central sector of the Main Ethiopian Rift. *Geological Society of America Bulletin*, **102**, 439–458.

WOLFENDEN, E., EBINGER, C., YIRGU, G., DEINO, A. & AYALEW, D. 2004. Evolution of the Main Ethiopian Rift: birth of a triple junction. *Earth and Planetary Science Letters*, **224**, 213–228.

# Heads and tails: 30 million years of the Afar plume

T. FURMAN[1], J. BRYCE[2], T. ROONEY[1], B. HANAN[3], G. YIRGU[4] & D. AYALEW[4]

[1]*Department of Geosciences, The Pennsylvania State University, University Park, PA 16802, USA (e-mail: furman@geosc.psu.edu)*

[2]*Department of Earth Sciences, University of New Hampshire, Durham, NH 03824, USA*

[3]*Department of Geological Sciences, San Diego State University, San Diego, CA 92182, USA*

[4]*Department of Geological and Geophysical Sciences, Addis Ababa University, Addis Ababa, Ethiopia*

**Abstract:** Primitive recent mafic lavas from the Main Ethiopian Rift provide insight into the structure, composition and long-term history of the Afar plume. Modern rift basalts are mildly alkalic in composition, and were derived by moderate degrees of melting of fertile peridotite at depths corresponding to the base of the modern lithosphere (*c.*100 km). They are typically more silica-undersaturated than Oligocene lavas from the Ethiopia–Yemen continental flood basalt province, indicating derivation by generally smaller degrees of melting than were prevalent during the onset of plume head activity in this region. Major and trace element differences between the Oligocene and modern suites can be interpreted in terms of melting processes, including melt-induced binary mixing of melts from the Afar plume and those from three mantle end-member compositions (the convecting upper mantle and two enriched mantle sources). The Afar plume composition itself has remained essentially constant over the past 30 million years, indicating that the plume is a long-lived feature of the mantle. The geochemical and isotopic compositions of mafic lavas derived from the Afar plume support a modified single plume model in which multiple plume stems rise from a common large plume originating at great depth in the mantle (i.e. the South African superplume).

Large igneous provinces, including both continental flood basalts (CFB) and oceanic plateaus, represent the largest outpourings of mafic igneous material on our planet. These provinces are attributed to melting associated with the initiation of mantle plume activity (Morgan 1972; White & McKenzie 1995; Richards *et al.* 1989; Hill *et al.* 1992). Several major flood basalt provinces (e.g. Deccan, Paraná–Etendeka, Greenland) have been studied intensively in recent years, leading to the recognition that flood basalt lavas often include contributions from both lithospheric mantle and deep mantle sources, in some cases overprinted by crustal contamination (Hawkesworth *et al.* 1998; Peate *et al.* 1992; Arndt *et al.* 1993; Baker *et al.* 1996*a*, 2000; Fram & Lesher 1997; Janney & Castillo 2001; Ewart *et al.* 2004). It is generally difficult to determine how these source components vary either temporally or spatially, as most flood basalt provinces are either (a) too old/altered to preserve sufficient fresh lavas for detailed geochemical analysis; or (b) deeply dissected by erosion and tectonism associated with the final stages of continental rifting. These difficulties make evaluation of temporal geochemical trends, particularly those associated with the change from plume head magmatism to plume tail activity (e.g. Farnetani *et al.* 2002), unfeasible. The lone exception to these limitations is the Cenozoic Ethiopia–Yemen continental flood basalt province (Fig. 1), the majority of which was emplaced at *c.* 30 Ma and has not yet been compromised by extensive faulting and alteration.

Most studies attribute Ethiopia–Yemen flood basalt activity to impact of the Afar plume head beneath the Ethiopian plateau (e.g. Baker *et al.* 1996*a*; Pik *et al.* 1999; Kieffer *et al.* 2004). The relationship between this copious volcanism at 30 Ma and sparser modern activity in the region is not entirely clear. Most problematic in this regard is the lack of an identifiable track linking the flood basalt province to the modern rift setting that corresponds to plate motion reconstructions. Specifically, it is unresolved whether Africa has been essentially stationary over the past 30 million years (Silver *et al.* 1998; Gripp & Gordon 2002) or whether it has moved north–northeastward over this time period (Gordon & Jurdy 1986; Burke 1996). If modern Afar volcanism is derived from the same plume that fed the Oligocene flood basalts, then geodynamic models must incorporate physical coupling between the plume and the overlying lithosphere (see Calais *et al.* 2006). This issue arises, in part, because Cenozoic volcanism in the East African rift is not limited to the CFB province

*From*: YIRGU, G., EBINGER, C.J., MAGUIRE, P.K.H. (eds) 2006. *The Afar Volcanic Province within the East African Rift System*. Geological Society, London, Special Publications, **259**, 95–119.
0305-8719/06/$15.00 

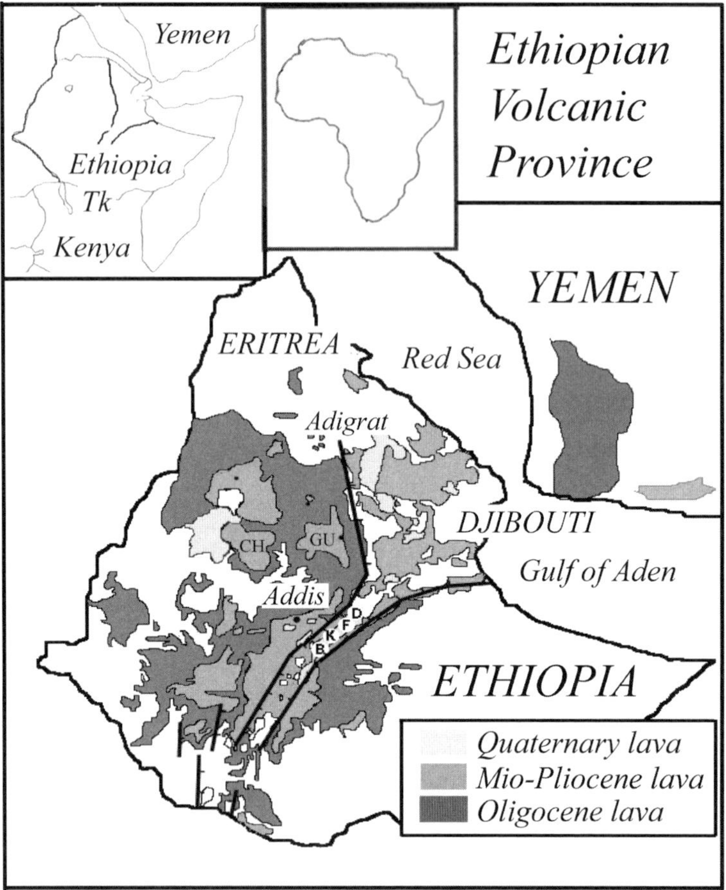


**Fig. 1.** Map of the Ethiopian Volcanic Province showing ages of volcanic rocks and sample localities mentioned in the text. Recent volcanic centres in the Main Ethiopian Rift are labelled: **D** = Dofan, **F** = Fantale, **K** = Kone, **B** = Boseti. **CH** and **GU** are the 23 Ma volcanic centres of Choke and Guguftu. On the inset map, **Tk** indicates the location of the Turkana province.

in Ethiopia and Yemen. In southernmost Ethiopia, voluminous basaltic activity began at *c.* 45 Ma, about 15 million years prior to the inferred arrival of the Oligocene plume head at the base of the Ethiopian lithosphere (Davidson & Rex 1980; Haileab *et al.* 2004; Vetel & Le Gall 2006). This dual-focus distribution of thick basalt sequences characterized by a variety of compositional differences has led to a range of geodynamic models involving both single (Ebinger & Sleep 1998), modified single (Furman *et al.* 2004, 2006) and double (George *et al.* 1998; Rogers *et al.* 2000; Pik *et al.* 2006; Rogers 2006) plumes, the validity of which remain unresolved. Most recently, volcanism in Afar and throughout East and South Africa has been attributed to the South African superplume (Janney *et al.* 2002; Kieffer *et al.* 2004; Furman *et al.* 2004, 2006), the largest seismic anomaly in the deep mantle (e.g. Ritsema *et al.* 1999; Gurnis *et al.* 2000; Nyblade *et al.* 2000; Zhao 2001; Ni *et al.* 2002). If our analysis indicates that the Afar plume has been a consistent feature of the deep mantle for 30 million years, it will provide a new window into the thermochemical structure of the South African superplume and its long-term evolution.

In this paper, we present new geochemical data on young mafic lavas from the Main Ethiopian Rift. These data permit characterization of the melting conditions and source components as they vary temporally. We then integrate our findings with the geochemical signatures from Oligocene and recent mafic lavas over the spatial extent of the CFB province. We combine data from several well-constrained regional studies from the Yemeni and Ethiopian portions of the Oligocene flood basalt sequences, and compare them to recent volcanics from Djibouti, the Afar region, the Main Ethiopian Rift and the Turkana province. The new data, taken in context with the CFB data, provide constraints on several outstanding questions relevant to flood basalt petrogenesis, namely: (a) What are the conditions of mantle melting at the onset of plume activity? (b) What source domains are involved in the genesis of the Oligocene magmas? (c) Do these same characteristics persist until the present time? (d) Is the Afar plume a long-lived feature of the mantle?

## Geodynamic setting

The Cenozoic Ethiopia–Yemen flood basalt province is located at the junction of three rifts: the Red Sea, the Gulf of Aden, and the East African rift system. Within central Ethiopia and Eritrea, the flood basalts cover an area of about 600 000 $km^2$, with an estimated total volume of about 350 000 $km^3$ (Mohr 1983; Mohr & Zanettin 1988). The lava pile thickness varies, reaching up to 2000 m along the margins of the southern Red Sea, and thinning to about 500 m towards both north and south. Most of the basalts and associated felsic rocks were erupted in a period of *c.* 1 million years at around 30 Ma (Zumbo *et al.* 1985; Baker *et al.* 1996*b*; Hofmann *et al.* 1997; Pik *et al.* 1998), concomitant with the onset of continental rifting by 29 Ma (Wolfenden *et al.* 2004). This continental flood basalt province contains the youngest and best-exposed sequence of mafic volcanic rocks associated with incipient continental break-up. It has been an area of episodic magmatic activity since the Oligocene, and is today a site of mafic volcanism attributed to the Afar plume (Vidal *et al.* 1991; Schilling *et al.* 1994; Marty *et al.* 1993, 1996; Deniel *et al.* 1994; Barrat *et al.* 1998). The Ethiopian continental flood basalt province thus provides an excellent opportunity to evaluate questions related to plume magmatism throughout the process of incipient continental rifting.

Detailed geochemical studies of modern hotspot volcanism in Hawaii (Lassiter & Hauri 1998; Blichert-Toft *et al.* 1999; DePaolo *et al.* 2001; Abouchami *et al.* 2005; Bryce *et al.* 2005), the Galápagos (Hoernle *et al.* 2000; Harpp & White 2001; Geldmacher *et al.* 2003; Thompson *et al.* 2004) and Iceland (Thirlwall 1995; Fitton *et al.* 1997; Hanan & Schilling 1997; Hanan *et al.* 1999; Kempton *et al.* 2000) have demonstrated convincingly that long-lived heterogeneities are preserved within mantle plumes located in ocean basins. These heterogeneities are expressed, to first order, spatially and/or temporally on these oceanic islands and thus are thought to represent melting of long-lived chemical and mineralogical domains located within the source region of the plume tail. No comparable study of spatial and temporal variations in plume volcanism has been attempted in a continental setting. The Afar plume provides an ideal laboratory to test these ideas using fairly young lavas ($<30$ Ma) from a tectonic environment that is well-constrained in both structural and geophysical terms (e.g. Mahatsente *et al.* 1999; Benoit *et al.* 2003; Keranen *et al.* 2004; Dugda *et al.* 2005; Kendall *et al.* 2005; Mackenzie *et al.* 2006; Bastow *et al.* 2005).

The mafic lavas in this study were collected from four volcanic centres within the Wonji Fault Belt (Mohr 1962; WoldeGabriel *et al.* 1990); from north to south they are Dofan, Fantale, Kone and Boseti (Fig. 1). Each volcano has a well-developed central silicic volcanic edifice, with basaltic lavas erupted primarily from cinder cones and along fissures towards the volcano flanks and between the volcanoes. At Kone, basaltic fissures cut through the nested Quaternary caldera structures and continue to the SW. These silicic volcanoes have been active for about three million years (WoldeGabriel *et al.* 1990; Chernet & Hart 1999) and have had several catastrophic eruptions which produced the pumice and ash that cover the rift valley floor.

## Sampling

Samples in this study were collected during field excursions in 2002 and 2003 and were chosen to obtain representative coverage of mafic lavas from the individual Quaternary volcanoes, augmented by post-Miocene syn-rift basalts emplaced along the border faults. This investigation is thus more comprehensive than previous studies that have broader regional coverage and/or include samples that range from Oligocene to Recent (Hart *et al.* 1989; Boccaletti *et al.* 1999; Chernet & Hart 1999). We focus here on fresh, young mafic lavas with $>6$ wt% MgO in order to elucidate the history of mantle melting and the nature of the source domains that contribute to this modern volcanism. The origin of more-evolved lavas (samples with $<6$ wt% MgO), including a discussion of shallow-level fractionation processes and its relationship to magmatic segmentation of the rift axis, will be presented elsewhere. The basalts are sparsely phyric, typically with $<5$ vol.% phenocrysts of olivine, accompanied in more-evolved basalts by clinopyroxene and plagioclase in variable proportions.

In order to assess the chemical heterogeneity within the modern volcanic province, we incorporate studies of mafic lavas from Djibouti (Vidal *et al.* 1991; Deniel *et al.* 1994), Erta 'Ale (Barrat *et al.* 1998; Ayalew *et al.* unpublished data) and the West Sheba Ridge in the Gulf of Aden (Schilling *et al.* 1992). To evaluate possible temporal evolution of the Afar plume, we compare these recent basalt suites to well-documented Oligocene continental flood basalts from Ethiopia (Pik *et al.* 1998, 1999; Kieffer *et al.* 2004) and Yemen (Chazot & Bertrand 1993; Baker *et al.* 1996*a*, 2000). There are limited data available for lavas emplaced between 30 and 3 Ma; we include analyses of the *c.*7 Ma Addis Ababa series (this study) and 23 Ma basalts from Choke and Guguftu volcanoes (Kieffer *et al.* 2004; Fig. 1). To broaden our regional scope, we compare the Afar province lavas to basalts from Turkana, nothern Kenya (Furman *et al.* 2004, 2006); geochemical characteristics of the Kenya rift are described elsewhere (Rogers 2006).

## Analytical methods

Centimetre-thick slabs were polished to remove saw marks, and then crushed into millimetre-sized chips. These chips were then powdered in a WC disc mill. Bulk rock analyses for major and trace elements were completed by DCP and ICP-MS analysis at Duke University and by ICP-MS at the Centre Récherches Pétrographiques et Géochimiques, Nancy, France (Table 1). Bulk rock Sr–Nd–Pb isotopic analyses (Table 2) were performed at UCLA and San Diego State University. Details of the analytical procedures are in the Appendix.

## Results

### *Major elements*

The modern Ethiopian rift lavas comprise primarily transitional tholeiitic basalts and trachybasalts (Fig. 2). Their bulk compositions are hypersthene- to mildly nepheline-normative, and evolved lavas tend towards silica-undersaturated compositions. The Ethiopian lavas are thus compositionally similar to contemporaneous mafic lavas from Turkana, northern Kenya (Furman *et al.* 2004) and 23 Ma basalts from Choke and Guguftu, but more alkali-rich than the Oligocene flood basalts which are dominated by quartz-normative tholeiites (Pik *et al.* 1999; Kieffer *et al.* 2004). The degree of differentiation of recent lavas varies between the individual volcanoes, with the most primitive basalts erupted from fissures immediately south of Kone. Each of the central volcanic complexes is characterized by low-MgO basalts through rhyolitic lavas.

The majority of lavas described in this study contain between 6.5 and 9.5 wt% MgO; no picritic lavas have been sampled in the modern rift setting although rare Fe-rich picrites (ferropicrites of Gibson *et al.* 2000) occur in both Ethiopian and Yemeni flood basalt sequences. Recent Ethiopian basalts have $TiO_2$ contents that range from 1.8–2.3 wt%, somewhat higher than values observed among Oligocene low-Ti lavas (1.1–2.1 wt%) but substantially lower than the primitive high-Ti suite (2.7–5.1 wt%). Pik *et al.* (1999) identified these coeval high- and low-titanium basalt series within the continental flood basalt sequence, and proposed a geographic control on eruptive composition. The presence of sparse high-Ti lavas among the Yemeni flood basalt sequence (Baker *et al.* 1996*b*), however, suggests that this lava type is more widespread than recognized previously.

### *Trace elements*

Abundances of compatible trace elements approach those expected for primary mantle melts (*c.* 200 ppm Ni, *c.* 750 ppm Cr, *c.* 33 ppm Sc; Table 1), suggesting that the most magnesian modern Ethiopian rift lavas have undergone only minor olivine fractionation during melt ascent and storage. Variation of these elements with MgO content among primitive Ethiopian basalts of all ages suggests a dominant role for olivine crystallization; in contrast, co-precipitation of olivine and clinopyroxene is inferred for Quaternary basalts from Turkana, nothern Kenya (Furman *et al.* 2004).

Chondrite-normalized rare earth element (REE) patterns of the Ethiopian rift basalts have moderate negative slopes and show little variation within or between eruptive centres (Fig. 3). Oligocene flood basalts from throughout Ethiopia (Pik *et al.* 1999) have lower overall abundances and consistently flatter rare earth element patterns, whereas primitive high-Ti samples have steeper patterns and substantially higher abundances, despite their higher MgO contents (Fig. 3). Two Oligocene high-Ti lavas from Adigrat (NW Ethiopia) have moderately steep, concave-down patterns that set them apart from the other flood basalts. In contrast to all of the Oligocene lavas, the 23 Ma lavas from Choke and Guguftu have rare earth element abundances that fall within the range of the modern rift lavas (Fig. 3), consistent with major element evidence (Fig. 2) suggesting these lavas are more closely related to Quaternary Turkana basalts than to the CFB and Main Ethiopian Rift products.

Incompatible trace element abundances in the young Ethiopian rift basalts are moderately

**Table 1.** *Major amd trace element and isotopic data for Ethiopian rift basalts. See supplementary data section in the Appendix for analytical techniques, precision and accuracy. The isotopic data presented for the 7 Ma Addis Ababa samples (denoted with*) represent initial values. The $\Delta 8/4$ values are computed according to Hart (1984); the NHRL model values are corrected to 7 Ma values using $\kappa = 2.5$.*

| Sample | 1027 | 1023 | 1018 | E99-2 | 1032 | N-18 | 1035 | 1035A | N-17 | E99-3 | 1030 |
|---|---|---|---|---|---|---|---|---|---|---|---|
| Location | Dofan | Dofan | Fantale | Fantale | Fantale | Fantale | Fantale | Fantale | Kone | Kone | Kone |
| Latitude (N) | 9.48 | 9.14 | 8.92 | | 8.87 | 8.91 | 8.80 | 8.80 | 8.88 | | 8.89 |
| Longitude (E) | 40.20 | 39.95 | 39.84 | | 39.91 | 39.83 | 39.68 | 39.68 | 39.79 | | 39.79 |
| $SiO_2$ | 46.78 | 46.15 | 46.88 | 46.52 | 47.48 | 46.26 | 47.28 | 46.41 | 46.64 | 46.68 | 47.00 |
| $Al_2O_3$ | 15.73 | 15.75 | 16.26 | 16.72 | 16.25 | 15.63 | 15.48 | 15.64 | 16.37 | 16.58 | 16.33 |
| $TiO_2$ | 2.25 | 2.20 | 2.58 | 2.46 | 2.47 | 2.24 | 2.31 | 2.26 | 2.71 | 2.41 | 2.33 |
| $Fe_2O_3$ | 12.77 | 12.45 | 13.46 | 13.14 | 12.88 | 12.06 | 12.52 | 12.34 | 14.27 | 13.23 | 13.00 |
| MgO | 7.35 | 8.08 | 6.23 | 6.35 | 6.65 | 7.35 | 7.68 | 7.80 | 6.27 | 6.65 | 6.83 |
| MnO | 0.19 | 0.18 | 0.21 | 0.20 | 0.20 | 0.19 | 0.19 | 0.18 | 0.21 | 0.19 | 0.20 |
| CaO | 11.00 | 11.60 | 9.13 | 9.48 | 9.62 | 10.78 | 11.14 | 10.87 | 9.04 | 10.41 | 10.00 |
| $Na_2O$ | 2.68 | 2.41 | 3.41 | 3.36 | 3.33 | 3.20 | 2.85 | 2.90 | 3.34 | 3.09 | 3.10 |
| $K_2O$ | 0.58 | 0.77 | 1.01 | 1.12 | 0.93 | 1.12 | 0.71 | 0.68 | 0.98 | 0.76 | 0.78 |
| $P_2O_5$ | 0.42 | 0.40 | 0.61 | 0.52 | 0.57 | 0.50 | 0.44 | 0.42 | 0.50 | 0.41 | 0.38 |
| Sum | 99.75 | 99.99 | 99.78 | 99.87 | 100.38 | 99.32 | 100.60 | 99.50 | 100.32 | 100.41 | 99.95 |
| Cs | 0.05 | 0.09 | 0.08 | 0.34 | 0.11 | 0.11 | 0.08 | 0.10 | 0.04 | 0.17 | 0.06 |
| Rb | 10.6 | 16.4 | 13.8 | 18.2 | 15.9 | 17.0 | 12.3 | 14.6 | 11.4 | 9.9 | 10.7 |
| Ba | 235 | 308 | 504 | 456 | 465 | 414 | 288 | 279 | 428 | 343 | 357 |
| Th | 1.59 | 1.91 | 1.46 | 2.09 | 1.56 | 2.01 | 1.29 | 1.45 | 1.61 | 1.56 | 1.43 |
| U | 0.4 | 0.4 | 0.5 | 0.5 | 0.4 | 0.5 | 0.4 | 0.4 | 0.4 | 0.2 | 0.4 |
| Nb | 27.4 | 28.8 | 33.7 | 29.8 | 29.7 | 32.2 | 22.7 | 23.2 | 32.9 | 22.0 | 23.2 |
| Ta | 1.7 | 1.7 | 2.0 | 2.0 | 1.8 | 1.9 | 1.4 | 1.4 | 2.0 | 1.5 | 1.5 |
| Pb | 1.7 | 1.6 | 2.4 | 2.4 | 2.3 | 2.2 | 2.0 | 2.1 | 2.0 | 1.5 | 1.6 |
| Sr | 382 | 591 | 411 | 539 | 522 | 497 | 415 | 462 | 479 | 487 | 448 |
| Zr | 190 | 124 | 157 | 168 | 150 | 155 | 121 | 125 | 169 | 129 | 132 |
| Hf | 4.4 | 2.9 | 3.6 | 3.3 | 3.5 | 3.6 | 2.8 | 2.9 | 4.2 | 2.8 | 3.4 |
| Y | 28.5 | 24.8 | 28.6 | 27.4 | 27.3 | 27.6 | 25.9 | 24.7 | 27.2 | 23.1 | 23.9 |
| Co | 66.6 | 50.0 | 47.6 | 47.9 | 45.3 | 46.7 | 47.6 | 50.3 | 47.6 | 48.2 | 65.6 |
| Cr | 172 | 221 | 38 | 58 | 64 | 214 | 218 | 232 | 30 | 106 | 89 |
| Ni | 91 | 86 | 44 | 43 | 64 | 82 | 84 | 82 | 49 | 57 | 62 |
| V | 350 | 298 | 287 | 286 | 273 | 277 | 313 | 306 | 300 | 300 | 283 |
| Sc | 32.1 | 26.5 | 20.7 | 25.4 | 31.2 | 27.9 | 30.9 | 30.3 | 27.5 | 31.0 | 31.3 |
| La | 20.3 | 18.9 | 20.8 | 27.3 | 22.2 | 24.6 | 15.3 | 17.1 | 23.1 | 19.7 | 19.2 |
| Ce | 44.9 | 41.3 | 50.6 | 59.4 | 50.6 | 52.5 | 35.8 | 38.6 | 51.9 | 45.2 | 41.7 |
| Pr | 3.0 | 5.3 | 6.2 | 6.9 | 6.7 | 6.6 | 4.5 | 5.1 | 6.5 | 5.2 | 5.5 |

*(continued)*

**Table 1.** *Continued*

| Sample<br>Location | 1027<br>Dofan | 1023<br>Dofan | 1018<br>Fantale | E99-2<br>Fantale | 1032<br>Fantale | N-18<br>Fantale | 1035<br>Fantale | 1035A<br>Fantale | N-17<br>Kone | E99-3<br>Kone | 1030<br>Kone |
|---|---|---|---|---|---|---|---|---|---|---|---|
| Nd | 25.4 | 21.9 | 25.9 | 30 | 28.5 | 28.4 | 19.4 | 21.3 | 28.5 | 23.3 | 23.5 |
| Sm | 5.67 | 4.82 | 5.50 | 6.34 | 6.20 | 5.97 | 4.31 | 4.69 | 6.24 | 5.17 | 5.26 |
| Eu | 2.02 | 1.74 | 2.09 | 2.36 | 2.32 | 2.13 | 1.54 | 1.73 | 2.27 | 1.89 | 1.98 |
| Gd | 5.96 | 5.10 | 5.88 | 6.17 | 6.35 | 6.46 | 4.56 | 5.09 | 6.68 | 4.79 | 5.34 |
| Tb | 0.96 | 0.73 | 0.84 | 0.89 | 0.92 | 1.01 | 0.68 | 0.75 | 1.07 | 0.72 | 0.84 |
| Dy | 5.34 | 4.02 | 4.47 | 4.84 | 5.05 | 5.05 | 3.73 | 4.16 | 5.35 | 4.17 | 4.58 |
| Ho | 1.07 | 0.78 | 0.87 | 0.93 | 0.97 | 0.99 | 0.71 | 0.82 | 1.02 | 0.78 | 0.93 |
| Er | 2.82 | 2.03 | 2.28 | 2.37 | 2.45 | 2.61 | 1.86 | 2.11 | 2.75 | 2.00 | 2.32 |
| Yb | 2.51 | 1.77 | 1.96 | 2.14 | 2.18 | 2.25 | 1.67 | 1.88 | 2.37 | 1.90 | 2.06 |
| Lu | 0.37 | 0.27 | 0.29 | 0.32 | 0.33 | 0.34 | 0.25 | 0.27 | 0.36 | 0.28 | 0.32 |
| $^{87}Sr/^{86}Sr$ | 0.703722 | 0.704102 | — | 0.704117 | — | 0.703961 | 0.704198 | — | — | — | — |
| $^{143}Nd/^{144}Nd$ | 0.512862 | 0.512817 | — | 0.512798 | — | 0.512799 | 0.512767 | — | — | — | — |
| $\varepsilon_{Nd}$ | 4.4 | 3.5 | — | 3.1 | — | 3.1 | 2.5 | — | — | — | — |
| $^{206}Pb/^{204}Pb$ | 18.827 | 18.752 | — | — | — | 18.732 | 18.627 | — | — | — | — |
| $^{207}Pb/^{204}Pb$ | 15.581 | 15.584 | — | — | — | 15.597 | 15.585 | — | — | — | — |
| $^{208}Pb/^{204}Pb$ | 38.896 | 38.782 | — | — | — | 38.834 | 38.748 | — | — | — | — |
| $\Delta 8/4$ | 50.72 | 48.38 | — | — | — | 56.00 | 60.10 | — | — | — | — |

| Sample<br>Location | N-16<br>Kone | N-22<br>Kone | KO-12<br>Kone | N-20<br>Kone | 1037<br>Kone | N-15<br>Kone | 1036<br>Kone | E99-11<br>Kone | KO-4B<br>Kone | N-21<br>Kone |
|---|---|---|---|---|---|---|---|---|---|---|
| Latitude (N) | 8.84 | 8.72 | | 8.83 | 8.86 | 8.78 | 8.84 | | | 8.78 |
| Longitude (E) | 39.70 | 39.68 | | 39.69 | 39.75 | 39.63 | 39.71 | | | 39.68 |
| $SiO_2$ | 46.14 | 45.42 | 46.53 | 45.86 | 47.14 | 46.35 | 46.50 | 46.28 | 46.43 | 46.86 |
| $Al_2O_3$ | 15.91 | 15.29 | 15.58 | 14.86 | 15.06 | 14.89 | 14.78 | 15.03 | 14.75 | 14.10 |
| $TiO_2$ | 2.46 | 2.31 | 2.27 | 2.21 | 1.95 | 2.22 | 2.20 | 2.14 | 2.17 | 1.90 |
| $Fe_2O_3$ | 12.82 | 12.43 | 12.65 | 12.46 | 11.92 | 12.18 | 12.56 | 12.42 | 12.55 | 11.73 |
| MgO | 6.83 | 7.97 | 8.14 | 8.33 | 8.35 | 8.64 | 8.51 | 8.78 | 8.85 | 10.44 |
| MnO | 0.19 | 0.19 | 0.18 | 0.19 | 0.19 | 0.19 | 0.20 | 0.17 | 0.19 | 0.18 |
| CaO | 10.95 | 11.19 | 10.78 | 10.58 | 10.17 | 10.91 | 10.69 | 11.37 | 10.74 | 11.02 |
| $Na_2O$ | 3.06 | 3.12 | 2.89 | 3.25 | 2.87 | 2.73 | 3.21 | 2.70 | 3.22 | 2.82 |
| $K_2O$ | 0.82 | 1.13 | 0.64 | 1.12 | 0.88 | 1.06 | 1.07 | 0.65 | 1.00 | 0.67 |
| $P_2O_5$ | 0.45 | 0.55 | 0.60 | 0.54 | 0.46 | 0.45 | 0.55 | 0.33 | 0.60 | 0.37 |
| Sum | 99.63 | 99.61 | 100.26 | 99.41 | 98.99 | 99.61 | 100.27 | 99.87 | 100.50 | 100.09 |
| Cs | 0.13 | 0.34 | 0.15 | 0.26 | 0.09 | 0.15 | 0.19 | 0.12 | 0.33 | 0.13 |

| | | | | | | | | | | |
|---|---|---|---|---|---|---|---|---|---|---|
| Rb | 13.1 | 28.2 | 12.2 | 23.6 | 12.7 | 18.4 | 22.2 | 11.2 | 26.7 | 12.7 |
| Ba | 299 | 405 | 272 | 391 | 401 | 324 | 411 | 241 | 357 | 246 |
| Th | 1.57 | 3.25 | 1.57 | 2.69 | 1.63 | 2.04 | 2.30 | 1.84 | 2.97 | 1.41 |
| U | 0.4 | 0.9 | 0.3 | 0.8 | 0.4 | 0.5 | 0.8 | 0.4 | 0.7 | 0.4 |
| Nb | 24.9 | 47.3 | 18.7 | 40.8 | 29.6 | 30.2 | 41.3 | 20.6 | 33.0 | 22.9 |
| Ta | 1.5 | 2.7 | 1.5 | 2.3 | 1.8 | 1.7 | 2.4 | 1.5 | 2.5 | 1.3 |
| Pb | 2.0 | 2.3 | 3.0 | 2.6 | 2.0 | 2.0 | 2.6 | 1.7 | 3.0 | 3.2 |
| Sr | 521 | 619 | 488 | 586 | 479 | 468 | 404 | 427 | 526 | 392 |
| Zr | 141 | 175 | 122 | 164 | 148 | 138 | 158 | 125 | 148 | 141 |
| Hf | 3.4 | 3.7 | 3.1 | 3.7 | 3.7 | 3.2 | 3.5 | 3.0 | 3.5 | 3.6 |
| Y | 25.7 | 27.1 | 21.4 | 25.5 | 27.0 | 24.1 | 27.8 | 21.2 | 23.2 | 23.5 |
| Co | 48.8 | 52.0 | 51.3 | 50.2 | 61.3 | 49.6 | 47.7 | 54.0 | 50.1 | 54.1 |
| Cr | 143 | 204 | 289 | 339 | 292 | 345 | 309 | 465 | 433 | 556 |
| Ni | 73 | 96 | 109 | 135 | 118 | 145 | 123 | 160 | 141 | 165 |
| V | 302 | 287 | 300 | 259 | 256 | 293 | 277 | 287 | 283 | 287 |
| Sc | 31.0 | 31.3 | 29.4 | 30.7 | 32.4 | 17.7 | 33.4 | 29.4 | 33.5 | 24.7 |
| La | 19.2 | 31.5 | 18.6 | 28.2 | 23.9 | 20.2 | 22.9 | 17.7 | 27.9 | 17.0 |
| Ce | 42.2 | 64.4 | 40.2 | 57.9 | 51.7 | 42.8 | 51.9 | 40.7 | 57.3 | 37.1 |
| Pr | 5.5 | 7.7 | 5.2 | 7.0 | 6.8 | 5.3 | 6.1 | 4.7 | 7.0 | 4.8 |
| Nd | 24.2 | 31.6 | 22.6 | 29.3 | 28.8 | 23.2 | 24.1 | 20.6 | 28 | 21.1 |
| Sm | 5.19 | 6.16 | 5.28 | 5.70 | 6.28 | 4.84 | 4.98 | 4.64 | 5.73 | 4.50 |
| Eu | 1.87 | 2.05 | 1.85 | 1.99 | 2.12 | 1.67 | 1.62 | 1.71 | 1.82 | 1.61 |
| Gd | 5.73 | 6.51 | 4.75 | 6.20 | 5.81 | 5.22 | 5.09 | 4.45 | 5.53 | 4.89 |
| Tb | 0.91 | 1.01 | 0.70 | 0.95 | 0.93 | 0.85 | 0.72 | 0.67 | 0.76 | 0.81 |
| Dy | 4.74 | 4.98 | 3.90 | 4.65 | 5.10 | 4.31 | 3.84 | 3.73 | 4.52 | 4.21 |
| Ho | 0.92 | 0.96 | 0.80 | 0.90 | 1.00 | 0.85 | 0.72 | 0.76 | 0.84 | 0.81 |
| Er | 2.38 | 2.53 | 1.91 | 2.44 | 2.63 | 2.27 | 1.91 | 2.01 | 2.30 | 2.21 |
| Yb | 2.17 | 2.09 | 1.73 | 2.05 | 2.34 | 1.96 | 1.67 | 1.78 | 1.90 | 2.04 |
| Lu | 0.31 | 0.32 | 0.31 | 0.31 | 0.35 | 0.30 | 0.24 | 0.28 | 0.31 | 0.29 |
| $^{87}Sr/^{86}Sr$ | — | 0.703768 | 0.704303 | 0.703996 | 0.704251 | 0.704203 | 0.703949 | — | 0.704015 | 0.704228 |
| $^{143}Nd/^{144}Nd$ | — | 0.512838 | 0.512765 | — | 0.512768 | 0.512800 | 0.512799 | — | 0.512801 | 0.512793 |
| $\varepsilon_{Nd}$ | — | 3.9 | 2.5 | — | 2.5 | 3.2 | 3.1 | — | 3.2 | 3.0 |
| $^{206}Pb/^{204}Pb$ | — | 18.945 | 18.658 | 18.811 | 18.557 | 18.680 | 18.808 | — | 18.801 | 18.624 |
| $^{207}Pb/^{204}Pb$ | — | 15.595 | 15.593 | 15.592 | 15.584 | 15.588 | 15.590 | — | 15.596 | 15.586 |
| $^{208}Pb/^{204}Pb$ | — | 38.896 | 38.826 | 38.873 | 38.632 | 38.879 | 38.862 | — | 38.872 | 38.794 |
| $\Delta 8/4$ | — | 36.45 | 64.15 | 50.35 | 56.96 | 66.79 | 49.61 | — | 51.46 | 65.06 |

*(continued)*

**Table 1.** *Continued.*

| Sample<br>Location | E99-6<br>Kone | N-02<br>Boseti | N-01<br>Boseti | E99-13<br>Boseti | E99-20<br>Boseti | N-12<br>E. Rift<br>Wall | N-09<br>E. Rift<br>Wall | W-04<br>E. Rift<br>Flank | E01-15A<br>W. Rift<br>Flank |
|---|---|---|---|---|---|---|---|---|---|
| Latitude (N) | | 8.44 | 8.39 | | | 8.28 | 8.26 | | 8.92 |
| Longitude (E) | | 39.40 | 39.33 | | | 39.41 | 39.50 | | 39.55 |
| $SiO_2$ | 47.11 | 47.14 | 46.78 | 47.37 | 46.87 | 50.59 | 48.53 | 45.05 | 47.52 |
| $Al_2O_3$ | 14.51 | 15.50 | 14.65 | 15.15 | 15.09 | 16.32 | 15.49 | 16.94 | 15.60 |
| $TiO_2$ | 1.82 | 2.20 | 2.15 | 2.16 | 2.20 | 2.20 | 1.54 | 2.57 | 2.42 |
| $Fe_2O_3$ | 11.59 | 11.98 | 11.96 | 12.09 | 12.16 | 11.83 | 11.42 | 12.63 | 14.54 |
| MgO | 10.66 | 7.60 | 8.46 | 8.46 | 9.13 | 5.95 | 8.63 | 7.25 | 5.74 |
| MnO | 0.18 | 0.18 | 0.18 | 0.18 | 0.18 | 0.18 | 0.19 | 0.18 | 0.21 |
| CaO | 11.16 | 10.31 | 11.10 | 11.35 | 11.25 | 8.97 | 9.62 | 11.49 | 8.50 |
| $Na_2O$ | 2.54 | 3.02 | 2.75 | 2.79 | 2.65 | 3.21 | 3.36 | 2.41 | 3.27 |
| $K_2O$ | 0.63 | 1.04 | 0.90 | 0.77 | 0.66 | 1.32 | 1.02 | 0.75 | 0.96 |
| $P_2O_5$ | 0.35 | 0.58 | 0.50 | 0.48 | 0.39 | 0.44 | 0.41 | 0.39 | 0.53 |
| Sum | 100.55 | 99.55 | 99.43 | 100.80 | 100.58 | 101.00 | 100.18 | 99.66 | 99.29 |
| Cs | 0.26 | 0.11 | 0.10 | 0.41 | 0.06 | 0.16 | 0.24 | 0.05 | 0.22 |
| Rb | 12.1 | 13.1 | 14.0 | 24.6 | 10.9 | 14.6 | 16.0 | 11.0 | 21.5 |
| Ba | 240 | 437 | 373 | 662 | 291 | 504 | 371 | 200 | 497 |
| Th | 1.80 | 1.94 | 1.68 | 3.93 | 1.90 | 1.98 | 1.69 | 1.62 | 2.52 |
| U | 0.4 | 0.4 | 0.5 | 0.9 | 0.3 | 0.5 | 0.4 | 0.5 | 0.5 |
| Nb | 19.7 | 30.5 | 27.0 | 42.0 | 20.2 | 27.9 | 22.4 | 24.4 | 22.0 |
| Ta | 1.4 | 1.8 | 1.6 | 2.8 | 1.5 | 1.6 | 1.3 | 1.5 | 1.3 |
| Pb | 1.9 | 2.7 | 2.5 | 4.7 | 2.2 | 4.3 | 5.8 | 1.6 | 4.2 |
| Sr | 399 | 501 | 470 | 730 | 471 | 587 | 523 | 468 | 440 |
| Zr | 117 | 176 | 146 | 215 | 125 | 210 | 131 | 148 | 190 |
| Hf | 2.7 | 3.9 | 3.2 | 4.4 | 2.8 | 4.8 | 3.2 | 3.5 | 4.5 |
| Y | 20.5 | 26.6 | 23.7 | 26.9 | 20.6 | 32.2 | 24.4 | 23.0 | 41.4 |
| Co | 54.7 | 46.7 | 49.5 | 61.8 | 50.3 | 44.7 | 44.0 | 47.1 | 49.0 |
| Cr | 757 | 240 | 342 | 721 | 540 | 60 | 336 | 40 | 49 |
| Ni | 209 | 97 | 109 | 202 | 141 | 47 | 135 | 72 | 64 |
| V | 276 | 260 | 278 | 250 | 279 | 227 | 217 | 267 | 322 |
| Sc | | 29.4 | 33.5 | | | 24.7 | 27.3 | 27.8 | 30.8 |
| La | 17.5 | 25.5 | 22.0 | 40.7 | 18.8 | 26.4 | 18.7 | 18.1 | 26.0 |
| Ce | 38.9 | 53.8 | 47.2 | 86.8 | 41.7 | 58.0 | 39.7 | 40.5 | 52.7 |
| Pr | 4.5 | 6.9 | 6.0 | 10.0 | 5.1 | 7.4 | 5.0 | 5.4 | 7.1 |
| Nd | 20.4 | 28.9 | 25.4 | 41.9 | 22 | 31.4 | 21.2 | 22.6 | 30.7 |
| Sm | 4.37 | 5.89 | 5.31 | 8.50 | 4.79 | 6.27 | 4.43 | 5.05 | 6.75 |
| Eu | 1.58 | 2.11 | 1.88 | 2.71 | 1.81 | 2.02 | 1.50 | 1.81 | 2.17 |

| | | | | | | | | | |
|---|---|---|---|---|---|---|---|---|---|
| Gd | 4.16 | 6.31 | 5.64 | 7.17 | 4.61 | 6.94 | 5.04 | 5.22 | 7.43 |
| Tb | 0.68 | 0.96 | 0.90 | 0.95 | 0.63 | 1.11 | 0.81 | 0.82 | 1.19 |
| Dy | 3.74 | 4.85 | 4.45 | 4.93 | 3.76 | 5.76 | 4.31 | 4.37 | 6.80 |
| Ho | 0.77 | 0.97 | 0.83 | 0.92 | 0.71 | 1.12 | 0.86 | 0.85 | 1.39 |
| Er | 1.90 | 2.48 | 2.21 | 2.14 | 1.89 | 3.03 | 2.32 | 2.18 | 3.81 |
| Yb | 1.78 | 2.20 | 1.97 | 2.01 | 1.76 | 2.79 | 2.16 | 1.87 | 3.40 |
| Lu | 0.27 | 0.33 | 0.28 | 0.30 | 0.24 | 0.42 | 0.31 | 0.28 | 0.52 |
| $^{87}Sr/^{86}Sr$ | — | — | 0.704485 | — | 0.704117 | 0.704982 | 0.704255 | — | — |
| $^{143}Nd/^{144}Nd$ | — | — | 0.512730 | — | — | 0.512585 | 0.512695 | — | — |
| $\varepsilon_{Nd}$ | — | — | 1.8 | — | — | −1.0 | 1.1 | — | — |
| $^{206}Pb/^{204}Pb$ | — | — | 18.518 | — | 18.694 | 17.853 | 18.836 | — | — |
| $^{207}Pb/^{204}Pb$ | — | — | 15.589 | — | 15.591 | 15.578 | 15.620 | — | — |
| $^{208}Pb/^{204}Pb$ | — | — | 38.686 | — | 38.791 | 38.181 | 38.854 | — | — |
| $\Delta 8/4$ | — | — | 67.07 | — | 56.30 | 96.97 | 45.43 | — | — |

| Sample<br>Location | E01-20<br>W. Rift<br>Flank | E01-32<br>Awash<br>Gorge | E-43<br>W. Rift<br>Flank | AA99-10*<br>Addis<br>Ababa | AA99-6*<br>Addis<br>Ababa | AA99-03*<br>Addis<br>Ababa | AA99-19<br>Addis<br>Ababa |
|---|---|---|---|---|---|---|---|
| Latitude (N) | 9.02 | 8.88 | 9.13 | | | | |
| Longitude (E) | 39.56 | 40.10 | 39.73 | | | | |
| $SiO_2$ | 48.93 | 47.09 | 47.74 | 45.12 | 46.14 | 47.67 | 48.49 |
| $Al_2O_3$ | 15.01 | 16.20 | 15.37 | 13.05 | 14.11 | 15.11 | 17.17 |
| $TiO_2$ | 2.50 | 2.22 | 3.09 | 1.50 | 1.61 | 2.26 | 1.94 |
| $Fe_2O_3$ | 13.60 | 13.06 | 13.05 | 11.25 | 12.79 | 12.53 | 11.59 |
| MgO | 5.97 | 6.49 | 7.20 | 14.38 | 12.14 | 8.40 | 6.36 |
| MnO | 0.21 | 0.23 | 0.18 | 0.16 | 0.18 | 0.18 | 0.16 |
| CaO | 10.35 | 11.09 | 9.25 | 11.06 | 9.11 | 9.80 | 9.69 |
| $Na_2O$ | 3.25 | 2.95 | 3.19 | 2.32 | 3.02 | 2.94 | 3.25 |
| $K_2O$ | 0.72 | 0.56 | 1.21 | 0.63 | 0.74 | 1.08 | 0.91 |
| $P_2O_5$ | 0.46 | 0.37 | 0.51 | 0.26 | 0.32 | 0.59 | 0.38 |
| Sum | 101.00 | 100.26 | 100.79 | 99.73 | 100.16 | 100.56 | 99.94 |
| Cs | 8.8 | 0.010.07 | 0.43 | 0.28 | 0.25 | 0.03 | 0.23 |
| Rb | 330 | 7.7 | 26.8 | 15.4 | 15.8 | 13.9 | 18.2 |
| Ba | 1.64 | 364 | 405 | 236 | 294 | 379 | 320 |
| Th | 0.1 | 1.14 | 3.39 | 2.20 | 1.73 | 1.88 | 2.33 |
| U | 24.7 | 0.3 | 0.9 | 0.5 | 0.3 | 0.2 | 0.4 |
| Nb | 1.5 | 17.9 | 44.6 | 23.6 | 27.4 | 32.8 | 22.7 |
| Ta | 2.2 | 1.1 | 2.8 | 1.7 | 2.0 | 1.9 | 1.5 |
| Pb | 457 | 1.8 | 3.4 | 1.9 | 2.4 | 3.2 | 3.2 |
| Sr | 154 | 470 | 694 | 454 | 361 | 544 | 636 |

(continued)

**Table 1.** *Continued*

| Sample<br>Location | E01-20<br>W. Rift<br>Flank | E01-32<br>Awash<br>Gorge | E-43<br>W. Rift<br>Flank | AA99-10*<br>Addis<br>Ababa | AA99-6*<br>Addis<br>Ababa | AA99-03*<br>Addis<br>Ababa | AA99-19<br>Addis<br>Ababa |
|---|---|---|---|---|---|---|---|
| Zr | 3.6 | 127 | 269 | 99 | 144 | 174 | 133 |
| Hf | 28.3 | 2.9 | 5.9 | 2.5 | 2.9 | 4.0 | 2.9 |
| Y | 45.1 | 24.7 | 30.8 | 17.9 | 26.2 | 28.4 | 22.4 |
| Co | 52 | 47.5 | 54.0 | 64.7 | 64.3 | 58.2 | 44.9 |
| Cr | 50 | 45 | 109 | 1136 | 663 | 391 | 64 |
| Ni | 313 | 73 | 72 | 410 | 389 | 124 | 35 |
| V | 32.5 | 281 | 281 | 234 | 218 | 258 | 272 |
| Sc | 19.8 | 29.9 | 23.9 | 30.8 | | 30.8 | |
| La | 45.0 | 15.9 | 33.9 | 17.4 | 19.6 | 27.4 | 20.7 |
| Ce | 5.9 | 35.3 | 47.8 | 36.7 | 36.1 | 61.5 | 44.2 |
| Pr | 25.8 | 4.8 | 9.7 | 4.3 | 4.4 | 7.5 | 5.3 |
| Nd | 8.53 | 20.9 | 39.7 | 17.8 | 18.3 | 30.9 | 23 |
| Sm | 2.01 | 4.80 | 8.27 | 3.67 | 3.93 | 6.33 | 4.86 |
| Eu | 6.06 | 1.76 | 2.71 | 1.27 | 1.33 | 2.11 | 1.63 |
| Gd | 0.96 | 5.12 | 8.15 | 4.01 | 3.94 | 6.54 | 4.53 |
| Tb | 5.21 | 0.82 | 1.17 | 0.57 | 0.61 | 0.99 | 0.65 |
| Dy | 1.02 | 4.40 | 5.93 | 3.06 | 3.47 | 5.21 | 3.92 |
| Ho | 2.69 | 0.90 | 1.13 | 0.70 | 0.67 | 1.02 | 0.74 |
| Er | 2.36 | 2.37 | 2.86 | 1.60 | 1.83 | 2.57 | 1.93 |
| Yb | 0.34 | 2.17 | 2.32 | 1.64 | 1.81 | 2.33 | 1.96 |
| Lu | 0.704237 | 0.32 | 0.33 | 0.21 | 0.28 | 0.35 | 0.28 |
| $^{87}Sr/^{86}Sr$ | 0.512753 | 0.704321 | 0.703841 | 0.703643 | 0.703880 | 0.704064 | — |
| $^{143}Nd/^{144}Nd$ | 2.2 | 0.512767 | 0.512842 | 0.512886 | 0.512780 | 0.512748 | — |
| $\varepsilon_{Nd}$ | | 2.5 | 4.0 | 4.8 | 2.8 | 2.1 | — |
| $^{206}Pb/^{204}Pb$ | — | 18.429 | 18.837 | 18.779 | — | 18.212 | — |
| $^{207}Pb/^{204}Pb$ | — | 15.580 | 15.597 | 15.594 | — | 15.570 | — |
| $^{208}Pb/^{204}Pb$ | — | 38.540 | 38.851 | 38.769 | — | 38.398 | — |
| $\Delta 8/4$ | — | 63.23 | 45.01 | 43.74 | — | 75.19 | — |

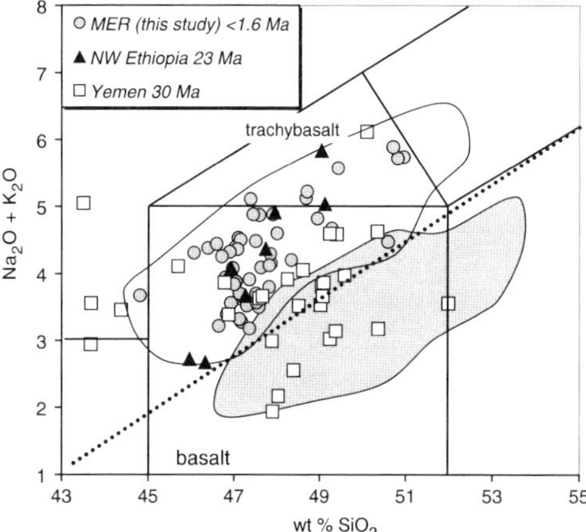


**Fig. 2.** The total alkalis–silica classification diagram for mafic volcanics of the Ethiopia–Yemen CFB province shows consistent temporal variations. Data are recalculated to 100% on an anhydrous basis with $Fe^{3+} = 15\%$ of total iron; the dotted line is the alkalic–tholeiitic division of Macdonald & Katsura (1964) for Hawaiian lavas. Mildly alkalic basalts from the modern Main Ethiopian Rift (MER samples from this study shown as data points; field includes data from Hart *et al.* 1989; Boccaletti *et al.* 1999; Chernet & Hart 1999) define a range distinct from that of tholeiitic 30 Ma Ethiopian lavas (shaded field; Hart *et al.* 1989; Pik *et al.* 1999; Kieffer *et al.* 2004). Yemeni CFBs (Chazot & Bertrand 1993; Baker *et al.* 1996*b*) generally overlap the Ethiopian suite but include samples with higher alkali contents. Note that 23 Ma basalts from NW Ethiopia (Kieffer *et al.* 2004) plot within the modern MER field.

enriched over primitive mantle (Sun & Mcdonough 1989), and generally overlap values measured in recent Djibouti lavas attributed to the Afar plume (Vidal *et al.* 1991; Deniel *et al.* 1994). Normalized incompatible trace element variation diagrams (Fig. 3) show generally parallel patterns within the Ethiopian rift suite: all samples have prominent positive Ba and Nb–Ta anomalies and weakly negative K anomalies. In detail, however, some consistent differences are found between lavas from individual volcanic centres that require distinct histories of melting and/or a limited degree of source heterogeneity along the rift axis. For example, MgO-rich Kone basalts have the highest abundances of most highly incompatible trace elements, suggesting these lavas are derived by a lower degree of melting than occurs beneath neighbouring volcanoes; this observation is consistent with the higher normative nepheline contents of those same lavas.

Regional and temporal differences observed among incompatible trace element abundances provide insight into the source regions of the mafic lavas through space and time (Fig. 3). Recent Djibouti basalts have slightly lower abundances of the most highly incompatible elements and P, but otherwise define patterns that parallel those of modern Main Ethiopian Rift lavas. Primitive Oligocene high-Ti lavas from Ethiopia and Yemen, including Adigrat, have abundances of the most highly incompatible trace elements that generally overlap the Ethiopian rift suite, but the older samples lack Ba anomalies and have consistently higher abundances of the middle and heavy REE, Zr, Hf, Ti and Y (Pik *et al.* 1999). In contrast, Oligocene lavas with <7 wt% MgO have lower contents of the more highly incompatible elements than the Ethiopian rift lavas, and overlapping abundances of the moderately incompatible elements (Fig. 4). Most of these suites are characterized by lavas with positive Ba anomalies (leading to elevated Ba/Rb) and negative K anomalies that suggest melting in the presence of minor amphibole.

Mafic Ethiopian rift lavas define a restricted range in La/Nb–Ba/Nb that is close to primitive mantle values, indicating a limited role for enriched mantle or mantle lithosphere sources (Fig. 4). Many of the Ethiopian rift basalts plot within the field defined by recent lavas derived from the Afar plume as sampled in Djibouti and Erta' Ale. Primitive high-Ti Oligocene lavas overlap these fields at low Ba/Nb values, but moderate- and low-Ti flood basalts from both Ethiopia and Yemen extend to much higher values of La/Nb and Ba/Nb, indicating a substantial contribution from enriched mantle sources. Similarly, Ce/Pb values of the Ethiopian rift and primitive high-Ti lavas plot primarily within the range of mantle-derived basalts, whereas the other suites trend towards both crustal and lithospheric end-members (Fig. 4). This ratio is influenced by crustal contributions, as Pb is elevated in crustal rocks relative to mantle lithologies, and the Ce/Pb value of mantle-derived liquids is well constrained ($25 \pm 5$; Hofmann *et al.* 1986). It is interesting to note that Yemen flood basalts that plot within the field of modern rift lavas in Fig. 2 have consistently low values of Ce/Pb (<10) that suggest crustal contamination has played a role in elevating their alkali abundances.

## *Radiogenic isotopic data*

Radiogenic isotope signatures of recent Ethiopian rift basalts (Table 1) indicate along-axis variations in source composition and/or melt generation processes, and probably also reflect variations in both crustal stretching (Maguire *et al.* 2006) and lithospheric thinning (Bastow *et al.* 2005). Lavas from

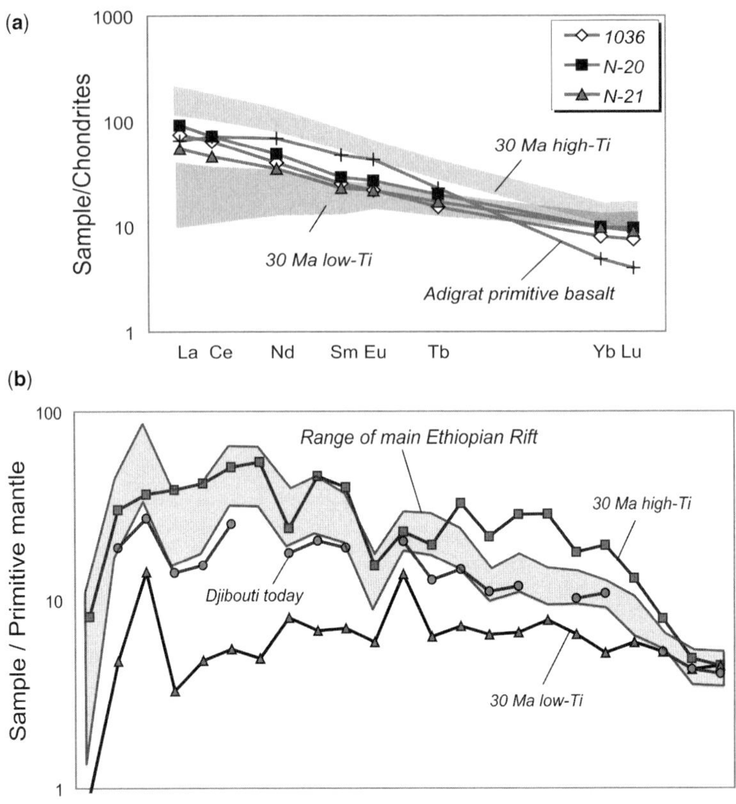


**Fig. 3.** Incompatible element abundances of selected Ethiopian lavas. (**a**) Rare earth elements. Modern rift basalts have abundances intermediate to those of low- and high-$TiO_2$ continental flood basalts (data from Pik *et al.* 1999; Kieffer *et al.* 2004); three representative samples are shown (see Table 1). A small number of primitive basalts from Adigrat have more highly concave REE patterns (Pik *et al.* 1999). Recent mafic lavas from Djibouti (Deniel *et al.* 1994) and 23 Ma Ethiopian basalts (Kieffer *et al.* 2004) have REE patterns (not shown) that fall within the range of the modern rift basalts (normalizing values of Boynton, 1984). (**b**) Primitive mantle-normalized incompatible trace element diagrams (normalizing values of Sun & McDonough 1989) for selected mafic lavas. The range of modern rift basalts is shaded; representative samples from Djibouti and the Oligocene low- and high-$TiO_2$ suites are shown for comparison. See text for discussion.

Dofan, the northernmost volcanic centre studied, have Sr–Nd–Pb isotope values that fall within the range of young plume-derived Djibouti and Erta 'Ale basalts (Fig. 5): $^{87}Sr/^{86}Sr \sim 0.7037$, $^{143}Nd/^{144}Nd \sim 0.51286$, $^{206}Pb/^{204}Pb \sim 18.83$, $^{208}Pb/^{204}Pb \sim 38.90$. Isotopic signatures trend slightly towards more enriched values from Dofan southward to Fantale and Kone. The most radiogenic values are seen at Boseti and Gedemsa, i.e. $^{87}Sr/^{86}Sr$ up to 0.7053. Post-rift basalts from the Addis Ababa series have isotopic compositions that overlap the Dofan lavas and extend to compositions similar to those observed at Fantale and Kone; syn-rift basalts sampled along the rift walls and away from the main rift axis have more enriched isotopic signatures similar to those at Boseti and Gedemsa.

At a regional scale, Sr–Nd isotope analyses do not allow individual source components to be distinguished reliably. Ethiopian rift basalts and the modern Afar plume both lie within the broad range defined by 30 million-year-old lavas from Ethiopia and Yemen; 23 Ma basalts from Choke and Guguftu plot within this field as well. The exceptions to this overall uniform picture are the two Oligocene high-Ti lavas from Adigrat noted previously to have unusual rare earth element abundances: these two samples have Sr–Nd isotopic compositions that are more depleted than any other Ethiopian lavas, and fall

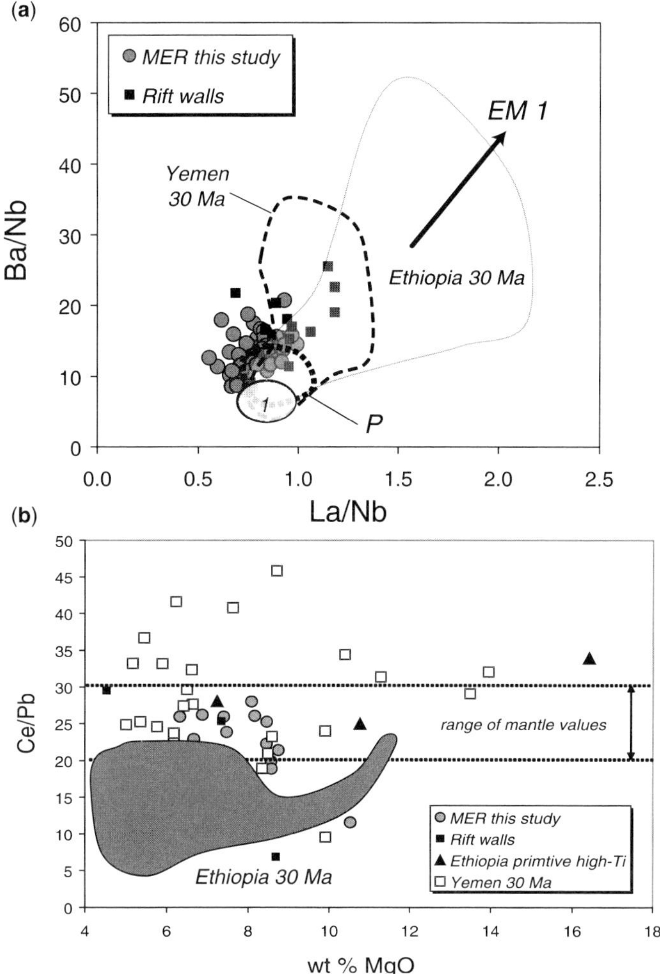


**Fig. 4.** Relative abundances of incompatible trace elements help constrain the source components that contribute to mafic volcanism. (**a**) La/Nb–Ba/Nb variations among modern rift basalts define a restricted range with La/Nb values similar to those of the primitive mantle, and Ba/Nb values somewhat elevated. Previously published analyses from the same region (Boccaletti *et al.* 1999, not shown) and our new data on syn-extensional basalts from the rift walls extend to higher values, requiring greater contributions from non-primitive source components. Most of the rift basalts plot near the field of young mafic lavas from Djibouti (Deniel *et al.* 1994) and Erta'Ale (Barrat *et al.* 1998), encircled by the dotted line and believed to represent the modern Afar plume. The Ethiopian (Pik *et al.* 1999; Kieffer *et al.* 2004) and Yemeni (Chazot & Bertrand 1993; Baker *et al.* 1996*a*) flood basalts define distinct ranges, but both groups are substantially enriched relative to the modern samples. The field labelled '1' encompasses primitive high-Ti Ethiopian lavas. (**b**) Ce/Pb–MgO variations show that most of the Main Ethiopian Rift lavas plot within the range of mantle-derived basalts, whereas the rift wall lavas and both Yemeni and Ethiopian flood basalts record contributions from lithospheric or crustal sources. Primitive high-Ti Ethiopian lavas plot within the range of mantle-derived basalts (sources of data: Baker *et al.* 1996*a*; Pik *et al.* 1999; Kieffer *et al.*).

near the MORB suite from the nearby West Sheba Ridge (Fig. 5).

Pb isotope signatures of the recent Ethiopian rift lavas form an array between the modern Afar plume and less radiogenic compositions broadly similar to those observed in 30-million-year-old continental flood basalts from both northern Ethiopia (including Adigrat) and Yemen (Fig. 5). These isotopic values are clearly distinct from compositions observed in Turkana, Kenya from 45 Ma through the present (Furman *et al.* 2004, 2006). Interestingly, the 23 Ma Ethiopian basalts from Choke and Guguftu with unusual major and trace element abundances have Pb isotopic

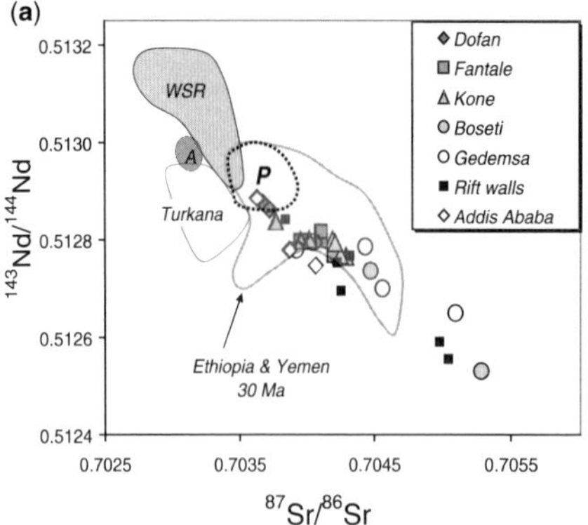


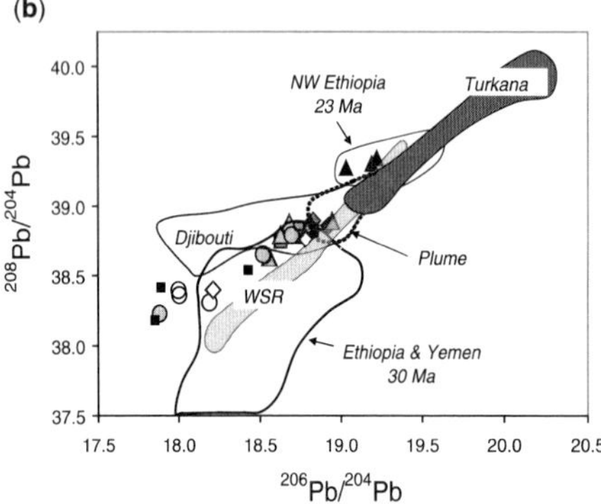


**Fig. 5.** Radiogenic isotope variations among Ethiopian lavas. (**a**) $^{87}Sr/^{86}Sr-^{143}Nd/^{144}Nd$ variations are strongly correlated among the recent rift lavas, and define a trend that extends from the Afar plume (labelled **P**), defined by mafic lavas from Djibouti (Deniel *et al.* 1994) and Erta' Ale (Barrat *et al.* 1998), towards radiogenic compositions more typical of crustal and lithospheric sources (Gedemsa data from Trua *et al.* 1999; Peccerillo *et al.* 2003). Lavas from Dofan, the northernmost volcano examined here, and the Addis Ababa series have isotopic compositions that overlap the Afar plume. Oligocene flood basalts (Chazot & Bertrand 1993; Baker *et al.* 1996*a*; Pik *et al.* 1999; Kieffer *et al.* 2004) define a broad range that encompasses many of the modern lavas but is distinct from Turkana basalts erupted 45 Ma through to the present (Furman *et al.* 2004, 2005). Two primitive high-Ti basalts from Adigrat (field labelled **A**) are compositionally distinct from all other basalts reported here. (**b**) The $^{206}Pb/^{204}Pb-^{208}Pb/^{204}Pb$ systematics of these sample suites further define the source components involved. Modern rift basalts vary from compositions overlapping the Afar plume to lower Pb isotope values, and generally overlap the range of 30 Ma flood basalts from both Ethiopia (including Adigrat) and Yemen. Lavas from Turkana have consistently higher Pb isotope compositions, ranging from a high-$\mu$ component sampled between 40–17 Ma to the modern Afar plume, and there is no overlap between the flood and rift basalts and the Turkana suite. Two 23 Ma lavas from Choke, Ethiopia (Kieffer *et al.* 2004) have elevated Pb isotope compositions relative to all other Ethiopian rocks (see text for discussion).

compositions that fall within the Turkana array rather than the long-term Ethiopia/Afar array (Fig. 5).

## Discussion: melting histories and basaltic source regions

### *Depth and extent of melting*

Modern Ethiopian rift basalts can be distinguished from the Oligocene continental flood basalts on the basis of major and trace element geochemistry, whilst the radiogenic isotopic compositions of the two groups show greater variability and, perhaps as a result, greater overlap. We first explore the extent to which the bulk geochemical differences reflect spatial and temporal variations in processes associated with melt extraction and emplacement, or whether they instead require distinct suites of lithospheric and sublithospheric source domains.

Primitive recent Ethiopian rift basalts have bulk compositions that can be derived by melting fertile spinel peridotite as suggested, for example, on the basis of their REE profiles (Fig. 3). The $SiO_2$ and FeO* contents of these lavas (back-corrected for olivine fractionation; Fig. 6) plot near the range of experimentally derived melts of fertile peridotite at a pressure of *c.* 30 kb (Baker & Stolper 1994; Kushiro 1996). The associated depth of *c.* 100 km is greater than the lithospheric thickness inferred on the basis of mantle tomography (e.g. Bastow *et al.* 2005; Tiberi *et al.* 2005). The melting pressures inferred for nepheline-normative lavas in this manner are slightly but consistently greater than those obtained for hypersthene-normative basalts (Fig. 6), although there is some overlap between the groups. A similar pattern is observed among young Djibouti lavas: samples with lower degrees of silica saturation indicate melting at greater depths than those inferred for hypersthene- and quartz-normative MgO-rich lavas (Fig. 6).

Among the Oligocene flood basalts, inferred pressures of melting range from 15 to $\geq$30 kb for hypersthene- and quartz-normative samples, and from 30 to >40 kb for nepheline-normative lavas. The more highly silica-saturated lavas also yield calculated primary melt compositions with higher $SiO_2$ and FeO* relative to the nepheline-normative lavas, consistent with their derivation by greater degrees of melting (cf. Kushiro 1996; Wasylenki *et al.* 2003). A more realistic picture of the melting process provides for polybaric melting, i.e. vertical aggregates of melts derived over a range of depths within a melting column that extends over tens of kilometres. The pressure

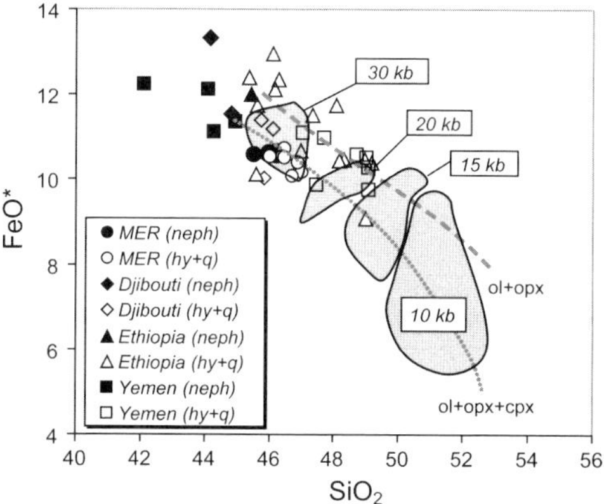


**Fig. 6.** $FeO^*$ versus $SiO_2$ for selected mafic lavas and experimental melts in equilibrium with mantle peridotite (Baker & Stolper 1994; Kushiro 1996). The dotted line connects experimental runs for melts coexisting with cpx + oliv + opx, and the dashed line connects experimental melts coexisting with oliv + opx after higher degrees of melt removal. Major element compositions were renormalized with total iron as FeO, then corrected for olivine fractionation to Mg# $\sim$ 72; only samples with Mg# > 62 were corrected using this procedure. Experimental data are plotted as reported, and are shown in the enclosed fields. Filled symbols indicate calculated primary melts for lavas with nepheline-normative compositions; open symbols indicate starting compositions that are quartz- or hypersthene-normative. Calculated primary melts for nepheline-normative lavas indicate consistently higher pressures and lower extents of melting than the more silica-saturated compositions. Sources of data: Deniel *et al.* 1994; Baker *et al.* 1996*a*; Pik *et al.* 1999; Kieffer *et al.* 2004.

estimates presented here provide simply a framework for comparison. Interestingly, there is no geographical control on the calculated melting depths, as lavas from Ethiopia and Yemen indicate comparable melting conditions. Most of the low-Ti Oligocene lavas are too evolved to be evaluated using this calculation scheme, as they have MgO contents below 7 wt.% and have undergone extensive fractionation of olivine $\pm$ clinopyroxene $\pm$ plagioclase feldspar (Pik *et al.* 1998, 1999).

## *Identifying lithospheric and sub-lithospheric source regions*

Prior studies of the Ethiopia–Yemen continental flood basalts and Plio-Quaternary lavas from the Afar region have identified several distinct source domains that contribute to magma genesis. To first order, these components include the Afar plume, one or more additional mantle reservoir(s), and a subordinate contribution from the continental crust or mantle lithosphere (Hart *et al.* 1984; Chazot & Bertrand 1993; Baker *et al.* 1996*a*, 2000; Pik *et al.* 1999; Kieffer *et al.* 2004). The Afar plume is broadly thought to include a component with $^{87}Sr/^{86}Sr$ $\sim$0.704, $^{143}Nd/^{144}Nd$ $\sim$0.51295, $^{206}Pb/^{204}Pb$ $\sim$18.8 and $^{3}He/^{4}He$ up to 19 $R/R_A$ (Hart *et al.* 1989; Schilling *et al.* 1992; Vidal *et al.* 1991; Marty *et al.* 1993, 1996; Deniel *et al.* 1994; Scarsi & Craig 1996; Pik *et al.* 1999), but the compositions of the remaining source regions are poorly constrained. For example, Hart *et al.* (1989) infer an enriched mantle component, while Pik *et al.* (1999) infer the presence of both MORB-like mantle and a second depleted component.

One reason for the lack of consensus is that Pb isotope compositions vary widely among Ethiopian plateau and rift lavas, and tend not to correlate well with Sr–Nd systematics (Fig. 5). Although Pb isotopic compositions of mafic melts are often heterogeneous at all scales, presumably reflecting melt segregation and aggregation processes (Saal *et al.* 1998; Kent *et al.* 2002; Bryce & DePaolo 2004), there are consistent spatial and temporal patterns among the lavas that correlate with both geographical and geochemical parameters, bringing us closer to an integrated understanding of the source regions.

Following Hart (1984), we use the dimensionless parameter $\Delta 8/4$ to systematize variations in Pb isotopic composition ($\Delta 8/4$ indicates the difference in $^{208}Pb/^{204}Pb$ of a sample relative to the Northern Hemisphere Reference Line; see Table 1); this parameter itself has no physical meaning but provides a framework within which to evaluate chemical distinctions among and between sample suites. Variations in $\Delta 8/4$- $^{87}Sr/^{86}Sr$ among the flood basalts and recent mafic lavas show clear regional trends that can be interpreted in terms of source regions sampled by the melting process (Fig. 7; Baker *et al.* 1996*a*). The modern Afar plume is well-constrained by the isotopic compositions of MgO-rich lavas from Erta 'Ale and Djibouti, and the remaining samples form linear arrays that radiate from this region of isotopic space.

Recent lavas from the Ethiopian rift and Djibouti define a trend towards high, EM1-like $^{87}Sr/^{86}Sr$ and $\Delta 8/4$. Samples in this group have La/Nb-Ba/Nb values that are not dramatically elevated relative to primitive mantle, so they indicate a radiogenic source region with only moderately enriched incompatible trace element abundances, interpreted as the lower crust (Baker *et al.* 1996*a*; Trua *et al.* 1999). A second well-defined trend among flood basalts from Ethiopia and Yemen extends towards high, EMII-like $^{87}Sr/^{86}Sr$ and low $\Delta 8/4$ values. This trend was interpreted by Baker *et al.* (1996a) as indicating an upper crustal component. The

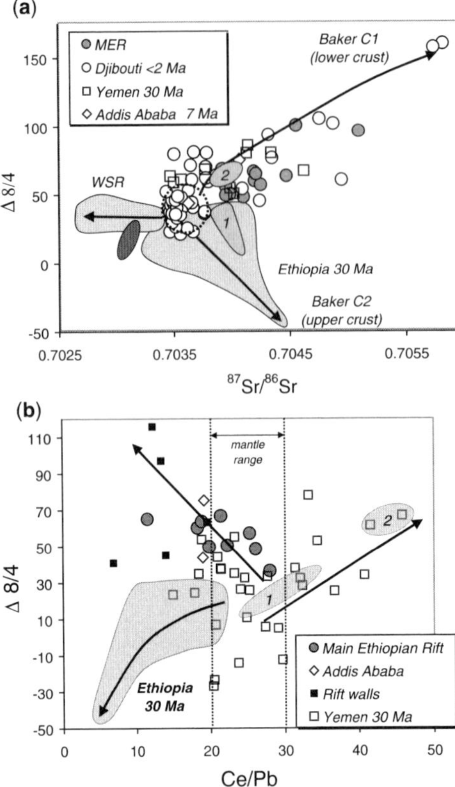


**Fig. 7.** $^{87}Sr/^{86}Sr$ and Ce/Pb vs. $\Delta 8/4$ require regional and temporal differences in source components contributing to mafic volcanism in the Afar region. Prior to plotting, samples from Yemen where crustal contamination could be demonstrated through anomalously high values of $^{207}Pb/^{204}Pb$ and/or $\delta^{18}O$ (Baker *et al.* 1996*a*, 2000) were removed. (**a**) Regional sample suites define tight arrays that extend from the Afar plume (dotted circle; data from Djibouti (Deniel *et al.* 1994) and Erta' Ale (Barrat *et al.* 1998) towards distinct additional source regions. Labelled components from Baker *et al.* (1996*a*) are discussed in the text. Fields labelled 1 and 2 indicate primitive high-Ti Oligocene lavas from Ethiopia (Pik *et al.* 1999) and Yemen (Baker *et al.* 1996a), respectively; field 3 encircles primitive Oligocene Adigrat basalts (Pik *et al.* 1999). Data on mid-ocean basalts from the nearby West Sheba Ridge (WSR) are from Schilling *et al.* (1992); Main Ethiopian Rift (MER) samples include lavas in Table 1 and data from Trua *et al.* (1999) and Peccarillo *et al.* (2003). (**b**) High $\Delta 8/4$ components recorded in young Main Ethiopian Rift (MER) lavas and Oligocene Yemen continental flood basalts (Baker *et al.* 1996*a*) can be resolved by their distinct Ce/Pb values. These data indicate distinct source components in each region that lie outside the range of mantle-derived liquids (Hofmann *et al.* 1996). All samples plotted here have >6.5 wt% MgO. Fields labelled 1 and 2 indicate primitive high-Ti Oligocene lavas from Ethiopia (Pik *et al.* 1999) and Yemen (Baker *et al.* 1996*a*), respectively. See text for discussion.

high Ba/Nb–La/Nb values of these samples suggest instead that the high $^{87}Sr/^{86}Sr$ component may reside in the enriched mantle lithosphere; low degrees of melting can also produce variations in these parameters, but the quartz- and hypersthene-normative compositions of these lavas are not consistent with low degrees of mantle melting. There is evidence that the source region was metasomatized to a small degree by hydrous fluids: the high Ba and low K contents of the lavas relative to Kone basalts suggest the presence of amphibole. Interestingly, the primitive high-Ti lavas interpreted by Pik *et al.* (1999) as representing the essential Oligocene plume (HT2 series) also plot within this array, suggesting they include a contribution from hydrated mantle lithosphere. The high degrees of melting inferred for both of these sample groups makes assimilation of surrounding lithologies during ascent physically plausible.

The third distinct trend observed among the mafic lavas examined here is defined by recent sea-floor basalts from the West Sheba Ridge near Djibouti (Schilling *et al.* 1992). These samples define an array trending from the modern plume composition towards lower $^{87}Sr/^{86}Sr$ at near-constant $\Delta 8/4$. We interpret this trend to indicate involvement of the convecting upper-mantle source region for mid-ocean ridge basalts. Note that the high-Ti lavas from Adigrat plot along this array and are clearly isotopically different from all other lavas observed in the Oligocene through Recent Ethiopian volcanic provinces. We emphasize that the volume of these lavas is small, and their occurrence rare, but they provide insight into a source component apparently sampled only during flood basalt volcanism.

To resolve the additional components involved in long-term Afar volcanism we incorporate trace element evidence, as relative abundances of incompatible trace elements are sensitive to the mineralogy of specific source regions. Variations in Ce/Pb against $\Delta 8/4$ (Fig. 7) show clearly the trend of crustal involvement for Ethiopian Oligocene basalts identified above. Interestingly, trends for modern Ethiopian rift lavas and Oligocene Yemen flood basalts can be distinguished from one another on the basis of their distinct Ce/Pb values. From this diagram, we conclude that the crustal source sampled by the modern Ethiopian rift lavas has been available throughout development of the rift, whereas the Yemen rocks record involvement of a separate crustal component that is absent in Ethiopia.

We can also combine our understanding of the melting conditions with the isotopic characteristics of the basalts. In each area, basalts derived at higher pressures and/or through lower degrees of melting (i.e. nepheline-normative compositions) have more depleted Sr and Nd isotopic signatures than corresponding lavas that are derived at more shallow depths and/or through greater extents of melting (Fig. 8). This observation reinforces the existence of an extended melting column within the mantle, and helps constrain the occurrence of different source domains in space and time.

The integrated geochemical evidence allows us to identify the source components involved in regional volcanism over the past 30 million years. First, Oligocene through Recent basalt volcanism in Ethiopia, Yemen and Djibouti includes a contribution from a common composition which we interpret as the Afar plume (Fig. 8). This component is identifiable on the basis of Sr–Nd–Pb isotopes and incompatible trace element abundances in basalts of all ages from throughout the region (Figs 4 and 5). Its persistence through time suggests that it forms the core of the upwelling mantle plume, and its occurrence in melts formed at high pressures indicates that it is a fertile composition. This plume composition appears to be nearly constant in both space and time, and that all of the regional trace element and isotopic data form binary arrays that radiate from this composition towards geochemically distinct end-members (cf. Schilling *et al.* 1992).

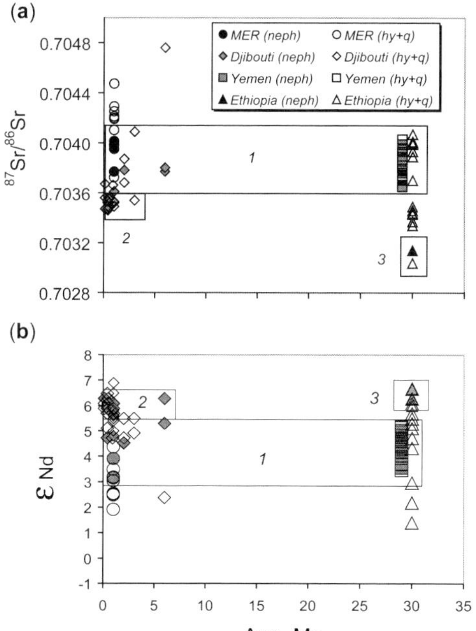


**Fig. 8.** Temporal and spatial variations in Sr–Nd isotopic compositions of primitive (>6.5 wt% MgO) lavas from Ethiopia, Yemen and the modern Afar plume indicate three source components present at low degrees of mantle melting. Each of these source components is isotopically unradiogenic relative to those recorded in lavas derived by higher degrees of melting. Box 1 encompasses a compositional range similar to bulk silicate earth that is found in lavas of all ages, interpreted here as representing the core of the Afar plume. Box 2 is a MORB-like composition found in ocean floor basalts and lavas from Djibouti (Deniel *et al.* 1994) and Erta' Ale (Barrat *et al.* 1998). Box 3 encloses lavas from Adigrat, Ethiopia (Pik *et al.* 1999). See text for discussion.

Three compositionally distinct source domains can be identified on the basis of the mixing trajectories. One is the depleted mantle (DMM), observed both in recent lavas from Djibouti and Erta 'Ale and in primitive Oligocene lavas from Adigrat, Ethiopia (Fig. 8). We also identify an enriched mantle (EMI-like) or crustal component in both low- and high-Ti mafic lavas from the entire Ethiopian plateau; these lavas are characterized by higher La/Nb–Ba/Nb values and require higher time-integrated parent–daughter ratios to account for their more radiogenic isotope compositions relative to the plume or DMM. A second enriched component (EMII-like) is recorded in within-rift and rift wall basalts from central Ethiopia and Djibouti, the off-rift volcanics from the Addis Ababa series, and the Oligocene Yemen basalts. We concur with Baker *et al.* (1996*a*) and Trua *et al.* (1999) that this source is located in the lower crust; alternatively the enriched source may represent mafic magmas emplaced during the earliest stage of flood basalt activity (e.g. Mackenzie *et al.* 2005; Maguire *et al.* 2006).

Two observations about the enriched source domains and their occurrence are significant. First, the enriched mantle contributions are found only in lavas derived from moderate degrees of mantle melting. Secondly, each enriched mantle composition forms a unique mixing array with the Afar plume. Together, these observations indicate that the plume records melting of chemically distinct source domains in the lithosphere and asthenosphere. This conclusion is consistent with numerical models that suggest an ascending plume will melt or displace, rather than entrain, surrounding and overlying mantle material (Farnetani *et al.* 2002).

## Geodynamic evolution of the region

### *Documenting the head-to-tail transition*

Numerical modelling of thermochemical plumes derived from a deep mantle thermal boundary layer predicts that geochemical domains present within a plume head will also be preserved in magmatism associated with the plume tail (Farnetani *et al.*

2002). This prediction results in part from the observation that ascending plume material displaces, rather than entrains, surrounding ambient upper mantle, so heterogeneities are thinned and stretched, preserving discrete flow paths within the plume core. We use this numerical prediction, derived for vigorous plumes emplaced beneath oceanic lithosphere, as a framework to assess geochemical heterogeneity in mafic lavas derived from the Afar plume throughout its 30 million-year history. Long-lived spatial variability in plume tails has been documented in Hawaii (e.g. Lassiter & Hauri 1998; DePaolo *et al.* 2001; Blichert-Toft *et al.* 2003; Mukhopadhyay *et al.* 2003; Kurz *et al.* 2004; Abouachami *et al.* 2005; Bryce *et al.* 2005) and the Galápagos (Geist 1992; White *et al.* 1993; Hoernle *et al.* 2000; Harpp & White 2001; Geldmacher *et al.* 2003; Thompson *et al.* 2004). There are good reasons to expect that the results from Afar may differ from those observed in the oceanic setting: the Afar plume has a substantially lower buoyancy flux than those calculated for Hawaii, Galápagos and Iceland (Ebinger & Sleep 1998) and was emplaced beneath thick continental lithosphere. In addition, the sampling density in the Ethiopia–Yemen flood basalt province is not yet sufficient to resolve consistent spatial variations in lava chemistry. Nonetheless, our results will help test the robustness of the numerical models developed for oceanic hot spots.

The new data presented here indicate temporal trends in source availability and mixing that in turn provide insight into the long-term evolution of the Afar plume system. The Oligocene flood basalts in Ethiopia form the thickest sequence, indicating the highest flux of both material and heat. Significantly, these lavas show the greatest and most uniform contribution from the lithospheric mantle, indicating high degrees of melting over a broad region. This lithospheric source region apparently does not play a major role in generating the recent MER basalts, suggesting it has been thinned and/or partially removed under the rift axis itself. In contrast, lithospheric geochemical signatures dominate the chemistry of lavas from highly extended portions of southwestern Ethiopia (Stewart & Rogers 1996; George *et al.* 1998; Rogers *et al.* 2000) and much of the Kenya rift (e.g. Macdonald 1994; Späth *et al.* 2001), indicating that lithospheric removal is not widespread.

The geochemical evidence suggests that the total magmatic and thermal flux of the Afar plume has waned through time. Modern mafic lavas from Djibouti, Erta' Ale and the Main Ethiopian Rift often show evidence of crustal interaction, presumably occurring in magma chambers within the crust (e.g. Ayalew *et al.* 2006). In the modern rift, low magma supply rates over the past 3 million years led to development of nested silicic calderas at, for example, Kone and Gedemsa. We suspect that the low magma flux calculated for the modern Afar plume relative to Hawaii and the Galápagos (Ebinger & Sleep 1998) means that it may not be feasible to correlate quantitatively any spatial variations in the chemical domains over the life of the Afar plume.

## *Geochemical evidence for one- and two-plume models*

The recognition of long-lived geochemically consistent contributions to magmatism in the Ethiopia–Yemen region supports models that call upon the Afar plume to be the source for both Oligocene (plume head) and recent (plume tail) magmatism. The broad distribution of long-lived volcanism in Ethiopia and portions of the East African rift has led to various models of plume structure. The suggestion that there are two discrete plumes (George *et al.* 1998; Rogers *et al.* 2000) was based on Sr–Nd isotopic compositions of lavas from southern Ethiopia and the Kenya rift. Although appealing in some ways, this model has two difficulties that have not been overcome: (1) it does not adequately account for the distribution of lavas within the EARS, as no viable plume track exists for the proposed Kenya plume; and (2) the south Ethiopian lavas were primarily derived from the lithosphere and thus provide little insight into sub-lithospheric processes (Furman *et al.* 2004).

The single-plume model proposed by Ebinger & Sleep (1998) places a plume head beneath southern Ethiopia at *c.* 45 Ma and attributes widespread Cenozoic magmatism (e.g. Hoggar and Tibesti massifs, the modern Afar area) to the lateral transport of plume material along zones of thinned lithosphere. This model benefits from physical simplicity, but is not immediately consistent with recent geophysical studies indicating possible plume stems beneath both Afar and the Tanzania craton (Debayle *et al.* 2001; Nyblade *et al.* 2000; Weeraratne *et al.* 2003; Montelli *et al.* 2004).

Similarly, the consistent Pb isotope differences between contemporaneous sub-lithospheric magmas erupted in Ethiopia and Turkana (Furman *et al.* 2004, 2005) are compatible with the geophysical evidence for separate plume stems beneath these two areas. Turkana lavas have a distinctive high $^{206}Pb/^{204}Pb$ signature that is not observed in mafic lavas associated with the Afar plume. This characteristic Pb isotopic signature is, however, found in other areas associated with the South African superplume (e.g. St Helena – Chaffey *et al.* 1989; South Africa – Janney *et al.* 2002). Over the long history of volcanism in Ethiopia and Turkana, only basalts from the 23 Ma Choke shield volcano have Pb isotope compositions that plot within the Turkana array (Fig. 5; Kieffer *et al.* 2004). These data are intriguing because they may indicate northward magma transport associated with volcano-tectonic activity at 23 Ma in the Turkana area (Furman *et al.* 2006).

The isotopic data thus cannot discriminate between models invoking two unrelated plumes (e.g. George *et al.* 1998) or one large heterogeneous plume with multiple stems as predicted by numerical and experimental results (Ishida *et al.* 1999; Davaille *et al.* 2003). The clear geophysical evidence for the South African superplume (Ritsema *et al.* 1999; Gurnis *et al.* 2000; Zhao 2001; Ni *et al.* 2002) appears to favour the latter scenario, although the shallow mantle structure cannot yet be resolved in sufficient spatial detail. The two small isotopically distinct plume stems may rise from the superplume and interact with regionally unique crustal and sublithospheric sources during ascent. Tomographic investigation of plume pairs (Azores and Canaries, Ascension and St Helena) indicates that two areas of upwelling can merge at mantle depths over 1000 km, thus tapping a common source at depth (Montelli *et al.* 2004). The large physical dimension of the South African superplume—it represents the largest-known heat and mass transfer from the lowermost portion of the mantle (e.g. Romanowicz & Gung 2002) — makes it likely that a variety of source materials coexist within the ascending region, and hence two or more chemically distinct plume heads could come from the same major deep mantle thermal and chemical anomaly.

## Conclusions

Mafic lavas from Ethiopia and Yemen record 30 million years of volcanism associated with the Afar plume. Modern Ethiopian rift basalts have major and trace element geochemical features that are distinct from those of the Oligocene continental flood basalts, whereas the isotopic characteristics of these groups show greater overlap. The geochemical differences can be interpreted in terms of melting processes (depth and degree of melting), as well as mixing of melts of the Afar plume and from mantle sources with three end-member compositions. Taken together, the geochemical evidence indicates that the Afar plume is a long-lived feature of the mantle and that its fundamental isotopic composition has effectively not changed during the transition from Oligocene plume head volcanism to modern plume tail activity. Temporal changes in Ethiopian volcanic geochemistry thus appear to reflect the evolving tectonic environment rather than a significant change in sub-lithospheric source composition.

Specifically, we find that:

1. Oligocene lavas are generally derived by greater degrees of melting than the recent basalts. Most of the flood basalts are hypersthene- and quartz-normative, while the younger lavas include a higher proportion of nepheline-normative compositions.
2. The isotopic compositions of Ethiopian and Yemeni basalts of all ages are linked with the depth and degree of melting: the Sr and Nd isotopic signatures of lavas derived from smaller degrees of melting and/or melting at higher pressure are more depleted than those signatures found in lavas derived at shallower depths and/or through greater extents of melting.
3. The Sr- and Pb-isotopic signatures of Oligocene through Recent mafic lavas define three distinct binary mixing arrays that radiate from a common composition interpreted as the Afar plume. The end-members involved in mixing include both depleted (DM, interpreted as the convecting upper mantle source region for mid-ocean ridge basalts) and enriched (EMI-like, inferred to be within the lower crust and EMII-like, inferred to reside within the upper crust) mantle components, each of which is observed in a geographically restricted area.
4. Lack of ternary mixing among these components suggests that the ascending plume melts chemically distinct local source domains in the overlying asthenosphere and lithosphere, rather than entraining and incorporating ambient mantle on a regional scale.

This research was funded by NSF-EAR 0207764 and a George H. Deike, Jr. grant from The Pennsylvania State University to T. Furman. J. Bryce acknowledges support from NSF-EAR 0338385 during preparation of the manuscript. We are grateful to E. Klein and G. Dwyer for performing major and trace element analyses. Special thanks to Kassahun Ejeta and Roeland Doust for masterful assistance during field work. Thanks also to Daniel and Elizabeth Larson for help with sampling and field logistics. We gratefully acknowledge the many thought-provoking conversations with members of EAGLE-US, UK and Ethiopia throughout this project. The manuscript benefited greatly from thoughtful reviews by G. WoldeGabriel and R. Macdonald; we thank C. Ebinger for careful editorial handling.

## Appendix

### *Supplemental material: analytical procedures*

E99- and KO-series lavas and samples AA99-6, -19 were analysed at the Centre Récherches

Pétrographiques et Géochimiques, Nancy, France by G. Yirgu (major elements by ICP-AES and trace elements by ICP-MS). All other lavas were analysed for major and minor elements (including Ba and Sr) by DCP on an ARL-Fisons Spectraspan 7 and for $P_2O_5$ and all trace elements by ICP-MS using a VG PlasmaQuad-3 at Duke University (M. Rudnicki and G. Dwyer, analysts). Precision based on replicate analyses of samples and natural basalt standards are generally $<1\%$ for $SiO_2$, Sr, Y, Zr, Nb, La, Ce; $<3\%$ for other major elements, Ba, Sr, Rb, Cs, Cr, Sc, V, Co, Pr, Nd, Sm, Eu, Gd, Tb, Dy, Ho, Er, Hf and Ta; $<5\%$ Ni, Yb, Lu, Pb, Th and $<8\%$ for U. All oxides are presented as wt %; trace elements are given as ppm.

All Pb analyses, a majority of the Nd analyses, and several Sr analyses were carried out in the clean laboratories and analytical facilities at San Diego State University. These Sr, Nd, and Pb isotope ratios were determined from a single HF-$HNO_3$ dissolution of approximately 0.4 g of rock powder. Pb was separated according to a two-column anion exchange chromatographic procedure described by Hanan & Schilling (1989). Nd and Sr elemental concentrations were subsequently carried out using procedures modified from Schilling *et al.* (1994), with Eichrom Ln-Spec resin for the rare earth separation, and Eichrom Sr-Spec as a final step for cleaning up the Sr. The total analytical blanks for these procedures were $<90$ pg Pb, $<200$ pg Nd, and $<300$ pg Sr. Thus no blank corrections were made to the data reported in Table 1. At SDSU, the Nd and Pb isotope ratios were measured on the Nu Plasma HR by multicollector inductively coupled plasma mass spectrometry (MC-ICP-MS). Sr isotope ratios were measured using VG Instruments Sector 54 seven-collector thermal ionization mass spectrometer. The $^{143}Nd/^{144}Nd$ and $^{87}Sr/^{86}Sr$ ratios were normalized to correct for mass fractionation using values of $^{146}Nd/^{144}Nd = 0.7219$ and $^{88}Sr/^{86}Sr = 0.1194$, respectively. The $^{143}Nd/^{144}Nd$ of AMES Nd standard $= 0.512\,118 \pm 0.000\,003$, and the NIST SRM 987 Sr standard $= 0.710\,219 \pm 0.000\,005$ (all $2\sigma/n$). Sr isotopic data in Table 1 are renormalized to NIST SRM 987 $= 0.710\,250$. The Nd reported in Table 1 is given relative to La Jolla Nd $= 0.511\,816$. Pb isotope compositions analysed by MC-ICP-MS on the Nu Plasma HR at SDSU using the Tl-doping technique with NIST SRM 997 Tl and $^{205}Tl/^{203}Tl = 2.3889$ (White *et al.* 2000; Thirlwall 2002; Albarède *et al.* 2004). NIST SRM981 was used to correct the Pb isotope data for fractionation and machine bias (White *et al.* 2000). The NIST 981 standard was run every second or third analysis. The final Pb data were adjusted to the Todt *et al.* (1996) values for NIST 981 relative to the bracketing standard runs. The NIST SRM 981 Pb standard averaged (with $2\sigma/n$) $^{206}Pb/^{204}Pb = 16.9372$ (0.0006), $^{207}Pb/^{204}Pb = 15.4915$ (0.0006), and $^{208}Pb/^{204}Pb = 36.6957$ (0.0016) during the analysis period. The uncertainties for the Pb isotope ratios, representing error propagation of the in-run $2\sigma/n$ analytical errors plus the $2\sigma/n$ reproducibility of the standards, were less than 0.002 for $^{206}Pb/^{204}Pb$ and $^{207}Pb/^{204}Pb$ and $<0.006$ for $^{208}Pb/^{204}Pb$.

A portion of the Sr and Nd isotopic measurements presented in Table 1 were carried out in the W. M. Keck Foundation Center for Isotope Geochemistry at UCLA. Approximately 200–250 mg aliquots of whole rock powders were digested in a HF–$HNO_3$ mixture. Samples for Sr isotopic study were prepared using standard ion chromatographic separation techniques employing cation exchange resins. Rare earth separations were carried out using a procedure involving a second cation ion exchange resin. Purified Sr salts were loaded in $Ta_2O_5$ on rhenium filaments for analysis. The strontium isotopic measurements were made via thermal ionization mass spectrometry (TIMS) on a VG sector-54 multicollector mass spectrometer operating in dynamic mode. Isotopic fractionation was corrected with an internal normalization of $^{86}Sr/^{88}Sr = 0.1194$. The average for NIST SRM 987 measured during the course of these analyses was $0.710\,238 \pm 0.000\,027$ ($2\sigma$). The $^{87}Sr/^{86}Sr$ values reported in Table 1 are normalized to NBS-987 values of 0.710 250. Neodymium isotopic measurements were also carried out TIMS on a Sector 54-30 multicollector operating in dynamic mode; Nd isotopic compositions were normalized to $^{146}Nd/^{144}Nd = 0.7219$ to correct for within-run fractionation. Standards run over the course of the analyses provided mean values (with $2\sigma$) of the La Jolla standard ($n = 5$) of $^{143}Nd/^{144}Nd = 0.511\,830 \pm 0.000\,017$ ($2\sigma$). Two replicate samples were measured for Nd isotopic compositions in both the UCLA and SDSU facilities. These samples (N22 and E43) agreed in $^{143}Nd/^{144}Nd$ values within 0.000 022, or well within the overlapping uncertainties. The values reported in the table for these two samples are those evaluated at SDSU. For the sake of consistency, the UCLA Nd isotopic data are normalized to the SDSU values for the La Jolla Nd standard (i.e. $^{143}Nd/^{144}Nd = 0.511\,816$).

## References

ABOUACHAMI, W., HOFMANN, A.W., GALER, S.J.G., FREY, F.A., EISELE, J. & FEIGENSON, M. 2005. Lead isotopes reveal bilateral asymmetry and vertical continuity in the Hawaiian mantle plume. *Nature*, **434**, 851–856.

ALBARÈDE, F., TÉLOUK, P., BLICHERT-TOFT, J., BOYET, M., AGRANIER, A. & NELSON, B. 2004. Precise and accurate isotopic measurements using multiple-collector ICPMS. *Geochimica et Cosmochimica Acta*, **68**, 2725–2744.

ARNDT, N.T., CZAMANSKE, G.K., WOODEN, J.L. & FEDORENKO, V.A. 1993. Mantle and crustal contributions to continental flood volcanism. *Tectonophysics*, **223**, 39–52.

AYALEW, D., EBINGER, C., BOURDON, E., WOLFENDEN, E., YIRGU, G. & GRASSINEAU, N. 2006. Temporal compositional variation of early syn-rift rhyolites along the southwestern Red Sea and northern Main Ethiopian Rift: implications for dyking of the crust. *In*: YIRGU, G., EBINGER, C.J. & MAGUIRE, P.K.H. (eds) *The Afar Volcanic Province within the East African Rift System*. Geological Society, London, Special Publications, **259**, 121–130.

BAKER, J.A., THIRLWALL, M.F. & MENZIES, M.A. 1996*a*. Sr–Nd–Pb isotopic and trace element evidence for crustal contamination of plume-derived flood basalts: Oligocene flood volcanism in western Yemen. *Geochimica et Cosmochimica Acta*, **60**, 2559–2581.

BAKER, J.A., SNEE, L. & MENZIES, M. 1996*b*. A brief Oligocene period of flood volcanism in Yemen: implications for the duration and rate of continental flood volcanism at the Afro-Arabian triple junction. *Earth and Planetary Science Letters*, **138**, 39–55.

BAKER, J.A., MACPHERSON, C.G., MENZIES, M.A., THIRLWALL, M.F., AL-KADASI, M. & MATTERY, D.P. 2000. Resolving crustal and mantle contributions to continental flood volcanism, Yemen; constraints from mineral oxygen isotope data. *Journal of Petrology*, **41**, 1805–1820.

BAKER, M.B. & STOLPER, E.M. 1994. Determining the composition of high-pressure mantle melts using diamond aggregates. *Geochimica et Cosmochimica Acta*, **58**, 2811–2827.

BARRAT, J.A., FOURCADE, S., JAHN, B.M., CHEMINÉE, J.L. & CAPDEVILA, R. 1998. Isotope (Sr, Nd, Pb, O) and trace element geochemistry of volcanics from the Erta' Ale range (Ethiopia). *Journal of Volcanology and Geothermal Research*, **80**, 85–100.

BASTOW, I.D., STUART, G.W., KENDALL, J.-M. & EBINGER, C.J., 2005. Upper-mantle seismic structure in a region of incipient continental breakup: northern Ethiopian Rift. *Geophysical Journal International*, **162**, 479–493.

BENOIT, M., NYBLADE, A., TUJI, M., AYELE, A., ASFAW, L., LANGSTON, C. & VANDECAR, J. 2003. Upper mantle seismic velocity structure beneath East Africa and the depth extent of thermal anomalies. *Geophysical Research Abstracts*, **5**, 07361.

BERTRAND, H., CHAZOT, G., BLICHERT-TOFT, J. & THORAL, S. 2003. Implications of widespread high-μ volcanism on the Arabian Plate for Afar mantle plume and lithosphere composition. *Chemical Geology*, **198**, 47–61.

BLICHERT-TOFT, J., FREY, F.A. & ALBARÈDE, F. 1999. Hf isotope evidence for pelagic sediments in the source of Hawaiian basalts. *Science*, **285**, 879–882.

BLICHERT-TOFT, J., WEIS, D., MAERSCHALK, C., AGRANIER, A. & ALBARÈDE, F. 2003. Hawaiian hot spot dynamics as inferred from the Hf and Pb isotope evolution of Manua Kea volcano. *Geochemistry, Geophysics, Geosystems*, **4**, 8704, doi:10.1029/2002GC000340, 2003.

BOCCALETTI, M., MAZZUOLI, R., BONINI, M., TRUA, T. & ABEBE, B. 1999. Plio-Quaternary volcanotectonic activity in the northern sector of the Main Ethiopian Rift: relationships with oblique rifting. *Journal of African Earth Sciences*, **29**, 679–698.

BRYCE, J.G. & DEPAOLO, D.J. 2004. Pb isotopic heterogeneity in basaltic phenocrysts. *Geochimica et Cosmochimica Acta*, **68**, 4453–4468.

BRYCE, J.G., DEPAOLO, D.J. & LASSITER, J.C. 2005. Geochemical structure of the Hawaiian plume: Sr, Nd, and Os isotopes in the 2.8 km HSDP-2 section of Mauna Kea volcano. *Geochemistry, Geophysics, Geosystems*, **6**, Q09G18, doi: 10.1029/2004GC000809.

BURKE, K. 1996. The African plate. *South African Journal of Geology*, **99**, 339–409.

CALAIS, E., EBINGER, C., HARTNADY, C. & NOCQUET, J.M. 2006. Kinematics of the East African Rift from GPS and earthquake slip vector data. *In*: YIRGU, G., EBINGER, C.J. & MAGUIRE, P.K.H. (eds) *The Afar Volcanic Province within the East African Rift System*. Geological Society, London, Special Publications, **259**, 9–22.

CHAFFEY, D.J., CLIFF, R.A. & WILSON, B.M. 1989. Characterization of the St. Helena magma source. *In*: SAUNDERS, A.D. & NORRY, M.J. (eds) *Magmatism in the Ocean Basins*. Geological Society, London, Special Publications, **42**, 257–276.

CHAZOT, G. & BERTRAND, H. 1993. Mantle sources and magma-continental crust interactions during early Red Sea–Gulf of Aden rifting in Southern Yemen; elemental and Sr, Nd, Pb isotope evidence. *Journal of Geophysical Research*, **98**, 1819–1835.

CHERNET, T. & HART, W.K. 1999. Petrology and geochemistry of volcanism in the northern Main Ethiopian Rift–southern Afar transition region. *Acta Vulcanologica*, **11**, 21–41.

DAVAILLE, A., LE BARS, M. & CARBONNE, C. 2003. Thermal convection in a heterogeneous mantle. *Comptes Rendus Geoscience*, **335**, 141–156.

DAVIDSON, A. & REX, D.C. 1980. Age of volcanism and rifting in southwestern Ethiopia. *Nature*, **283**, 657–658.

DEBAYLE, E., LEVEQUE, J.-J. & CARA, M. 2001. Seismic evidence for a deeply rooted low-velocity anomaly in the upper mantle beneath the northeastern Afro/Arabian continent. *Earth and Planetary Science Letters*, **193**, 423–436.

DENIEL, C., VIDAL, P., COULON, C., VELLUTINI, P.-J. & PIGUET, P. 1994. Temporal evolution of mantle sources during continental rifting: the volcanism of Djibouti. *Journal of Geophysical Research*, **99**, 2853–2869.

DEPAOLO, D.J., BRYCE, J.G., DODSON, A., SHUSTER, D. & KENNEDY, B. 2001. Isotopic evolution of Mauna Loa and the chemical structure of the Hawaiian plume. *Geochemistry, Geophysics, Geosystems*, **2**(7), doi:10.1029/2000GC000133.

DUGDA, M.T., NYBLADE, A.A., JULIA, J., LANGSTON, C.A., AMMON, C.J. & SIMIYU, S. 2005. Crustal structure in Ethiopia and Kenya from receiver function analysis. *Journal of Geophysical Research*, **110**, B01303.

DUNCAN, R.A. 1981. Hotspots in the southern ocean—An absolute framework of reference for motion of the Gondwana continents, *Tectonophysics*, **74**, 29–42.

EWART, A., MARSH, J.S., MILNER, S.C., DUNCAN, A.R., KAMBER, B.S. & ARMSTRONG, R.A. 2004. Petrology and geochemistry of Early Cretaceous bimodal continental flood volcanism of the NW Etendeka, Namibia; Part 1, Introduction, mafic lavas and re-evaluation of mantle source components. *Journal of Petrology*, **45**, 59–105.

EBINGER C.J. & SLEEP N.H. 1998. Cenozoic magmatism throughout east Africa resulting from impact of a single plume. *Nature*, **395**, 788–791.

FARNETANI, C.G., LEGRAS, B. & TACKLEY, P.J. 2002. Mixing and deformations in mantle plumes. *Earth and Planetary Science Letters*, **196**, 1–15.

FITTON, J.G., SAUNDERS, A.D., NORRY, M.J., HARDARSON, B.S. & TAYLOR, R.N. 1997. Thermal and chemical structure of the Iceland plume: new insights from the alkaline basalts of the Snaefell volcanic centre. *Journal of the Geological Society*, London, **153**, 197–208.

FRAM, M.S. & LESHER, C.E. 1997. Generation and polybaric differentiation of East Greenland early Tertiary flood basalts. *Journal of Petrology*, **38**, 231–275.

FURMAN, T., BRYCE, J.G., KARSON, J. & IOTTI, A. 2004. East African Rift System (EARS) plume structure: insights from Quaternary mafic lavas of Turkana, Kenya. *Journal of Petrology*, **45**, 1069–1088.

FURMAN, T., KALETA, K.M. & BRYCE, J.G. 2006, Tertiary mafic lavas of Turkana, Kenya: constraints on temporal evolution of the EARS and the occurrence of HIMU volcanism in Africa. *Journal of Petrology* (doi: 10.1093/petrology/eg1009).

GEIST, D.J. 1992. An appraisal of melting processes and the Galápagos Hotspot: major- and trace-element evidence. *Journal of Volcanology and Geothermal Research*, **52**, 65–82.

GELDMACHER, J., HANAN, B.B., BLICHERT-TOFT, J., HARPP, K., HOERNLE, K., HAUFF, F., WERNER, R. & KERR, A.C. 2003. Hafnium isotopic variations in volcanic rocks from the Caribbean Large Igneous Province and the Galápagos hot spot tracks. *Geochemistry, Geophysics, Geosystems*, **4**, doi: 10.1029/2002GC000477.

GEORGE, R., ROGERS, N. & KELLEY, S. 1998. Earliest magmatism in Ethiopia: evidence for two mantle plumes in one flood basalt province. *Geology*, **26**, 923–926.

GIBSON, S.A., THOMPSON, R.N. & DICKIN, A.P. 2000. Ferropicrites: geochemical evidence for Fe-rich streaks in upwelling mantle plumes. *Earth and Planetary Science Letters*, **174**, 355–374.

GORDON, R.G. & JURDY, D.M. 1986. Cenozoic global plate motions. *Journal of Geophysical Research*, **91**, 12,389–12,406.

GRIPP, A.E. & GORDON, R.G. 2002. Young tracks of hotspots and current plate velocities. *Geophysical Journal International*, **150**, 321–361.

GURNIS, M., MITROVICA, J.X., RITSEMA, J. & VAN HEIJST, H.-J. (2000). Constraining mantle density structure using geological evidence of surface uplift rates: the case of the African Superplume. *Geochemistry, Geophysics, Geosystems*, **2**, Paper no. 1999GC000035.

HAILEAB, B., BROWN, F.H., MCDOUGALL, I. & GATHOGO, P.N. 2004. Gombe Group basalts and initiation of Pliocene deposition in the Turkana depression, northern Kenya and southern Ethiopia. *Geological Magazine*, **141**, 41–53.

HANAN, B.B. & SCHILLING, J.-G. 1997. The dynamic evolution of the Iceland mantle plume; the lead isotope perspective. *Earth and Planetary Science Letters*, **151**, 43–60.

HANAN, B.B. & SCHILLING, J.-G. 1989. Easter microplate evolution: Pb isotope evidence. *Journal of Geophysical Research*, **94**, 7432–7448.

HANAN, B., BLICHERT-TOFT, J., KINGSLEY, R. & SCHILLING, J.-G. 1999. Depleted Iceland mantle plume geochemical signature: Artifact of multicomponent mixing? *Geochemistry, Geophysics, Geosystems*, **1**, doi:10.1029/1999GC000009.

HARPP, K.S. & WHITE, W.M. 2001. Tracing a mantle plume: isotopic and trace element variations of Gal'pagos seamounts. *Geochemistry, Geophysics, Geosystems*, **2**, Paper no. 2000GC000137.

HART, S.R. 1984. A large-scale isotopic anomaly in the southern hemisphere. *Nature*, **309**, 753–757.

HART, W.K., WOLDEGABRIEL, G., WALTER, R.C. & MERTZMAN, S.A. 1989. Basaltic volcanism in Ethiopia: constraints on continental rifting and mantle interactions. *Journal of Geophysical Research*, **94**, 7731–7748.

HAWKESWORTH, C., KELLY, S., TURNER, S. & LE ROEX, A. 1998. Mantle processes during Gondwana breakup and dispersal. *Journal of African Earth Sciences*, **27**, 108–109.

HILL, R.I., CAMPBELL, I.H., DAVIES, G.F. & GRIFFITHS, R.W. 1992. Mantle plumes and continental tectonics. *Science*, **256**, 186–193.

HOERNLE, K., WERNER, R., MORGAN, J.P., GARBE-SCHÖNBERG, C.-D., BRYCE, J. & MRAZEK, J. 2000. Existence of complex spatial zonation in the Galápagos plume for at least 14 m.y. *Geology*, **28**, 435–438.

HOFMANN, A.W., JOCHUM, K.P., SEUFERT, M. & WHITE, W.M. 1986. Nb and Pb in oceanic basalts: new constraints on mantle evolution. *Earth and Planetary Science Letters*, **79**, 33–45.

HOFMANN, C., COURTILLOT, V., FÉRAUD, G., ROCHETTE, P., YIRGU, G., KETEFO, E. & PIK, R. 1997. Timing of the Ethiopian flood basalt event and implications for plume birth and global change. *Nature*, **389**, 838–841.

ISHIDA, M., MARUYAMA, S., SUETSUGU, D., MATSUZAKA, S. & EGUCHI, T. 1999. Superplume

Project: Towards a new view of whole Earth dynamics. *Earth Planets Space*, **51**, i–v.

JANNEY, P.E. & CASTILLO, P.R. 2001. Geochemistry of the oldest Atlantic oceanic crust suggests mantle plume involvement in the early history of the central Atlantic Ocean. *Earth and Planetary Science Letters*, **192**, 291–302.

JANNEY, P.E., LE ROEX, A.P., CARLSON, R.W. & VILJOEN, K.S. 2002. A chemical and multi-isotope study of the Western Cape melilitite province, South Africa: implications for the sources of kimberlites and the origin of the HIMU signature in Africa. *Journal of Petrology*, **43**, 2339–2370.

KEMPTON, P.D., FITTON, J.G., SAUNDERS, A.D., NOWELL, G.M., TAYLOR, R.N., HARDARSON, B.S. & PEARSON, G. 2000. The Iceland plume in space and time: a Sr–Nd–Pb–Hf study of the North Atlantic rifted margin. *Earth and Planetary Science Letters*, **177**, 255–271.

KENDALL, J.-M., STUART, G.W., EBINGER, C.J., BASTOW, I.D. & KEIR, D. 2005. Magma-assisted rifting in Ethiopia. *Nature*, **433**, 146–148.

KENT, A.J.R., BAKER, J.A. & WIEDENBECK, M. 2002. Contamination and melt aggregation processes in continental flood basalts: constraints from melt inclusions in Oligocene basalts from Yemen. *Earth and Planetary Science Letters*, **202**, 577–594.

KERANEN, K., KLEMPERER, S.L., GLOAUGEN, R. and the EAGLE working group (2004). Three-dimensional seismic imaging of a protoridge axis in the Main Ethiopian rift. *Geology*, **32**, 949–952.

KIEFFER, B., ARNDT, N., ET AL. 2004. Flood and shield basalts from Ethiopia: magmas from the African Superswell. *Journal of Petrology*, **45**, 793–834.

KURZ, M.D., CURTICE, J., LOTT, III, D.E. & SOLOW, A. 2004. Rapid helium isotopic variability in Mauna Kea shield lavas from the Hawaiian Scientific Drilling Project. *Geochemistry, Geophysics, Geosystems*, **5**, Q04G14, doi:10.1029/2002 GC000439.

KUSHIRO, I. 1996. Partial melting of a fertile mantle peridotite at high pressures: an experimental study using aggregates of diamond. *In*: BASU, A. & HART, S. (eds) *Earth Processes: Reading the Isotopic Code*. American Geophysical Union, 109–122.

LASSITER, J.C. & HAURI, E.H. 1998. Osmium-isotope variations in Hawaiian lavas: evidence for recycled oceanic lithosphere in the Hawaiian plume. *Earth and Planetary Science Letters*, **164**, 483–496.

MACDONALD, R. 1994. Petrological evidence regarding evolution of the Kenya Rift Valley. *Tectonophysics*, **236**, 373–390.

MACKENZIE, G., THYBO, H. & MAGUIRE, P. 2005. Crustal velocity structure across the Main Ethiopian Rift: results from two-dimensional wide-angle seismic modeling. *Geophysical Journal International*, **163**, 994–1006.

MAGUIRE, P.K.H., KELLER, G.R. ET AL. 2006. Crustal structure of the northern Main Ethiopian Rift from the EAGLE controlled-source survey: a snapshot of incipient lithospheric break-up. *In*: YIRGU, G., EBINGER, C.J. & MAGUIRE, P.K.H. (eds) *The Afar Volcanic Province within the East African Rift System*. Geological Society, London Special Publications, **259**, 269–291.

MAHATSENTE, R., JENTZSCH, G. & JAHR, T. 1999. Crustal structure of the Main Ethiopian Rift from gravity data: 3-dimensional modeling. *Tectonophysics*, **313**, 363–382.

MARTY, B., APPORA, I., BARRAT, J.-A., DENIEL, C., VELLUTINI, P. & VIDAL, P. 1993. He, Ar, Sr, Nd and Pb isotopes in volcanic rocks from Afar: evidence for a primitive mantle component and constraints on magmatic sources. *Geochemical Journal*, **27**, 223–232.

MARTY, B., PIK, R. & YIRGU, G. 1996. Helium isotopic variations in Ethiopian plume lavas: nature of magmatic sources and limit on lower mantle contribution. *Earth and Planetary Science Letters*, **144**, 223–237.

MOHR, P.A. 1962. The Ethiopian rift system. *Bulletin of the Geophysical Observatory*, **3**, 33–62.

MOHR, P. 1983. Ethiopian flood basalt province. *Nature*, **303**, 577–584.

MOHR, P. & ZANETTIN, B. 1988. The Ethiopian flood basalt province. *In*: MACDOUGALL, J.D. (ed.) *Continental Flood Basalts*. Kluwer Academic, Dordrecht, 63–110.

MONTELLI, R., NOLET, G., DAHLEN, F.A., MASTERS, G., ENGDAHL, E.R. & HUNG, S.-H. 2004. Finite-frequence tomography reveals a variety of plumes in the mantle. *Science*, **303**, 388–343.

MORGAN, W.J. 1972. *In*: SHAGAM, R. ET AL. (eds) *Studies in Earth and Space Science*. Geological Society of America Memoir, **132**.

MUKHOPADHYAY, S., LASSITER, J.C., FARLEY, K.A. & BOGUE, S.W. 2003. Geochemistry of Kauai shield-stage lavas: Implications for the chemical evolution of the Hawaiian plume. *Geochemistry, Geophysics, Geosystems*, **4**(1), 1009, doi:10.1029/2002 GC000342.

NI, S., TAN, E., GURNIS, M. & HELMBERGER, D. 2002. Sharp sides to the African superplume. *Science*, **296**, 1850–1852.

NYBLADE, A.A., OWENS, T.J., GURROLA, H., RITSEMA, J. & LANGSTON, C.A. 2000. Seismic evidence for a deep upper mantle thermal anomaly beneath east Africa. *Geology*, **28**, 599–602.

PEATE, D.W., HAWKESWORTH, C.J. & MANTOVANI, M.S.M. 1992. Chemical stratigraphy of the Paraná lavas (South America): classification of magma types and their spatial distribution. *Bulletin of Volcanology*, **55**, 119–139.

PECCERILLO, A., BARBERIO, M.R., YIRGU, G., AYALEW, D., BARBIERI, M. & WU, T.W. 2003. Relationship between mafic and peralkaline silicic magmatism in continental rift settings: a petrological, geochemical and isotopic study of the Gedemsa volcano, central Ethiopian rift. *Journal of Petrology*, **44**, 2003–2032.

PIK, R., DENIEL, C., COULON, C., YIRGU, G., HOFMANN, C. & AYALEW, D. 1998. The northwestern Ethiopian Plateau flood basalts: classification and spatial distribution of magma types. *Journal of Volcanology and Geothermal Research*, **81**, 91–111.

PIK, R., DENIEL, C., COULON, C., YIRGU, G. & MARTY, B. 1999. Isotopic and trace element signatures of Ethiopian flood basalts; evidence for plume-lithosphere interactions. *Geochimica et Cosmochimica Acta*, **63**, 2263–2279.

PIK, R., MARTY, B. & HILTON, D.R. 2006. How many mantle plumes in Africa? The geochemical point of view. *Chemical Geology*, **226**, 100–114.

RICHARDS, M.A., DUNCAN, R.A. & COURTILLOT, V.E. 1989. Flood basalts and hot-spot tracks: plume heads and tails. *Science*, **246**, 103–107.

RITSEMA, J., VAN HEIJST, H.J., & WOODHOUSE, J.H. 1999. Complex shear wave velocity structure imagedc beneath Africa and Iceland. *Science*, **286**, 1925–1928.

ROGERS, N., MACDONALD, R., FITTON, J.G., GEORGE, R., SMITH, M. & BARREIRO, B. 2000. Two mantle plumes beneath the East African rift system: Sr, Nd and Pb isotope evidence from Kenya Rift basalts. *Earth and Planetary Science Letters*, **176**, 387–400.

ROGERS, N. 2006. Basaltic magmatism and the geodynamics of the African Rift. *In*: YIRGU, G., EBINGER, C.J. & MAGUIRE, P.K.H. (eds) *The Afar Volcanic Province within the East African Rift System*. Geological Society, London Special Publications, **259**, 77–93.

ROMANOWICZ, B. & GUNG, Y. 2002. Superplumes from the core-mantle boundary to the lithosphere; implications for heat flux. *Science*, **296**, 513–516.

RUDNICK, R.L., MCDONOUGH, W.F. & CHAPPELL, B.W. 1993. Carbonatite metasomatism in the northern Tanzanian mantle: petrographic and geochemical characteristics. *Earth and Planetary Science Letters*, **114**, 463–475.

SAAL, A.E., HART, S.R., SHIMIZU, N., HAURI, E.J. & LAYNE, G.D. 1998. Pb isotopic variability in melt inclusions for oceanic island basalts, Polynesia. *Science*, **282**, 1481–1484.

SAMUEL, H. & FARNETANI, C.G. 2003. Thermochemical convection and helium concentrations in mantle plumes. *Earth and Planetary Science Letters*, **207**, 39–56.

SCARSI, P. & CRAIG, H. 1996. Helium isotope ratios in Ethiopian Rift basalts. *Earth and Planetary Science Letters*, **144**, 505–516.

SCHILLING, J.-G., HANAN, B.B., MCCULLY, B., KINGSLEY, R.H. & FONTIGNIE, D. 1994. Influence of the Sierra Leone mantle plume on the equatorial Mid-Atlantic Ridge: a Nd–Sr–Pb isotopic study. *Journal of Geophysical Research*, **99**, 12005–12028.

SCHILLING, J.-G., KINGSLEY, R.H., HANAN, B.B. & MCCULLY, B.L. 1998. Nd–Sr–Pb isotopic variations along the Gulf of Aden: evidence for Afar mantle plume-continental lithosphere interaction. *Journal of Geophysical Research*, **97**, 10927–10966.

SHAW, J.E., BAKER, J.A., MENZIES, M.A., THIRLWALL, M.F. & IBRAHIM, K.M. 2003. Petrogenesis of the largest intraplate volcanic field on the Arabian Plate (Jordan): a mixed lithosphere–asthenosphere source activated by lithospheric extension. *Journal of Petrology*, **44**, 1657–1679.

SILVER, P., RUSSO, R.M. & LITHGOW-BERTELLONI, C. 1998. Coupling of South American and African plate motion and plate deformation. *Science*, **279**, 60–63.

SPÄTH, A., LE ROEX, A.P. & OPIYO-AKECH, N. 2001. Plume-lithosphere interaction and the origin of continental rift-related alkali volcanism – the Chyulu Hills volcanic province, southern Kenya. *Journal of Petrology*, **42**, 765–787.

STEWART, K. & ROGERS, N. 1996. Mantle plume and lithosphere contributions to basalts from southern Ethiopia. *Earth and Planetary Science Letters*, **139**, 195–211.

SUN, S.-S. & MCDONOUGH, W.F. 1989. Chemical and isotopic systematics of oceanic basalts: implications for mantle composition and processes. *In*: SAUNDERS, A.D. & NORRY, M.J. (eds) *Magmatism in the Ocean Basins*. Geological Society, London, *Special Publications*, **42**, 313–345.

THIRLWALL, M.F. 1995. Generation of the Pb isotopic characteristics of the Iceland plume. *Journal of the Geological Society*, London, **152**, 991–996.

THIRLWALL, M.F. 2002. Multicollector ICP-MS analysis of Pb isotopes using a $^{207}Pb$–$^{204}Pb$ double spike demonstrates up to 4000 ppm/amu systematic errors in Tl-normalization. *Chemical Geology*, **184**, 255–279.

THOMPSON, P.M.E., KEMPTON, P.D., WHITE, R.V., KERR, A.C., TARNEY, J., SAUNDERS, A.D., FITTON, J.G. & MCBIRNEY, A. 2004. Hf–Nd isotope constraints on the origin of the Cretaceous Caribbean plateau and its relationship to the Galápagos plume. *Earth and Planetary Science Letters*, **217**, 59–75.

TIBERI, C., EBINGER, C., BALLU, V., STEWART, G. & OLUMA, B. 2005. Inverse models of gravity data from the Red Sea–Aden–East African rifts triple junction zone. *Geological Society of America Bulletin*, **163**, 775–787.

TIBERI, C., EBINGER, C., BALLU, V., STUART, G. & OLUMA, B. 2005. Inverse models of gravity data from the Red Sea-Aden-East African Rifts triple junction zone. *Geophysical Journal International*, **163**, 775–787.

TODT, W., CLIFF, R.A., HANSER, A. & HOFMANN, A.W. 1996. Evaluation of a 202Pb-205Pb double spike for high precision lead isotope analysis. *In*: BASU, A. & HART, S. (eds) *Earth Processes: Reading the Isotopic Code*. American Geophysical Union, 429–437.

TRUA, T., DENIEL, C. & MAZZUOLI, R. 1999. Crustal control in the genesis of Plio-Quaternary bimodal magmatism of the Main Ethiopian Rift (MER): geochemical and isotopic (Sr, Nd, Pb) evidence. *Chemical Geology*, **155**, 201–231.

VETEL, W. & LE GALL, B. 2006. Dynamics of prolonged continental extension in magmatic rifts: the Turkana Rift case study (Northern Kenya). *In*: YIRGU, G., EBINGER, C.J. & MAGUIRE, P.K.H. (eds) *The Afar Volcanic Province within the East African Rift System*. Geological Society, London, *Special Publications*, **259**, 209–233.

VIDAL, P., DENIEL, C., VELLUTINI, P.J., PIGUET, P., COULON, C., VINCENT, J. & AUDIN, J. 1991.

Changes of mantle sources in the course of a rift evolution: the Afar case. *Geophysical Research Letters*, **18**, 1913–1916.

WASYLENKI, L.E., BAKER, M.B., KENT, A.J.R. & STOLPER, E. 2003. Near-solidus melting of the shallow upper mantle: partial melting experiments on depleted peridotite. *Journal of Petrology*, **44**, 1163–1191.

WEERARATNE, D.S., FORSYTH, D.W., FISHER, K.M. & NYBLADE, A.A. 2003. Evidence for an upper mantle plume beneath the Tanzanian Craton from Rayleigh wave tomography. *Journal of Geophysical Research*, **108**, doi: 10.1029/2002JB002273.

WHITE, W.M., MCBIRNEY, A.R. & DUNCAN, R.A. 1993. Petrology and geochemistry of the Galápagos Islands: portrait of a pathological mantle plume. *Journal of Geophysical Research*, **98**, 19,533–19,563.

WHITE, W.M., ALBARÈDE, F. & TÉLOUK, P. 2000. High-precision analysis of Pb isotopic ratios using multicollector ICP-MS. *Chemical Geology*, **167**, 257–270.

WHITE, R.S., & MCKENZIE, D.M. 1995. Mantle plumes and flood basalts, *Journal of Geophysical Research*, **100**, 17,543–17,585.

WOLDEGABRIEL, G., ARONSON, J.L. & WALTER, R.C. 1990. Geology, geochronology, and rift basin development in the central sector of the Main Ethiopian Rift. *Geological Society of America Bulletin*, **102**, 439–458.

WOLFENDEN, E., EBINGER, C., YIRGU, G., DEINO, A. & AYALEW, D. 2004. Evolution of the northern Main Ethiopian Rift: birth of a triple junction. *Earth and Planetary Science Letters*, **224**, 213–228.

ZHAO, D. 2001. Seismic structure and origin of hotspots and mantle plumes. *Earth and Planetary Science Letters*, **192**, 251–265.

ZUMBO, V., FÉRAUD, G., VELLUTINI, P., PIGUET, P. & VINCENT, J. 1985. First $^{40}Ar/^{39}Ar$ dating on Early Oligocene to Plio-Pleistocene magmatic events of the Afar–Republic of Djibouti. *Journal of Volcanology and Geothermal Research*, **65**, 281–295.

# Temporal compositional variation of syn-rift rhyolites along the western margin of the southern Red Sea and northern Main Ethiopian Rift

D. AYALEW[1], C. EBINGER[2], E. BOURDON[2,3], E. WOLFENDEN[2], G. YIRGU[1] & N. GRASSINEAU[2]

[1]*Department of Earth Sciences, Addis Ababa University, PO Box 729/1033 Addis Ababa, Ethiopia (e-mail: dereayal@geol.aau.edu.et)*

[2]*Department of Geology, Royal Holloway University of London, Egham, TW20 0EX, UK*

[3]*Now at: Institut de Geologie, Université de Neuchatel, Rue Emile Argand 11, CP 2, 2007 Neuchatel, Switzerland*

**Abstract:** Structural and geochronological relations indicate that the felsic rocks at the top of the Oligocene flood basalt sequences in the Afar volcanic province were erupted coevally with the initial rifting in the Red Sea and Gulf of Aden. In this study, we use the newly established volcanic–tectonic history to examine the geochemical evolution with time of felsic volcanics as rifting has progressed to seafloor spreading in the southern Red Sea and northern Main Ethiopian Rift. Geochemical analyses (major and trace elements; Sr, Nd and O isotopic compositions) of syn-rift rhyolites ranging in age from 28 to 2.5 Ma indicate that the rhyolites can be derived from mantle-sourced basaltic magma through fractional crystallization accompanied by variable amounts of crustal contamination (e.g. $^{87}Sr/^{86}Sr = 0.70489–0.70651$; $^{143}Nd/^{144}Nd = 0.51254–0.51283$; $\delta^{18}O = +4.5$ to $+6.4‰$). The input of crust tends to increase with time, which suggests the weakening and heating of the crust in response to lithospheric thinning and magma injection in the past *c.* 30 Ma. These results support earlier structural and thermomechanical models for rift formation in the southern Red Sea rift and the younger, less-evolved northern Main Ethiopian Rift system.

The initiation of continental flood basalt (CFB) magmatism is spatially and temporally related to continental break-up, which results in the formation of oceanic crust, at least within the past 200 Ma. However, the relationships between the timing of CFB formation and rifting leading to ocean-floor formation are complex, and can vary in time and space (see also Courtillot *et al.* 1999; Hawkesworth *et al.* 1999; Menzies *et al.* 2002). Magmatism can predate continental extension by several million years (e.g. Ethiopia–Yemen), magmatism and break-up can be synchronous (e.g. Paraná–Etendeka, North Atlantic Tertiary volcanic province/Greenland–UK), or magmatism can postdate break-up by several million years (e.g. Australia–India). This is not unusual, in that plumes may interact with continental lithosphere independent of plate-driving forces. Although the details of plume melting remain controversial, there is agreement that lithospheric thinning or pre-existing thin zones are needed to produce the huge volumes of melt seen in CFB provinces.

In Ethiopia, Oligocene–Miocene felsic volcanic strata (rhyolites and minor trachytes) capping the flood basalt sequences are well exposed along the western margin of the southernmost sector of the southern Red Sea and the northernmost Main Ethiopian Rift (MER). The rhyolites are preferentially localized on or near the border fault, indicating that rifting preceded or at least was coeval with the felsic volcanism in both the southern Red Sea rift and the northernmost Main Ethiopian Rift (Wolfenden *et al.* 2004, 2005). This shows a clear relationship between the emplacement of large volumes of rhyolite and formation of the large offset border fault systems. The timespan for silicic volcanism shows southward younging along the border fault and the silicic volcanism migrates from the wider border faults towards narrow (*c.* 10 km-wide) magmatic segments within the rift (Wolfenden *et al.* 2005). The overall observations show multiple episodes of riftward migration of the locus of magmatism and faulting through time.

Samples of felsic volcanic rock (pyroclastic flow and lava) from the western margin (Fig. 1) of the southern Red Sea and northern Main Ethiopian Rift, where precise geochronological data are available, were analysed for chemical (major and trace elements) and isotopic (Sr, Nd and O) determinations. Geochemical studies on precisely dated

*From*: YIRGU, G., EBINGER, C.J. & MAGUIRE, P.K.H. (eds) 2006. *The Afar Volcanic Province within the East African Rift System*. Geological Society, London, Special Publications, **259**, 121–130.
0305-8719/06/$15.00 

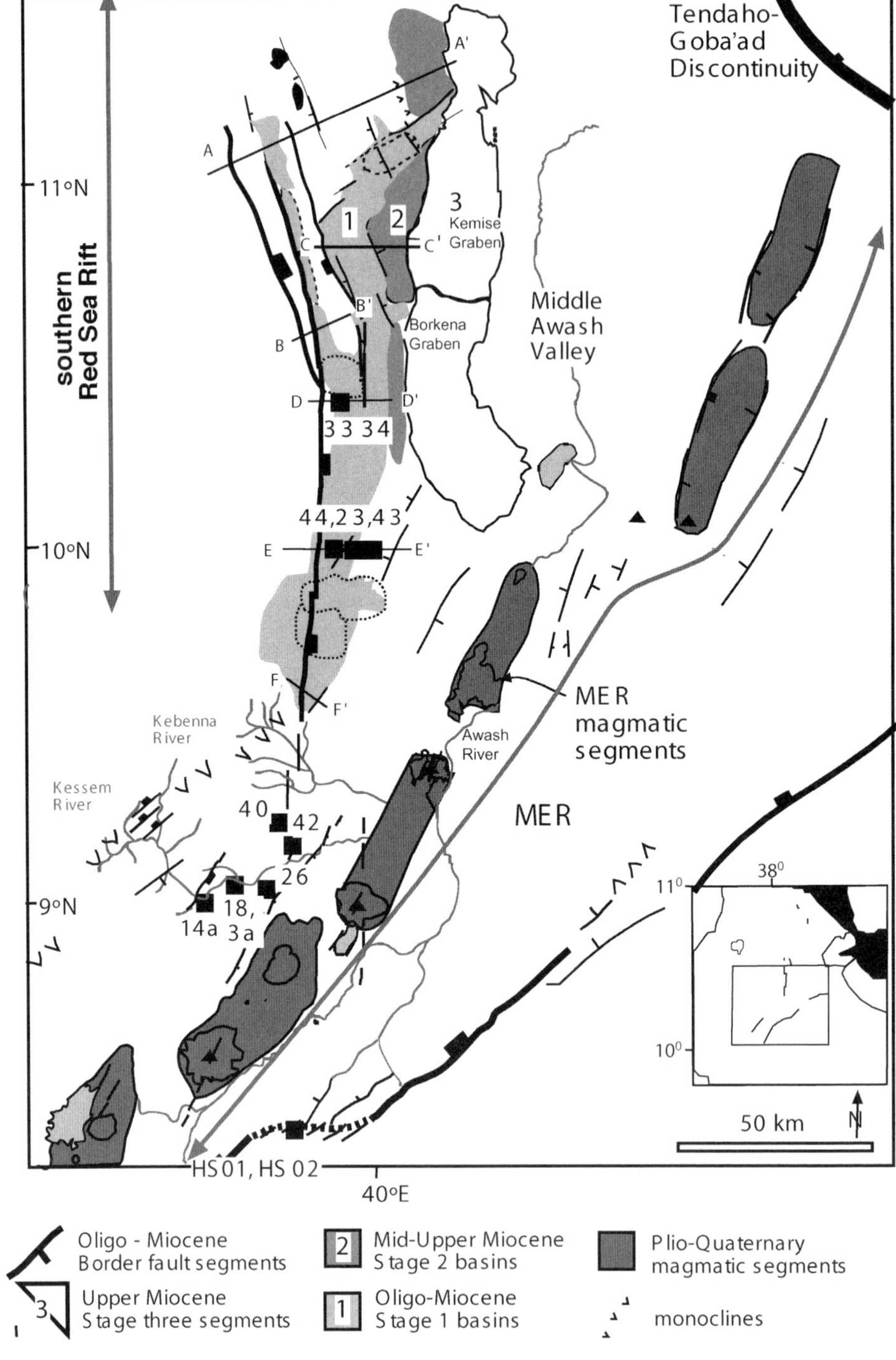


**Fig. 1.** Stages of basin development in the southernmost sector of the southern Red Sea and in the northern Main Ethiopian Rift (MER). BF = border fault. Numbering refers to stages of basin development in the Kemise, Ataye, Baso-Werena and Adama basins. The thin dashed lines enclose inter-basin volcanic complexes. Filled squares represent sample locations. The inset map shows the position of the study area with respect to the Red Sea–Afar–Main Ethiopian Rift triple junction. Adapted from Wolfenden *et al.* 2005.

samples are very sparse for the Ethiopian igneous province and this is the first comprehensive study. The main objective of this research is to constrain the geochemical evolution of the silicic volcanic rocks, which mark the onset of rifting in the southern Red Sea, through time in the light of existing and new data. We interpret our results in the light of new geophysical observations for the northern MER and uplifted plateau region that demonstrate significant modification of the crust and mantle lithosphere by magmatic processes (e.g. Kendall *et al.* 2005; Keranen *et al.* 2004; Mackenzie *et al.* 2005).

## Tectonic setting

The Red Sea–Aden–Main Ethiopian Rift triple junction lies on the broad Ethiopian plateau, believed to have developed above a Palaeogene mantle plume (Marty *et al.* 1996; Pik *et al.* 1999; Schilling & Kingsley 1992). Compilations of $^{40}Ar/^{39}Ar$ data show that flood basalts and associated rhyolites were erupted across a *c.* 1000 km diameter region between 31 and 29 Ma (Ayalew *et al.* 2002; Ayalew & Yirgu 2003; Baker *et al.* 1996; Coulié *et al.* 2003; Hofmann *et al.* 1997; Rochette *et al.* 1998; Ukstins *et al.* 2002), roughly coeval with the initiation of NE-directed extension in the southern Red Sea (Wolfenden *et al.* 2005) and the Gulf of Aden (Watchorn *et al.* 1998). An enigmatic province in southwestern Ethiopia comprising *c.* 1 km of basalts capped by rhyolites dated at 45 Ma may signal initial plume–lithosphere interactions (e.g. Ebinger *et al.* 1993; Ebinger & Sleep 1998) or mark a separate plume province (e.g. George *et al.* 1998).

Wolfenden *et al.* (2005) integrate geochronological data (Ukstins *et al.* 2002) and structural data, and show that rift formation had commenced by 29 Ma in the southern Red Sea rift, with faulting and basin subsidence propagating southward to *c.* 10° N. The Gulf of Aden rift had also initiated by 30 Ma (e.g. d'Acremont *et al.* 2005). Thus, rifting was coeval with the rhyolitic volcanism at the close of the flood basalt cycle. Initial rifting in the third arm of the Afar triple junction, the northernmost Main Ethiopian Rift (MER), initiated around *c.* 11 Ma (Chernet *et al.* 1998; Wolfenden *et al.* 2004). Initial crustal extension within the southern and central MER commenced between 18 and 15 Ma (Ebinger *et al.* 2000; WoldeGabriel *et al.* 1990). Thus, the MER, the northernmost sector of the East African rift system, propagated northward from the *c.* 25 Ma Turkana rift, a region of anomalously thinned lithosphere first rifted during Mesozoic time. The Red Sea–Aden–Main Ethiopian Rift triple junction is, therefore, a relatively young feature, developed only during the past 11 Ma, or 20 million years after the flood volcanism.

## Age of silicic volcanism along the western margin

Along the western rift margin of the Afar depression, the flood volcanic sequences overlie marine sedimentary strata deposited on a passive continental margin in Mesozoic time (Hunegnaw *et al.* 1998). The volcanic packages are up to 2000 m thick and comprise basaltic lava flows overlain by rhyolites such as ignimbrites, airfall tuffs and lavas with interbedded basalts. The flood volcanic stratigraphy commonly shows a repetitious basalt–rhyolite succession. Silicic volcanics commonly constitute up to 50% of the preserved volcanic thickness and lie towards the top of the flood volcanic sequences. The rhyolitic centres comprise massive glassy effusive volcanic rocks and phenocryst-rich crystalline intrusions, and form irregular to dome-shaped flows.

The first silicic ignimbrite in the western margin of the Afar depression was erupted at 30.2 Ma and bimodal basalt–rhyolite volcanism continued in parts of the broad flood basalt province until 10 Ma (Kieffer *et al.* 2004; Wolfenden *et al.* 2004). The ages of the silicic volcanic rocks outside the faulted rift valley range from about 30 Ma in the north of the province to 10.8 Ma at Mount Guna (Kieffer *et al.* 2004). Silicic volcanism occurred early during the main basaltic episode and continued to erupt with Quaternary volcanism occurring in the rift valleys and central plateau. The range of dates for the studied rhyolites is from 30.2 to 2.5 Ma (Fig. 2).

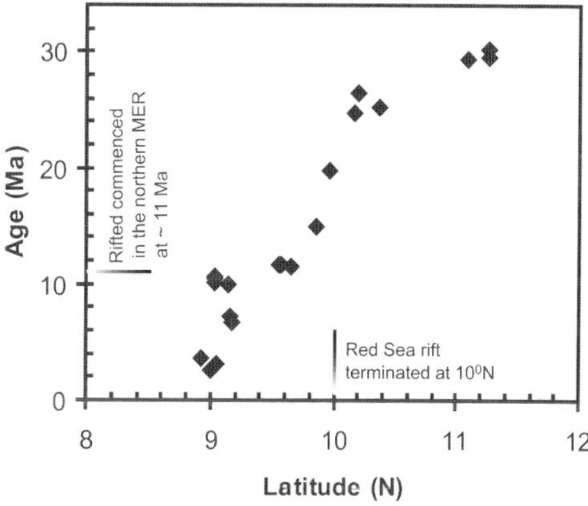


**Fig. 2.** Ages of silicic volcanic rocks in the western margin of the Afar depression, with migration of the centre of eruption from north to south. Note the Red Sea rift terminated at 10° N until linkage of the MER and southern Red Sea occurred at ~11 Ma.

## Petrological characters of the silicic volcanic rocks

The Ethiopian volcanism (both plateau and rift) is bimodal with respect to $SiO_2$ content (basic–acidic distribution), a feature shared with most CFB. Intermediate rocks are scarce or absent. In the total alkalis–silica (TAS) classification, the volcanic rocks range from basalt, through basaltic andesite, trachybasalt, basaltic trachyandesite to trachyte and rhyolite (Ayalew *et al.* 2002; Kieffer *et al.* 2004; Peccerillo *et al.* 2003; Pik *et al.* 1998).

The silicic extrusive rocks are dominantly pyroclastic flows (welded and non-welded ignimbrites) with some lava that comprise phenocrysts (with up to 20 vol.% mode) and microphenocrysts largely of sanidine and quartz with minor amounts of opaques and clinopyroxene. The phenocrysts sometimes tend to occur in cumulophyric clots and are often broken and partially resorbed. The groundmass is commonly glassy showing eutaxitic texture and it is partially devitrified with dark brown stains and shows microcrystalline felsitic and spherulitic textures. The felsic rocks show evidence of hydrothermal alteration with surface colours of orange, pink, yellow, brown and red.

Silicic volcanic rocks consist mainly of rhyolite and minor trachyte (Table 1) and show mildly peralkaline affinity. There is one intermediate rock (Sample E01–3a). Sample E01–26 has anomalously high CaO (24%) and LOI (18%). For clarity, we simply refer to these rocks as rhyolites. The petrogenesis of the silicic volcanic rocks, including sample localities, petrography, stratigraphy, age and tectonic evolution, was presented in detail elsewhere (Ayalew *et al.* 2002; Ayalew & Yirgu 2003; Wolfenden *et al.* 2004, 2005) and are omitted from the forthcoming discussion. The geochemical character of the silicic rocks varies from north to south and within each region the character of the silicic rocks matches that of the underlying flood basalts.

## Sr–Nd–O isotopes

Table 2 reports the Sr–Nd–O isotopic compositions of the rhyolites and an intermediate lava (sample E01–3a) along the western margin. The studied rhyolites have low Sr (<76 ppm, except sample E01–3a with Sr = 272 ppm) and high Rb (97–169 ppm) concentrations; consequently, they have high Rb/Sr ratios (1–11). The high Rb/Sr ratios are susceptible to even slight modification/alteration. The feldspar separates, on the other hand, have relatively low Rb/Sr ratios compared to the whole rock, so that their Sr isotopic compositions are more reliable than whole rock data. For this reason, Sr isotopic compositions of rhyolites were carried out on feldspar separates and the measured $^{87}Sr/^{86}Sr$ ratios are corrected to initial Sr isotope ratios. Alkali feldspars separated from the whole rocks have $^{87}Sr/^{86}Sr$ ratios that are markedly lower than their hosts (Table 2). Given the low Sm/Nd ratios of the studied rhyolites, the measured $^{143}Nd/^{144}Nd$ ratios are essentially considered to be the same as the initial $^{143}Nd/^{144}Nd$ ratios.

Figure 3 illustrates the Sr–Nd isotopic compositions of the study rhyolites compared to that of the plateau (Oligo-Miocene) and rift (Quaternary) rhyolites. This comparison reveals that Oligocene rhyolites, erupted probably under thick and perhaps cool crust, have the most mantle-like Sr and Nd isotopic ratios. The marginal rhyolites display lower $^{143}Nd/^{144}Nd$ and higher $^{87}Sr/^{86}Sr$ ratios compared to the plateau Oligocene rhyolites. They generally plot in the field of syn-rift Miocene and Quaternary rhyolites, indicating the link between faulting/rifting and silicic magma formation.

The oxygen isotopic compositions ($\delta^{18}O$ values) measured on quartz and alkali feldspar separates of the rhyolites range from +4.1 to +6.4‰. There are three values around +5.5‰, but the three other values are quite scattered. These values are substantially lower for rhyolites relative to the average values of most silicic volcanic rocks constrained to be between +7 and +8‰ (Eiler 2001), and may reflect that meteoric fluid circulation under high temperature conditions produced the $^{18}O$-depletion (Criss & Taylor 1986). Meteoric water involved more progressively in the superficial magma chamber probably due to tectonic deformations.

## Origin of marginal rhyolites

The west marginal rhyolites of the Afar depression cannot be derived from partial melting of the local crust, as they have relatively low initial $^{87}Sr/^{86}Sr$ ratios ~0.70549 and high $^{143}Nd/^{144}Nd$ ~0.51272, on average. Most Precambrian Ethiopian crustal rocks have $^{143}Nd/^{144}Nd$ ratios lower than those observed in the studied rocks (Asrat *et al.* 2004; Teklay *et al.* 1998).

Silicic partial melts of juvenile basaltic underplates cannot be distinguished from that derived by differentiation of basaltic magmas on major element variations, because both processes involve crystal–liquid equilibria. One possibility to discriminate partial melting from fractional crystallization trends is the variation patterns of compatible vs. incompatible elements (Cameron *et al.* 1996; Halliday *et al.* 1991; Hanson 1989). Figure 4 shows the variation of Sr (compatible element) vs. Zr (incompatible element) for west marginal rhyolites,

**Table 1.** *Major (wt %) and trace (ppm) elements for selected rhyolites along the western margin of the Afar depression*

| Sample | EO1–14A | EO1–26 | E01–18 | EO1–3A | E01–42 | E01–40 | EEW00–23 | EEW00–43 | EEW00–44 | EEW02–21 | EEW00–34 | EEW00–33 | HS–01 | HS–02 |
|---|---|---|---|---|---|---|---|---|---|---|---|---|---|---|
| Age (Ma) | 3.555 | 2.54 | 10.144 | 10.56 | 10.008 | 6.619 | 20–26 | ~ 19 | 15–19 | 11.673 | ~25 | 20–25 | 7.807 | 7.98 |
| Lat (°N) | 8.921 | 9 | 9.029 | 9.039 | 9.138 | 9.178 | 9.927 | 9.964 | 10.001 | 10.690 | 10.331 | 10.356 | 8.34 | 8.35 |
| Long (°E) | 39.5501 | 39.6325 | 39.5675 | 39.5653 | 39.7253 | 39.5428 | 39.981 | 39.937 | 39.869 | 40.025 | 39.847 | 39.832 | 39.6729 | 39.6699 |
| $SiO_2$ | 75.23 | 52.71 | 73.96 | 67.24 | 73.67 | 74.37 | 73.51 | 72.58 | 71.55 | 73.55 | 74.78 | 75.20 | 70.32 | 71.70 |
| $TiO_2$ | 0.41 | 0.36 | 0.46 | 1.11 | 0.37 | 0.35 | 0.38 | 0.48 | 0.66 | 0.41 | 0.42 | 0.30 | 0.34 | 0.43 |
| $Al_2O_3$ | 10.72 | 8.52 | 11.22 | 15.51 | 11.02 | 11.07 | 10.01 | 13.70 | 14.48 | 10.81 | 13.14 | 12.18 | 13.45 | 10.37 |
| $Fe_2O_3^*$ | 4.99 | 4.89 | 3.73 | 5.19 | 4.42 | 4.45 | 5.49 | 2.92 | 1.95 | 5.22 | 1.50 | 2.34 | 4.72 | 7.72 |
| MnO | 0.12 | 0.19 | 0.08 | 0.15 | 0.09 | 0.10 | 0.18 | 0.03 | 0.08 | 0.17 | 0.04 | 0.05 | 0.11 | 0.27 |
| MgO | 0.22 | 0.38 | 0.14 | 2.08 | 0.03 | 0.21 | 0.22 | 0.39 | 0.35 | 0.13 | 0.06 | 0.12 | 0.11 | 0.08 |
| CaO | 0.41 | 24.28 | 0.29 | 2.17 | 0.13 | 0.29 | 0.30 | 0.52 | 0.37 | 0.25 | 0.32 | 0.19 | 0.35 | 0.47 |
| $Na_2O$ | 4.13 | 2.79 | 4.59 | 2.48 | 4.08 | 2.45 | 4.65 | 1.11 | 5.05 | 4.63 | 4.21 | 3.67 | 5.08 | 3.65 |
| $K_2O$ | 4.61 | 4.51 | 4.40 | 4.36 | 4.82 | 5.65 | 4.56 | 7.48 | 5.49 | 4.51 | 4.84 | 4.95 | 4.75 | 4.38 |
| $P_2O_5$ | 0.04 | 0.14 | 0.05 | 0.22 | 0.04 | 0.03 | 0.04 | 0.03 | 0.07 | 0.03 | 0.04 | 0.02 | 0.03 | 0.03 |
| Total | 100.87 | 98.78 | 98.93 | 100.51 | 98.66 | 98.96 | 99.34 | 99.25 | 100.05 | 99.70 | 99.36 | 99.02 | 99.25 | 99.09 |
| LOI | 2.67 | 17.98 | 0.59 | 7.15 | 0.65 | 3.14 | 0.77 | 2.18 | 0.66 | 1.00 | 0.38 | 1.19 | 1.02 | 4.26 |
| Ni | 3.2 | 7.1 | 4.5 | 5.7 | 3.4 | 6.4 | 5.9 | 3.9 | 3.4 | 5.8 | 6.9 | 5.8 | 7.0 | 4.1 |
| Cr | 2.3 | 4.6 | 2.6 | 3.5 | 2.5 | 6.3 | 7.2 | 3.1 | 3.2 | 3.0 | 4.6 | 3.7 | 2.8 | 4.1 |
| V | 6.8 | 5.1 | 11.2 | 21.9 | 6.5 | 10.8 | 11.9 | 10.4 | 9.5 | 9.2 | 5.3 | 7.1 | 4.8 | 1.3 |
| Sc | 1.5 | 1.0 | 8.3 | 10.1 | 6.4 | 5.3 | 5.4 | 5.5 | 5.1 | 2.2 | 5.7 | 3.3 | 2.6 | 2.1 |
| Zn | 226.2 | 185.5 | 131.8 | 130.5 | 178.8 | 282.5 | 278.6 | 195.9 | 103.0 | 203.1 | 156.1 | 123.3 | 181.0 | 290.6 |
| Ga | 29.0 | 20.4 | 28.3 | 25.7 | 29.4 | 32.8 | 32.2 | 29.6 | 29.2 | 31.4 | 27.9 | 28.1 | 30.3 | 31.4 |
| Pb | 21.0 | 13.4 | 10.6 | 16.2 | 13.8 | 22.5 | 21.7 | 13.8 | 17.4 | 17.1 | 14.6 | 15.6 | 19.5 | 23.4 |
| Sr | 16.6 | 75.5 | 17.0 | 271.9 | 12.5 | 16.0 | 16.0 | 64.6 | 40.9 | 10.7 | 72.0 | 35.1 | 14.5 | 22.0 |
| Rb | 168.6 | 96.7 | 107.4 | 81.2 | 121.3 | 150.3 | 160.6 | 140.4 | 130.3 | 120.1 | 128.4 | 150.0 | 166.5 | 145.1 |
| Ba | 90.8 | 306.1 | 331.8 | 817.2 | 247.3 | 49.2 | 67.0 | 417.3 | 432.5 | 80.7 | 579.2 | 305.5 | 145.3 | 304.0 |
| Zr | 957.2 | 855.5 | 863.3 | 663.8 | 1181.8 | 1736.2 | 1564.2 | 982.9 | 1035.7 | 906.4 | 737.9 | 696.4 | 1404.8 | 1152.3 |
| Nb | 118.8 | 116.0 | 108.1 | 74.3 | 137.4 | 197.2 | 168.7 | 120.5 | 166.4 | 105.8 | 111.5 | 125.7 | 151.7 | 159.1 |
| Th | 20.0 | 12.8 | 15.5 | 11.8 | 17.4 | 23.1 | 28.0 | 14.1 | 23.1 | 13.9 | 17.0 | 20.3 | 19.0 | 18.4 |
| Y | 118.7 | 103.8 | 69.4 | 72.0 | 88.8 | 200.0 | 159.7 | 97.4 | 86.1 | 101.5 | 92.0 | 74.3 | 110.3 | 127.9 |
| La | 129.8 | 91.6 | 82.7 | 91.3 | 83.8 | 156.2 | 190.8 | 112.2 | 130.6 | 85.3 | 141.8 | 88.9 | 102.6 | 148.5 |
| Ce | 220.6 | 179.2 | 166.8 | 189.4 | 178.6 | 294.0 | 349.4 | 226.4 | 252.9 | 204.0 | 155.0 | 147.1 | 207.6 | 276.6 |
| Nd | 116.2 | 83.4 | 78.8 | 87.6 | 81.2 | 168.6 | 169.6 | 100.4 | 99.0 | 84.7 | 130.6 | 86.0 | 78.5 | 122.4 |

The data are presented in geographical order. Sample E01–26 has anomalously high CaO (24%) and LOI (18%). Age data are from Ukstins *et al.* (2002); Wolfenden *et al.* (2004, 2005).
*Total Fe was analysed as $Fe_2O_3$

**Table 2.** *Sr–Nd–O isotopic compositions of rhyolites from along the western margin of Afar depression*

| Sample | Age (Ma) | Phase | $^{87}Sr/^{86}Sr$ | $^{87}Sr/^{86}Sr_i$ | $^{143}Nd/^{144}Nd$ | Qz/AF$\delta^{18}O$ (‰) |
|---|---|---|---|---|---|---|
| E 01–3a | 10.56 | WR | 0.705537 ± 10 | 0.705411 | 0.512654 ± 5 | |
| | | AF | 0.705445 ± 10 | 0.705363 | | |
| E 01–14a | 3.56 | WR | | | 0.512569 ± 6 | |
| E-01–26 | 2.54 | WR | | | 0.512778 ± 5 | +5.69 |
| E 01–18 | 10.14 | WR | | | 0.512797 ± 6 | |
| | | WR | | | 0.512794 ± 4 | |
| | | | | | ***mean 0.512796*** | |
| | | AF | 0.704893 ± 16 | 0.704814 | | |
| E 01–40 | 6.62 | WR | | | 0.512815 ± 4 | +4.56 |
| | | AF | 0.706508 ± 10 | 0.706429 | | |
| E 01–42 | 10.01 | WR | | | 0.512834 ± 5 | |
| | | WR | | | 0.512826 ± 4 | |
| | | WR | | | 0.512834 ± 4 | |
| | | | | | ***mean 0.512831*** | |
| EEW 02–21 | 11.67 | WR | | | 0.512658 ± 4 | +6.36 |
| | | WR | | | 0.512652 ± 5 | |
| | | | | | ***mean 0.512655*** | |
| EEW 00–23 | 19.76 | WR | | | 0.512762 ± 4 | +4.69 |
| EEW 00–33 | 25.52 | WR | 0.708968 ± 10 | 0.704594 | 0.512768 ± 5 | |
| | | AF | 0.705355 ± 9 | 0.705156 | | |
| EEW 00–34 | 25.52 | WR | 0.706879 ± 12 | 0.705054 | 0.512769 ± 4 | +5.57 |
| | | AF | 0.705275 ± 10 | 0.705076 | | |
| EEW 00–43 | 19.76 | AF | 0.70579 ± 10 | 0.705636 | | +5.37 |
| EEW00–44 | 19.76 | WR | 0.708496 ± 12 | 0.705971 | 0.512720 ± 4 | |
| HS01 | 7.81 | WR | | | 0.512538 ± 5 | |
| HS02 | 7.98 | WR | | | 0.512708 ± 5 | |

WR, whole rock; AF, alkali feldspar; Qz, quartz.
Sr and Nd isotope ratios were determined on the VG354 Thermal Ionization Spectrometer at Royal Holloway University of London using multidynamic techniques procedures and normalization described in Thirlwall (1991a and b). For Nd, powder samples were leached for 1 h in hot 6 M HCl. For Sr on alkali feldspar, crystals were handpicked to eliminate oxidation and glass or matrix, and leached for 1/2 h in hot 2.5 M HCl. In both case, leachate was removed by multiple rising in ultra-pure water. Not all samples contained alkali feldspars. Some whole rocks Sr analyses were performed on leached powders used for Nd, for comparisons (see text). All Nd were loaded on Re filaments and run as oxide over a period of 6 months during which the mean $^{143}Nd/^{144}Nd$ value of our Aldrich laboratory standard was 0.511415 ±8 (n = 17) and fully consistent with the long-term Aldrich mean of 0.511418 (Thirlwall, 1991a,b). All Sr were loaded on Re filaments between two layers of $TaF_5$. They are given relative to a $^{87}Sr/^{86}Sr$ value of 0.710248 for SRM987 with an external precision of 0.000011 (n = 26 over 6 months). $^{87}Sr/^{86}Sr$ ratios were corrected for initial values using a Rb/Sr of 0.55 on average (the Rb/Sr of alkali feldspar is in the range 0.1–1 according to published values of solid-liquid partition coefficients for Rb and Sr in alkali feldspar). WR, whole rock; AF, alkali feldspar.
Quartz and feldspar (samples E01–26 and EEW0043) minerals have been analysed for $\delta^{18}O$ using a VG Isotech (now VG instruments) Optimal dual inlet IRMS on line to the Laser Prep system (Mattey, 1997). Quantities of 1.7 mg for each hand-picked pure phase mineral have been combusted via a synrad $CO_2$ laser with presence of $BrF_5$ in excess. The released $O_2$ was directly analysed in the mass spectrometer, after cryogenic cleaned up. Three mineral standards were analysed during these runs: two are internal, GMG II, a garnet of +5.7 ‰ and a quartz of +8.8 ‰, which are calibrated against the international NBS-30 biotite at +5.1‰. All $\delta^{18}O$ are reported relative to V-SMOW. The overall precision on standards and sample replicates is better than ± 0.1‰.

compared with those of Gedemsa rhyolites thought to be generated by fractional crystallization of basaltic magma. It is apparent that the marginal rhyolites cannot be obtained by a single stage partial melting of basalts or their derivatives. The observed array of data is closely modelled by variable degrees of fractional crystallization of basalt.

Field observations show that the felsic rocks are always found in association with basalts and occur towards the upper part of an individual volcanic cycle, although there is a local recurrence of the basalt–rhyolite succession. This systematic stratigraphic relationship suggests that the silicic magmas are intimately associated with mafic magma, and the rhyolite marks the final phase of each magma intrusion cycle.

## Crustal involvement on silicic magma through time

The focus of this study is to understand the compositional variation of the rhyolites through time, in an attempt to explain the role of rifting on the creation of large crustal magma reservoirs where mafic magma differentiate to give rise to silicic magmas. As observed in the MER, magma chambers occur where there is a dense network of faults, i.e. two or more trends of fault (Keranen *et al.* 2004).

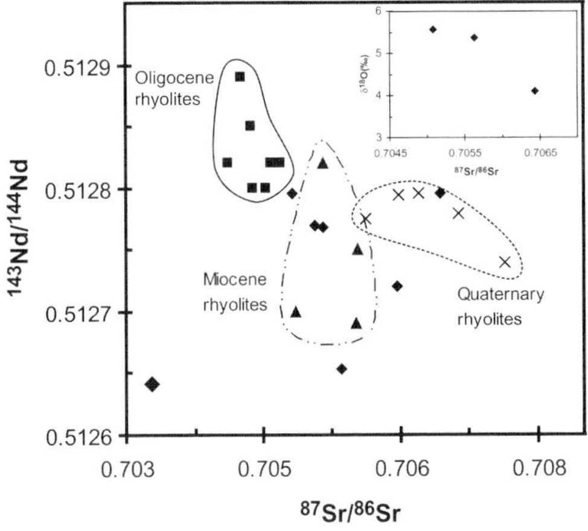


**Fig. 3.** Variation of $^{87}Sr/^{86}Sr$ vs. $^{143}Nd/^{144}Nd$ for rhyolites from the western margin of the Afar depression compared to the field of plateau and rift rhyolites. Note that most marginal rhyolites show substantial overlap with Miocene and Quaternary rhyolites. The inset displays variation of $^{87}Sr/^{86}Sr$ vs. $\delta^{18}O$ for rhyolites.

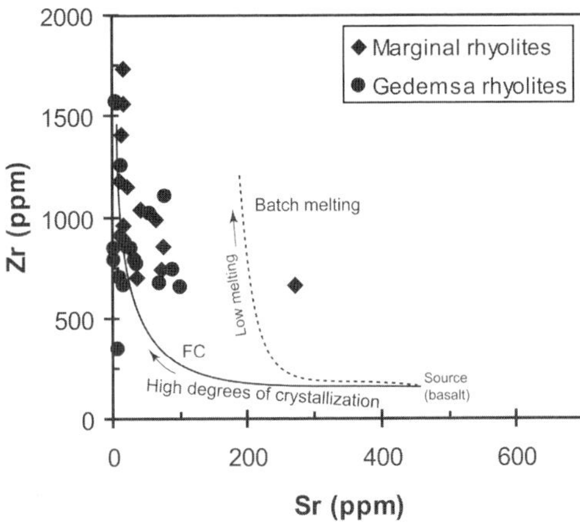


**Fig. 4.** Variation patterns of compatible (e.g. Sr) vs. incompatible (e.g. Zr) elements for west marginal rhyolites, illustrating that they can be modelled by fractional crystallization of basaltic magma. Also shown is the field of Gedemsa rhyolites thought to be generated by fractional crystallization of basaltic magma, and partial melting and fractional crystallization trends (calculated after Peccerillo *et al.* 2003).

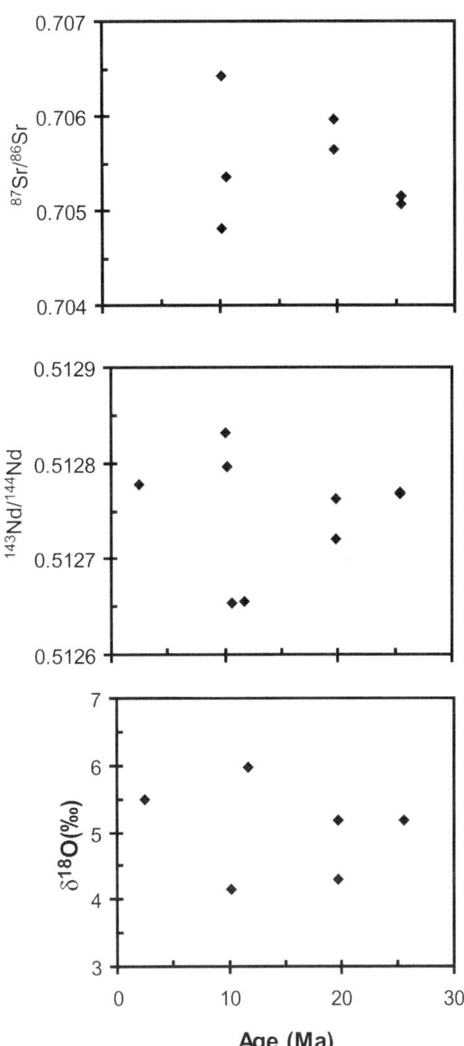


**Fig. 5.** Temporal variations of Sr, Nd and O isotope values for west marginal rhyolites. There is a trend between sample age and $^{87}Sr/^{86}Sr$, $^{143}Nd/^{144}Nd$ and $\delta^{18}O$.

In a plot of Sr–Nd–O isotope values with sample age (Fig. 5), there is a discernible trend. There is a negative correlation between sample age and $^{87}Sr/^{86}Sr$ values. $^{143}Nd/^{144}Nd$ ratios are roughly positively correlated with sample age. $\delta^{18}O$ decrease with time. There is no such systematic variation between major and trace elements, and sample age (not shown). The observed Sr–Nd–O isotope variations with time clearly reflect increasing participation of the basement over time on the genesis of silicic magma. This is most likely caused by progressive thermal and mechanical weakening of the crust within the faulted rift valley, and a concomitant shallowing in depth to the brittle–ductile transition in response to lithospheric thinning and magma injection in the past *c.* 30 Ma (Bastow *et al.* 2005; Stuart *et al.* 2006; Maguire *et al.* 2006).

The temporal variation of Sr–Nd–O isotopes of the western margin rhyolites indicates that the input of crust (or the rate of crustal contamination) on silicic magma generation is controlled by the stage of rifting and rate of magma addition. In the earliest stage (*c.* 30 Ma), the crust was unstretched and cold, and consequently the amount of assimilation of colder crustal rocks was limited. The repeated addition of large volumes of melt into the lithosphere during rifting would lead to weakening and heating of the crust, resulting in an increase in the rate of contamination.

$^{18}O$ depletion in the marginal rhyolites suggests that meteoric fluids were involved in the magmatic process, under high temperature conditions. Crustal rocks have generally both $^{87}Sr/^{86}Sr$ and $\delta^{18}O$ high and hence crustal contamination will increase the $\delta^{18}O$, not deplete it, therefore fluid circulation must have occurred. Hence, contamination of the silicic magma must have occurred at shallow levels where hydrothermal fluids are normally found.

## Conclusions

The timespan for silicic volcanism along the western margin of the southernmost sector of the southern Red Sea and the northernmost Main Ethiopian Rift indicates southward younging along the border fault and migrates from the wider border faults towards narrow (*c.* 10 km-wide) magmatic segments within the central rift. This indicates riftward migration of the locus of magmatism and faulting through time.

The felsic volcanic strata, capping the flood basalt sequences, are preferentially localized on or near the border fault, indicating that the age of the rhyolites actually dates the onset of rifting in the southern Red Sea and northern Main Ethiopian Rift. This clearly implies that rifting provides a mechanism for generation of large volumes of felsic magmatism.

The Sr–Nd–O isotopic data illustrate that the input of crust on the genesis of silicic magmas tends to increase with time, which suggests that the crust thermally weakens in response to lithospheric thinning and magma injection in the past ~30 Ma. The repeated addition of large volumes of melt into the lithosphere during rifting would lead to weakening and heating of the crust. Meteoric fluids were involved in the $^{18}O$ depletion of the marginal rhyolites.

Dereje Ayalew was supported by a travel grant from the Royal Society. Analyses were funded by the RHUL Research Committee. E. Bourdon was supported by EC Marie Curie Training and Mobility award. C. Ebinger and E. Wolfenden acknowledge support from University of London Research Funds, National Geographic Society, and NERC grant NER/A/S/2000/1004. A. Peccerillo and T. Furman are gratefully acknowledged for their insightful reviews.

## References

Asrat, A., Barbey, P., Ludden, J.N., Reisberg, L., Gliezes, G. & Ayalew, D. 2004. Petrology and isotope geochemistry of the Pan-African Negash pluton, northern Ethiopia: mafic–felsic magma interactions during the construction of shallow-level calc-alkaline plutons. *Journal of Petrology*, **45**, 1147–1179.

Ayalew, D. & Yirgu, Y. 2003. Crustal contribution to the genesis of Ethiopian plateau rhyolitic ignimbrites: basalt and rhyolite geochemical provinciality. *Journal of the Geological Society, London*, **160**, 47–56.

Ayalew, D., Barbey, P., Marty, B., Reisberg, L., Yirgu, G. & Pik, R. 2002. Source, genesis and timing of giant ignimbrite deposits associated with Ethiopian continental flood basalts. *Geochimica et Cosmochimica Acta*, **66**, 1429–1448.

Baker, J., Snee, L. & Menzies, M. 1996. A brief Oligocene period of flood volcanism in Yemen: implications for the duration and rate of continental flood volcanism at the Afro-Arabian triple junction. *Earth and Planetary Science Letters*, **138**, 39–55.

Bastow, I., Stuart, G.W., Kendall, J.M. & Ebinger, C. 2005. Upper mantle seismic structure in a region of incipient continental breakup: northern Ethiopian rift. *Geophysical Journal International*, **162**, 479–493.

Cameron, K.L., Parker, D.F. & Sampson, D.E. 1996. Testing crustal melting models for the origin of flood rhyolites: a Nd–Pb–Sr isotopic study of the Tertiary David Mountains volcanic field, west Texas. *Journal of Geophysical Research*, **101**, 20407–20422.

Chernet, T., Hart, W.K., Aronson, J.L. & Walter, R.C. 1998. New age constraints on the timing of volcanism and tectonism in the northern Main Ethiopian Rift–southern Afar transition zone (Ethiopia). *Journal of Volcanological and Geothermal Research*, **80**, 267–280.

Coulié, E., Quidelleur, X., Gillot, P.Y., Courtillot, V., Lefèvre, J.C. & Chiesa, S. 2003. Comparative K–Ar and Ar–Ar dating of Ethiopian and Yemenite Oligocene volcanism: implications for timing and duration of the Ethiopian traps. *Earth and Planetary Science Letters*, **206**, 477–492.

Courtillot, V., Jaupart, C., Manighetti, I., Tapponnier, P. & Besse, J. 1999. On causal links between flood basalts and continental breakup. *Earth and Planetary Science Letters*, **166**, 177–195.

Criss, R.E. & Taylor, H.P. 1986. Meteoric–hydrothermal systems. *In*: Valley, J.W., Taylor, H.P. & O'Neil, J.R. (eds) *Reviews in Mineralogy: Stable Isotopes in High Temperature Geological Processes*. Mineralogical Society of America, Vol. 16 Ch. 11, 373–424.

D'ACREMONT, E., LEROY, S., BESLIER, M.O., BELLAHSEN, N., FOURNIER, M., ROBIN, C., MAIA, M. & GENTE, P. 2005. Structure and evolution of the eastern Gulf of Aden conjugate margins from seismic reflection data. *Geophysical Journal International*, **160**, 869–890.

EBINGER, C.J. & SLEEP, N.H. 1998. Cenozoic magmatism throughtout east Africa resulting from the impact of a single plume. *Nature*, **395**, 788–791.

EBINGER, C.J., YEMANE, T., WOLDEGABRIEL, G., ARONSON, J.L., & WALTER, R.C. 1993. Late Eocene–Recent volcanism and rifting in the southern Main Ethiopian Rift. *Journal of the Geological Society, London*, **150**, 99–108.

EBINGER, C.J., YEMANE, T., HARDING, D., TESFAYE, S., REX, D. & KELLEY, S. 2000. Rift deflection, migration and propagation: linkage of the Ethiopian and Eastern rifts, Africa. *Geological Society of America Bulletin*, **102**, 163–176.

EILER, J.M. 2001. Oxygen isotope variations of basaltic lavas and upper mantle rocks. *In*: VALLEY, D.R. & COLE, D. (eds) *Reviews in Mineralogy: Stable Isotope Geochemistry*. Mineralogical Society of America, Vol. 43 Ch. 5, 319–364.

GEORGE, R., ROGERS, N. & KELLY, S. 1998. Earliest magmatism in Ethiopia: evidence for two mantle plumes in one flood basalt province. *Geology*, **26**, 923–926.

HALLIDAY, A.N., DAVIDSON, J.P., HILDRETH, W. & HOLDEN, P. 1991. Modelling the petrogenesis of high Rb/Sr silicic magmas. *Chemical Geology*, **92**, 107–114.

HANSON, G.N. 1989. An approach to trace element modeling using a simple igneous as an example. *In*: LIPIN, B.R. & MCKAY, G.A. (eds) *Reviews in Mineralogy: Geochemistry and mineralogy of the rare earth elements.* Mineralogical Society of America, Vol. 21, Ch. 4, 79–97.

HAWKESWORTH, C., KELLEY, S., TURNER, S., LE ROEX, A. & STOREY, B. 1999. Mantle processes during Gondwana break-up and dispersal. *Journal of African Earth Sciences*, **28**, 239–261.

HOFMANN, C., COURTILLOT, V., FÉRAUD, G., ROCHETTE, P., YIRGU, G., KETEFO, E. & PIK, R. 1997. Timing of the Ethiopian flood basalt event and implications for plume birth and global change. *Nature*, **389**, 838–841.

HUNEGNAW, A., SAGE, L. & GONNARD, R. 1998. Hydrocarbon potential of the intracratonic Ogaden basin SE Ethiopia. *Journal of Petroleum Geology*, **21**, 401–425.

KENDALL, J.M., STUART, G., EBINGER, C., BASTOW, I. & KEIR, D. 2005. Magma-assisted rifting in Ethiopia. *Nature*, **433**, 146–148.

KERANEN, K., KLEMPERER, S., GLOAGUEN, R. & EAGLE Working Group. 2004. Imaging a protoridge axis in the Main Ethiopian Rift. *Geology*, **39**, 949–952.

KIEFFER, B., ARNDT, N., *ET AL.* 2004. Flood and Shield basalts from Ethiopia: magmas from the African superswell. *Journal of Petrology*, **45**, 793–834.

MACKENZIE, G.D., THYBO, H. & MAGUIRE, P.K.H. 2005. Crustal velocity structure across the main Ethiopian Rift: results from two-dimensional wide-angle seismic modelling. *Geophysical Journal International*, **162**, 994–1006.

MARTY, B., PIK, R. & YIRGU, G. 1996. Helium isotopic variations in Ethiopian plume lavas: nature of magmatic sources and limit on lower mantle contribution. *Earth and Planetary Science Letters*, **144**, 223–237.

MATTEY, D.P. 1997. LaserPrep: An Automatic Laser-Fluorination System for Micromass 'Optima' or 'Prism' Mass Spectrometers. *Micromass Application Note*, **107**, 8pp.

MENZIES, M.A., KLEMPERER, S.L., EBINGER, C.J. & BAKER, J. 2002. Characteristics of volcanic rifted margins. *In*: MENZIES, M.A., KLEMPERER, S.L., EBINGER, C.J. & BAKER, J., (eds) *Volcanic Rifted Margins*, p. 1–14. *Geological Society of America Special Paper*, **362**.

PECCERILLO, A., BARBERIO, M.R., YIRGU, Y., AYALEW, D., BARBERI, M. & WU, T.W. 2003. Relationships between mafic and acid peralkaline magmatism in continental rift settings: a petrological, geochemical and isotopic study of the Gedemsa volcano, central Ethiopian rift. *Journal of Petrology*, **44**, 2003–2032.

PIK, R., DENIEL, C., COULON, C., YIRGU, G., HOFMANN, C. & AYALEW, D. 1998. The northwestern Ethiopian plateau flood basalts: classification and spatial distribution of magma types. *Journal of Volcanology and Geothermal Research*, **81**, 91–111.

PIK, R., DENIEL, C., COULON, C., YIRGU, G. & MARTY, B. 1999. Isotopic and trace element signatures of Ethiopian flood basalts: evidence for plume–lithosphere interactions. *Geochimica et Cosmochimica Acta*, **63**, 2263–2279.

ROCHETTE, P., TAMRAT, E., FÉRAUD, G., PIK, R., COURTILLOT, V., KETEFO, E., COULON, C., HOFMANN, C., VANDAMME, D. & YIRGU, G. 1998. Magnetostratigraphy and timing of the Oligocene Ethiopian traps. *Earth and Planetary Science Letters*, **164**, 497–510.

SCHILLING, J.G. & KINGSLEY, R.H. 1992. Nd–Sr–Pb isotopic variations along the Gulf of Aden: evidence for Afar mantle–plume–continental lithosphere interaction. *Journal of Geophysical Research*, **97**, 10,927–10,966.

TEKLAY, M., KRÖNER, A., MEZGER, K. & OBERHÄNSLI, R. 1998. Geochemistry, Pb–Pb single zircon ages and Rb–Sr isotope composition of Precambrian rocks from southern and eastern Ethiopia: implication for crustal evolution in east Africa. *Journal of African Earth Sciences*, **26**, 207–227.

THIRLWALL, M.F. 1991*a*. Long-term reproducibility of multicollector Sr and Nd isotope ratio analysis. *Chemical Geology*, **94**, 85–104.

THIRLWALL, M.F. 1991*b*. High-precision multicollector isotopic analysis of low levels of Nd as oxide. *Chemical Geology*, **94**, 13–22.

THIRLWALL, M.F., JENKINS, C., VROON, P.Z. & MATTEY, D. 1997. Crustal interaction during construction of ocean islands: Pb–Sr–Nd–O isotope geochemistry of the shield basalts of Gran Canaria, Canary Islands. *Chemical Geology*, **135**, 233–262.

Ukstins, I., Renne, P., Wolfenden, E., Baker, J., Ayalew, D. & Menzies, M. 2002. Matching conjugate volcanic rifted margins: $^{40}Ar/^{39}Ar$ chronostratigraphy of pre- and syn-rift bimodal flood volcanism in Ethiopia and Yemen. *Earth and Planetary Science Letters*, **198**, 289–306.

Watchorn, F., Nichols, G. & Bosence, D. 1998. Rift-related sedimentation and stratigraphy, southern Yemen (Gulf of Aden). *In*: Purser, B. & Bosence, D. (eds) *Sedimentary and Tectonic Evolution of Rift Basins*. Chapman & Hall, pp. 165–189.

WoldeGabriel, G., Aronson, J.L., & Walter, R.C. 1990. Geology, geochronology and rift basin development in the central sector of the Main Ethiopian Rift. *Geological Society of America Bulletin*, **102**, 439–485.

Wolfenden, E., Ebinger, C., Yirgu, G., Deino, A. & Ayalew, D. 2004. Evolution of the northern Main Ethiopian Rift: birth of a triple junction. *Earth and Planetary Science Letters*, **224**, 213–228.

Wolfenden, E., Ebinger, C., Yirgu, G., Renne, P. & Kelley, S.P. 2005. Evolution of the southern Red Sea rift: birth of a magmatic margin. *Geological Society of America Bulletin*, **117**, 846–864.

# Part 3: Rifting in the Afar volcanic province: Modelling and kinematics

The four papers in this section describe the results of seismicity, remote sensing, palaeomagnetic and geodetic studies of rift structure and kinematics along the length of the Ethiopian rift and Red Sea rifts.

**Ayele *et al.*** determine locations and focal mechanisms for a swarm of earthquakes that occurred in the southeastern Afar depression in 2000. The relocated cluster of events does not coincide with any major surface fault trace, but detailed geological maps are lacking from this remote region. Source depths are about 7 km. Focal mechanisms indicate movement along WNW–ESE to east–west striking normal faults; these mechanisms are highly oblique to the approximately north–south strike of faults in the Main Ethiopian Rift. Ayele *et al.* (2006) interpret this activity as evidence for continued movement along faults that initiated during the Oligocene opening of the Gulf of Aden; these faults strike approximately east–west.

**Casey *et al.*** analyse faults and eruptive centres within the $<2$ Ma magmatic segments of the Main Ethiopian Rift, which are the present-day locus of strain. Their aim is to determine the relative importance of faulting and magmatism in strain accommodation as rifting proceeds to seafloor spreading. Field-calibrated remote sensing imagery reveals that up to half of the faults have eruptive centres or fissural flows along their length, confirming the importance of magmatism in late-stage extension. Both the orientation of faults and elongation of Quaternary calderas indicate a $106^\circ$ E extension direction, consistent with independent seismicity and geodetic studies. Casey *et al.* simulate the shape of the largest fault in the population with an elastic plate, and predict that faults are restricted to the upper 10 km of the crust. The absence of large offset faults along the length of the magmatic segments and the coincidence of faults and eruptive centres indicate that magma intrusion to the lower crust and dyke intrusion to the upper crust accommodate extension near the transition between continental rifting and seafloor spreading.

**Kidane *et al.*** report the first palaeomagnetic results from the Main Ethiopian Rift. Measurements were made at 20 sites to test plate kinematic models for the right-stepping, en echelon arrangement of Quaternary magmatic segments within the rift. Data also include five K–Ar age determinations. Both felsic and basltic flows were sampled, with natural remanent magnetization values similar to those found in earlier studies of the Afar depression. The mean palaeomagnetic direction determined from these new samples is statistically identical to the predicted Apparent Polar Wander Path reference curve for Africa (Besse & Courtillot 2003). There is no evidence for rotations about vertical axes as predicted by oceanic transform models or transtensional models for the en echelon arrangement of magmatic segments in the Main Ethiopian Rift. Kidane *et al.* (2006) attribute the Quaternary rift segmentation to magma intrusion, rather than large offset faults.

**Pizzi *et al.*** utilize field, geodetic and geomorphological analyses to examine the kinematics of rifting in the central Main Ethiopian Rift. Their work focuses on the partitioning of strain between the Asela border fault and Quaternary faults in the Wonji fault belt ($\sim$magmatic segments). They estimate slip rates of around 2 mm $a^{-1}$ in the direction N95$^\circ$ E within the magmatically active Wonji belt. Geomorphological and fault analyses document 300–400 m of displacement along the Asela border fault since mid-Pleistocene time. They conclude that strain is accommodated by a combination of magmatic deformation of the rift floor and slip along the rift border fault. Their model of shallow detachment faulting can explain both the rotation of fault blocks and geodetic data.

## Reference


BESSE, J. & COURTILLOT, V. 2003. Apparent and true polar wander and the geometry of the geomagnetic field in the last 200 My. *Journal of Geophysical Research*, **108**, 10.1029/2003JB002684.



*From*: YIRGU, G., EBINGER, C.J. & MAGUIRE, P.K.H. (eds) 2006. *The Afar Volcanic Province within the East African Rift System*. Geological Society, London, Special Publications, **259**, 131.
0305-8719/06/$15.00

# New evidence for Afro-Arabian plate separation in southern Afar

ATALAY AYELE[1], ANDREW A. NYBLADE[2], CHARLES A. LANGSTON[3], MICHEL CARA[4] & JEAN-J. LEVEQUE[4]

[1]*Geophysical Observatory, Addis Ababa University, PO Box 1176 Addis Ababa, Ethiopia (e-mail: atalay@geobs.aau.edu.et)*

[2]*Department of Geosciences, Pennsylvania State University, University Park, Pennsylvania, USA*

[3]*Center for Earthquake Research and Information, University of Memphis, Memphis, Tennessee, USA*

[4]*École et Observatoire des Sciences de la Terre, 5 rue René Descartes, F-67084 Strasbourg Cedex, France*

**Abstract:** The May 2000 earthquake cluster, around 10° N and 41° E in southern Afar, has been studied using high quality data from 12 temporary and permanent broadband seismic stations deployed in the area. 140 earthquakes have been located using P- and S-wave arrival times, a well-constrained velocity model, and a double-difference location algorithm. Source mechanisms and moment magnitudes for the four largest events ($M > 4$) have been obtained from moment tensor inversion. There is no clear alignment of the epicentres along a fault zone; however, the events are clustered slightly southeast of Mount Amoissa along WNW–ESE extension of the Ayelu–Amoissa (Abida/Dabita) lineament. Focal mechanisms show fault motion along WNW–ESE to east–west striking normal faults, with extension oblique to the orientation of the Main Ethiopian Rift. The non-double-couple components of the source mechanisms range from 18–25%, suggesting that the seismic activity is of tectonic origin and not volcanic. Source depths are $\leq 7$ km, in good agreement with estimates of the elastic thickness of the Afar lithosphere. We suggest that the Gewane earthquake swarm represents remnant strain accommodation along a previous line of weakness in southern Afar related to the separation of Arabia from Africa because the focal mechanisms show north–south extension similar to many of the events in central Afar at the triple junction where Arabia is presently rifting away from Africa.

The Afar depression is a diffuse triple junction where the Red Sea, Gulf of Aden and Main Ethiopian Rifts join (Fig. 1). The fastest rate of plate separation (1.6 cm $a^{-1}$) is between Africa and Arabia (Chu & Gordeon 1998; Walpersdorf *et al.* 1999). The Main Ethiopian Rift (MER) developed over the past 11 Ma and is the youngest of the three rift arms (Wolfenden *et al.* 2004). Opening of the Gulf of Aden and Red Sea rifts began *c.* 20–30 Ma (e.g. Manighetti *et al.* 1998). Different rates of extension across the three rift arms, coupled with weakening of the lithosphere from a Palaeogene mantle plume beneath Afar (Hart *et al.* 1989; White & Mckenzie 1989; Ebinger & Sleep 1998), may have contributed to making the location of the triple junction unstable (McKenzie & Morgan 1969; Garfunkel & Beyth 2004). The present location of the Afar triple junction is within the Lake Abhe area (Manighetti *et al.* 1997, 1998), but may have migrated to this position from an initial location in southern Afar (Tesfaye *et al.* 2003).

An earthquake swarm occurred in May 2000 about 50 km SE of Gewane town in southern Afar, an area where there has been little reported seismicity in recent years (Figs 1 & 2). Several earthquakes were felt as far away as 120 km. The sesimic activity was recorded on 12 seismic stations, and locations of the three largest events, with magnitudes between 4 and 5, were reported by the National Earthquake Information Center (NEIC). Hofstetter & Beyth (2003) calculated fault plane solutions for the three events in the NEIC catalogue by inverting waveforms from nearby permanent stations FURI and ATD (Fig. 1, Table 1a). Two of the mechanisms show strike–slip motion and one shows oblique motion with predominantly north–south extension.

The three events in the NEIC catalogue indicate that the swarm is located east of the northern

*From*: YIRGU, G., EBINGER, C.J. & MAGUIRE, P.K.H. (eds) 2006. *The Afar Volcanic Province within the East African Rift System*. Geological Society, London, Special Publications, **259**, 133–141.
0305-8719/06/$15.00 

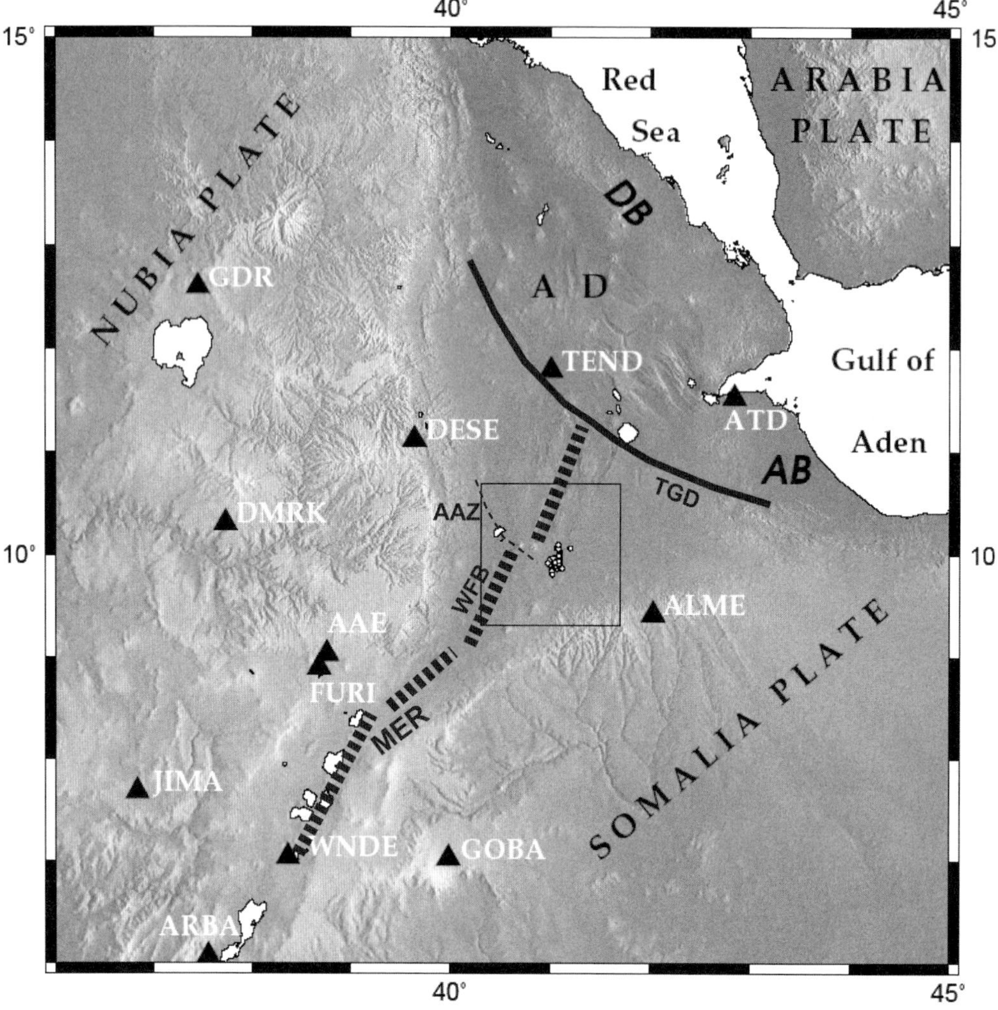


**Fig. 1.** Shaded relief map of Afar and surrounding areas. Black triangles represent station locations, and station names are in white letters. The outlined rectangle represents our area of interest in this study and shown in Fig. 2, and small white dots show the epicentres of the Gewane earthquakes. AD, Afar Depression; DB, Danakil Block; AB, Ali Sabhieh block; MER, Main Ethiopian Rift. Bold dashed line shows approximate location of the Wonji Fold Belt (WFB); thin dashed line shows approximate location of the arcuate accommodation zone (AAZ) marking location of the proposed initial triple junction (Tesfaye *et al.* 2003), and bold solid line shows approximate location of the Tendaho–Goba'ad discontinuity (TGD).

MER and south of the Afar triple junction (Mohr 1967; Ebinger & Casey 2001, Fig. 1). This location, away from the main zones of active extension, is tectonically puzzling, as is the north–south extension indicated by one of the focal mechanisms from Hofstetter & Beyth (2003), which is orthogonal to the direction of opening in the MER. In this paper, we investigate further the Gewane earthquake swarm and discuss its geodynamic implications for the tectonic development of southern Afar. Locations of 140 events within the swarm are presented along with new focal mechanisms for the four largest events.

## Background

Southern Afar is the transition zone between the north–south striking southern Red Sea rift faults and the NE-trending Main Ethiopian Rift (MER)

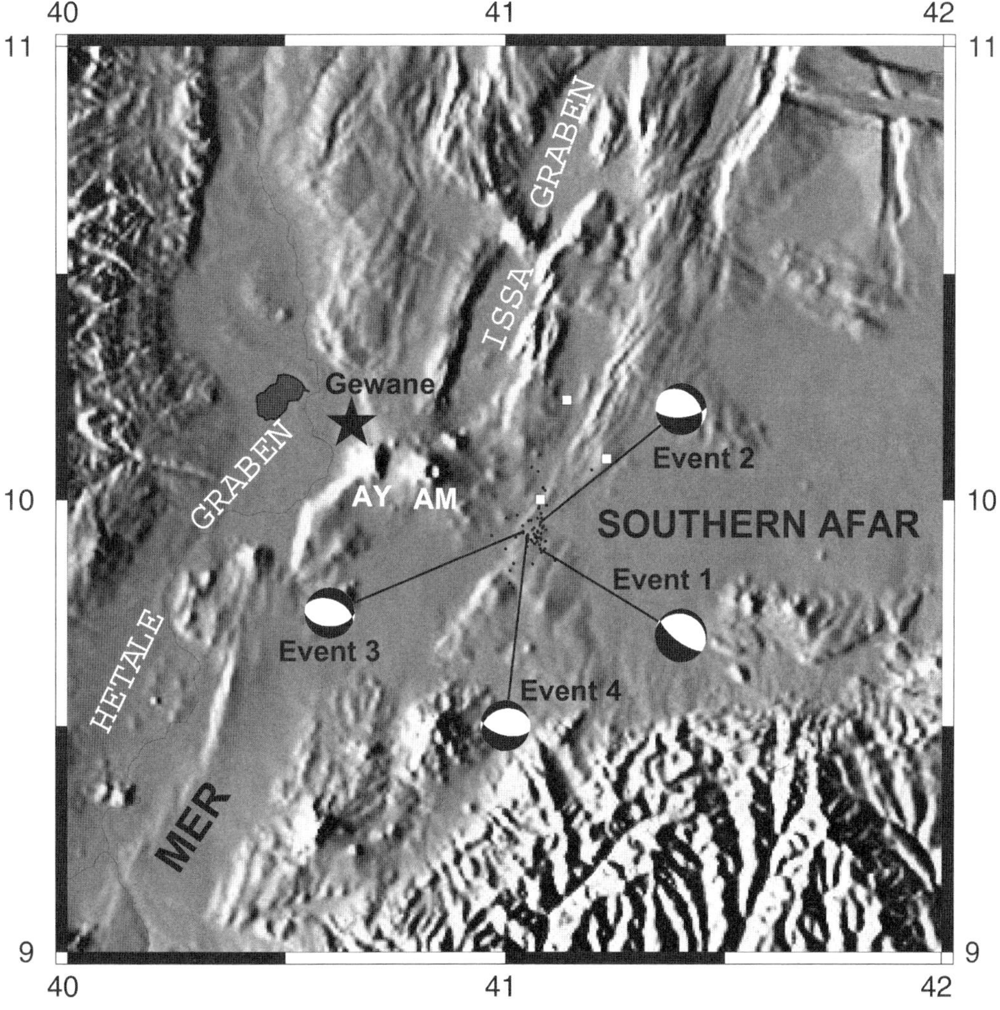


**Fig. 2.** Shaded relief map of the study area in southern Afar. Black dots represent our event locations and white rectangles the NEIC locations for the three events used by Hofstetter & Beyth (2003) (see Table 1a). AY, Mount Ayelu; AM, Mount Amoissa. 'Beach-balls' show focal mechanisms obtained in this study (lower hemisphere projections), and the black star shows the location of Gewane town.

**Table 1.** *List of earthquakes with hypocentres and source parameters*

| Event | Date | Origin time | Lat. | Lon. | H(km) | *Mw* | Strike | Dip | Rake |
|---|---|---|---|---|---|---|---|---|---|
| **(a) NEIC catalogue and Hofstetter & Beyth (2003)** | | | | | | | | | |
| 1 | 2000/05/10 | 08:43:31.8 | 10.22 | 41.14 | 10 | 4.6 | 2;269 | 79;73 | 17;168 |
| 2 | 2000/05/12 | 17:54:03.8 | 10.00 | 41.08 | 10 | 4.2 | 5;272 | 78;76 | 12;165 |
| 3 | 2000/05/16 | 20:47:51.9 | 10.09 | 41.23 | 10 | 4.4 | 113;241 | 60;43 | −57; −133 |
| **(b) This study** | | | | | | | | | |
| 1 | 2000/05/10 | 08:43:28.4 | 09.89 | 41.08 | 7 | 4.4 | 126;288 | 71,20 | −84; −107 |
| 2 | 2000/05/10 | 23:17:52.9 | 09.94 | 41.08 | 5 | 4.3 | 86;310 | 33;65 | −111; −51 |
| 3 | 2000/05/12 | 17:54:02.1 | 09.93 | 41.04 | 7 | 4.2 | 91;290 | 45;46 | −104; −76 |
| 4 | 2000/05/16 | 20:47:48.6 | 09.91 | 41.05 | 6 | 4.2 | 103;257 | 59;33 | −76; −112 |

(Tesfaye *et al.* 2003; Wolfenden *et al.* 2004), but few details are known about its geodynamic development because much of the region is covered by the *c.* 2 Ma Afar stratoid series (Audin *et al.* 2004). The northern terminus of the MER is believed to be the Tendaho–Goba'ad discontinuity, near Lake Abhe (Tesfaye *et al.* 2003; Acton *et al.* 2000, Figs 1, 2). The MER is characterized by NE–SW oriented Miocene border faults and the Wonji Fault Belt (WFB), a north–south to N20° E-trending system of en echelon right-stepping Quaternary to recent faults cutting the rift floor (Mohr 1967). The WFB coincides with right-stepping en echelon magmatic segments which are the locus of present-day extension in the MER (Ebinger & Casey 2001). The largest displacement of magmatic segments within southern Afar occurs between the Issa and Hertale grabens along the Ayelu–Amoissa line (Mohr 1967, Fig. 2).

The cause and depth extent of the along-axis segmentation in the MER is poorly known. Di Paola (1970) postulated a WNW–ESE trending cross-rift structure to explain the Ayelu–Amoissa volcanic alignment. There is evidence for an early phase of rifting in southern Afar, which may have been influenced by the propagation of the Gulf of Aden ridge (Mohr 1967; Audin *et al.* 2004). An arcuate accommodation zone close to the Ayelu–Amoissa volcanic alignment may mark a previous location of southwestern Arabia prior to the northward migration of the triple junction to its present location in central Afar (Tesfaye *et al.* 2003, Fig. 1). The last episode of rifting within the accommodation zone took place between 1.8 and 1.6 Ma adjacent to the Ayelu–Amoissa volcanic complex, which marks the intersection of the WFB and the WNW–ESE trending Ayelu–Amoissa (Abida/Dabita) lineament (Berhe 1986, Figs 1, 2).

The northern part of the MER around Gewane has not been seismically active before May 2000, except for some poorly located earthquake activity in September and October 1938 and December 1971 (Gouin 1979). The apparent lack of seismicity in southern Afar is in marked contrast to the occurrence of seismicity elsewhere in the Afar, for example in central and eastern Afar and the border faults of western Afar (e.g. Hofstetter & Beyth 2003). Focal mechanisms for events in central Afar mainly show north–south extension, while focal mechanisms for events along the western border of Afar show predominantly east–west extension (see Fig. 6).

## Data and methods

Data for this study come from several sources. Two temporary broadband seismic experiments were operating in Ethiopia in 2000, the Ethiopia broadband seismic experiment (Nyblade & Langston 2002), which consisted of six stations at that time (DMRK, JIMA, ARBA, GOBA, AAE, TEND; Fig. 1), and the French 'Institut National des Sciences de l'Univers'(INSU) broadband seismic experiment, which consisted of three stations (ALME, DESE, GDR; Fig. 1). Data from two permanent global stations, FURI (IRIS/USGS) and ATD (GEOSCOPE), were also used for this study, as well as travel times from a short-period Ethiopian station (WNDE).

Onset times of P and S phases were hand-picked to within 0.1 sec after removing instrument responses from the waveforms. Preliminary epicentral locations for 140 earthquakes were obtained from the P- and S-wave arrival times, a well-constrained 1D velocity model from the EAGLE-controlled source experiment (Daly *et al.* 2004), and the HYPOINVERSE 2000 program (Klein 2002). The events were then relocated using a double-difference earthquake location algorithm (Waldhauser & Ellsworth 2000), and the results before and after relocation are shown in Fig. 3.

To obtain the fault plane solutions and moment magnitudes for the four largest events, moment tensor inversion was performed on three-component displacement waveforms from four or more stations using a time domain linear inversion algorithm (Langston 1981; Ammon *et al.* 1998). Green's functions for the inversion were calculated using the reflection-matrix method (Kennett 1983) as implemented by Randall (1994) and a 1D crustal structure model from the EAGLE-controlled source experiment (Daly *et al.* 2004). Waveforms were bandpass-filtered from 20 to 65 seconds and the entire wavetrain was included in the inversion. Within this passband, almost all the data from the broadband stations have good signal-to-noise ratios both for body and surface waves (Fig. 4). The inversion was performed for source depths between 4 and 20 km at 1 km depth increments, and the fault parameters reported are from the inversion at the depth which yielded the best fit to the data (Fig. 5; Table 1b).

## Results and discussion

Uncertainties in event locations are less than about 10 km (Figs 2 & 3). The scatter in the locations is reduced by applying the double-differencing, but there is no clear alignment of the epicentres that would define a fault zone. The locations of the three largest events in the NEIC catalogue fall outside of the cluster at about 41° E longitude and 10° N latitude, but after location and relocation in this study they all lie within the centre of the cluster. The cluster is located slightly SE of

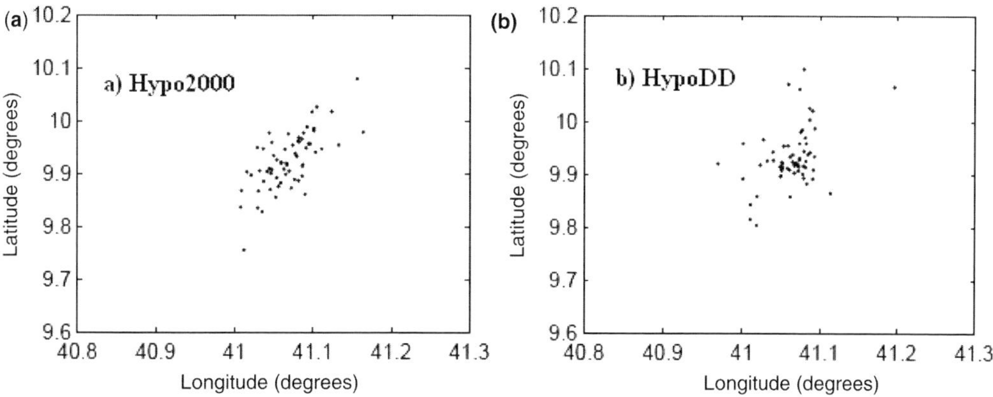


**Fig. 3.** Plots of event epicentres (**a**) before and (**b**) after relocation using the HypoDD program.

Mount Amoissa, along WNW–ESE extension of the Ayelu–Amoissa (Abida/Dabita) lineament proposed by Di Paola (1970).

The results from the moment tensor inversion show fault motion along WNW–ESE to east–west striking normal faults. The extension direction is oblique to the orientation of the MER (Fig. 2). The non-double-couple components range from 18–25%, suggesting that the seismic activity is of tectonic origin and not volcanic. Source nucleation $\leq$7 km depth (Table 1b) is in good agreement with estimates of the elastic thickness of the Afar lithosphere (Ebinger & Hayward 1996; Keranen *et al.* 2004; Bilham *et al.* 1999; Ayele *et al.* 2004).

Focal mechanims for events 1, 3 and 4 in Table 1b were also reported by Hofstetter & Beyth (2003) (Table 1a). The mechanism obtained in this study for event 4 is similar to the one reported by Hofstetter & Beyth (2003) (event 3, Table 1a), but the mechanisms for events 1 and 3 are different (normal fault with north–south extension vs. strike-slip). We attribute these differences to the better data coverage provided by the data set used in this study.

What are the tectonic implications of these results for the geodynamic setting of southern Afar? The location of the swarm along the Ayelu–Amoissa (Abida/Dabita) lineament suggests a kinematic link to this cross-rift structure. We suggest that the Gewane earthquake swarm represents remnant strain accommodation along a previous line of weakness in southern Afar related to the separation of Arabia from Africa associated with this lineament. This suggestion is based on several lines of reasoning.

First, the focal mechanisms for the four largest events in the swarm show north–south extension similar to many of the events in central Afar at the triple junction where Arabia is presently rifting away from Africa (Figs 2 & 6). In May 1980, there were also two earthquakes in southeastern Afar in the Ali-Sabieh block that have similar focal mechanisms (i.e. north–south extension) to the Gewane swarm (Fig. 6).

Next, deformation throughout Afar over the past 2 Ma has been characterized by widespread crustal deformation resulting in inhomogeneous extension across the region (Acton *et al.* 2000). The Afar depression acts as a soft link for the three arms of the triple junction, its location is unstable (McKenzie & Morgan 1969; Garfunkel & Beyth 2004), and the boundaries of the Arabian, Nubian and Somalian plates are not yet clearly developed (Williams *et al.* 2004). Consequently, the pattern of rifting may not show a simple geometry (Sykes 1970), and north–south extension related to the Nubia–Arabia plate motion south of the present-day triple junction location should not necessarily be unexpected.

Finally, present-day extension within the northern MER has localized to <20 km-wide zones of dyking, faulting and volcanism (Mohr 1967; Bilham *et al.* 1999; Ebinger & Casey 2001; Keranen *et al.* 2004), and many of these Quaternary magmatic segments are cut by en echelon fault zones, possibly forming proto-ridge axes for future sea floor spreading. There are several dextral offsets of the magmatic segments in the MER that align with structures as old as Mesozoic extending under the entire Ethiopian swell. The ancient linear structures are manifested at the rift margins and within the rift itself, including the elliptical orientation of Quaternary calderas (Mohr 1967; Acocella *et al.* 2002). The Ayelu–Amoissa volcano-tectonic lineament is a good example of one of these structures with dextral displacement oblique to the MER axis. It has been proposed that structures like the Ayelu–Amoissa

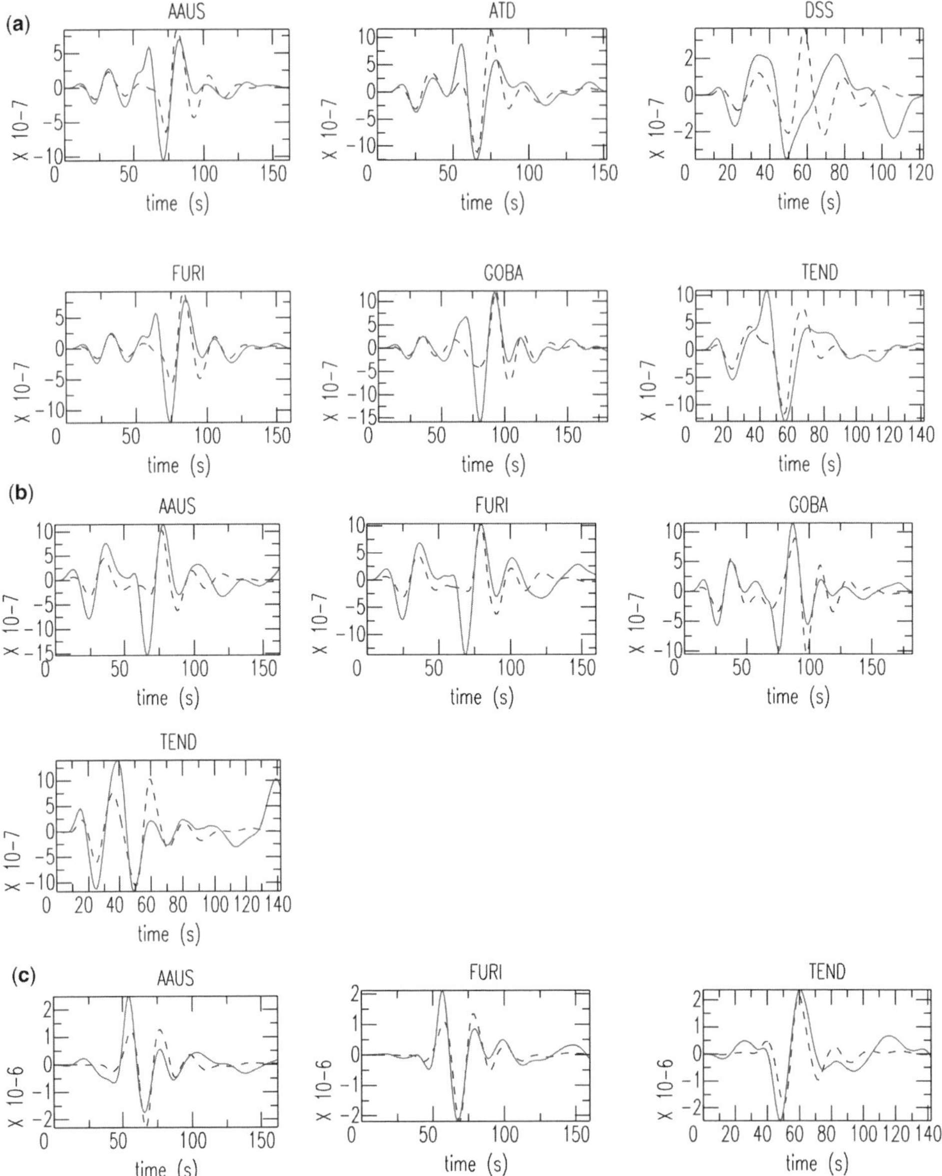


**Fig. 4.** Waveform fits of the observed (solid lines) and synthetic (dashed lines) seismograms for event 1 (Fig. 2, Table 1b). (**a**) Vertical, (**b**) radial and (**c**) transverse components of ground displacement. Station locations are shown in Fig. 1.

volcano-tectonic lineament were once transform fault zones (Mohr 1967; Barberi & Varet 1977). However, the focal mechanisms for the four largest events in the swarm do not show strike–slip motion. Thus, the Gewane swarm probably does not represent proto-ridge development along the MER within the southern Afar, and having ruled out this possibility, the only remaining plausible explanation for the north–south extension in the Gewane swarm is strain release along

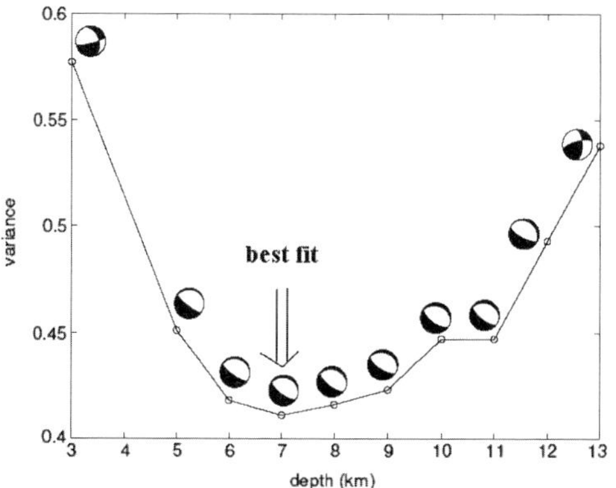


**Fig. 5.** Plot showing variance reduction in waveform fits vs. source depth for event 1 (Fig. 2, Table 1b). Focal mechanisms obtained from the moment tensor inversion are shown for each source depth. A best fit to the data is obtained for a focal depth of 7 km.

the Ayelu–Amoissa volcano-tectonic lineament related to the separation of the Nubian and Arabian plates.

## Summary and conclusions

We have located 140 earthquakes from the May 2000 swarm near Gewane in southern Afar, and obtained focal mechanisms and moment magnitudes for the four largest events within this swarm. There is no clear alignment of the epicentres along a fault zone, however, the events are clustered slightly southeast of Mount Amoissa along the Ayelu–Amoissa (Abida/Dabita) lineament. The results from the moment tensor inversion show fault motion along WNW–ESE to east–west striking normal faults. The extension direction is oblique to the orientation of the MER. The non-double-couple components of the source mechanisms range from 18–25%, suggesting that the seismic

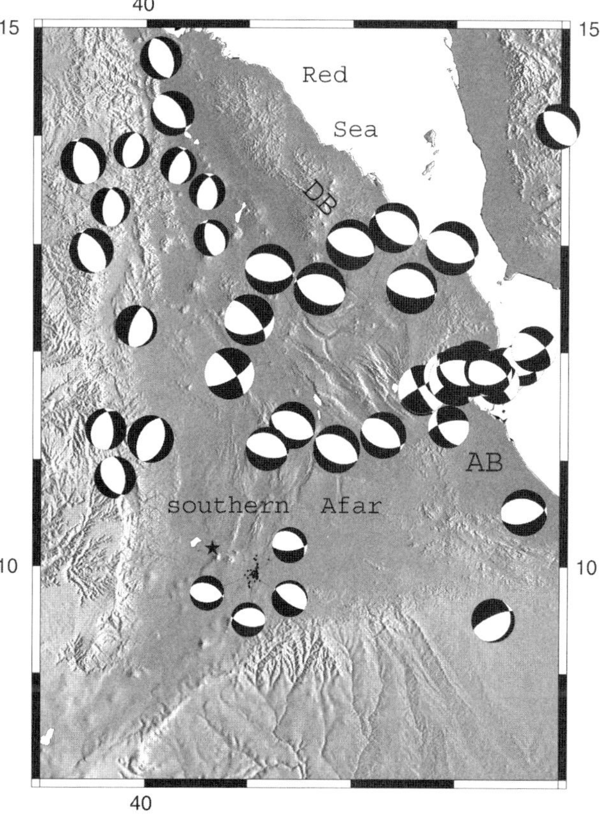


**Fig. 6.** Shaded relief map of Afar and surrounding areas showing focal mechanism plots for the four events used in this study (Table 1b) in southern Afar, plus focal mechanisms from Ayele *et al.* (2004) and the Harvard CMT solutions between 1977 and 2004. DB, Danakil Block; AB, Ali Sabhieh block.

activity is of tectonic origin and not volcanic. Source depths of ≤7 km are in good agreement with estimates of the elastic thickness of the Afar lithosphere.

We suggest that the Gewane earthquake swarm represents remnant strain accommodation along a previous line of weakness in southern Afar related to the separation of Arabia from Africa primarily because the focal mechanisms show north–south extension similar to many of the events in central Afar at the triple junction where Arabia is presently rifting away from Africa. Deformation in Afar over the past 2 Ma has been widespread, resulting in inhomogeneous crustal extension, and therefore north–south extension related to the Nubia–Arabia plate motion south of the present-day triple junction should not necessarily be unexpected.

We thank Chuck Ammon for his assistance with the moment tensor inversions, George Randall for use of his reflectivity code, Pöl Wessel and Walter H.F. Smith for Generic Mapping Tools (GMT), and Derek Keir and an anonymous reviewer for constructive comments. This research has been funded by the National Science Foundation (grants EAR 993093 and 0003424).

## References

ACOCELLA, V., KORME, T. & FUNICIELLO, R. 2002. Elliptic calderas in the Ethiopian Rift: control of pre-existing structures. *Journal of Volcanology and Geothermal Research*, **119**, 189–203.

ACTON, G.D., TESSEMA, A., JACKSON, M. & BILHAM, R. 2000. The tectonic and geomagnetic significance of paleomagnetic observations from volcanic rocks from central Afar, Africa. *Earth and Planetary Science Letters*, **180**, 225–241.

AMMON, C.J., HERRMANN, R.B., LANGSTON, C.A. & BENZ, H. 1998. Faulting parameters of the January 16, 1994 Wyomissing Hills, Pennsylvania earthquakes. *Seismological Research Letters*, **69**, 261–269.

AUDIN, L., QUIDELLEUR, X., COULIE, E., COURTILLOT, V., GILDER, S., MANIGHETTI, I., GILLOT, P.-Y., TAPPONIER, P. & KIDANE, T. 2004. Paleomagnetism and K–Ar and Ar /Ar ages in the Ali Sabieh area (Republic of Djibouti and Ethiopia): Constraints on the mechanism of Aden ridge propagation into southern Afar during the last 10 Myr. *Geophysical Journal International*, **158**, 327–345.

AYELE, A., STUART, G., BASTOW, I., KENDALL, J.-M. & KEIR, D. 2004. Tectonic implication of the August 2002 Earthquake Sequence at the Western Danakil Microplate in northwest Afar. *In*: YIRGU, G. *et al.* (eds) *Proceedings of International Conference on the East African Rift System*, June 20–24 2004. Addis Ababa, Ethiopia: Ethiopian Geoscience and Mineral Engineering Association, pp. 20–22.

BARBERI, F. & VARET, J. 1977. Volcanism of Afar: Small-scale plate tectonics implications. *Geological Society of America Bulletin*, **88**, 1251–1266.

BERHE, S.M. 1986. Geologic and geochronologic constraints on the evolution of the Red Sea–Gulf of Aden and Afar Depression. *Journal of African Earth Science*, **5**, 101–117.

BILHAM, R., BENDICK, R., LARSON, K., MOHR, P., BRAUN, J., TESFAYE, S. & ASFAW, L. 1999. Secular and tidal strain across the Main Ethiopian Rift. *Geophysical Research Letters*, **26**, 2789–2792.

CHU, D. & GORDON, R.G. 1998. Current plate motions across the Red Sea. *Geophysical Journal International*, **135**, 313–328.

DALY, E., KEIR, D., EBINGER, C.J., STUART, G., AYELE, A. & WALTHAM, D. 2004. Crustal structure of the northern Main Ethiopian Rift from a tomographic inversion of local earthquakes. *In*: YIRGU, G. *et al.* (eds) *Proceedings of International Conference on the East African Rift System*, June 20–24 2004. Addis Ababa, Ethiopia: Ethiopian Geoscience and Mineral Engineering Association, pp. 51–52.

DI PAOLA, G.M. 1970. Geological–Geothermal report on the central part of the Ethiopian rift valley. *Addis Ababa, Ministry of Mines*, 46 pp.

EBINGER, C.J. & CASEY, M. 2001. Continental breakup in magmatic provinces: An Ethiopian example. *Geology*, **29**, 527–530.

EBINGER, C.J. & HAYWARD, N.J. 1996. Soft plates and hot spots: Views from Afar. *Journal of Geophysical Research*, **101**(B10), 21,859–21,876.

EBINGER, C.J. & SLEEP, N.H. 1998. Cenozoic magmatism throughout east Africa resulting from impact of single plume. *Nature*, **395**, 788–791.

GARFUNKEL, Z. & BEYTH, M. 2004. Constraints on the structural development of Afar imposed by the major surrounding plates. *In*: YIRGU, G. *et al.* (eds) *Proceedings of International Conference on the East African Rift System*, June 20–24 2004. Addis Ababa, Ethiopia: Ethiopian Geoscience and Mineral Engineering Association, p. 74.

GOUIN, P. 1979. Earthquake history of Ethiopia and the Horn of Africa. *International Development Research Center (IDRC)*, Ottawa, Ontario.

HART, W.K., WOLDEGABRIEL, G., WALTER, R.C. & MERTZMAN, S.A. 1989. Basaltic volcanism in Ethiopia: Constraints on continental rifting and mantle interactions. *Journal of Geophysical Research*, **94**(B6), 7731–7748.

HOFSTETTER, R. & BEYTH, M. 2003. The Afar Depression: interpretation of the 1960–2000 earthquakes. *Geophysical Journal International*, **155**, 715–732.

KENNET, B.L.N. 1983. *Seismic Wave Propagation in Stratified Media.* Cambridge University Press, Cambridge, England.

KERANEN, K., KLEMPERER, S.L., GLOAGUEN, R. & EAGLE Working Group. 2004. Three-dimensional seismic imaging of a protoridge axis in the Main Ethiopian Rift. *Geological Society of America*, **32**, 949–952.

KLEIN, F.W. 2002. User's Guide to Hypoinverse-2000. A Fortran program to solve for earthquake locations and magnitudes. *USGS open file report*, **02–171**, 1–171.

LANGSTON, C.A. 1981. Source inversion of seismic waveforms: the Koyna, India, earthquakes of 13 September, 1967. *Bulletin Seismological Society of America*, **71**, 1–24.

MANIGHETTI, I., TAPPONIER, P., COURTILLOT, V., GRUSZOW, S. & GILLOT, P.Y. 1997. Propagation of rifting along the Arabia–Somalia plate boundary: The gulfs of Aden and Tadjoura, *Journal of Geophysical Research*, **102**, 2681–2710.

MANIGHETTI, I., TAPPONIER, P., GILLOT, P.T., JACQUES, E., COURRTILLOT, V., ARMIJO, R., RUEGG, J.C. & KING, G. 1998. Propagation of rifting along the Arabia–Somalia plate boundary into Afar, *Journal of Geophysical Research*, **103**, 4947–4974.

MCKENZIE, D.P. & MORGAN, W.J. 1969. Evolution of triple junctions. *Nature*, **224**, 125–133.

MOHR, P.A. 1967. Major volcano-tectonic lineament in the Ethiopian Rift System. *Nature*, **213**(5077), 664–665.

NYBLADE, A.A. & LANGSTON, C. A. 2002. Broadband seismic experiments probe the East African Rift. *EOS Transactions AGU*, **83**, 405–408.

RANDALL, G.E. 1994. Efficient calculation of differential seismograms for laterally homogeneous Earth models. *Geophysical Journal International*, **118**, 245–254.

SYKES, L.R. 1970. Seismicity of the Indian Ocean and possible nascent Island Arc between Ceylon and Australia. *Journal of Geophysical Research*, **75**, 5041–5055.

TESFAYE, S., HARDING, D.J. & KUSKY, T.M. 2003. Early continental breakup boundary and migration of the Afar triple junction. *Geological Society of America*, **115**, 1053–1067.

WALDHAUSER, F. & ELLSWORTH, W.L. 2000. A double-difference earthquake location algorithm: method and application to the northern Hayward fault, California. *Bulletin Seismological Society America*, **80**, 1548–1368.

WALPERSDORF, A., VIGNY, C., RUEGG, J.C., HUCHON, P., ASFAW, L.M. & AL KHIRBASH, S. 1999. Five years of GPS observations on the Arta-Sana'a baseline across the Afar triple junction. *Journal of Geodynamics*, **28**, 225–236.

WHITE, R. & MCKENZIE, D. 1989. Magmatism at rift zones: The generation of volcanic continental margins and flood basalts. *Journal of Geophysical Research*, **94**(B6), 7685–7729.

WILLIAMS, F.M., WILLIAMS, M.A.J. & AUMENTO, F. 2004. Tensional fissures and crustal extension rates in the northern part of the Main Ethiopian Rift. *Journal of African Earth Sciences*, **38**, 183–197.

WOLFENDEN, E., EBINGER, C.J., YIRGU, G., DEINO, A. & AYALEW, D. 2004. Evolution of the northern Main Ethiopian Rift: birth of a triple junction. *Earth and Planetary Science Letters*, **224**, 213–228.

# Strain accommodation in transitional rifts: extension by magma intrusion and faulting in Ethiopian rift magmatic segments

M. CASEY[1], C. EBINGER[2], D. KEIR[2], R. GLOAGUEN[2,3] & F. MOHAMED[2]

[1]*Earth Sciences, School of Earth and Environment, University of Leeds, Leeds LS2 9JT, UK (email: m.casey@earth.leeds.ac.uk)*

[2]*Department of Geology, Royal Holloway University of London, Egham TW20 0EX, UK*

[3]*TU Bergakademie Freiberg, Institut für Geologie, B. von Cotta 2, D-09599 Freiberg, Germany*

**Abstract:** Active deformation within the northern part of the Main Ethiopian Rift (MER) occurs within approximately 60 km-long, 20 km-wide 'magmatic segments' that lie within the 80 km-wide rift valley. Geophysical data reveal that the crust beneath the <1.9 Ma magmatic segments has been heavily intruded; magmatic segments accommodate strain via both magma intrusion and faulting. We undertake field and remote sensing analyses of faults and eruptive centres in the magmatic segments to estimate the relative proportion of strain accommodated by faulting and magma intrusion and the kinematics of Quaternary faults. Up to half the ≤10 km-long normal faults within the Boset–Kone and Fantale–Dofen magmatic segments have eruptive centres or extrusive lavas along their length. Comparison of the deformation field of the largest Quaternary fault and an elastic half-space dislocation model indicates a down-dip length of 10 km, coincident with the seismogenic layer thickness and the top of the seismically imaged mafic intrusions. These relations suggest that Quaternary faults are primarily driven by magma intrusion into the mid- to upper crust, which triggers faulting and dyke intrusion into the brittle upper crust. The active volcanoes of Boset, Fantale and Dofen all have elliptical shapes with their long axes in the direction N105, consistent with extension direction derived from earthquake focal mechanisms. Calderas show natural strains ranging from around 0.30 for Boset, 0.55 for Fantale, and 0.94 for Dofen. These values give extension strain rates of the order of 0.3 microstrain per year, comparable to geodetic models. Structural analyses reveal no evidence for transcurrent faults linking right-stepping magmatic segments. Instead, the tips of magmatic segments overlap, thereby accommodating strain transfer. The intimate relationship between faulting and magmatism in the northern MER is strikingly similar to that of slow-spreading mid-ocean ridges, but without the hard linkage zones of transform faults.

Fault and magmatic structures within the transition zone between continental and oceanic rifting are masked by thick wedges of seaward-dipping lavas on the majority of passive continental margins worldwide (e.g. Korenaga *et al.* 2003; Holbrook & Kelemen 1993). As a result, there are few quantitative data on strain distribution immediately prior to continental breakup, nor on the along-axis variations of a clearly 3D process. Thus, we cannot discriminate between the increasingly more complex models for rupture of rheologically layered continental lithosphere (e.g. Buck 2004). The role of magmatism in the continental breakup process is also poorly understood (e.g. Buck 2004).

Oceanic rift zone studies provide important insights into the relative importance of faulting and magmatism. In Iceland, tensile stresses from far-field plate motions accumulate over decades before being released during relatively short time periods by faulting and dyke intrusions (e.g. Bjornsson 1985; Jonsson *et al.* 1997). The magma injection is expressed at the surface by surface uplift due to increased pressure in the underlying magma source and, locally, by shallow seismicity (e.g. Sigmundsson *et al.* 1997; Feigl *et al.* 2000). In the Southern Iceland Seismic Zone and Aden rift, magma injection at the base of the brittle crust may induce stresses that exceed the Coulomb failure threshold, thereby triggering earthquakes (Feigl *et al.* 2000; Cattin *et al.* 2005). On the Upper East Rift Zone of the Kilauea volcano in Hawaii, seismic swarms with events of magnitude 3 and less are accompanied by dyke intrusions at shallow depth (Cervelli *et al.* 2002). The along-strike propagation of dykes is further facilitated by the local increase of tensile stresses at the tips of the advancing magma-filled fissures (Rubin & Pollard 1988; Rubin 1992). A separate group of models considers strain accommodation by faulting processes without magma intrusion. Within the incipient seafloor spreading centres of the Afar depression, Acocella *et al.* (2003), Manighetti *et al.* (1997) and Gupta & Scholz (2000) proposed that Quaternary normal faults nucleate from

*From*: YIRGU, G., EBINGER, C.J. & MAGUIRE, P.K.H. (eds) 2006. *The Afar Volcanic Province within the East African Rift System*. Geological Society, London, Special Publications, **259**, 143–163.
0305-8719/06/$15.00 

tensile stress-induced fractures as they grow downward and along strike, with a mechanism similar to that proposed for oceanic ridges in Iceland (Gudmundsson 1992) and at the East Pacific Rise (Wright *et al.* 1995). Alternatively, Grant & Kattenhorn (2004) propose that en echelon fissures in Iceland formed above upward-propagating, dipping normal faults.

New and existing data from the seismically and volcanically active Main Ethiopian Rift, East Africa, demonstrate its similarities to mid-ocean ridges, yet within a continental setting. Uniquely, we have the opportunity to capture processes in a rift transitional between continental and oceanic rifting. Field studies in the MER and Afar depression have documented that dyking and faulting, mostly localized in <20 km-wide magmatic segments, accommodate extension (Hayward & Ebinger 1996; Ebinger & Casey 2001). The relative proportion and process, however, remain debated (e.g. Acocella *et al.* 2003). The aims of our field and remote sensing studies are to (1) characterize structural patterns in four magmatic segments along the length of the rift; (2) estimate strain accommodation by faulting and magma intrusion in two magmatic segments; and (3) integrate these results with those of complementary geophysical studies in the rift to propose a model for the transition from continental to oceanic rifting. Our work, integrated with the EAGLE geophysical data, indicates that the along-axis segmentation of this transitional rift is largely controlled by magmatic intrusion, rather than segmented extensional fault systems.

## Tectonic setting

The Main Ethiopian Rift system developed within an Oligocene flood basalt province, and it overlies a broad region of anomalously low velocity mantle (e.g. Montelli *et al.* 2004; Benoit *et al.* 2003; Bastow *et al.* 2005) (Fig. 1). A synthesis of $^{40}Ar/^{39}Ar$ data shows that up to 2 km of flood basalts and rhyolites were erupted throughout the southernmost Red Sea region at around 31 Ma (e.g. Hofmann *et al.* 1997; Pik *et al.* 2003), roughly coeval with the separation of Arabia from Africa (Wolfenden *et al.* 2005). Less widespread and probably related volcanism affected SW Ethiopia at 45–39 Ma (e.g. Ebinger *et al.* 2000), and volcanism continued outside the developing rifts until around 11 Ma (Kieffer *et al.* 2004) when volcanism was localized to the rift valleys. The volcanic units overlying uplifted Mesozoic marine sequences now form a 2500 m-high plateau on both sides of the southern Red Sea and western Gulf of Aden (Fig. 1). Thus, the long-lived volcanism throughout the 1000 km-radius region suggests the asthenospheric hot zone is also long-lived.

The triple junction between the Red Sea, Gulf of Aden and the Main Ethiopian Rift, the northernmost sector of the East African rift system, lies within the Afar depression west of the Danakil microplate (e.g. Manighetti *et al.* 1997; Tesfaye *et al.* 2003) (Fig. 1). Oligocene extension in both the Red Sea and Aden rifts was northeast-directed, and linked to the separation of Arabia from Africa (e.g. Courtillot *et al.* 1987; Bellahsen *et al.* 2003). WNW-directed extension within the third arm of the triple junction, the Main Ethiopian Rift (MER), commenced at around 18 Ma in southwestern Ethiopia, and propagated northward into the Afar depression after 11 Ma (Wolfenden *et al.* 2004; WoldeGabriel *et al.* 1990). Thus, the MER is a much younger, and less evolved, rift than the Red Sea and Gulf of Aden rifts.

Seismic and geochemical data provide constraints on melting and melt emplacement beneath the MER. P- and S-wave tomography models indicate that the lithosphere–asthenosphere boundary lies at around 70 km subsurface with anomalously low velocity zones in the upper asthenosphere attributed to a combination of higher temperatures and the presence of partial melt (Bastow *et al.* 2005). Major element compositions of Quaternary mafic lavas from the MER are derived from parental melts generated at depths of 60–75 km (Rooney *et al.* 2005; Furman *et al.* 2006), consistent with the tomographic estimates of the lithosphere–asthenosphere boundary. Analyses of SKS shear wave splitting observations indicate that partial melt accumulates in vertically oriented dykes that cross cut the lithosphere (Kendall *et al.* 2005; Keir *et al.* 2006). Thus, there is, and probably has been, a ready supply of melt to the northern MER.

Geophysical and geochemical data show magmatic modification to the crust throughout the plateau region, and its relation to fault-controlled extension. Crustal thickness is greatest (40–50 km) beneath the uplifted plateau flanking the rift valley (Dugda *et al.* 2005; Stuart *et al.* 2006), with evidence for 10–15 km of magmatic underplate beneath the uplifted plateau where volcanism spans >40 Ma (Mackenzie *et al.* 2005) (Fig. 2). Beneath the MER, crustal thickness decreases from ~38 km in the south to 26 km beneath Fantale volcano in the southern Afar depression, with a significant decrease north of Boset volcano (Dugda *et al.* 2005; Maguire *et al.* 2006). Refraction, receiver function and gravity data suggest that (a) crustal stretching is minor south of Fantale volcano (9° N; Figs 2 & 3) and (b) that the crust comprises an increasingly larger proportion of new igneous material as one moves from south to north (Mackenzie *et al.* 2005; Tiberi *et al.* 2005; Dugda *et al.* 2005; Cornwell *et al.* 2006). Ayale *et al.* (2006) interpreted systematic

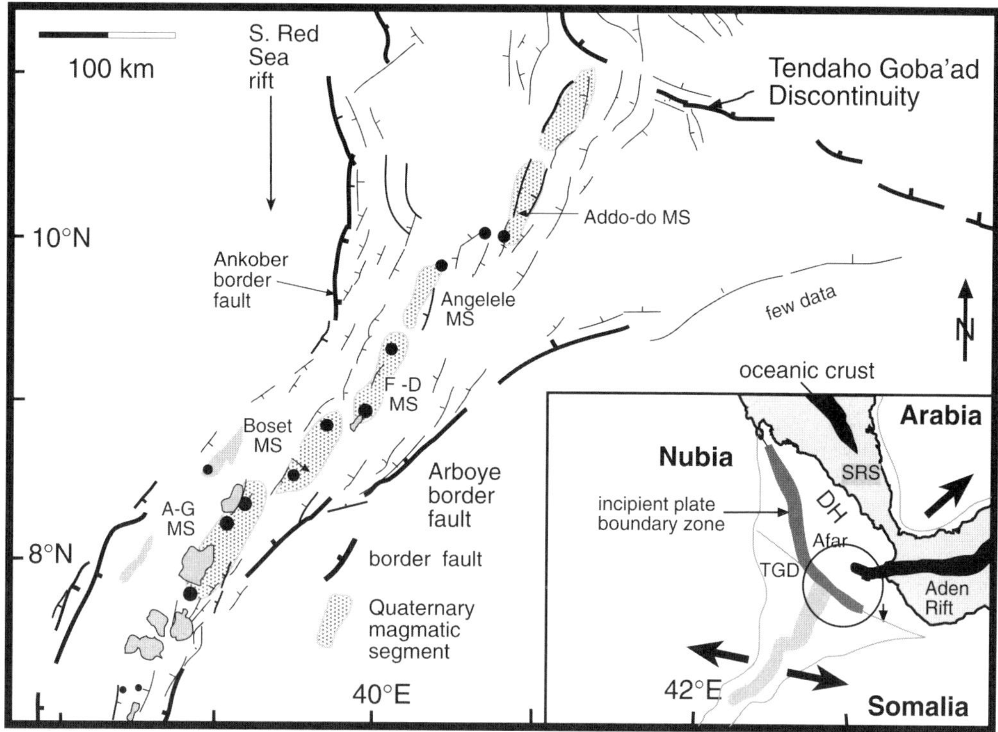


**Fig. 1.** Zones of Quaternary faulting and magmatism (magmatic segments) with respect to border faults of the Main Ethiopian Rift (MER) and the southern Red Sea rift. Inset shows the plate kinematic relation of the MER to the southern Red Sea and Gulf of Aden rifts, and opening directions between the Nubian, Somalian and Arabian plates. A–G: Aluto–Gedemsa, F–D: Fantale–Dofen, SRS: Southern Red Sea, DH: Danakil Horst, TGD: Tendaho Goba'ad Discontinuity. Black dots indicate locations of mahir shield complexes After Wolfenden *et al.* (2004).

changes in the chemistry of felsic lavas as increasing amounts of crustal melting over the period 31 Ma to present, consistent with the spatial localization of strain.

In the northernmost Main Ethiopian Rift, a system of aligned Quaternary eruptive centres forms a right-stepping, en echelon array within the rift valley, which is bounded by mid-Miocene border faults (Ebinger & Casey 2001; Wolfenden *et al.* 2004). These Quaternary magmatic segments are superposed on the older Red Sea and Gulf of Aden rift structures within the Afar depression (Hayward & Ebinger 1996; Boccaletti *et al.* 1999; Acocella *et al.* 2003). Historic and local seismicity patterns indicate that N45-trending border faults are inactive and that strain is accommodated within the less than 20 km-wide magmatic segments (Keir *et al.* 2006). A discrete cluster of activity coincides with the oblique intersection of the Red Sea and MER rifts (Keir *et al.* 2006) (Figs 1 & 2).

Geodetic, fault slip and seismicity data place strong observational constraints on the current opening direction along the MER, and seafloor spreading patterns constrain motions back to about 3.2 Ma. Fernandes *et al.* (2004) and Calais *et al.* (2006) predict 6–7 mm $a^{-1}$ opening in a direction N95 $\pm$ 5 based on models of instantaneous GPS and fault slip data. Pizzi *et al.* (2006) analysed fault slip indicators in dated Quaternary lavas, and found an extension direction of N95. Keir *et al.* (2006) inverted focal mechanism solutions for earthquakes in the northern MER occurring over a 15-month period, and they estimate an extension direction of N109 $\pm$ 12, which is identical to the opening direction from a single geodetic profile across the rift (Bilham *et al.* 1999). Chu & Gordon (1999) find an opening direction of N105 with an opening velocity of 6 mm $a^{-1}$ with plate kinematic data averaged over the past 3.2 Ma. The similarity of these independent estimates suggests that opening directions of N135 based on fault kinematic indicators alone (e.g. Acocella & Korme 2002) may be averaging pre-3.2 Ma (N135E) and present day (N105E) opening directions.

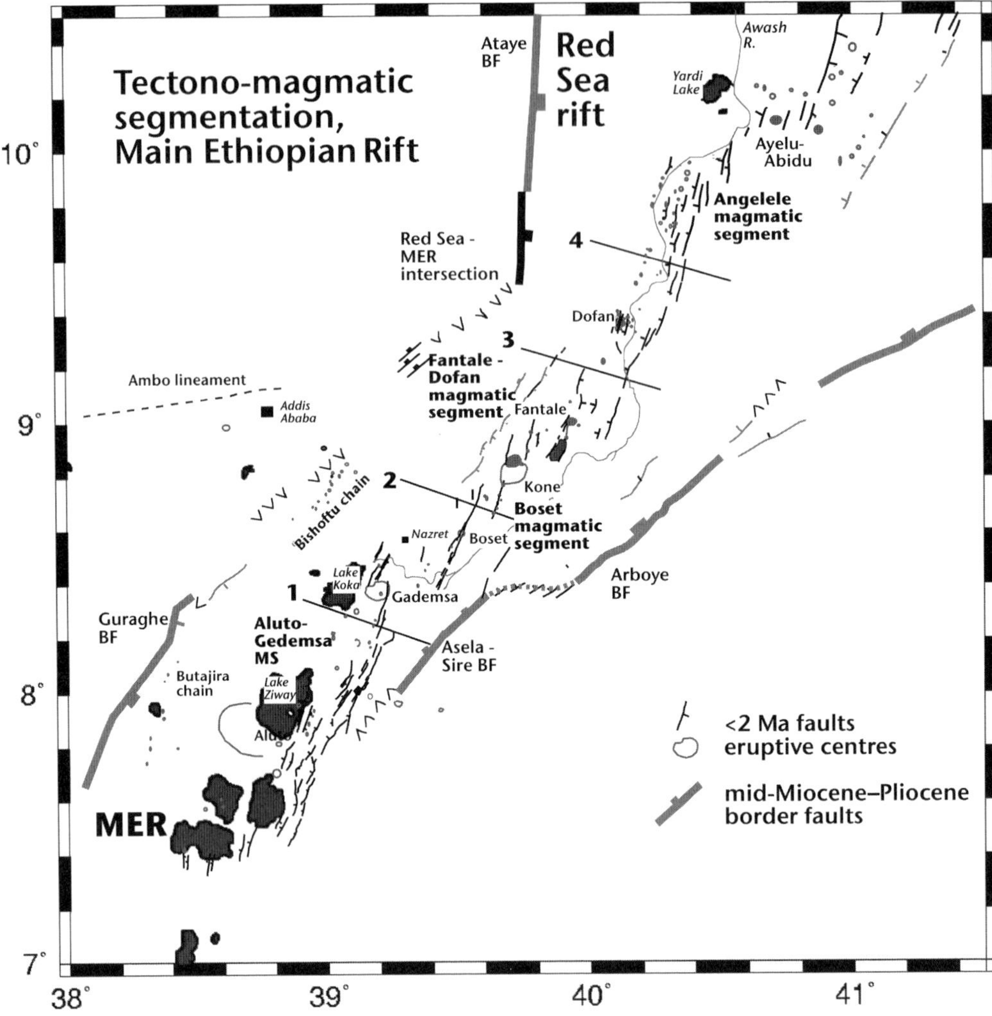


**Fig. 2.** Locations of Quaternary fault systems and eruptive magmatic centres with respect to the large offset (2–5 km) Miocene–Recent border faults of the northern MER. Note also the Butajira and Debre Zeit chains of eruptive volcanic centres to the west of the Aluto–Gedemsa and Boset–Kone magmatic segments. There are no faults with measurable throws along these aligned chains, unlike the magmatic segments. Lines 1–4 are locations of topographic profiles shown in Fig. 4.

## Magmatic segments

Right-stepping en echelon faults, fissures and chains of Quaternary eruptive centres were collectively referred to as the Wonji belt, which Mohr (1962) and Meyer *et al.* (1975) interpreted as a riftward migration of strain from border faults. With better geochronology data and imagery, Ebinger & Casey (2001) delineated a Quaternary along-axis segmentation of faults and eruptive centres, and described a series of magmatic segments within the loosely defined Wonji belt. This narrow, heavily faulted zone fades out south of 7.5° N where collapse calderas and faults occur across a much broader zone within the rift, and border faults are seismically active (Asfaw 1992; Keir *et al.* 2005). The N10-striking faults and aligned volcanic chains within the Quaternary magmatic segments of the Wonji belt are oblique to the N45 strike of the large offset Miocene border faults bounding the MER, which themselves are oblique to the N20 trend of the southern MER (e.g. Meyer *et al.* 1975; Pizzi *et al.* 2006) (Figs 2 & 3).

The Wonji belt is segmented at the surface and at depth. Tomographic images reveal an along-axis segmentation in crust and upper mantle velocity patterns that follows the same right-stepping pattern as the magmatic segments of the Wonji belt. Bastow *et al.* (2005) image segmented, NE-trending low-velocity zones beneath the magmatic segments, which they interpret as discrete zones of enhanced melt production in the mantle. The low-velocity zones are offset slightly from the rift axis towards the NW. The magmatic segments are characterized by positive Bouguer anomalies (Tiberi *et al.* 2005; Cornwell *et al.* 2006) and underlain by anomalously high-velocity material interpreted as cooled mafic intrusions (Keranen *et al.* 2004). Keranen *et al.* (2004) tomographically image a clear right-stepping pattern of high velocities between 20 km and 7 km in the crust beneath magmatic segments.

There are also two chains of Quaternary magmatism within the MER: the Debre Zeit (Bishoftu) lineament, and the Butajira lineament (e.g. Mohr 1968; Rooney *et al.* 2005) (Fig. 2). Unlike the Wonji belt magmatic segments, these chains are largely aseismic and show little structural or morphological evidence of active strain (Keir *et al.* 2006). They are underlain by hot asthenosphere (Bastow *et al.* 2005), but they lack the large relative positive Bouguer anomalies of the magmatic segments (Cornwell *et al.* 2005). Keir *et al.* (2006) suggest that these are either unfavourably oriented 'failed' magmatic segments, or incipient zones of strain.

Several models have been proposed for the along-axis segmentation of the northern MER. The obliquity of Miocene border faults and Quaternary faults in the Wonji belt led Bonini *et al.* (1997) to propose a model of oblique extension along the border faults after a change in extension direction, which probably occurred at around 3.5 Ma to accommodate along-axis propagation in the Gulf of Aden (Wolfenden *et al.* 2004). The Quaternary development, therefore, of the Wonji belt (our magmatic segments) has been attributed to oblique extension, with a left-lateral component of motion along the rift floor (e.g. Mohr 1968; Boccaletti *et al.* 1998).

Alternatively, Meyer *et al.* (1975) recognized the Wonji belt as a tectonic and magmatic rift stage transitional between continental and oceanic rifting, with a narrow zone of magma intrusion. Ebinger & Casey (2001) and Keranen *et al.* (2004) likened the 'magmatic segments' to along-axis segmentation seen on slow-spreading mid-ocean ridges, based on interpretations of geophysical and geological data. The earlier kinematic models focused on mechanical deformation across the rift, and considered the magmatism a consequence of the brittle surface deformation.

We briefly describe the characteristics of this along-axis segmentation between Aluto volcano in the south and Ayelu–Abidu in the north summarizing existing constraints and illustrating with cross-rift topographic profiles extracted from 90 m SRTM digital elevation models (Figs 3 & 4). This region coincides with the coverage of the EAGLE experiment, which provides critical subsurface and kinematic information to constrain surface observations (e.g. Maguire *et al.* 2006). Segment names are derived from the peralkaline silicic centres in each segment, rather than the geographic locales used in Ebinger & Casey (2001), which have both Amharic and Oromiya names, leading to confusion. Segment terminations are delineated at right-steps in the zones of magmatic construction. The boundary between the partly sediment-covered Fantale–Dofen and Angelele magmatic segments is less clear. There two right steps in the new border fault systems east of Dofen that mark the tip of the Fantale–Dofen segment. The Angelele segment to the north is marked to its east by an elliptical border fault system (Figs 2 & 3).

### *Aluto–Gedemsa magmatic segment*

The 20 km-wide, 70 km-long Aluto–Gedemsa magmatic segment corresponds to an approximately 50 m-high dome marked by numerous collapse calderas and volcanoes (Fig. 4a). This magmatic segment shows only minor crustal thinning with respect to crustal thickness beneath the flanks (Stuart *et al.* 2006; Maguire *et al.* 2006; Tiberi *et al.* 2005). The 0.85–0.29 Ma Gademsa caldera lies near the northern end of this magmatic segment (Morton *et al.* 1979; Boccaletti *et al.* 1999). Pizzi *et al.* (2006) provide detailed maps and interpretations of faults and eruptive centres in the Aluto–Gedemsa segment, as well as the more diffuse zone of faulting and magmatism to the south.

The morphologically youthful faults cut 1.97 Ma basalt flows, which are overlain by a 1.6–0.6 Ma package of ignimbrites, and Holocene flows (Pizzi *et al.* 2006). Individual faults within the magmatic segment are up to 20 km in length, and bound narrow horsts and grabens (e.g. Fig. 4a). Pizzi *et al.* (2006) estimate minimum slip rates of around 1 mm $a^{-1}$.

### *Boset–Kone magmatic segment*

The *c.* 65 km-long Boset magmatic segment steps 15 km to the east, and lies about 500 m lower than, the Aluto–Gedemsa magmatic segment (Figs 3 & 4a). The Keradi–Balchi fault system lying to the west of the magmatic segment developed between 3.2 and 2.5 Ma, but morphology and seismicity patterns indicate that it is now inactive (Wolfenden *et al.* 2004; Keir *et al.* 2006).

The *c.* 1.9 Ma Wonji series marks the base of the magmatic segment (WoldeGabriel *et al.* 1990;

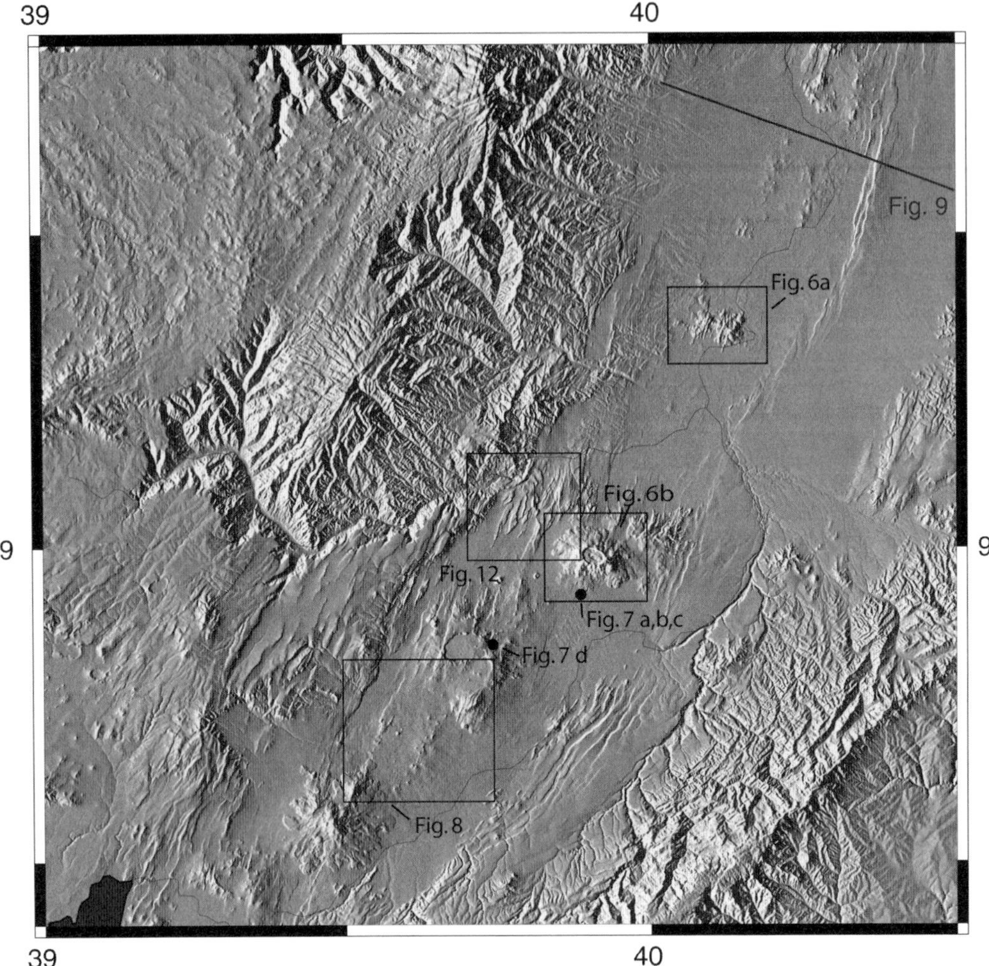


**Fig. 3.** 90 m SRTM greyscale digital elevation model showing the locations of the areas corresponding to figures. Note the consistent decrease in elevation from south to north along the length of the rift.

Boccaletti *et al.* 1998; Kidane *et al.* 2006). Boset (Bosetti) volcano lies near the southern tip of the segment, and the elongate, multiphase Kone caldera lies near its northern end (Figs 2 & 3). Morton *et al.* (1979) report an age of 0.79 Ma for the basal trachytes of Boset. Williams *et al.* (2004) constrain late stage eruptions to around 14 000–6800 BP. Along the Boset–Kone magmatic segment, two 30–40 km-long fault arrays and aligned chains of cinder cones are themselves arranged in a right-stepping en echelon pattern (Figs 2, 3 & 4a).

## *Fantale–Dofen magmatic segment*

Faults and aligned eruptive centres in the Fantale–Dofen magmatic segment developed within the zone of intersection of the *c.* 25 Ma Red Sea rift, and the MER (Wolfenden *et al.* 2004). Unlike the Aluto–Gedemsa and Boset–Kone magmatic segments to the south, the Fantale–Dofen magmatic segment lies within a half-graben bounded along its eastern side by a series of 60–80 m-high scarps (Fig. 3; km 38, Fig. 4b). This graben lies 550 m below the *c.* 1300 m-high Boset–Kone magmatic segment, consistent with the >5 km decrease in crustal thickness north of between Boset and Fantale volcano (Maguire *et al.* 2006; Stuart *et al.* 2006).

This magmatic segment is marked by felsic shield volcanoes with elongate calderas at both its southern (Fantale) and northern (Dofen) ends (Figs 2, 3 & 6). Lava flows originating from a fissure located on Fantale's southern flank erupted between 1810 and 1820 (e.g. Gibson 1967;

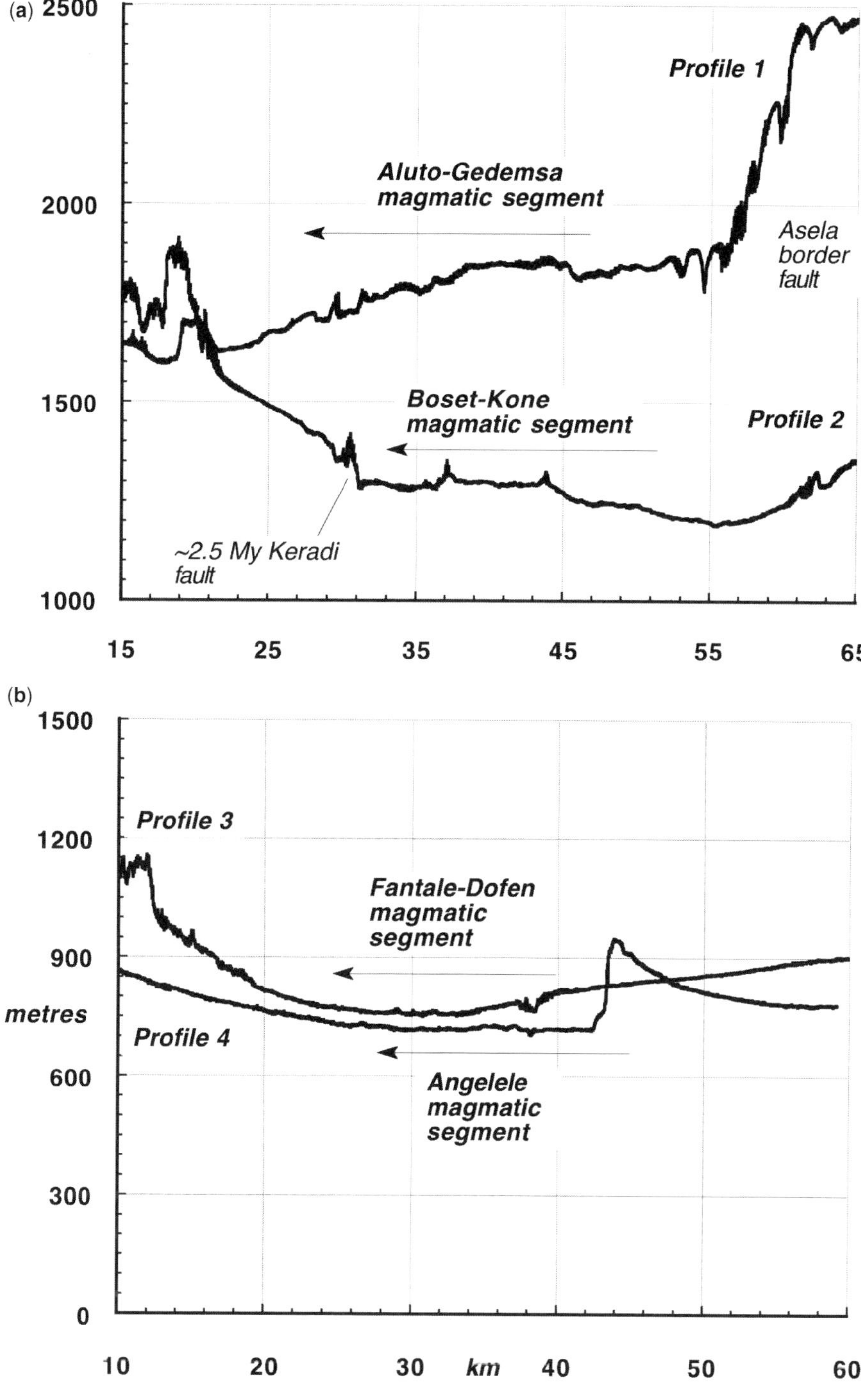


**Fig. 4.** Vertically exaggerated topographic profiles extracted from 90 m SRTM topography database to illustrate tectono-magmatic relations and their variations along the length of the rift. Profiles 1–4 are marked on Fig. 2. (**a**) Profiles 1 and 2 crossing the Aluto–Gedemsa and Boset–Kone magmatic segments, respectively. (**b**) Profiles 3 and 4 crossing the Fantale–Dofen and Angelele magmatic segments, respectively. Note the change from low dome to half-graben morphology of the active tectonic zone as one moves from south to north.

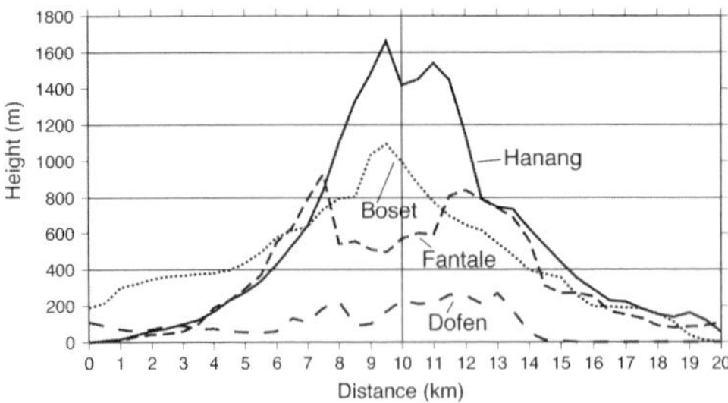


**Fig. 5.** Profiles through the volcanic edifices of Boset, Fantale and Dofen. The unfaulted profile of Hanang volcano (Tanzania) is shown for comparison. Geological and geophysical data show no evidence of extensional strain across Hanang (e.g. LeGall *et al.* 2004). The MER volcanoes are all broader and lower than Hanang; flattening increases from south to north.

Williams *et al.* 2004), and they are now cut by N10° E-striking faults (Figs 2, 3 & 6a). Open fissures, and fissures with lavas, are common throughout this belt (e.g. Asfaw 1982; Williams *et al.* 2004). Although some K–Ar dates for the extended apron of the Dofen volcanic complex provide dates of 1.7 Ma, historic flows and sulfur extrusions attest to ongoing activity within the central part of the rifted Dofen shield (Chernet 2005). Normal faults with >100 m throw downdrop the Dofen shield complex, producing a narrow rift valley in the central caldera complex (Figs 3, 5 & 6a). The Dofen region experienced intense seismicity during the period 2001–2003, with an approximately $M_L = 4.8$ event located SE of Dofen volcano (Keir *et al.* 2006).

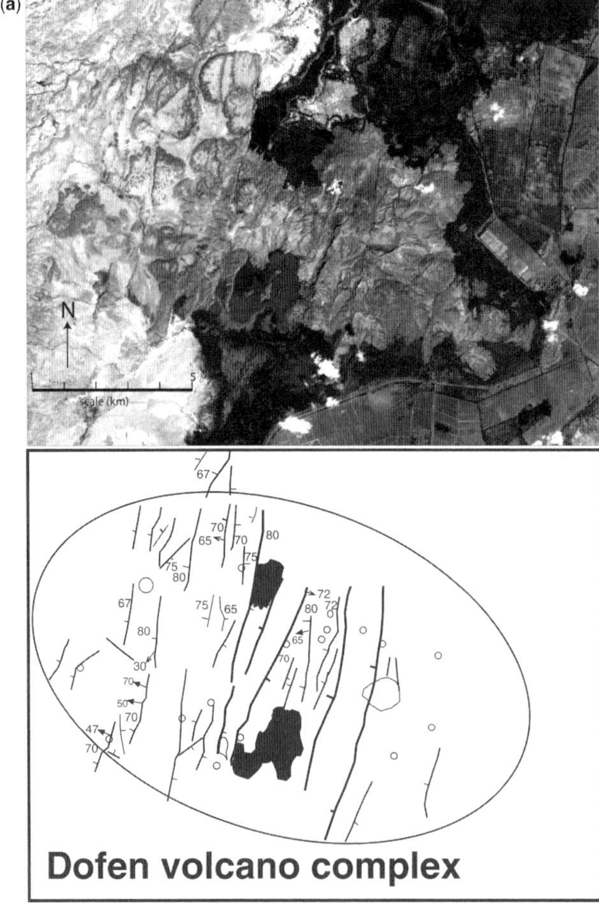


**Fig. 6.** (**a**) Band 2 of Aster 15 m resolution imagery showing the Dofen volcanic edifice. The elliptical region with roughly east–west long axis is taken to be the strained volcanic edifice (ellipse above shows regional extension direction). Fault dips and fault slip directions from field studies in 1990 shown in line drawing below image. Open circles indicate eruptive centres.

### *Angelele magmatic segment*

The 60 km-long Angelele magmatic segment has the form of a half-graben bounded on its eastern side by a 230 m-high fault system, the largest active fault in the area (Figs 2, 3 & 4b). This 'new' border fault has an en echelon soft linkage zone with the fault system bounding the eastern side of the Fantale–Dofen magmatic segment; the individual fault segments form a right-stepping system between 9.5 and 10° N (Figs 2 & 3). It is also a shallow depression like the Fantale–Dofen segment to the south (Figs 3 & 4b).

Geological and geophysical information is sparse in this remote region; only ASTER imagery and DEM data were used. Faults and aligned eruptive centres do not show the right-stepping pattern of magmatic segments to the south, perhaps owing to superposition of the MER on Oligocene Red Sea rift structures (e.g. Tesfaye *et al.* 2003). Chernet *et al.* (1998) provide a lone K–Ar date of $0.83 \pm 0.02$ Ma for a basalt flow midway down the magmatic segment. Faults bounding the segment show greater throws than to the south. Keir *et al.* (2006) locate a number of seismic events in this magmatic segment, which lies at the limit of the EAGLE array.

## Structural observations

Field sites within the Boset–Kone and Fantale–Dofen magmatic segments were chosen to field check satellite imagery and to examine the relation, in time and space, between faulting and magmatism. In this regional study of the Boset–Kone and Fantale–Dofen magmatic segments, we measured the lengths and orientations of faults that cut <1.9 Ma lava flows, as well as the position of scoria cones, fissural or vent flows, and dykes. We measure fault lengths and orientations using field-calibrated high-resolution imagery, 1 : 50 000 topography sheets, and 90 m-resolution SRTM. Landsat Thematic Mapper imagery (15–30 m resolution) and ASTER imagery (15 m resolution) provide a regional context, and field traverses were made on all but the Angelele magmatic segment. We confined our study to faults of length greater than 200 m that were clearly distinguishable from felsic and basaltic flow fronts (e.g. Figs 6 & 8). Fault displacements can only be estimated from topographic relief, as vertical resolution of seismic data is about 500 m in the shallow subsurface. We therefore do not examine length-displacement relations in this regional study. Faults immediately outside the Quaternary magmatic

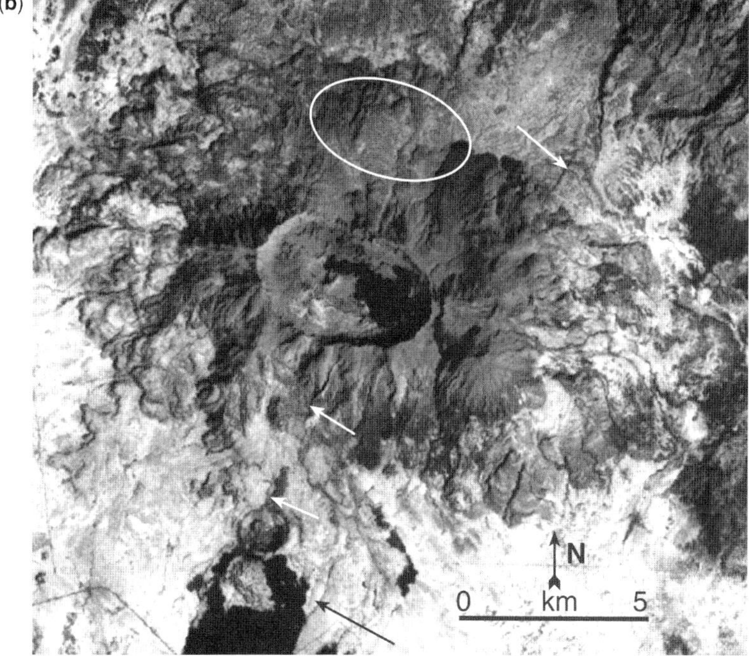


**Fig. 6.** Continued. (**b**) Landsat-7 Thematic Mapper (30 m resolution) image of Fantale volcanic edifice. Arrows point to active N10E fault scarps.

segments show little or no seismicity and are more deeply incised. These faults are also excluded from our database.

An example of the observations used to interpret Quaternary faults is shown later in Fig. 11. The fault scarp in the foreground shows the relay zone between two segments. The segments seen here are typical of their regularity in the Quaternary faults. The faults are often marked at the surface by a vertical fissure that may have very little movement (Fig. 7a). The fissure sometimes separates a rotated block of rock and the upthrow side of the fault (Fig. 7c). Figure 7b is a view onto the surface of the rotated block, which forms a smooth monoclinal ramp from upthrown to downthrown side. Figure 7d is a view from a relay zone towards one of the segments.

Measurements of fault, dyke, and fissure geometries were made throughout the Boset–Kone and Fantale–Dofen magmatic segments. Faults in the basalts rapidly degrade, splitting along individual columnar basalts, leading to near-vertical scarps that rarely preserve the fault slip plane. Fault kinematic indicators were only measured in the Dofen caldera complex, and contribute to our evaluation of Quaternary extension direction (Fig. 6).

There are no active large offset faults along the length of the magmatic segments, nor across the Boset–Kone magmatic segments; we can infer that there are no buried faults, as only the 2–4 km throws on the border faults are imaged in EAGLE seismic data (Mackenzie *et al.* 2005; Maguire *et al.* 2006). The cumulative frequency plots of all identified faults within the Boset–Kone magmatic segment ($N = 123$) and the Fantale–Dofen magmatic segment ($N = 164$) are shown in Fig. 9. The limited length range of observations makes it speculative to interpret the completeness of the database. Clearly, many faults may be covered by sediments and/or lavas, and faults with small surface displacements will be difficult to identify in DEM and imagery.

The orientations of individual segments forming linked faults is shown in Fig. 10. We also estimate strain direction and magnitude as given by the elliptical shape of Boset, Fantale and Dofen volcanoes (e.g. Fig. 6b).

**Fig. 7.** Faults and fissures from the overlap zone between the Boset–Kone and Fantale–Dofen magmatic segments. (**a**) Fissure with very little shear movement beside Lake Beseka; downthrow is to the left. (**b**) View onto the rotated block of a normal fault, Lake Beseka. (**c**) Wide fissure and rotated block (to left) Lake Beseka. (**d**) Fault, relay zone and fissure, near the Kone caldera complex. The left side of the hill on the skyline gives the orientation of the fault at depth.

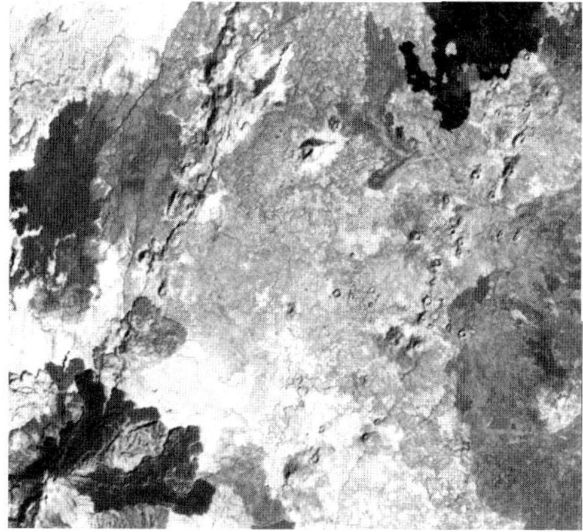

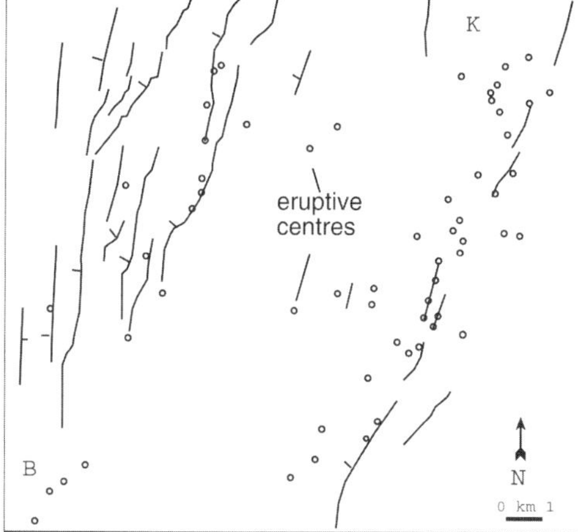


**Fig. 8.** Landsat TM imagery showing the close spatial coincidence of faults and eruptive centres (fissures, cinder cones, shields) on the central Boset–Kone magmatic segment. Location shown in Fig. 3. Note the parallel faults and aligned scoria cones, as well as fissural flows from some faults. Boset (B) and Kone (K). Most of the relief in this zone is magmatic construction.

The positive relief of the Aluto–Gedemsa and Boset–Kone magmatic segments deflect rivers away, whereas the shallow depressions of the Fantale–Dofen and Angelele magmatic segments capture the Awash River and its tributaries. Faults of the northern two segments, therefore, expose Pliocene–Recent ephemeral lake deposits (e.g. WoldeGabriel *et al.* 1990; Asfaw *et al.* 1992), striae and other slip indicators along faults in fluvio-lacustrine and volcaniclastic sequences. These field observations and the fault and eruptive centre analyses of imagery from the Boset–Kone and Fantale–Dofen magmatic segments reveal

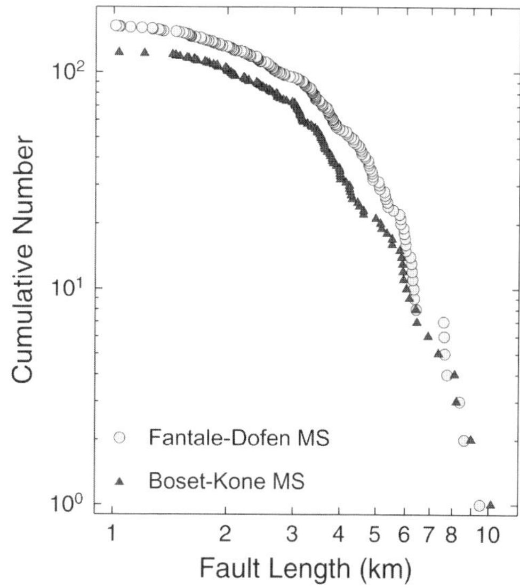


**Fig. 9.** Cumulative frequency plots of faults from the Boset–Kone magmatic segment and Fantale–Dofen magmatic segment.

several consistent patterns (online supplementary imagery and Table 1).

(1) Faulting and magmatism occur in *c.* 20–40 km-long right-stepping arrays that lie on either side of the magmatic segment (e.g. Figs 2 & 8). Within the Boset–Kone magmatic segment, these arrays emanate from the Boset and Kone shield complexes, and faults curve slightly into these volcanoes. Faults are more numerous around the multiple collapse calderas of Kone and Dofen, suggesting some formed to accommodate magma withdrawal, rather than tectonic strain (e.g. Asfaw *et al.* 2006). Owing to the right-stepping nature of the chains in the Boset–Kone segment, there is no clear 'spreading axis'. The Fantale–Dofen segment, however, shows a low relief, around 10 km-wide central graben striking north–south, with most faults and eruptive centres along the eastern side of this depression. The felsic shield volcanoes lie near the tips of the magmatic segments, suggesting the rise of magma is inhibited at segment tips, and crustal reservoirs form (e.g. Keranen *et al.* 2004).

(2) 51% of the faults in the Boset–Kone magmatic segment have aligned cones and/or fissures along their lengths. For example, small basaltic cones and fissural eruptions mark a 10 km-long, <100 m-high, N10° E-striking fault array south of Kone caldera (Fig. 8). Fewer faults and eruptive centres are exposed in the depression of the Fantale–Dofen magmatic segment where 41% of

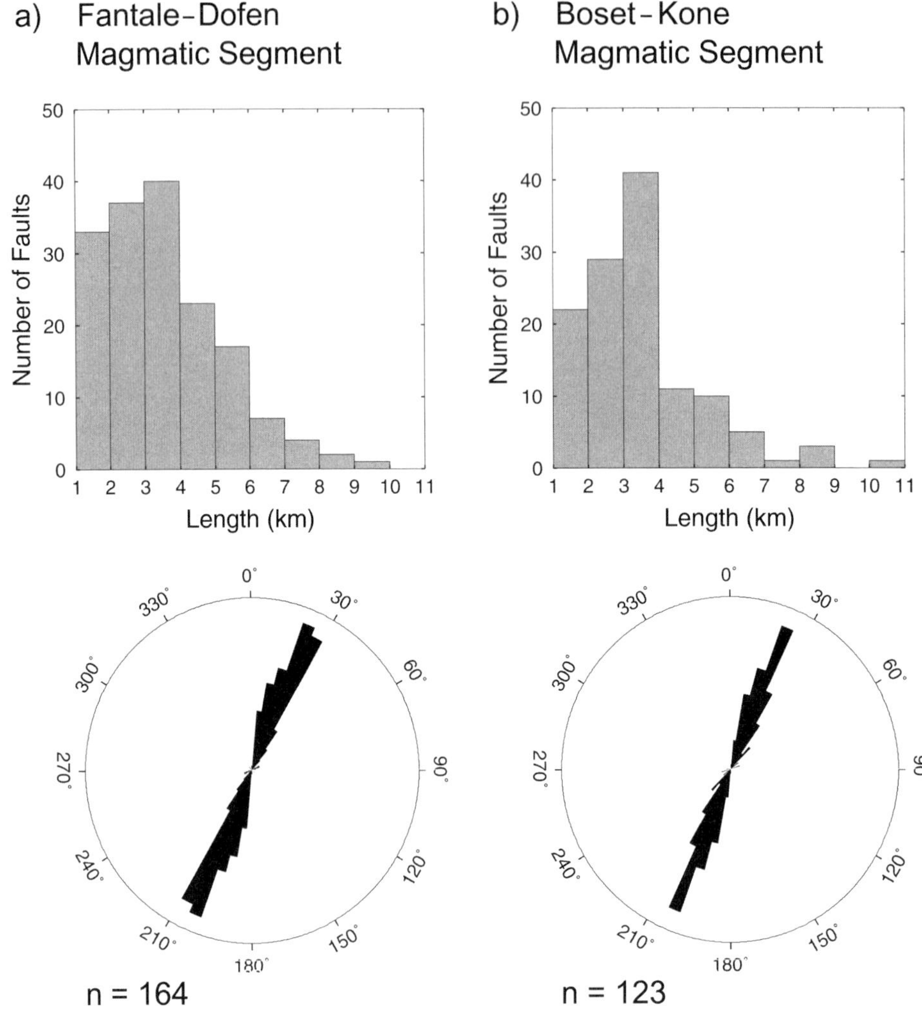


**Fig. 10.** Histograms of fault lengths and rose diagrams of fault azimuth for the fault populations shown in Fig 9. (**a**) Fantale–Dofen magmatic segment and (**b**) Boset–Kone magmatic segment.

the faults have eruptive centres along their lengths, but where much of the magmatic segment is buried by recent sediments.

(3) Where the interaction between faults and cones could be clearly defined either in the field or on imagery, we found that basalts were located preferentially at the tips of the faults and open cracks (Figs 8 & 11).

(4) Maximum fault length is less than 11 km within both the Boset–Kone magmatic segments (Figs 9 & 10). Most faults are $\leq$6 km-long, with more faults identified in the Fantale–Dofen segment than in the Boset–Kone magmatic segment (Fig. 10). The largest fault in the Fantale–Dofen magmatic segment has a displacement of 230 m, but faults within the Boset–Kone magmatic segment have <80 m relief. More faults are detected in the Fantale–Dofen ($N = 164$) than the Boset–Kone ($N = 123$) magmatic segment.

(5) The majority of faults and fissures strike N10–N30° E (Fig. 10). A small subset of short faults striking N45–70° E link arrays within the magmatic segments (e.g. Figs 6b, 8 & 12). Similar patterns are found in the Aluto–Gedemsa magmatic segment (Pizzi *et al.* 2006).

(6) Faults outside the magmatic segments show both N15° E and N45° E trends. Figure 12 shows an area of faulting between the western

**Table 1.** *Axial ratio, natural strain and trend of long axis estimates for some Ethiopian volcanoes*

| Volcanic centre | Data source | Method | Height, m | Axial ratio | Natural strain | Trend of long axis (°) |
|---|---|---|---|---|---|---|
| Dofen edifice | Landsat | Measurement | 800 | 2.56 | 0.948 | 113 |
| Dofen edifice | SRTM | Moments | 800 | 1.34 | 0.293 | 110 |
| Fantale crater | Landsat | Measurement | 1900 | 1.67 | 0.512 | 114 |
| Fantale crater | Landsat | Measurement | 1100 | 1.77 | 0.571 | — |
| Fantale crater | Map | Measurement | 1900 | 1.61 | 0.476 | 111 |
| Fantale edifice | SRTM | Moments | 1200 | 1.36 | 0.308 | 109 |
| Fantale edifice | SRTM | Moments | 1450 | 1.55 | 0.438 | 107 |
| Fantale edifice | SRTM | Moments | 1580 | 1.95 | 0.668 | 106 |
| Boset edifice | SRTM | Moments | 1950 | 1.31 | 0.270 | 107 |
| Boset edifice | Map | Measurement | 2100 | 1.48 | 0.392 | 127 |

edge of the Fantale–Dofen magmatic segment and the rift border faults, immediately north of the tip of the Boset–Kone magmatic segment. In the area labelled 'inactive faults', there are two directions of faults with rhomb-shaped regions between. The rhomb-shaped regions are often pull-apart basins into which river courses have been diverted and which contain ephemeral floodplains. The region labelled 'active faults' lies within the Fantale–Dofen magmatic segment. These faults cut 2.5 Ma and older sequences but predate the magmatic segment formation (Wolfenden *et al.* 2004).

(7) There are no fault systems linking the right-stepping magmatic segments between the Aluto–Gedemsa and Boset–Kone magmatic segments, and between the Boset–Kone and Fantale–Dofen magmatic segments. Some short, N45–70° E-striking fault segments are noted along the length of the magmatic segments; these short faults allow linkage of en echelon faults, and may utilize Miocene–Pliocene faults (e.g. Figs 6a, 8 & 12). Normal faults occur across the breadth of the overlapping magmatic segment tips, rather than in narrow arrays seen in the central magmatic segments.

(8) Eruptive centres are elongate in a direction parallel to the extension direction. Figure 5 compares the morphology of Boset, Fantale and Dofen edifices, which show a progressive flattening from south to north along the length of the rift.

The small offsets on the younger faults and the irregularity of the surface caused by lava flows and other magmatic constructs means that faults with small offsets cannot be interpreted in terms of fault deformation alone. An exception is the 30 km-long, 250 m-high fault escarpment bounding the east side of the Angelele magmatic segment. (Figs 4 & 13). The easterly dipping side of the ridge produced by the escarpment shows a section with a discernible curvature running for 10 km east from the ridge.

## Explanation of structures

### *Down-dip fault length from footwall uplift*

The displacement difference boundary element method of Crouch & Starfield (1983) is a simple and convenient means of calculating the displacement fields around faults, which are idealised as normal fault dislocations in an elastic half-space. This idealization is scale independent. The only points of comparison between model and observations are the dimension of the fault and the distance to which the deformation field is perturbed by the presence of the fault; simulation of the footwall and hanging wall shapes would require additional assumptions beyond the scope of this study. The magnitude of the displacement on the fault has no influence on the extent of the perturbation, except that for small displacements in nature the roughness of the ground surface will mask the perturbation. A model was set up with a line of boundary elements 100 units long representing the Earth's surface: a boundary condition of zero shear and normal stress was applied. A fault dipping at 60° was defined by a 10-unit-long line of elements dipping from the mid-point of the line of elements representing the surface. A condition of zero relative normal displacement was used to hold the fault surfaces in contact. The relative shear displacement on the fault was given a linear increase from zero at the lower end to an arbitrary value of 0.01 units at the surface.

The vertical displacement of the surface on the footwall scaled horizontally and vertically to the topographic profile is given in Fig. 13. The horizontal scaling was applied to the down-dip length of the fault. From this it may be estimated that the down-dip length of the largest fault in the study area is around 10 km (cf. Fig. 9), consistent with the 7–10 km seismogenic layer thickness estimated

**Fig. 11.** Quaternary faults in the Asela region. The view direction is towards the SE, with a series of step faults leading up to the border fault in the background. The faults in the foreground show the regular segmentation of the seismically active faults. Extrusive lavas occur at segment linkage zone. Photo: Derek Keir.

by Keir *et al.* (2006). The topographic profile used is the one that has the longest horizontal range of footwall uplift and so the down-dip length of the fault can be taken as a maximum estimate for the faults in the region.

## *Cinder cones and extrusions*

The concentration of cinder cones close to the magmatic segment axes had been ascribed to their being fed along axis from relatively small magma chambers under the Quaternary volcanoes (Ebinger & Casey 2001). Geochemical and geophysical data from the EAGLE project indicate that some dykes and fissural flows are directly fed from upper mantle sources rather than from crustal reservoirs (Kendall *et al.* 2005; Rooney *et al.* 2005; Furman *et al.* 2006). We further propose that the regular tracks of cinder cones are the surface expression of dyke segments resulting from the variation of stress axis orientation towards the surface (Delaney & Pollard 1981).

## *Volcanoes as strain indicators*

The volcanic centres have elliptical forms with the long axis perpendicular to the younger normal faults, suggesting that they are the locus of strain, and record both the magnitude and the duration of strain (Fig. 6a, b). The axial ratios of three volcanoes, Dofen, Fantale and Boset, were estimated in two ways: measuring the long and short axes plus the orientation of the long axis from digital images, and by the method of De Chabalier & Avouac (1994) (Table 1). The latter method uses the second moments of the topography, with the assumption that the extension is perpendicular to the normal faults. Deviations of the actual long axis from this direction is taken to be the result of shear strains. In this study, these deviations were so small as to be negligible. The shortening direction is taken to be vertical and this is supported by the flattened nature of these volcanic edifices, especially visible for Dofen (Figs 3 & 5). The ratio given is thus the extension strain component. The ratio is also given as its natural logarithm, or natural strain, a strain measure that increases linearly with time under constant strain rate.

The data in Table 1 are quite variable according to method of obtaining the data and location of the measurements on each volcano. However, they show two trends: an increase in ratio from south to north, i.e. from Boset to Dofen; and an

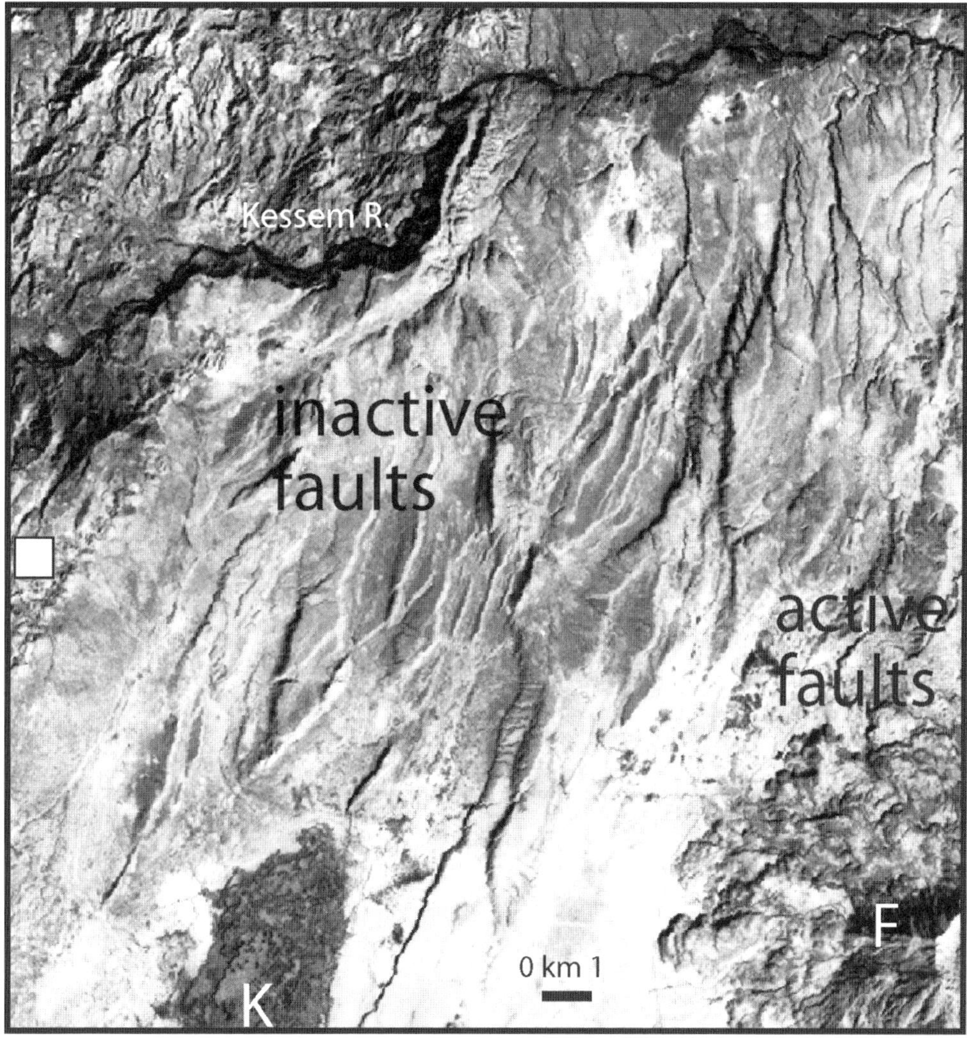


**Fig. 12.** Landsat TM imagery showing fault patterns outside the Quaternary Fantale–Dofen magmatic segment showing both N10E and N35E-striking, aseismic fault sets. The two fault sets intersect to form rhombohedral blocks. The active faults to the east cut the apron of Fantale volcano. The square on the left edge of the figure marks the position of a felsic ignimbrite marker bed dated at 2.5 Ma, which was used to estimate timing of fault movements (Wolfenden *et al.* 2004).

increase in ratio from measurements taken low to high on the volcanic edifices. The long axis orientation is generally consistent at around $110 \pm 3^{\circ}$ except for the measurement taken from a high contour of Boset.

Figure 5 shows a comparison of the profiles of the Ethiopian rift volcanoes with an elliptical shape (e.g. Fig. 6) and Hanang volcano from Tanzania with no detectable stretching. The Ethiopian volcanoes, especially Dofen, have lower profiles than Hanang. This observation supports the hypothesis that the elliptical shape of the MER volcanoes is the result of horizontal extension and vertical shortening, in a narrow zone of magma injection (magmatic segments).

For a simple extension deformation, the strain rate can be obtained from the finite strain ratio by the following:

$$d = \frac{\ln(l/l_0)}{\Delta t}$$

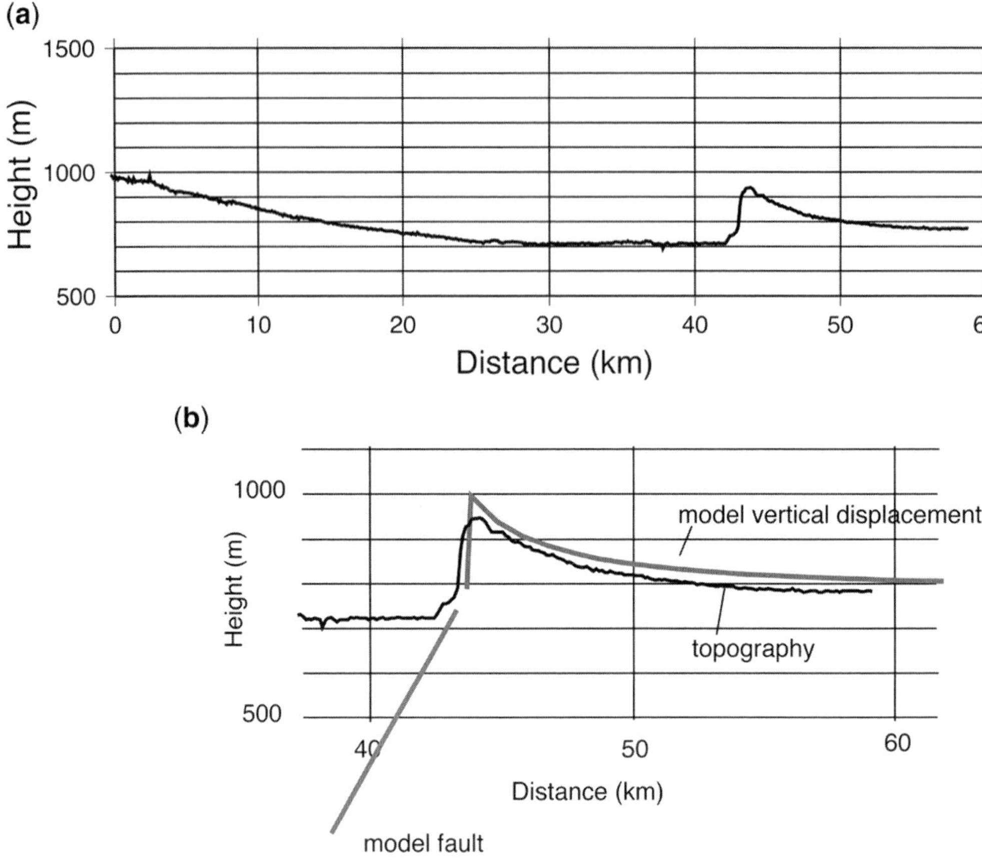


**Fig. 13.** (**a**) Detailed portion of the topographic profile through the Angelele magmatic segment (Profile 4, Figs 3 & 4b). (**b**) Vertical displacement of a model fault dipping at 60° scaled to fit the topographic profile shown in (a). The model displacement profile has been moved upwards to allow comparison. The model fault, scaled by the same amount as its displacement profile is also shown in the figure.

where $d$ is the strain rate, $l$ is the deformed state length of a line of $l_0$ in the undeformed state and $\Delta t$ is the time over which the deformation builds up. At the spatially averaged strain rate of 0.1 μ strain per year estimated by Bilham *et al.* (1999), the strains would take from 2–9 Ma to build up, a little longer than the 1.8 Ma age of the magmatic segments. The main sources of error in comparing the strain rate estimated from the time to reach the strains observed and that from present-day geodetic measurements come from the variability of the latter in space and time. The relatively short times of geodetic studies raise the possibility that the measurements vary considerably from the average. The degree to which the strain is localized also has an important influence on the result. If strain is more localized than was assumed by Bilham *et al.* (1999), then our estimates concur within error.

## Discussion

Seismicity data over the period 2001–2003, historic seismicity and geomorphology all indicate that the Miocene border faults are inactive and strain is localized to the *c.* 20 km-wide magmatic segments. Although the planform of the MER between 7.5° N and 9° N is typical of analogue models of 'transtensional rifting', the localization of strain to the Wonji belt magmatic segments means that border faults and en echelon magmatic segments developed diachronously; differences in orientation represent the lithospheric response to changes in plate forces. Two additional lines of evidence argue against transtensional deformation for the deformation localized in the magmatic segments. Palaeomagnetic data presented by Kidane *et al.* (2006) show negligible rotations about vertical axes either within or between magmatic segments, as would be expected if the region were in fact transtensional. Focal mechanism solutions of Keir *et al.* (2006) show primarily dip-slip movement on N20E-striking fault planes. The minor left-lateral slip mechanisms are consistent with slip on pre-existing N45° E-striking planes within a N110E extension regime (Ayele *et al.* 2000; Keir *et al.* 2006).

The zone of rhombohedra resulting from the intersection of the two fault orientations observed between the rift border faults and the Fantale–Dofen magmatic segment is interpreted as a phase of transtensional deformation between 3.5 Ma and 2.0 Ma, as noted by deformation of an ignimbrite marker bed (Fig. 12). Bonini *et al.* (1997) also come to this conclusion based on their model results, but these authors concluded that the transtension persists to the present day. In our interpretation, the duration of this phase of rifting is too short to accumulate measurable block rotations observable by palaeomagnetic investigations. Both models and indirect evidence from seismicity data indicate that the small offset normal faults, monoclines and open fissures form from upward propagation of both faults and faulted dykes. Magma intrusions to mid-crustal levels accommodate strain in the lower crust, and dykes and faults form in the brittle crust above intrusions (e.g. Fig. 9, Keir *et al.* 2006). Thus, patterns in the upper crust are similar to the model of Grant & Kattenhorn (2004), but with magma intrusion the trigger at depth. Bull *et al.* (2006) show that monoclines develop beyond the lateral tips of faults in Iceland showing the same morphology as those we observed in the MER (e.g. Fig. 7b); these relations have been noted by Rowland (pers. comm., 2005). This comparison with faults in Iceland also demonstrates the similarity of process, albeit at slower rates of strain and magma production. In studying the faults on a length scale of more than around 10 metres, the presence of a vertical fissure and monocline at the surface is unimportant. Throw should be measured from the undisturbed rock on the footwall down to the undisturbed rock on the hangingwall. The orientation of the fissure (sub-vertical) should not be taken as the dip of the fault, the plane of which is some tens of metres in the subsurface.

Buck (2004) demonstrated that extension of the lithosphere can take place at a lower tectonic stress when magma is present to form dykes. If the intrusion occurs at intermediate depths (5–10 km), the released load will be transferred upwards, driving extensional normal faults propagating towards the surface.

The deformation processes proposed here are very similar to those in the newly formed seafloor Asal–Ghoubbet rift. Cattin *et al.* (2005) used a visco-elastic lithospheric deformation model to simulate the observed geodetic and seismicity patterns in the Asal–Ghoubbet rift, Djibouti and propose that a dyking event at depth triggered normal faulting. They found a very good fit of model predictions with seismogenic layer thickness and heat flow observations. Their models allow >50% extension by magmatic intrusion.

The magmatic segments of the Ethiopian rift show a trend of increasing extension towards the north, as indicated by the deformed volcanic edifices and increase in number of identified faults from Boset–Kone ($N = 123$) to Fantale–Dofen ($N = 164$) magmatic segments (see Fig. 10). Seismic (Bastow *et al.* 2005; Maguire *et al.* 2006) and geochemical (Furman *et al.* 2006) data also show a south to north increase in the amount of crustal and mantle lithospheric thinning.

Using remote sensing, seismic, geochemical and geodetic data, and by analogy to the Asal–Ghoubbet rift zone, we can estimate the relative contribution of dyke intrusion and faulting to the total deformation. Synthesizing constraints on distribution of strain from field, the number of faults and their displacements do not appear adequate to account for all the strain recorded in the shapes of the volcanoes, nor predicted from geodetic models of Nubia–Somalia separation. One possibility is that there are large offset faults buried by lava flows, but one would expect flows to show the shape of the footwall uplift of these faults. Seismic data in Mackenzie *et al.* (2005) show no evidence for buried magmatic segment faults with offsets >500 m, the resolution in the upper 5 km. Bendick *et al.* (2006) compared seismic moment release and geodetic strain across the MER along a line through the Boset–Kone magmatic segment, and concluded that faulting alone could not account for plate boundary deformation over the past two decades. All the faults appear to have dips of 60° or greater and large displacements on normal faults rotates them to shallower dips. In addition, a large amount of extension on normal faults would lead to thinning detectable in EAGLE seismic data. Instead, only one fault with >200 m throw is observed; it marks a basin, rather than the topographic highs seen in the Aluto–Gedemsa and Boset–Kone magmatic segments. The larger faults are matched by apparently higher magma production rates to the north (Bastow *et al.* 2005; Furman *et al.* 2006; Pik *et al.* 2003).

The right-stepping en echelon arrangement of magmatic segments, as well as the fault zones and aligned chains of eruptive centres along the length of magmatic segments, mark a segmentation distinct from the Miocene border fault segmentation (e.g. Ebinger & Casey 2001). Our observations indicate that magma intrusion processes largely control the extension and rupture processes in the MER, as was seen by Cattin *et al.* (2005) in the

Asal–Ghoubbet rift. The segmentation in low-velocity zones near the lithosphere–asthenosphere boundary beneath the MER implies localized melt production zones along the length of the rift (Bastow *et al.* 2005), but crustal contamination makes it difficult to evaluate along-axis trends in geochemical data from the Quaternary centres (e.g. Ayele *et al.* 2006; Furman *et al.* 2006). We find no evidence for cross-rift structures linking en echelon segments at the surface, or at depth. Gravity and crustal seismic data mirror the surface expression of the magmatic segments; crust beneath narrow magmatic segments is denser and high velocity, whereas intersegment zones are not anomalous (Keranen *et al.* 2004; Tiberi *et al.* 2005; Cornwell *et al.* 2006). Likewise, the lack of rotation about vertical axes in intersegment zones argues against distributed strain in these zones.

We suggest that the en echelon along-axis segmentation of the MER represents the interplay between Miocene–Pliocene rift structures and the now-focused, localized magma intrusion zones marking the transition to seafloor spreading. The East African rift is linked to the Red Sea and Gulf of Aden rifts via the NE-striking northern MER, which itself is oblique to the remainder of the MER (e.g. Fig. 1). Maguire *et al.* (2006) and Bastow *et al.* (2005) propose an ancient suture zone of pre-rift structure cross-cutting the northern MER, possibly explaining the kink in the MER. The Miocene border faults are also oblique to the current extension direction, with a Pliocene transtensional episode recorded in fault patterns (N10° E extension along N45° E faults; e.g. Bonini *et al.* 1998). Finally, the deep-seated thermal anomaly beneath the rift trends NE to connect with the Afar plume centre imaged in global tomographic models (e.g. Montelli *et al.* 2004). The right-stepping en echelon pattern of magmatic segments is attributed to the combination of a NE-trending magma source zone at the lithosphere–asthenosphere boundary and an artifact of the pre-Quaternary geological history.

Our results, supported by comparison to the recently ruptured Asal–Ghoubbet seafloor spreading centre, suggest that the along-axis segmentation of mid-ocean ridges initiates prior to lithospheric rupture in areas with a ready supply of magma.

## Conclusions

Field, remote sensing and modelling studies integrated with EAGLE seismic results indicate that extension has localized to approximately 20 km-wide zones of small offset faults and magmatism since Quaternary time. Extension within these 60 km-long, right-stepping magmatic segments is accommodated by both faulting and magmatism. Although extrusive volcanism masks some fault offsets, geophysical data indicate that stretching by faulting across the rift zone is minor, and the only large offset faults are the inactive Miocene border faults outboard from the magmatic segments. The elongation of eruptive centres, the localization of seismicity and magmatism, and the absence of large offset faults all indicate that magma intrusion and dyking play a larger role than faulting in strain accommodation as rifting progresses to seafloor spreading.

The right-stepping, en echelon pattern of magmatic segments, and the obliquity of Quaternary and Miocene–Pliocene structures, record changes in extension direction through time. NE-striking, Miocene–Pliocene faults were replaced by N10–30E-striking faults in late Pliocene time, resulting in a transtensional regime. With the localization of strain to the narrow magmatic segments after 2 Ma, extension was normal to the strike of normal faults and aligned eruptive centres; the older NE-striking faults act as oblique-slip linkage segments. The coincidence of volcanic edifice elongation directions, Quaternary fault slip indicators, and focal mechanism solutions (Keir *et al.* 2005) all indicate a N105–110° E extension direction, in agreement with geodetic and plate kinematic models (e.g. Calais *et al.* 2006). The right-stepping en echelon pattern of magmatic segments is attributed to the combination of a NE-trending magma source zone at the lithosphere–asthenosphere boundary and the pattern of extension along border faults. The rates of extension inferred from volcano elongation estimates lie within the range of rates estimated from geodetic studies, assuming strain is localized to the 20 km-wide magmatic segments. Total strain increases from south to north, consistent with the observed increase in crustal and mantle lithospheric thinning (e.g. Bastow *et al.* 2005; Maguire *et al.* 2006).

Observations of Quaternary faults in the northern MER indicate that faults propagate upward, probably from mid- to upper crustal intrusion zones. Monoclines with open fissures mark a vertical crack above an upward-propagating, dipping fault plane. The stress concentrations at the propagating fault tips promote the extrusion of magma at these points. The deformation patterns observed in the northern MER are strikingly similar to patterns in the Asal–Ghoubbet and Iceland seafloor spreading centres where magma intrusion accommodates more strain than faulting.

We thank Laike Asfaw, Atalay Ayele, Bekele Abebe, Tesfaye Kidane, Tesfaye Korme, Ian Bastow, Ellen Wolfenden, Eve Daly, K. McClay, and Rosie Fletcher for discussion of field observations, and J.R. Rowland, J. Bull, and P.K.H. Maguire for very helpful reviews.

## References


ABEBE, B., BOCCALETTI, M., MAZZUOLI, R., BONINI, M.L. & TRUA, T. 1998. *Geological map of the Lake Ziway–Asela region, Main Ethiopian Rift.* ARCA-DB Map.

ACOCELLA, V. & KORME, T. 2002. Holocene extension direction along the Main Ethiopian Rift, East Africa. *Terra Nova*, **14**, 191–197.

ACOCELLA, V., KORME, T. & SALVINI, F. 2003. Formation of normal faults along the axial zone of the Ethiopian Rift. *Journal of Structural Geology*, **25**, 503–513.

ASFAW, L.M. 1982. Development of earthquake induced fissures. *Nature*, **286**, 551–553.

ASFAW, L.M. 1992. Seismic risk at a site in the East-African Rift System. *Tectonophysics*, **209**, 301–309.

ASFAW, L.M. 2006. Vertical deformation in the Main Ethiopian Rift: Levelling results in its northern part, 1995–2004. *In*: YIRGU, G., EBINGER, C.J. & MAGUIRE, P.K.H. (eds) *The Afar Volcanic Province within the East African Rift System.* Geological Society, London, Special Publications, **259**, 185–190.

AYELE, A. 2000. Normal left-oblique fault mechanisms as an indication of sinistral deformation between the Nubia and Somalia plates in the Main Ethiopian Rift. *Journal of African Earth Sciences*, **31**, 359–367.

AYELE, D., BOURDON, E., EBINGER, C., WOLFENDEN, E., YIRGU, G. & GRASSINEAU, N. 2006. Temporal compositional variation of syn-rift rhyolite along the western margin of the Red Sea and northern Main Ethiopian Rift. *In*: YIRGU, G., EBINGER, C.J. & MAGUIRE, P.K.H. (eds) *The Afar Volcanic Province within the East African Rift System.* Geological Society, London, Special Publications, **259**, 121–130.

BASTOW, I., STUART, G., KENDALL, M. & EBINGER, C. 2005. Upper mantle seismic structure in a region of incipient continental breakup: northern Main Ethiopian Rift, *Geophysical Journal International*, **162**, 479–493.

BELLAHSEN, N., FACCENNA, C., FUNICIELLO, C., DANIEL, J.M. & JOLIVET, L. 2003. Why did Africa separate from Africa? Insights from 3D laboratory experiments. *Earth and Planetary Science Letters*, **216**, 365–381.

BENDICK, R., BILHAM, R., KLEMPERER, S. & ASFAW, L.M. 2006. Distributed Nubia–Somalia relative motion and dike intrusion in the Main Ethiopian Rift. *Geophysical Journal International*, (in press).

BENOIT, M.H., NYBLADE, A.A., VANDECAR, J.C. & GURROLA, H. 2003. Upper mantle P velocity structure and mantle transition zone thickness beneath the Arabian Shield. *Geophysical Research Letters*, **30**, 1153.

BILHAM, R., BENDICK, R., LARSON, K., BRAUN, J., TESFAYE, S., MOHR, P. & ASFAW, L. 1999. Secular and tidal strain across the Ethiopian rift. *Geophysical Research Letters*, **27**, 2789–2984.

BJORNSSON, A. 1985. Dynamics of crustal rifting in Iceland. *Journal of Geophysical Research*, **90**, 151–162.

BOCCALETTI, M., BONINI, M., MAZZUOLI, R., ABEBE, B., PICCARDI, L. & TORTORICI, L. 1998. Quaternary oblique extensional tectonics in the Ethiopian Rift (Horn of Africa). *Tectonophysics*, **287**, 97–116.

BOCCALETTI, M., MAZZUOLI, R., BONINI, L., TRUA, T. & ABEBE, B. 1999. Plio-Quaternary volcanotectonic activity in the northern sector of the Main Ethiopian Rift: Relationships with oblique rifting. *Journal of African Earth Science*, **29**, 679–698.

BONINI, M., SOURIOT, T., BOCCALETTI, M. & BRUN, J.P. 1997. Successive orthogonal and oblique extension episodes in a rift zone: Laboratory experiments with application to the Ethiopian rift. *Tectonics*, **16**, 347–362.

BUCK, W.R. 2004. Consequences of asthenospheric variability on continental rifting. *In*: KARNER, G., TAYLOR, B., DRISCOLL, N. & KOHLSTEDT, B. (eds) *Rheology and Deformation of the Lithosphere at Continental Margins.* Columbia University Press, pp. 92–137,

BULL, J.M., MINSHULL, T., MITCHELL, N., DIX, J.K. & HARDARDOTTIR, L. 2006. Magmatic and Tectonic History of Icelands Western Rift at Lake Thingvallavatn. *Geological Society of America Bulletin*, **117**, 1451–1465.

CALAIS, E., EBINGER, C., HARTNADY, C. & NOCQUET, J.M. 2006. Mantle-driven plate motions in the East African Rift? *In*: YIRGU, G., EBINGER, C.J. & MAGUIRE, P.K.H. (eds) *The Afar Volcanic Province within the East African Rift System.* Geological Society, London, Special Publications, **259**, 9–22.

CATTIN, R., DOUBRE, C., DE CHABALIER, J.-B., KING, G., VIGNY, C., AVOUAC, J.-P. & RUEGG, J.-C. 2005. Numerical modelling of Quaternary deformation and post-seismic displacement in the Asal–Ghoubbet rift (Djibouti, Africa). *Earth and Planetary Science Letters*, **239**, 352–367.

CERVELLI, P., SEGALL, P., AMELUNG, F., GARBEIL, H., MEERTENS, C., OWEN, S., MIKLIUS, A. & LISKOWSKI, M. 2002. The 12 September 1999 upper East Rift Zone dike intrusion at Kilauea Volcano, Hawaii. *Journal of Geophysical Research*, **107**, 1029/2001JB000602.

CHERNET, T., HART, W., ARONSON, J. & WALTER, R.C. 1998. New age constraints on the timing of volcanism and tectonism in the northern Main Ethiopian Rift–southern Afar transition zone (Ethiopia). *Journal of Volcanology and Geothermal Research*, **80**, 267–280.

CHERNET, T. 2005. Geological and hydrothermal alteration mapping of the Dofen geothermal prospect and adjacent western escarpment (Ethiopia), *Proceedings of the World Geothermal Congress, Antalya, Turkey*, 24–29 April 2005.

CHU, D.H. & GORDON, R.G. 1999. Evidence for motion between Nubia and Somalia along the Southwest Indian Ridge. *Nature*, **398**, 64–67.

CORNWELL, D.G., MACKENZIE, G., MAGUIRE, P.K.H., ENGLAND, R., ASFAW, L.M. & OLUMA, B. 2006.

Northern Main Ethiopian Rift crustal structure from new high precision gravity data. *In*: YIRGU, G., EBINGER, C.J. & MAGUIRE, P.K.H. (eds) *The Afar Volcanic Province within the East African Rift System*. Geological Society, London, Special Publications, **259**, 307–321.

COURTILLOT, V., ARMIJO, R. & TAPPONNIER, P. 1987. The Sinai triple junction revisited. *Tectonophysics*, **141**, 181–190.

CROUCH, S.L. & STARFIELD, A.M. 1983. *Boundary Element Methods in Solid Mechanics*. Allen & Unwin, London.

DE CHABALIER, J.-B. & AVOUAC, J.-P. 1994. Kinematics of the Asal Rift (Djibouti) determined from the deformation of Fieale Volcano. *Science*, **265**, 1677–1681.

DELANEY, P.T. & POLLARD, D. 1981. Deformation of host rocks and flow of magma during growth of minette dykes and breccia-bearing inclusions near Ship Rock, New Mexico. *US Geological Survey Professional Paper*, 1202.

DUGDA, M., NYBLADE, A., JULIA, J., LANGSTON, C.A., AMMON, C. & SIMIYU, S. 2005. Crustal structure in Ethiopia and Kenya from receiver function analyses: Implication for rift development in eastern Africa. *Journal of Geophysical Research*, **110**, doi:10.1029/2004JB003065.

EBINGER, C.J. & CASEY, M. 2001. Continental breakup in magmatic provinces: An Ethiopian example. *Geology*, **29**, 527–530.

EBINGER, C., YEMANE, T., HARDING, D., TESFAYE, S., REX, D. & KELLEY, S. 2000. Rift deflection, migration, and propagation: Linkage of the Ethiopian and Eastern rifts, Africa, *Geological Society of America Bulletin*, **102**, 163–176.

FEIGL, K., GASPERI, J., SIGMUNDSSON, F. & RIGO, A. 2000. Crustal deformation near Hengill volcano Iceland 1993–1998: Coupling between magmatic activity and faulting inferred from elastic modelling of satellite radar interferometry. *Journal of Geophysical Research*, **105**, 25655–25670.

FERNANDES, R.M.S., AMBROSIUS, B.A.C., NOOMEN, R., BASTUS, L., COMBRINCK, L. MIRANDA, J.M. & SPAKMAN, W. 2004. Angular velocity of Nubia and Somalia from continuous GPS data: implications on present-day relative kinematics. *Earth and Planetary Science Letters*, **222**, 197–208.

FURMAN, T., BRYCE, J., ROONEY, T., HANAN, B., YIRGU, G. & AYALEW, D. 2006. Heads and tails: 30 million years of the Afar Plume. *In*: YIRGU, G., EBINGER, C.J. & MAGUIRE, P.K.H. (eds) *The Afar Volcanic Province within the East African Rift System*. Geological Society, London, Special Publications, **259**, 95–119.

GIBSON, I. 1967. Preliminary account of the volcanic geology of Fantale, Shoa. *Bulletin. Geophysical Observatory Addis Ababa*, **10**, 59–67.

GRANT, J.V. & KATTENHORN, S.A. 2004. Evolution of vertical faults at an extensional plate boundary, southwest Iceland. *Journal of Structural Geology*, **26**, 537–557.

GUDMUNDSSON, A. 1992. Formation and growth of normal faults at the divergent plate boundary in Iceland. *Terra Nova*, **4**, 464–471.

GUPTA, A. & SCHOLZ, C.H. 2000. Brittle strain transition in the Afar Depression: Implications for fault growth and seafloor spreading. *Geology*, **28**, 1087–1090.

HAYWARD, N. & EBINGER, C. 1996. Variations in along-axis segmentation of the Afar rift system. *Tectonics*, **15**, 244–257.

HOFMANN, C., COURTILLOT, V., FERAUD, G., ROCHETTE, P., YIRGU, G., KETEFO, E. & PIK, R. 1997. Timing of the Ethiopian flood basalt event: implications for plume birth and global change. *Nature*, **389**, 838–841.

HOLBROOK, W. & KELEMEN, P. 1993. Large igneous province on the US Atlantic margin and implications for magmatism during breakup. *Nature*, **364**, 433–436.

JONSSON, S., EINARSSON, P. & SIGMUNDSSON, F. 1997. Extension across a divergent plate boundary, the eastern Volcanic Rift Zone, south Iceland, 1967–1994, observed with GPS and electronic distance measurements. *Journal of Geophysical Research*, **102**, 11913–11929.

KEIR, D., EBINGER, C., STUART, G., DALY, E. & AYELE, A. 2006. Strain accommodation by magmatism and faulting as rifting proceeds to breakup: Seismicity of the northern Ethiopian Rift. *Journal Geophysical Research* (in press).

KENDALL, J.-M., STUART, G., EBINGER, C., BASTOW, I. & KEIR, D., 2005. Magma-assisted rifting in Ethiopia. *Nature*, **433**, 146–148.

KERANEN, K., KLEMPERER, S., GLOAGUEN, R. & EAGLE WORKING GROUP 2004. Imaging a protoridge axis in the Main Ethiopian Rift. *Geology*, **39**, 949–952.

KIDANE, T., EBINGER, C., ABEBE, B., PLATZMAN, E., KEIR, D., LAHITTE, P. & ROCHETTE, P. 2006. Palaeomagnetic constraints on continental breakup processes: Observations from the Main Ethiopian Rift. *In*: YIRGU, G., EBINGER, C.J. & MAGUIRE, P.K.H. (eds) *The Afar Volcanic Province within the East African Rift System*. Geological Society, London, Special Publications, **259**, 165–183.

KIEFFER, B., ARNDT, N., *ET AL.* 2004. Flood and shield basalts from Ethiopia: Magmas from the African superswell. *Journal of Petrology*, **45**, 793–834.

KORENAGA, J., HOLBROOK, W.S., *ET AL.* 2000. Crustal structure of the southeast Greenland margin from joint refraction and reflection seismic tomography. *Journal of Geophysical Research*, **105**, 21591–21614.

MAGUIRE, P.K.H., KELLER, G. *ET AL.* 2006. Crustal structure of the northern MER from the EAGLE controlled source survey: Snapshot of incipient lithospheric breakup. *In*: YIRGU, G., EBINGER, C.J. & MAGUIRE, P.K.H. (eds) *The Afar Volcanic Province within the East African Rift System*. Geological Society, London, Special Publications, **259**, 269–291.

MACKENZIE, G., THYBO, H. & MAGUIRE, P.K.H. 2005. Crustal velocity structure across the Main Ethiopian Rift: Results from two-dimensional wide-angle seismic modelling. *Geophysical Journal International*, **162**, 994–1006.

MANDL, G. 1987. Discontinuous fault zones. *Journal of Structural Geology*, **9**, 105–110.

MANIGHETTI, I., TAPPONNIER, P., COURTILLOT, V., GRUSZOW, S. & GILLOT, P.-Y. 1997. Propagation of rifting along the Arabia–Somalia plate boundary: The gulfs of Aden and Tadjoura. *Journal of Geophysical Research*, **102**, 2681–2710.

MEYER, W., PILGER, A., ROSLER, A. & STETS, J. 1975. Tectonic evolution of the northern part of the Main Ethiopian Rift. *In*: PILGER, A. & RÖSLER, A. (eds) *Afar Depression of Ethiopia*, Schweizerbart, Stuttgart.

MOHR, P.A. 1962. The Ethiopian Rift System. *Bulletin of the Geophysical Observatory*, Addis Ababa, **5**, 33–62.

MOHR, P.A. 1968. Transcurrent faulting in the Ethiopian Rift System. *Nature*, **218**, 938–941.

MONTELLI, R., NOLET, G., DAHLEN, F.A., ENGDAHL, E.R. & HUNG, S.H. 2004. Finite-frequency tomography reveals a variety of plumes in the mantle. *Science*, **303**, 338–343.

MORTON, W.H., REX, D.C., MITCHELL, J.G. & MOHR, P.A. 1979. Riftward younging of volcanic units in the Addis Ababa region, Ethiopian Rift valley. *Nature*, **280**, 284–288.

PIK, R., MARTY, B., CARIGNAN, J. & LAVE, J. 2003. Stability of the Upper Nile drainage network (Ethiopia) deduced from (U–Th)/He thermochronometry: Implications for uplift and erosion of the Afar plume dome. *Earth and Planetary Science Letters*, **215**, 73–88.

PIZZI, A., COLTORTI, M., ABEBE, B., DISPERATI, L., SACCHI, G. & SALVINI, R. 2006. The Wonji fault belt (Main Ethiopian Rift): Structural and geomorphological constraints and GPS monitoring. *In*: YIRGU, G., EBINGER, C.J. & MAGUIRE, P.K.H. (eds) *The Afar Volcanic Province within the East African Rift System*. Geological Society, London, Special Publications, **259**, 191–207.

ROONEY, T., FURMAN, T., YIRGU, G. & AYALEW, D. 2005. Structure of the Ethiopian lithosphere: Evidence from mantle xenoliths. *Geochimica et Cosmochimica Acta*, **69**, 3889–3910.

RUBIN, A. 1992. Dike-induced faulting and graben subsidence in volcanic rift zones. *Journal of Geophysical Research*, **97**, 1839–1858.

RUBIN, A.M. & POLLARD, D.D. 1988. Dike induced faulting in rift zones in Iceland and Afar. *Geology*, **16**, 413–317.

SIGMUNDSSON, F., EINARSSON, P., ROGNVALDSSON, S.T., FOULGER, G.R., HODGKINSON, K.R. & THORBERGSSON, G. 1997. The 1994–1995 seismicity and deformation at the Hengill triple junction, Iceland: Triggering of earthquakes by minor magma injection in a zone of horizontal shear stress. *Journal of Geophysical Research*, **102**, 15151–15161.

STUART, G., BASTOW, I. & EBINGER, C. 2006. Crustal structure of the northern Main Ethiopian Rift from receiver function studies. *In*: YIRGU, G., EBINGER, C.J. & MAGUIRE, P.K.H. (eds) *The Afar Volcanic Province within the East African Rift System*. Geological Society, London, Special Publications, **259**, 253–267.

TESFAYE, S., HARDING, D.J. & KUSKY, T. 2003. Early continental breakup boundary and migration of the Afar Triple Junction, Ethiopia. *Geological Society of America Bulletin*, **115**, 1053–1067.

TIBERI, C., EBINGER, C., BALLU, V., STUART, G. & OLUMA, B. 2005. Inverse models of gravity data from the Red Sea–Aden–East African Rifts triple junction zone. *Geophysical Journal International*, **163**, 775–787.

WILLIAMS, F., WILLIAMS, M. & AUMENTO, F. 2004. Tensional fissures and crustal extension rates in the northern part of the Main Ethiopian Rift. *Journal of African Earth Sciences*, **38**, 183–197.

WOLDEGABRIEL, G., ARONSON, J. & WALTER, R. 1990. Geology, geochronology, and rift basin development in the central sector of the Main Ethiopian Rift. *Geological Society of America Bulletin*, **102**, 439–458.

WOLFENDEN, E., EBINGER, C., YIRGU, G., DEINO, A. & AYALEW, D. 2004. Evolution of the northern Main Ethiopian Rift: Birth of a triple junction. *Earth and Planetary Science Letters*, **224**, 213–228.

WOLFENDEN, E., EBINGER, C., YIRGU, G., RENNE, P. & KELLEY, S.P. 2005, Evolution of the southern Red Sea rift: Birth of a magmatic margin. *Bulletin of the Geological Society of America*, **117**, 846–864.

WRIGHT, D.J., HAYMON, R.M. & FORNARI, D.J. 1995. Crustal fissuring and its relationship to magmatic and hydrothermal processes on the East Pacific Rise crest (9-degrees-12′ to 54′ N), *Journal of Geophysical Research*, **100**, 6097–6120.

# Palaeomagnetic constraints on continental break-up processes: observations from the Main Ethiopian Rift

TESFAYE KIDANE[1*], E. PLATZMAN[2,3], C. EBINGER[4], B. ABEBE[1] & P. ROCHETTE[5]

[1]*Department of Earth Sciences, AAU, PO Box 1176, Addis Ababa, Ethiopia (e-mail: tesfkida@geol.aau.edu.et)*

[2]*Department of Earth Sciences, University College London, Gower Street, London WC1E 6BT, UK*

[3]*Now at Department of Earth Sciences, University Southern California, 3651 Trousdale Parkway, Los Angeles, California, 90089, USA*

[4]*Department of Geology, Royal Holloway University of London, Egham, TW20 0EX, UK*

[5]*CEREGE University of Aix-Marseille 3, BP80, Europole de l'Arbois, 13545 Aix en Provence Cedex 4, France*

**Abstract:** We report the first palaeomagnetic results from the Main Ethiopian Rift (MER), the northernmost sector of the East African rift system. This part of the MER shows an along-axis tectono-magmatic segmentation pattern similar to that of slow-spreading mid-ocean ridges, which developed during the past 1.9 Ma. The aims of our palaeomagnetic, structural and geochronological studies are to test plate kinematic models for the right-stepping, en echelon 60–80 km-long magmatic segments. Twenty palaeomagnetic sites were sampled on either basalt or ignimbrite outcropping in the region adjacent to, and within, the $<1.9$ Ma-old tectono-magmatic segments of Gademsa–Koka, Boset and Fentale–Dofan. Five K–Ar age determinations were made to bracket the age of units studied in the palaeomagnetic analyses. The natural remanent magnetization intensity possibly exhibited a unimodal distribution with a value of 6.6 A/m ($\sigma = 5.6$ A/m) for the basalts and a bimodal distribution with magnetization intensity of 0.69 A/m ($\sigma = 0.55$ A/m) and 0.03 A/m ($\sigma = 0.02$ A/m), statistically similar to values from previous studies in the Afar triple junction zone (e.g. Kidane *et al.* 1999, 2002). Progressive heating, alternating field analysis, and susceptibility vs. temperature measurements indicated unblocking temperature ranging between 300 °C–600 °C for basalts and between 500 °C–660 °C for ignimbrites, suggesting the magnetic mineralogy to be titanomagnetite and magnetite for the former and magnetite and titanohematite for the latter. Palaeomagnetic measurements using both TH and AF technique revealed quasi-single component of magnetization with viscous remanent magnetization (VRM) on a few samples. Principal component analysis and statistical averaging resulted in an overall mean palaeomagnetic direction of ($D_s = 2.3°$, $I_s = 7.8°$, $\alpha_{95} = 7$, $K = 26.9$, $N = 17$) which is statistically identical to the expected direction ($D = 1.9°$, $I = 13.5°$, $\alpha_{95} = 2.5$, $K = 105.6$, $N = 32$) from the Apparent Polar Wander Path reference curve for Africa at 1.5 Ma (Besse & Courtillot 2003). The angular difference between the observed and expected directions above with their uncertainty is calculated to be $0.4° \pm 7.5°$. These results indicate that the Late Pliocene–Pleistocene rocks of the MER in the studied region do not suffer vertical axis rotation, arguing against transtensional and seafloor-spreading–transform kinematic models. We suggest that magma intrusion, rather than large offset faults, produce the right-stepping, en echelon magmatic segments of the MER, which is at the transition from continental to oceanic extension.

The Main Ethiopian Rift (MER) forms the northernmost sector of the East African rift system stretching from Lake Abbaya in the south to Gabilema volcano in the north where it forms the third arm of the Afar triple junction (Fig. 1). The MER is the 700 km-long, 80 km-wide volcanically active rift situated between the Ethiopian and Somalian plateaus, which are covered by Oligocene–Miocene lavas (e.g. Hofmann *et al.* 1997). It is one of the few places in the world where one can observe the process of continental rifting and study the patterns of faulting and eruptive volcanic centres immediately prior to continental break-up. Geological field observations and historic–recent seismic activity attests to ongoing extension in the MER (Gouin 1979; Asfaw 1992;

*From*: YIRGU, G., EBINGER, C.J. & MAGUIRE, P.K.H. (eds) 2006. *The Afar Volcanic Province within the East African Rift System*. Geological Society, London, Special Publications, **259**, 165–183.
0305-8719/06/$15.00 

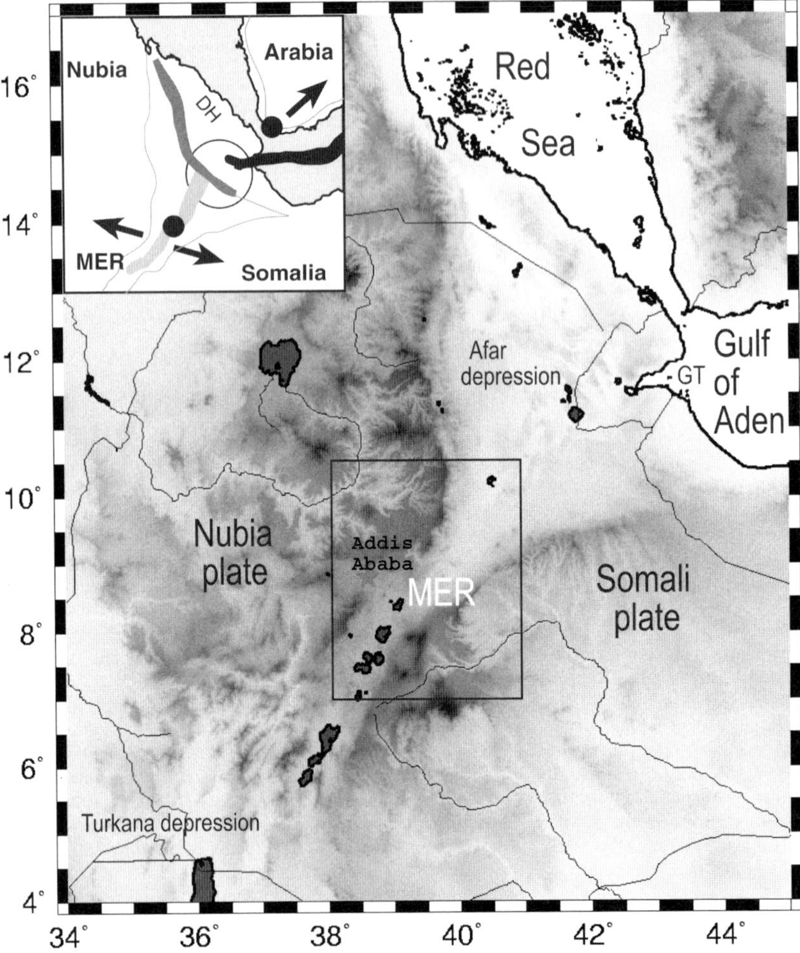


**Fig. 1.** Generalized elevation model diagram of the East African rift system. The inset area on the figure shows the studied region.

Keir *et al.* in press). Low seismic velocities and low degree of seismic anisotropy recorded beneath the rift indicate the presence of melt (Bastow *et al.* 2005; Kendall *et al.* 2005). Despite the importance of the MER, until recently most research has focused on the zones of incipient break-up near the Afar triple junction (e.g. Manighetti *et al.* 1998; Lahitte *et al.* 2003).

Oligocene–Recent volcanic rocks including basalts, ignimbrites and rhyolites crop out along the rift margin and also cover the floor of the MER (Fig. 1). In some places, these flows form thick piles, making them suitable for palaeomagnetic investigations and providing a good opportunity to study the geology, tectonic evolution as well as features of the geomagnetic field.

Along the flanks of the rift, large offset border faults are characterized by riftward en echelon right-stepping normal faults with a dominant orientation of NNE–SSW to NE–SW, having characteristic structural styles on the eastern and western margins (Di Paola 1972; Casey *et al.* 2006) (Fig. 2). The eastern margin fault systems are characterized by multiple faults each with >100 m of cumulative displacement, whereas the western margin is marked by a long erosional escarpment or faulted monoclines (e.g. WoldeGabriel *et al.* 1990; Abebe *et al.* 1998) (Fig. 2). Between these Miocene–Pliocene border faults, the central rift valley is marked by approximately 20 km-wide right-stepping, en echelon chains of eruptive volcanic centres and fissural flows cut by N15E-striking,

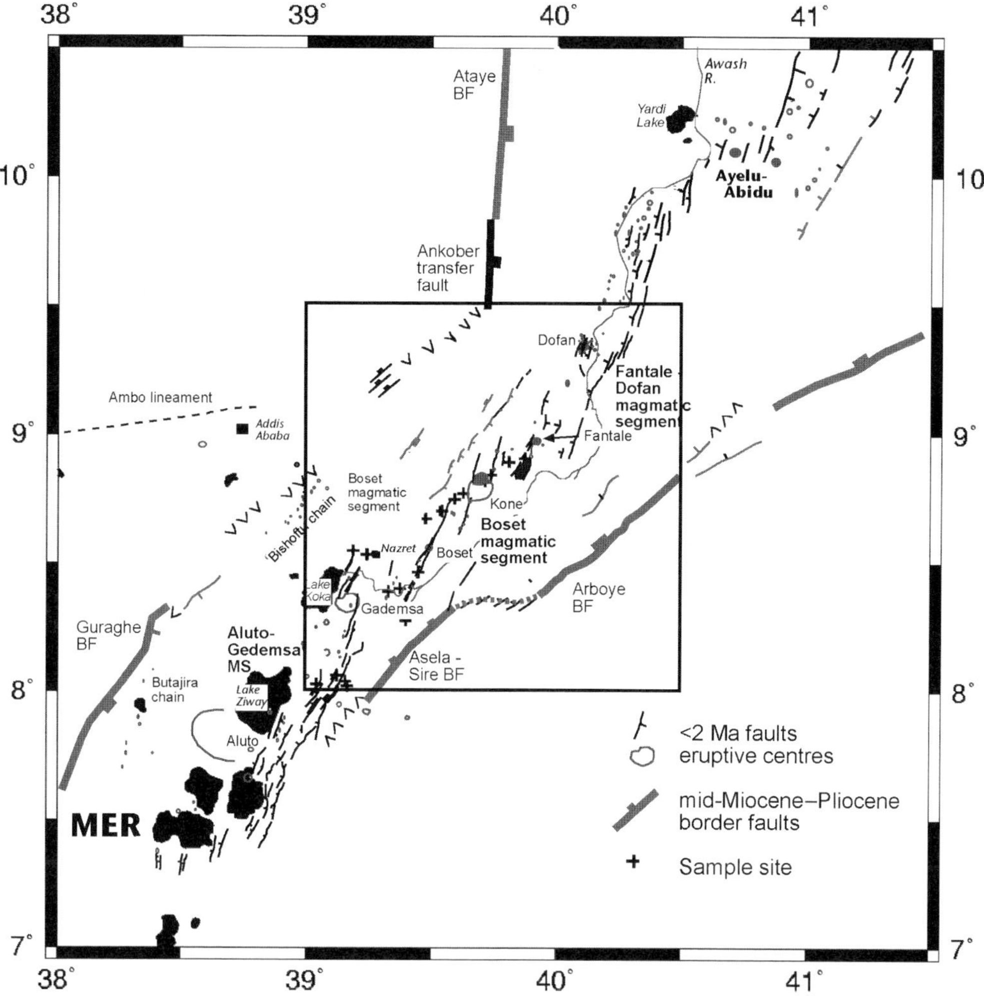


**Fig. 2.** Tectonic map of the MER with all palaeomagnetic sites reported as black crosses (from Casey *et al.* 2006). Square box shows area shown in Fig. 11.

small offset normal faults (e.g. Boccaletti *et al.* 1998; Ebinger & Casey 2001; Abebe *et al.* 1998). Where dates are available, the bimodal (rhyolitic and basaltic composition) lavas in the magmatic segments of the Wonji belt have ages <1.8 Ma (e.g. Boccaletti *et al.* 1998). The along-axis segmentation of the Wonji belt is apparent down to mid- to lower crustal levels: high-velocity, high-density material, interpreted as cooled mafic intrusions, rises to about 8 km depth beneath the *c.* 50 km-long 'magmatic segments' (Mahatsente *et al.* 1999; Keranen *et al.* 2004).

A number of interpretations of the kinematics of the along-axis segmentation of the Wonji belt have been suggested, with implications for fault block rotations. Bonini *et al.* (1997) compare analogue models of a change in extension direction by 45°, and interpret the segmentation as a manifestation of successive episodes of orthogonal and oblique-slip movement. Ebinger & Casey (2001) and Keir *et al.* (2006) suggest that dyke intrusion and faulting accommodate melt intrusions into the plate beneath the magmatic segments, and that the Miocene border fault systems are largely inactive. Keranen *et al.* (2004) interpret these magmatic segments as incipient seafloor spreading centres, which implies some rotations within the overlapping intersegment zones, if new oceanic crust is being created. Alternatively, Chernet *et al.* (1998) suggest that oblique-slip transform faults, or accommodation zones, cross-cut the rift valley linking the segmented border fault system in a pattern similar to seafloor-spreading centres.

The aim of our studies is to determine whether a change in palaeostress field has occurred since the development of the magmatic segments. To do so, one has to unambiguously observe the consequences on the geometry of older faults and the orientation of the new structures developing under the new stress field. However, one has to be cautious in identifying, measuring and interpreting the kinematic data, ensuring that one combines mesoscopic observations with subsurface data prior to making interpretation on the palaeostress field change. Data from the EAGLE (Ethiopian Afar Geoscientific Lithospheric Experiment) project provide the framework for our study. Kinematic information also has to be supported by reliable geochronological information on the outcrops where the data are obtained. In the absence of a good age control, the kinematic information could be misleading and contradictory. For example, in the MER, a dextral (Chorowicz *et al.* 1994) and sinistral (Boccaletti *et al.* 1999) movement has been reported on the same faults.

## Plate kinematic setting

Studies of the rifting history, fault geometry, palaeostress orientations and the tectonic evolution have led to a number of interpretations of the direction of opening between Nubia and Somalia across the MER. Local geological studies show a range of opening directions. An east–west extension direction during the entire rift evolution was suggested by Di Paola (1972) and WoldeGabriel *et al.* (1990) based on the orientation of structures within the Wonji belt while a NW–SE extension direction has been proposed by Chorowicz *et al.* (1994), Korme *et al.* (1997) and Acocella & Korme (2002), based on the interpretation of fault slip indicators across the rift zone. With improved age control, Boccaletti *et al.* (1992, 1994, 1998), Abebe (1993) and Bonini *et al.* (1997) proposed a change in the extension direction from N135 during the Miocene to N100-N115 during the Quaternary. They argue that a change in the stress field occurred in the MER from an original NW–SE direction of extension to an east–west trending extension. Wolfenden *et al.* (2004) find a change in extension direction from N130 to N100 occurred in the northern MER between 6.6 and 3.2 Ma, but age constraints are too sparse to extrapolate results over a larger area. Keir *et al.* (2006) estimate a N109° extension direction from inversion of focal mechanism solutions of earthquakes recorded during the period 2001–3, with most seismicity restricted to the Quaternary magmatic segments.

Estimates of Nubia–Somalia opening vectors from marine geophysical data spanning the past 3.2 Ma (Jestin *et al.* 1994: Lemaux *et al.* 2002) and GPS spanning 1997–present (Fernandes *et al.* 2004) show an opening direction of N90-109, with rates of 6–7 mm $a^{-1}$. Finally, Bilham *et al.* (1999) estimated about 3 mm $a^{-1}$ velocity over the period 1969–1997.

## Palaeomagnetism

### *Sample collection and laboratory treatment*

The studied sites were selected using 1 : 50 000 topographic maps in conjunction with satellite images. The distribution of these sites is indicated in Fig. 2. From the eastern rift margin around Assela through Nazreth to Metehara area along the segmented Wonji belt, 20 palaeomagnetic sites were collected in two field seasons. The first field season was in March 2003 where ten sites marked AS01–04 and Naz01–06 were collected. The second season took place in January 2004 and sites BF01–10 were collected. Palaeomagnetic sites (lava flows) were either from mafic lava flows (basalts) or felsic pyroclastic flows (ignimbrites). These rocks cropped out along fault scarps, road cuts, and river valley exposures.

During the sampling, care was taken to collect cores from the central part of the flow in order to avoid the possibility that the sample had been remagnetized during emplacement of the next flow. A total of about 160 cores were collected for palaeomagnetic analysis. Of these cores, 126 were oriented using both a magnetic and sun compass. The mean difference between the two readings was $1.8^\circ \pm 3.9^\circ$, which is statistically the same as the local declination value (1.4°, with a secular rate of increase of 0.2° per year) predicted from the 2004 Reference Field (IGRF) for the mean latitude and longitude ($\lambda = 8.4^\circ$, $\phi = 39.3^\circ$). Therefore, the effect of local anomalies due to the possible lightning strikes and high-intensity basalts is not significant and averages out in the calculated mean. Eight samples were also collected for geochronology purposes from the same localities as the palaeomagnetic samples. Eight to ten cores were drilled from each site, and the palaeomagnetic study was undertaken in four different laboratories following the state-of-the-art of routine palaeomagnetic procedures (e.g. Kidane *et al.* 1999): in the palaeomagnetic laboratory of the Department of Geology and Geophysics, Addis Ababa; CEREGE and IPGP, France; and UCL, United Kingdom. In the laboratories, samples were demagnetized to isolate the stable characteristic component of remanent magnetization (ChRM). From each core, at least two twin specimens were obtained. One was subjected to progressive alternating field (AF) demagnetization and the other to progressive heating or thermal (TH) demagnetization in order to detect possible overlapping components of remanent magnetization. Analysis in Addis Ababa and CEREGE were done only using AF where as at IPGP and UCL thermal treatments were also done.

## *Palaeomagnetic results*

*Natural Remanent Magnetization (NRM).* The histogram plot of the natural remanent magnetizations (NRMs) of 263 analysed specimens gave log normal distribution with three modes. When the basalts and the ignimbrites are plotted separately, the histogram for the NRMs of the basalts form a unimodal log normal distribution (Fig. 3a) while that for the ignimbrites form a bimodal log normal distribution (Fig. 3b). The mean NRM intensity of the basalts is found to be 6.6 A/m ($\sigma = 5.6$ A/m). Within the uncertainty, this value is the same as that reported by Prévot & Grommé (1975) for continental and subaerial basalts (3.1–4.1 A/m). It is also within the range of what has been reported previously in Afar (2.4–4.7–6.5) by Kidane *et al.* (1999, 2002) and the references therein.

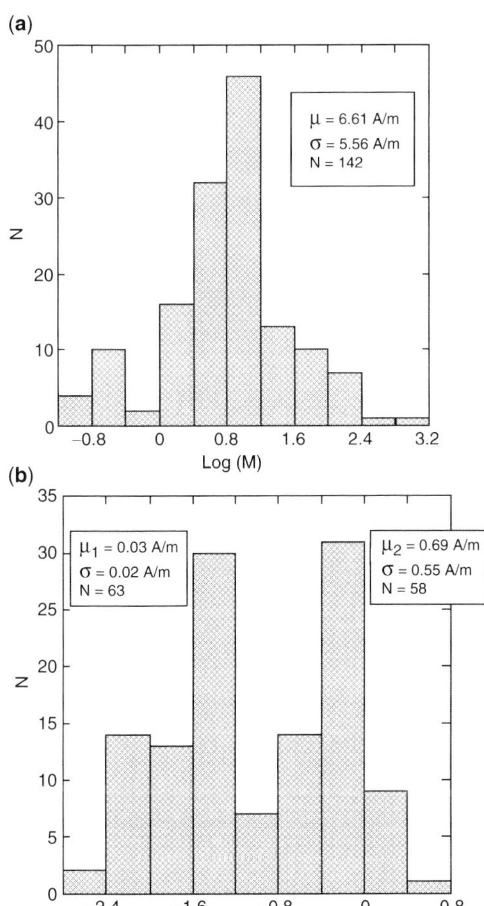


**Fig. 3.** Histogram of the NRM of the analysed samples: (**a**) basalts and (**b**) ignimbrite samples.

The bimodal distribution of remanent intensities observed in the ignimbrites with means of 0.03 A/m, $\sigma = 0.02$ A/m and 0.69 A/m, $\sigma = 0.55$ A/m is the first of its kind reported from Ethiopia. The two means, with a difference of one order of magnitude, are very well defined, indicative of a variation in composition and/or grain size of the magnetic carriers in the ignimbrite. The ignimbrites in the MER are known either to be pantelleritic or comenditic in composition (Ayalew *et al.* 2006), both of which are very rich in iron compared to the normal rhyolites. The difference in the magnetic behaviour of the two types of ignimbrites could have arisen from grain size differences related to their mode of eruptions (fissural or volcanic vent). However, we can say that there are two different kinds of ignimbrites at least from the viewpoint of the NRM intensity. Both ignimbrite types exhibit stable behaviour during demagnetization. They both have very well-defined linear orthogonal vector component diagrams.

It appears that the two modes could also be related to whether the samples are collected from the bottom or top part of the ignimbrite flow. Where samples are collected from the base of the flow, NRM intensity is 10 times higher than that at the top of the flow. Though better explanation for this variations will come from detailed palaeomagnetic and rock-magnetic studies, our preliminary interpretation is that it is related to the fast downward settling of heavy minerals including iron-oxides towards the base during the original cooling from high temperature.

*Alternating field (AF) and thermal (TH) demagnetization.* Samples from both the ignimbrites and basalts were subjected to progressive alternating field (at least 10 steps) and or thermal (15–19 steps) demagnetization. For most of the cores, twin specimens were produced: one of the twins was demagnetized with AF while the other was demagnetized thermally. In the rare cases where the cores did not produce two samples, some of them were treated with AF while the rest were treated with the thermal technique.

AF demagnetization reveals that the median destructive fields for basalts and ignimbrites are variable. Figure 4 indicates the median destructive

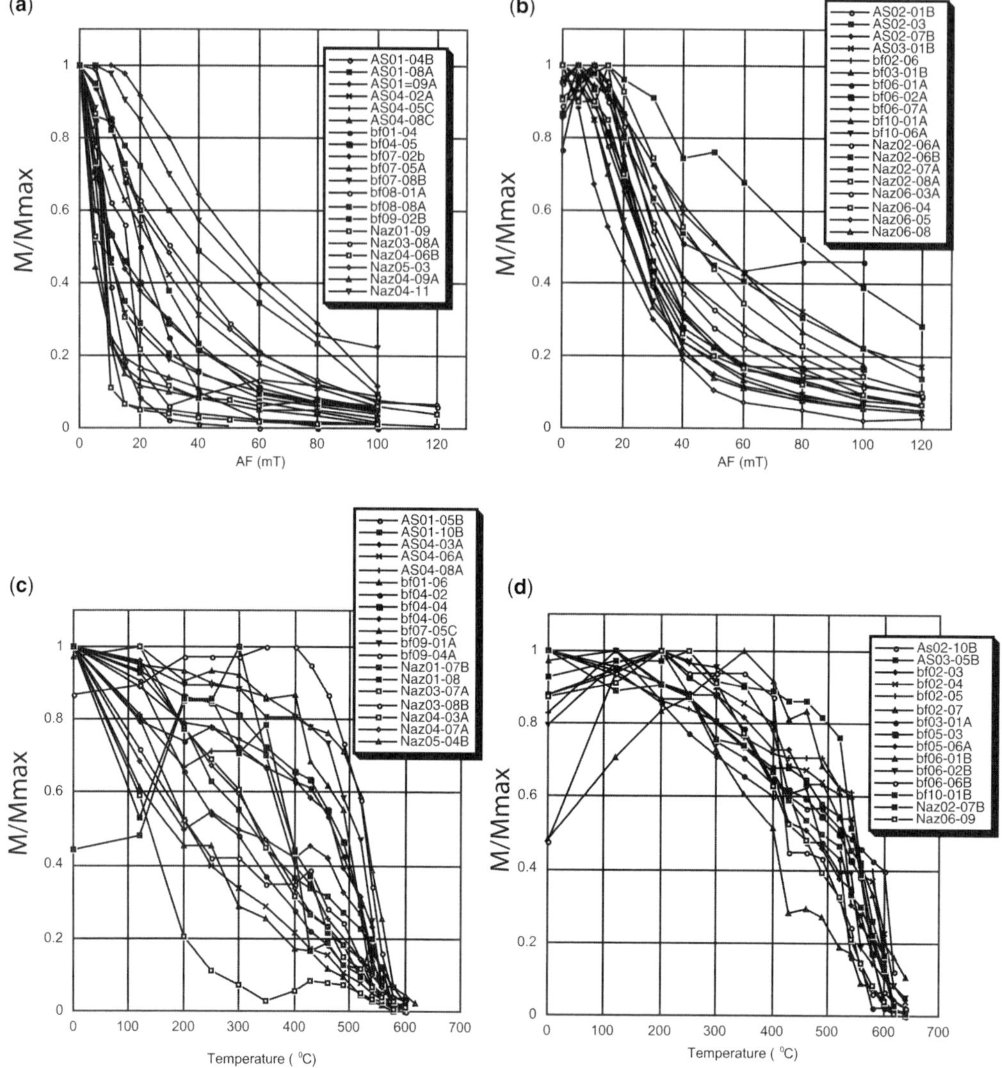


**Fig. 4.** Normalized magnetization intensity curves for AF (upper) demagnetization and TH (lower) demagnetization, (**a**) and (**c**) for basalt samples, (**b**) and (**d**) for ignimbrite samples.

field for some representative samples from basalts and ignimbrites (Fig. 4a, b). For the basalts, it ranged from 10 mT to 50 mT (Fig. 4a) while for the ignimbrites it is in the range of 20 mT to 80 mT (Fig. 4b). The magnetization decay curves in Figure 4 depict two different behaviours in that the maximum magnetization intensity is attained at the NRM values for the former and at values different from NRM for the latter, which would suggest that the ignimbrites are mostly remagnetized in a direction opposite to the ChRM.

The AF or thermally treated specimens of different samples did not show any systematic variations between laboratories. Results from the four laboratories indicated that the AF treatment identified one or two components of remanent magnetization. In general, an alternating field of between 0–20 mT isolates the first component (Fig. 5a, b, c). In a few cases, the two components showed a complete overlap so that the orthogonal vector components have a curved shape (Fig. 5d). The direction of this overprinted component is variable, ranging in direction from the current geomagnetic field to

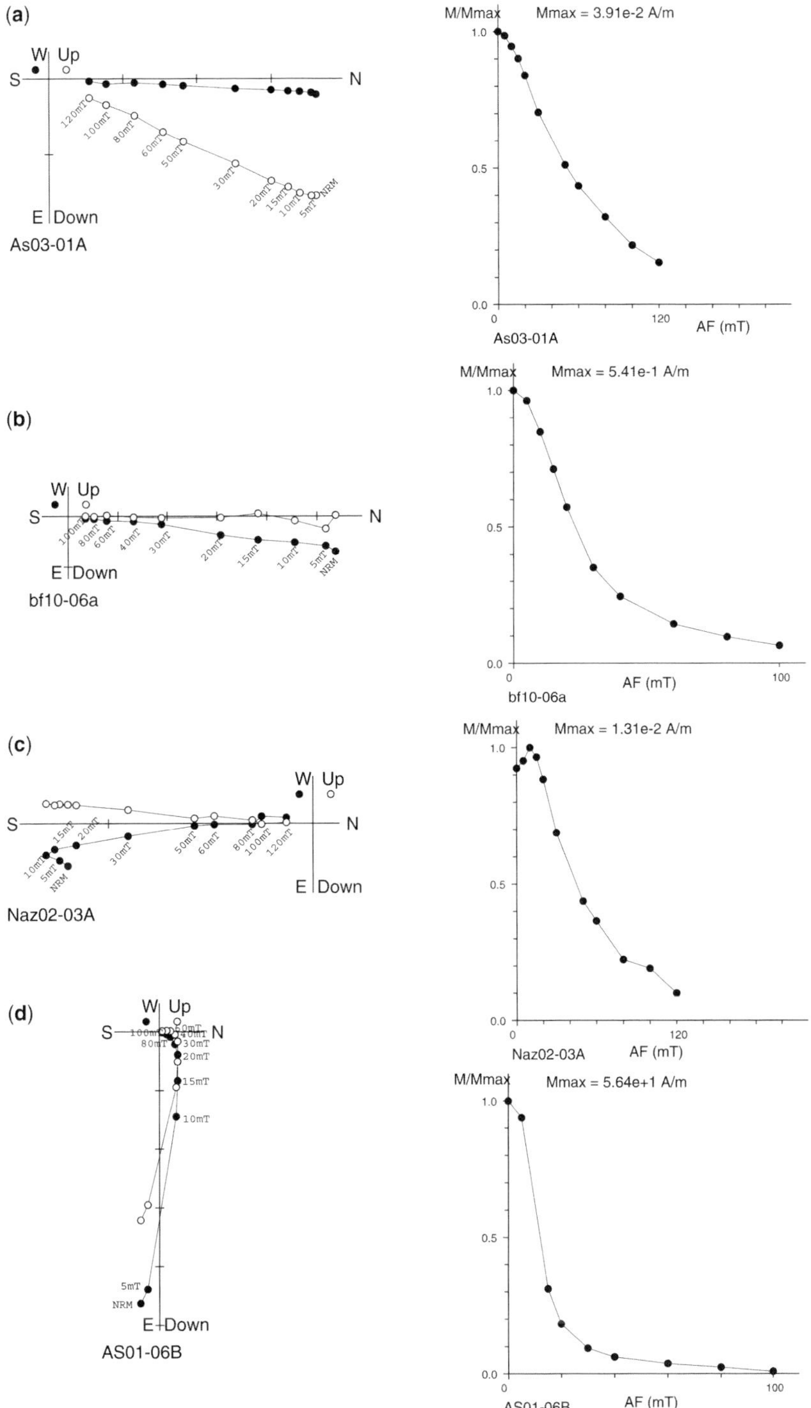


**Fig. 5.** Typical examples of vector component diagrams for the specimens treated by AF (**a**, **b**, **c**) representing single or quasi-single component and (**d**) representing strongly overlapping components.

some meaningless directions, suggesting viscous remagnetization (VRM) in the present Earth's magnetic field, and remagnetization related to effects of lightning, drilling and transportation processes. The second component of remanent magnetization, isolated above 20 mT in most of the samples, defined a straight-line segment directed towards the origin of the plot. This component is interpreted as the Characteristic Remanent Magnetization (ChRM) acquired during the cooling of the lava flows.

The thermal treatment of samples was done in the shielded room of IPGP and UCL. Samples studied in both these laboratories indicated that the VRM is cleaned by heating the specimen to a temperature of between 0–300 °C (Fig. 6a, b, c). For some of the samples, the VRM persists up to a temperature of 490 °C (Fig. 6d). Demagnetization is complete by 600 °C for the basalts whereas for some of the ignimbrites the stepwise heating went until a temperature of 660 °C, indicating the presence of significant amounts of titanohematite or hematite. Above about 400 °C, samples decayed in a straight-line segment directed towards the origin of the vector diagram. This direction is interpreted to be the ChRM, acquired during the cooling of the flows.

Representative decay curves for thermally treated samples indicate that, in basaltic samples, the ChRM has an unblocking temperature spectrum between 300 °C to 600 °C with 50% of the magnetization intensity cleaned by a temperature of 150 °C to 550 °C (Fig. 4c). These would suggest that the magnetic minerals carrying the magnetization are dominantly magnetites and titanomagnetite with higher Ti content in samples with low unblocking temperature and lower Ti content in the samples with high unblocking temperatures. The occurrence of some goethite or iron sulphides cannot be ruled out. For most of the samples, the maximum remanent intensity is attained at the NRM value except for three samples where the maximum intensity is attained after heating to 120 °C, 300 °C and 400 °C for the three samples.

For the ignimbrite samples, there is a narrower unblocking temperature range from 500 °C to 600 °C (Fig. 4d). The decay curves indicate that 50% of the magnetization is removed between 400 °C and 550 °C (Fig. 4d). Most of the samples attain the maximum magnetization intensity at fields or temperatures higher than the NRM, suggesting remagnetization in opposite direction to the ChRM. These results indicate that the magnetization in these samples is probably dominated by magnetite and titanohematite.

*K–T (susceptibility vs. temperature) experiments.* High temperature K–T experiments were undertaken at the UCL palaeomagnetism laboratory on a representative number of the samples obtained on the second sampling mission. Results indicate that all the samples are characterized by very simple, reversible curves without any major phase transitions. The basalts (Fig. 7a–f) have high initial total susceptibility ($K_{tot}$) with values ranging from 80*E-6 (Fig. 7c) to about 800*E-6 (Fig. 7e). Three behaviours are observed for the basalts; the first group shows an initial sharp decrease in the $K_{tot}$ between room temperature to 100 °C and then a smooth increase in $K_{tot}$ until a temperature of about 400 °C. From 400 to 600 °C, the susceptibility decreases sharply with the curves steepening further between 500 °C–600 °C (Fig. 7a, b); the second group shows similar behaviour to the first group except that it doesn't have the initial sharp decrease in the $K_{tot}$ (Fig. 7c, f); the third group is characterized by initial increase in $K_{tot}$ below 100 °C and then a sharp decrease in $K_{tot}$ until a temperature of 150 °C–200 °C, then a gentle increase (Fig. 7d) or gentle decrease (Fig. 7e) in $K_{tot}$, and finally a sharp decrease between 500 °C–600 °C.

The ignimbrites are generally characterized by negative susceptibility (Fig. 7g, h). In these samples, $K_{tot}$ gradually increases from room temperature to 500 °C–550 °C and then sharply decreases to a final temperature of 660 °C (Fig. 7g, h). For all the experiments done, the final $K_{tot}$ is lower than the initial $K_{tot}$ with identical heating and cooling curves, indicating no chemical transformation had occurred. Results from these experiments indicate that titanomagnetite and magnetite are the dominant carriers of the magnetization in the basalts. In the ignimbrites, titanomagnetite, magnetite and titanohematite are suggested to be the magnetization carriers.

*Palaeomagnetic directions.* On the orthogonal vector component diagrams where linear vector components are identified such as those displayed on Figs 5 and 6, the ChRM is determined by least square technique (Kirchvink 1980). On average, the maximum angular deviation of the ChRM from a straight line is less than 1°. However, where stable end points and linear segments could not be isolated, either because of overlaps of one or more components or due to lightning strikes, the technique of remagnetization circle has been used (Halls 1976, 1978).

For most of the samples analysed and for those that have quasi-single component magnetization, the ChRM identified by AF and TH techniques are identical for the twin specimens obtained from the same core samples (Figs 5a, b & 6a, b). Both AF and TH treatments were done on the expectations that for some cases one or the other

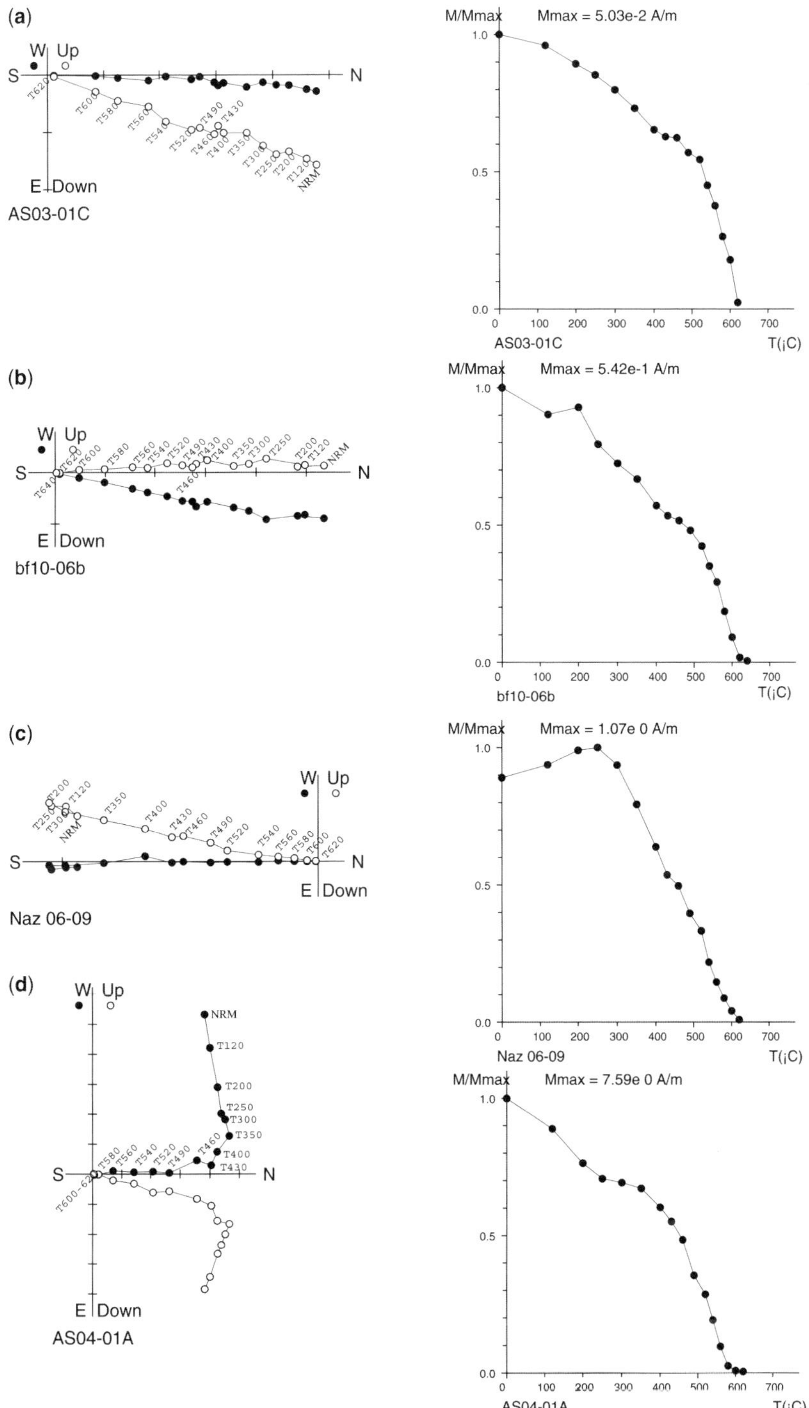


**Fig. 6.** Typical examples of vector component diagrams for the specimens treated by thermally (**a**, **b**, **c**) representing single or quasi-single component and (**d**) representing overlapping components.

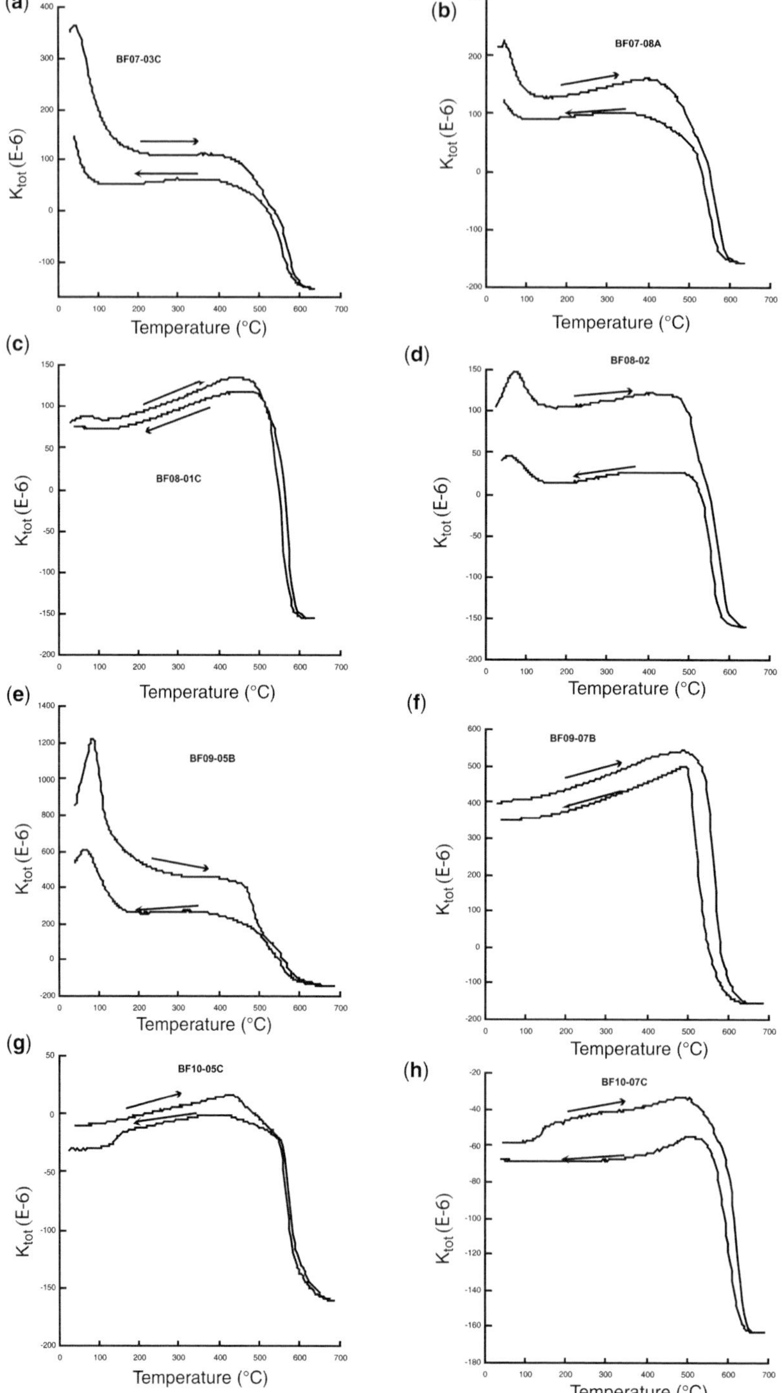


**Fig. 7.** Susceptibility vs. temperature ($K$–$T$) curves for some representative samples from basalts (**a–f**) and ignimbrites (**g**) and (**h**). The heating and cooling curves are shown by the arrows.

technique would better identify the vector components present. In this study, however, both AF and TH generally resulted in identical Zijderveld diagrams (Figs 5 & 6). These happen because for the most part the vector components of various specimens are characterized by non-overlapping unblocking temperatures and non-overlapping coercivity spectra.

The ChRM values determined at specimen level then were used to calculate the core and site (flow) averages using Fischer (1953) statistics or McFadden & McElhinny (1988) statistics for combined analysis of remagnetization circles and stable linear segments. The palaeomagnetic data analysis, from principal component determination to calculation of overall mean direction, was done using the 'palaeomac' programme (Cogne 2003).

The site averages calculated for all the 20 palaeomagnetic sites are reported in Table 1. Out of the 20 flows, 12 sites have a normal polarity and six sites have a reversed polarity with one site characterized by anomalous inclination. The remaining two sites, however, appear to give well-defined site mean directions in terms of declinations; they have anomalously high inclination and lower intensity, which suggest that they could be transitional directions. The confidence interval ($\alpha_{95}$) for all of the samples is less than 9°, indicating good quality palaeomagnetic data.

Figure 8 illustrates all the mean palaeomagnetic directions from all 20 sites, excluding the three sites that were eliminated. In this figure, the sites sampled during the first field season are shown in circles and those during the second field season are shown with inverted triangles. Both of these populations form random distributions about the mean, indicating that there is no bias due to the sampling at two different seasons. The normal and reversed polarities are 180° apart, and also they pass the palaeomagnetic stability reversal test (McFadden & McElhinny 1988). When an overall mean direction is calculated for the 17 sites with stable polarities, excluding the two transitional and the one with anomalous inclination (shown in italics, Table 1), the following values are obtained: $D_s = 2.3°$, $I_s = 7.8°$, $N = 17$, $K = 26.9$, $\alpha_{95} = 7°$. Because there was no tilting of all the young flows investigated in this study, there is no difference in the site mean directions in geographic and tectonic coordinates.

**Table 1.** *Palaeomagnetic site mean directions for all the analysed rocks are reported. All lava flows are laid down horizontal. The overall mean directions calculated for the 17 sites considered are also given and the 2 Ma reference curve from Besse & Courtillot (2003) for a mean latitude and longitude of the study area is shown. Italic numbers indicate the three sites not used for mean calculation*

| Sample name | Coordinates | | N | $D_s$ | $I_s$ | K | $\alpha_{95}$ |
|---|---|---|---|---|---|---|---|
| | Lat. | Lon. | | | | | |
| AS01 | 08°, 03.8′ | 039°, 07.4′ | 12 | 11.4 | −20.3 | 104.2 | 4.4 |
| AS02 | 08°, 01.2′ | 039°, 09.8′ | 16 | 357.9 | 6.8 | 170.1 | 2.8 |
| AS03 | 08°, 02.5′ | 039°, 09.5′ | 18 | 3.5 | 12.4 | 167.5 | 2.7 |
| AS04 | 08°, 01.7′ | 039°, 02.7′ | 12 | 3.6 | 9.1 | 109.1 | 4.3 |
| Naz01 | 08°, 32.1′ | 039°, 14.6′ | 7 | 186 | −26.7 | 131.7 | 5.3 |
| Naz02 | 08°, 32.1′ | 039°, 14.5′ | 15 | 178.2 | −21.2 | 40.8 | 6.1 |
| *Naz03* | *08°, 32.0′* | *039°, 14.5′* | *8* | *182.5* | *38.7* | *35.9* | *9.8* |
| Naz04 | 08°, 40.5′ | 039°, 28.7′ | 9 | 188.3 | 5.9 | 56.2 | 7.6 |
| Naz05 | 08°, 42.2′ | 039°, 32.5′ | 4 | 14.9 | −11.2 | 1393.8 | 2.5 |
| Naz06 | 08°, 24.1′ | 039°, 22.4′ | 13 | 178.8 | −2.6 | 58.5 | 5.5 |
| BF01 | 08°, 27.9′ | 039°, 26.8′ | 2 | 356.7 | 21.5 | 794.1 | 8.9 |
| BF02 | 08°, 33.0′ | 039°, 11.2′ | 7 | 174.4 | −28.2 | 432.8 | 2.9 |
| BF03 | 08°, 54.4′ | 039°, 52.2′ | 6 | 355.7 | 20.6 | 295.6 | 3.9 |
| BF04 | 08°, 53.4′ | 039°, 48.4′ | 5 | 354.5 | 0.9 | 658.9 | 3 |
| *BF05* | *08°, 50.5′* | *039°, 44.0′* | *6* | *240.6* | *−39* | *430* | *3.3* |
| *BF06* | *08°, 49.2′* | *039°, 42.7′* | *8* | *231* | *−42.1* | *682.6* | *2.1* |
| BF07 | 08°, 46.4′ | 039°, 37.4′ | 12 | 10.6 | −12.8 | 275.5 | 2.7 |
| BF08 | 08°, 45.0′ | 039°, 35.4′ | 10 | 357.6 | 15.9 | 240.8 | 3.1 |
| BF09 | 08°, 23.3′ | 039°, 19.8′ | 6 | 358.4 | 13 | 118.4 | 6.2 |
| BF10 | 08°, 16.6′ | 039°, 23.9′ | 6 | 8.5 | 2.5 | 113.7 | 6.3 |
| All mean | 08°, 30′ | 039°, 30′ | 17 | 2.3 | 7.8 | 26.9 | 7.0 |
| BC03 (2 Ma) | 08°, 30′ | 039°, 30′ | 32 | 1.9 | 13.5 | 105.6 | 2.5 |

Sample name; location coordinates (latitude and longitude); N, no. of samples; $D_s$ and $I_s$, declination and inclination in stratigraphic coordinates; K, Fischer precision parameter; $\alpha_{95}$, 95% confidence interval.

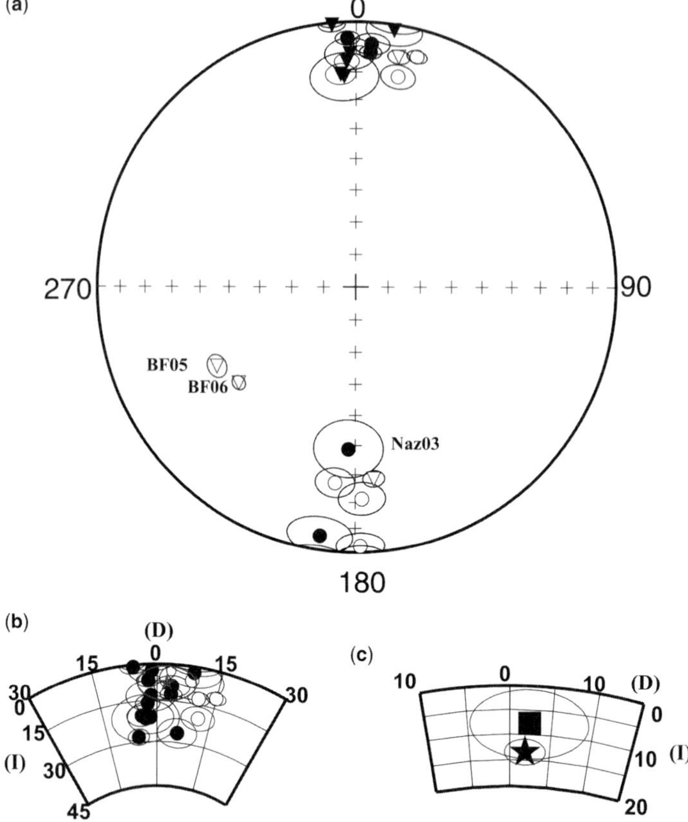


**Fig. 8.** Stereographic projection of all the 20 sites analysed. Symbols used distinguish sites from the first mission (circles) and those from the second mission (inverted triangles). The three transitional sites are also labelled. The remaining 17 sites with the overall mean is shown in (**b**) after the reversed polarities are changed into their normal counterparts. (**c**) Comparison of the overall mean direction (rectangular symbol) with the expected direction (star) from the theoretical Apparent Polar Wander Path Curve for Africa (Besse & Courtillot 2003).

*Geochronology.* From the 20 palaeomagnetic sites investigated, six representative sites were selected for K/Ar age determinations. The ages of the rest of the sites were then deduced either by their lateral continuity with the sites chosen for K/Ar analysis, stratigraphic positions, or lithological similarities with those dated sites.

The dating was performed using the unspiked K/Ar (Cassignol–Gillot) technique (Cassignol & Gillot 1982; Gillot & Cornette 1986). The details of the dating results are reported in Table 2. Five samples have been measured each in two replicates of separate aliquots. For one site where the result was thought very reliable, a single measurement was made. The age results obtained varied from $1.945 \pm 0.029$ Ma for the porphyritic basalt at the base of the eastern margin at Deneba area (giving the lower age limit of the studied rocks) to $0.025 \pm 0.01$ Ma for the young basalt along the recent fissures of the rift axis (giving the upper age limit) (see Table 2). Geological investigation indicates that the rest of the palaeomagnetic sites analysed are likely to have ages within the above range. Comparison of obtained ages and magnetic polarities (Table 2) with the Geomagnetic Polarity Time Scale (GPTS) of Cande & Kent (1995) indicate that they are in agreement.

## Interpretation and discussions

### *Tectonic rotation*

The palaeomagnetic observation on the 20 sites of Quaternary volcanics resulted in an overall mean direction of $D_s = 2.3°$, $I_s = 7.8°$, $\alpha_{95} = 7$, $K = 26.9$, $N = 17$ (see also Table 1 and discussion above). When the obtained overall mean

**Table 2.** *Results of K-Ar dating of six volcanic rocks (basalts and ignimbrites) in the studied region of the MER*

| Sample name | K% | $^{40}Ar^*$(%) | $^{40}Ar^*$(at/g) $*10^{12}$ | Age (Ma) | Uncertainty ($\pm 1\sigma$) | Observed Magnetic Polarity | Polarity scale of Cande & Kent (1995) |
|---|---|---|---|---|---|---|---|
| NAZ01 | 1.139 | 21.89 | 1.969 | 1.654 | 0.025 | | |
| Basalt | 1.139 | 24.37 | 1.975 | 1.659 | 0.024 | | |
| Mean ($n = 2$) | | | | **1.657** | **0.06** | **Reversed** | **Reversed (Matuyama)** |
| NAZ02 | 4.062 | 5.22 | 3.979 | 0.938 | 0.04 | | |
| Ignimbrite | 4.062 | 5.44 | 3.930 | 0.926 | 0.04 | | |
| Mean ($n = 2$) | | | | **0.932** | **0.023** | **Reversed** | **Reversed (Matuyama)** |
| AS04 | 1.067 | 0.68 | 0.0665 | 0.060 | 0.009 | | |
| Basalt | 1.067 | 0.75 | 0.0718 | 0.064 | 0.009 | | |
| Mean ($n = 2$) | | | | **0.062** | **0.009** | **Normal** | **Normal (Brunhes)** |
| AS02 | 4.536 | 33.05 | 3.059 | **0.646** | **0.01** | **Normal** | **Normal (Brunhes)** |
| Ignimbrite | | | | | | | |
| AS01 | 1.092 | 0.15 | 12.01 | 0.011 | 0.01 | | |
| Basalt | 1.092 | 0.56 | 43.23 | 0.038 | 0.01 | | |
| Mean ($n = 2$) | | | | **0.025** | **0.01** | **Normal** | **Normal (Brunhes)** |
| Deneba | 2.385 | 28.05 | 4.845 | 1.944 | 0.029 | | |
| Basalt | 2.385 | 31.90 | 4.853 | 1.947 | 0.029 | | |
| Mean ($n = 2$) | | | | **1.945** | **0.029** | **No palaeomagnetic measurement** | |

Sample name, sample identification; K%, total percentage of potassium in sample; $^{40}Ar^*$ (%), percentage of radiogenic argon in sample; $^{40}Ar^*$ (at/g), number of atoms of radiogenic argon 40 per gram of sample (in $10^{12}$at/g); Age (Ma), ages determined from two replicate argon measurements for five of the samples (means of these replicate determination indicated in **bold**) and in one case only the age is determined from single argon measurement; Uncertainty ( $\pm$ $1\sigma$), error margin is at the $1\sigma$ level; Magnetic polarity, the polarity of the site mean palaeomagnetic direction at same site; Polarity scale of Cande & Kent (1995), Magnetic polarity predicted by Cande & Kent, are shown in the table.

palaeomagnetic direction is compared with the expected direction (D = 1.9°, I = 13.5°, $\alpha_{95} = 2.5$, K = 105.6, N = 32) calculated from the Apparent Polar Wander Path reference curve for Africa at 1.5 Ma (Besse & Courtillot 2003), they are not statistically different (Fig. 9).

Therefore, there has been no tectonic rotation that can be detected by palaeomagnetism. The investigated region is located within and overlaps the <1.9 My magmato-tectonic segments of Gademsa–Koka, Boset and Fentale–Dofan (Figs 2 & 11). Given the 20 km overlap and right-stepping nature of the individual fault and fissure belts lying within each magmato-tectonic segment, counterclockwise rotation of about 11° would be predicted if the oblique rifting model was responsible for the Pleistocene evolution of the MER (Boccaletti *et al.* 1992, 1999; Bonini *et al.* 1997). We can extrapolate present-day extension velocities to estimate 12 km of opening across the magmatic segments during the past 2 Ma, which is probably a high estimate. This amount of opening would cause significant rotations in a transtensional setting as predicted by models of Bonini *et al.* (1997). Our palaeomagnetic studies in these areas argue against any detectable vertical axis rotation and therefore against this model for rift formation and evolution. Observations on recent structures in the rift indicated that, although shear surfaces are produced by cohesionless material such as pyroclastic fall deposits, most fractures open as Mode I cracks first and then evolve into Mode III cracks later (e.g. Accocella & Korme 2002). Thus, shear surfaces may not represent tectonic stresses, making it difficult to find true kinematic indicators on ignimbrites and basalts and explaining the contradictory kinematic results of Boccaletti *et al.* (1992) and Chorowicz *et al.* (1994).

Instead, the absence of significant rotations about vertical axes in magmatic segments and sites outside the magmatic segments is consistent with (a) strain transfer from border faults to magmatic segments at *c.* 2 Ma and (b) models of extension by dyke injection and minor faulting. Numerical models of Buck (2004, 2006) show that the presence of magma at the lithosphere–asthenosphere boundary reduces the tectonic forces required to rift continental lithosphere, with progressive melt

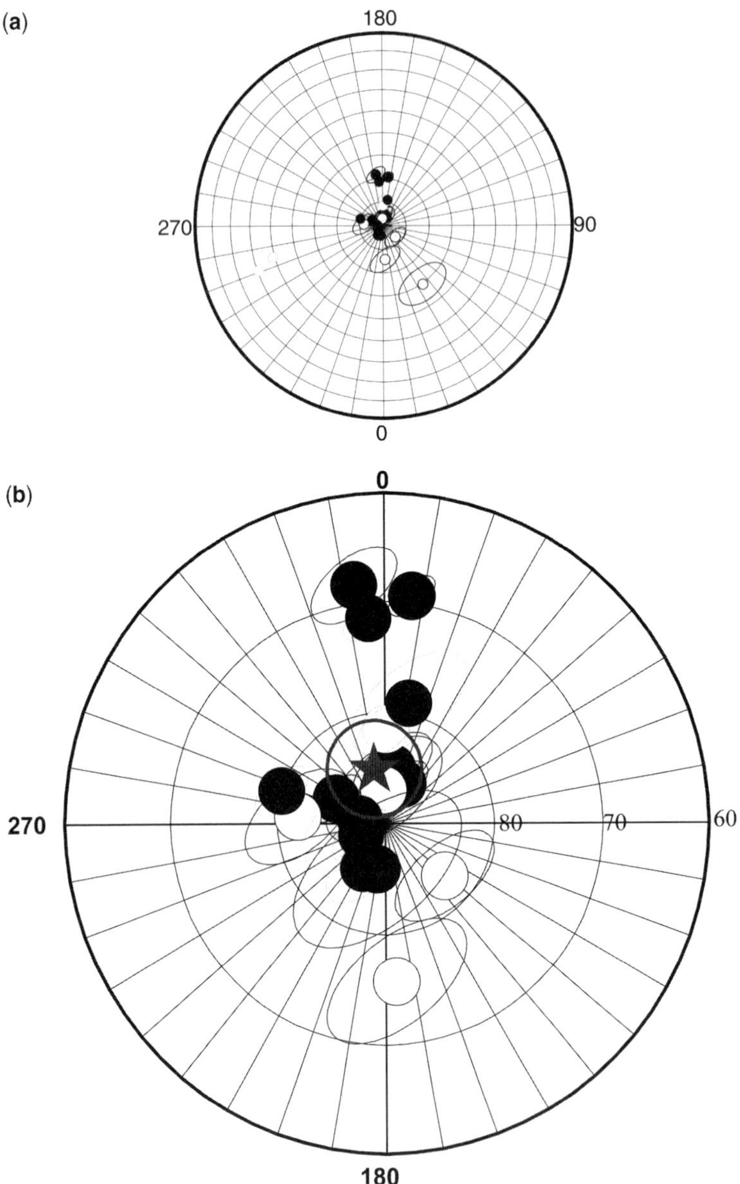


**Fig. 9.** Stereographic projection showing (**a**) the VGP scatter of all the studied sites with the three transitional sites shown in grey, and (**b**) the zoom-in of (a) with the three sites discarded and the mean VGP plotted for the remaining 17 sites considered stable.

injection localizing strain. Keir *et al.* (2006) propose that magma injection to <10 km subsurface beneath the magmatic segments triggers fault slip in the crust above the intrusions, leading to the formation of small offset faults. Field observations also show dykes reaching the surface, and open, magma-filled fissures in the magmatic segments (e.g. Accocella and Korme 2002; Casey *et al.* 2006). Thus, we propose that strain is accommodated largely by magma injection into the crust, rather than by movement along large offset fault border faults, or by large faults in the Quaternary magmatic segments. The lack of rotations in inter-segment zones suggests that no through-going transform faults have developed. Rather, the overlapping en echelon segments may mark a NE-trending melt supply zone linking the MER to the on-land continuation of the Red sea and Gulf of Aden ridge in the Afar depression.

### *Palaeosecular variation studies (PSV) and virtual geomagnetic poles (VGP)*

Table 3 and Fig. 11 present the VGPs of individual sites together with the mean palaeomagnetic pole for the studied region calculated from the above site mean directions for a mean position at 8.5°, 39.5° respectively. From the 17 sites considered, a mean pole at $\phi = 191.9°$, $\lambda = 85°$, $N = 17$, $K = 67.7$, $A_{95} = 4.4$ is calculated. This value is similar to the value $\phi = 171.5°$, $\lambda = 87.1°$, $N = 32$, $K = 105.6$, $A_{95} = 2.5$ predicted from the APWP reference curve of Besse & Courtillot (2003).

The studied region experienced only minor rotation so the individual VGPs calculated can be used to determine the VGP scatter with respect to the North Pole. For the 17 sites considered to have stable polarities, a VGP scatter of 8.9° (confidence interval 7.3°–11.4°) is obtained (Cox 1969). If site Naz 03, which was discarded due to anomalous inclination value, is included, the VGP scatter increases to 10.1° (confidence interval 8.3°–12.9°). Because the age ranges of the studied lava flows is between 1.9 Ma–0.025 Ma and includes both reversed and normal polarities in different locations, the VGP scatter should reflect the palaeosecular variation of the geomagnetic field. The obtained VGP scatter values are in good agreement with the recent 0–5 Ma model of McFadden *et al.* (1990). Figure 10 shows a comparison between the calculated VGP scatter for the MER and the reference curve of McFadden *et al.* (1990) for the period between 0–5 Ma. The calculated result is somewhat smaller than the predicted value from the reference curve. This could be due to the comparatively small number of sites in the MER. To see the sensitivity of this value, we include the remaining two sites (BF05 and BF06) which were eliminated for being transitional directions increases the VGP scatter to 14.6° (confidence interval 12.4°–18.7°), a value higher compared to the predicted scatter.

## Conclusions

This first palaeomagnetic investigation undertaken in the MER has provided the first quantitative data about the fault block motion in a region of continental break-up. In the study, the basalts and the ignimbrites showed stable demagnetization behaviour. Most of the samples showed one or two

**Table 3.** *Site mean Virtual Geomagnetic Poles (VGPs) for all analysed sites and mean palaeomagnetic pole for 17 sites considered. The mean VGP from Apparent Polar Wander Path Curve of Besse & Courtillot (2003) is also given as BC03. The non-considered three sites because of their transitional directions are shown in italics*

| Sample Name | N | $\phi_s$ | $\lambda_s$ | K | $A_{95}$ |
|---|---|---|---|---|---|
| AS01 | 12 | 187.5 | 68.3 | 2.4 | 4.7 |
| AS02 | 16 | 243.8 | 84.9 | 1.4 | 2.9 |
| AS03 | 18 | 155.9 | 86.1 | 1.4 | 2.7 |
| AS04 | 12 | 172.5 | 85 | 2.2 | 4.4 |
| Naz01 | 7 | 265.5 | −81.9 | 3.1 | 5.7 |
| Naz02 | 15 | 183.2 | −87 | 3.4 | 6.4 |
| *Naz03* | *8* | *34.6* | *−59.6* | *6.9* | *11.6* |
| Naz04 | 9 | 3.8 | −75.7 | 3.9 | 7.7 |
| Naz05 | 4 | 173 | 69.3 | 1.3 | 2.5 |
| Naz06 | 13 | 48.8 | −82.8 | 2.7 | 5.5 |
| BF01 | 2 | 349.1 | 85.8 | 4.9 | 9.4 |
| BF02 | 7 | 179.5 | −81.5 | 1.7 | 3.2 |
| BF03 | 6 | 332.4 | 85.5 | 2.2 | 4.1 |
| BF04 | 5 | 252.9 | 80 | 1.5 | 3 |
| *BF05* | *6* | *289.3* | *−30.5* | *2.3* | *3.9* |
| *BF06* | *8* | *285.5* | *−39.1* | *1.6* | *2.6* |
| BF07 | 12 | 184.7 | 71.4 | 1.4 | 2.7 |
| BF08 | 10 | 295.2 | 87.5 | 1.7 | 3.2 |
| BF09 | 6 | 260.4 | 87.6 | 3.2 | 6.3 |
| BF10 | 6 | 168.6 | 79 | 3.2 | 6.3 |
| Mean VGP | 17 | 191.9 | 85 | 67.7 | 4.4 |
| Mean VGP BC03 | 32 | 171.5 | 87.1 | 105.6 | 2.5 |

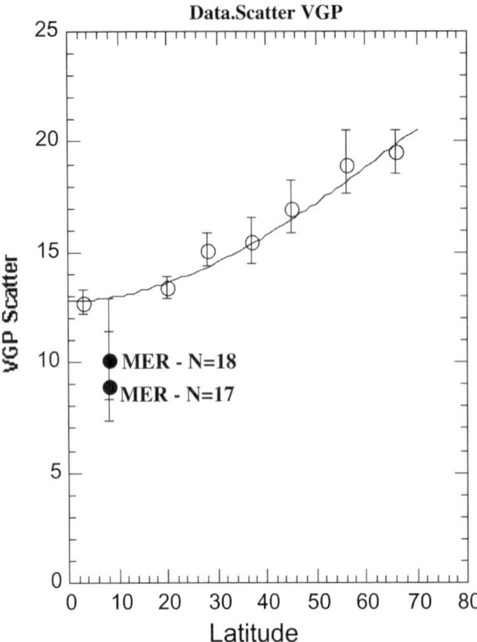


**Fig. 10.** Comparison of observed VGP scatter for the Pleistocene rocks of the MER ($N = 17$ and $N = 18$) with the reference curve of McFadden *et al.* (1991) for the period between 0–5 Ma.

components of magnetization with the first one being cleaned by AF fields of 15–20 mT and temperatures of up to 400–550 °C. Both the TH and AF techniques revealed identical orthogonal vector component diagrams for most samples. The NRM intensity is characterized by a unimodal distribution in the basalts with a magnitude similar to the one suggested for sub-aerial basalts (Prévot & Grommé 1975), and a bimodal distribution in the ignimbrites. The magnetization intensity values are similar to those reported in Afar (e.g. Kidane *et al.* 1999, 2002). Progressive heating and alternating field analysis and susceptibility vs. temperature measurements suggest that the magnetic mineralogy is titanomagnetite and magnetite for the basalts and magnetite and titanohematite for the ignimbrites.

In the majority of samples, palaeomagnetic measurements using both TH and AF techniques revealed quasi-single component of magnetization with a viscous remanent magnetization (VRM) observable on few samples. Principal component analysis and statistical averaging resulted in an overall mean palaeomagnetic direction which is statistically identical to the expected direction for stable Africa

A geochronological investigation performed on the six selected samples using the unspiked K/Ar (Cassignol–Gillot) technique (Cassignol & Gillot 1982; Gillot & Cornette 1986) confirmed that rocks analysed have Late Pliocene–Quaternary ages (1.945–0.025 Ma).

The VGPs calculated for 17 stable sites from site mean palaeomagnetic direction provide a mean pole at $\phi = 191.9°$, $\lambda = 85°$, $N = 17$, $K = 67.7$, $A_{95} = 4.4$, which is similar to the value $\phi = 171.5°$, $\lambda = 87.1°$, $N = 32$, $K = 105.6$, $A_{95} = 2.5$ predicted from the APWP reference curve of Besse & Courtillot (2003). Given that the region has not suffered any vertical axis rotation, a VGP scatter of 8.9° (confidence interval 7.3°–11.4°) is obtained. For $N = 17$, a VGP scatter of 10.1° (confidence interval 8.3°–12.9°) (Cox 1969). This VGP scatter when the uncertainties are considered is in good agreement with the recent 0–5 Ma model of McFadden *et al.* (1991). These palaeomagnetic results indicate that the palaeosecular variation has been adequately sampled.

The declination results from all the sites and the overall mean direction is reported as small arrows and big arrow respectively; at each site, the corresponding north direction is also indicated by a long straight line on a blown-up digital elevation map of the studied site (Fig. 11). Comparison of the overall mean direction of $D_s = 2.3°$, $I_s = 7.8°$, $\alpha_{95} = 7$, $K = 26.9$, $N = 17$ with the expected direction ($D = 1.9°$, $I = 13.5°$, $\alpha_{95} = 2.5$, $K = 105.6$, $N = 32$) calculated from the Apparent Polar Wander Path reference curve for Africa at 1.5 Ma (Besse & Courtillot 2003) resulted in a declination difference of only ($\Delta D = 0.4° \pm 7.5°$). In the oblique rifting model, however, using simple geometrical relations for an opening rate of about 5 mm $a^{-1}$, one expects a counterclockwise rotation of about 11° since 1.9 Ma (see Fig. 3 Boccaletti *et al.* 1999). The above data show a a statistically significant difference between our palaeomagnetically obtained observations and the counterclockwise rotations predicted by the oblique rifting model; instead, our results show no vertical axis rotation.

The absence of detectable rotations about vertical axes from sites located between tectono-magmatic segments, and between the magmatic segments and the border faults, discounts transtenstional kinematic models, and those of spreading segments separated by transform faults. Our results interpreted in light of structural and geophysical data indicate that extensional strain is accommodated by magma injection in a NE-trending zone beneath the MER which may serve to link seafloor spreading in the southernmost Red Sea and easternmost Gulf of Aden to the East African rift system.

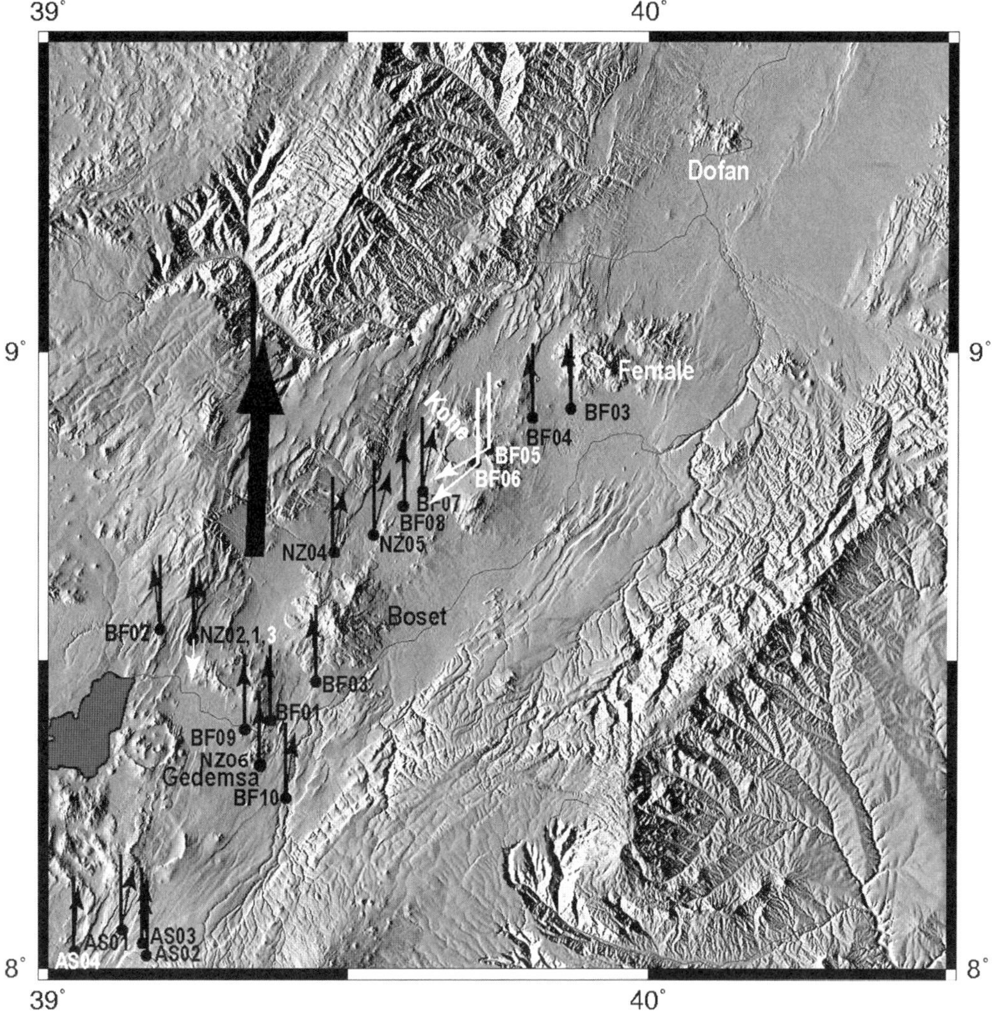


**Fig. 11.** SRTM digital elevation model of the study region (see Fig. 2) with all the site mean palaeomagnetic directions shown as black arrows. At each site, the north direction is indicated by a longer dark line. The three transitional directions are indicated by white arrows. The overall mean palaeomagnetic direction in the region is given by the large, bold arrow beside the north direction. The sites with stable reversed polarities are changed to their stable counterparts in this figure.

The Science and Technology Commission, FDRE, is highly appreciated for the grant that the first author used for field sampling and some part of the analysis. The French Embassy in Addis Ababa and the Royal Society funded laboratory visits to France and the UK, respectively. C. Ebinger acknowledges NERC grant NER/A/S/2000/1004. The authors would like to thank V. Courtillot, P.Y. Gillot and P. Lahitte for their kind help for using their lab facilities during the preliminary part of the work. We are grateful to D. Keir for the invaluable discussions with the first author. Finally, the authors would like to express their gratitude to Mike Fuller and anonymous reviewers for their critical review and suggestions that helped to improve the paper.

## References

Abebe, B. 1993. *Studio geologico-structurale del Rift Etiopico a sud di Assela*. Ph.D. Thesis, University of Firenze, Florence, Italy, 153pp.

Abebe, B., Boccaletti, M., Mazzuoli, R., Bonini, M., Tortorici, L. & Trua, T. 1998. Geological map of the Lake Ziway–Asela region, Main Ethiopian rift, ARCA-DB Map.

ACCOCELLA, V. & KORME, T. 2002. Holocene extension direction along the Main Ethiopian Rift, East Africa. *Terra Nova*, **14**, 191–197.

ASFAW, L.M. 1992. Constraining the African pole of rotation. *Tectonophysics*, **209**, 55–63.

AYALEW, D., EBINGER, C., BOURDON, E., WOLFENDEN, E., YIRGU, G. & GRASSINEAU, N. 2006. Temporal compositional variation of syn-rift rhyolites along the western margin of the southern Red Sea and northern Main Ethiopian Rift. *In*: YIRGU, G., EBINGER, C.J. & MAGUIRE, P.K.H. (eds) *The Afar Volcanic Province within the East African Rift System*. Geological Society, London, Special Publications, **259**, 123–132.

BASTOW, I., STUART, G.W., KENDALL, J.-M. & EBINGER, C. 2005. Upper mantle seismic structure in a region of incipient continental breakup: northern Ethiopian rift. *Geophysical Journal International*, **162**, 479–493.

BESSE, J. & COURTILLOT, V. (2003). Apparent and True polar wander and the geometry of the geomagnetic field in the last 200 million years. *Journal of Geophysical Research*, **108**(B10), 2469 doi 10.1029/2003JB002684.

BILHAM, R., BENDICK, R., LARSON, K., MOHR, P., BRAUN, J., TESFAYE, S., ASFAW, L. 1999. Secular and tidal strain across the Main Ethiopian Rift. *Geophysics Research Letters*, **26**, 2789–2792.

BOCCALETTI, M., GETANEH, A. & TORTORICI, L. 1992. The Main Ethiopian Rift: an example of oblique rifting. *Annales Tectonicae*, **6**, 20–25.

BOCCALETTI, M., MAMMO, T., BONINI, M. & ABEBE, B. 1994. Seismotectonics of East African Rift System: evidence of active oblique rifting. *Annales Tectonicae*, **8**, 87–99.

BOCCALETTI, M., BONINI, M., MAZZUOLI, R., ABEBE, B., PICCARDI, L. & TORTORICI, L. 1998. Quaternary oblique extensional tectonics in the Ethiopian Rift (Horn of Africa). *Tectonophysics*, **287**, 97–116.

BONINI, M., SOURIOT, T., BOCCALETTI, M. & BRUN, J.P. 1997. Successive orthogonal and oblique extension episodes in a rift zone: Laboratory experiments with application to the Ethiopian rift. *Tectonics*, **16**, 347–362.

BOSWORTH, W. & STRECKER, M.R. 1997. Stress field changes in the Afro-Arabian rift System during the Miocene to Recent periods. *Tectonophysics*, **278**, 47–62.

BUCK, W.R. 2004. Consequences of asthenospheric variability on continental rifting. *In*: KARNER, G.D., TAYLOR, B., DRISCOLL, N.W. & KOHLSTEDT, D.L. (eds), *Rheology and Deformation of the Lithosphere at Continental Margins*. Columbia University Press, New York, 1–30.

CANDE, S.C. & KENT, D.V. 1995. Revised calibration of the geomagnetic polarity timescale for the Late Cretaceous and Cenozoic. *Journal of Geophysical Research*, **100**, 6093–6095.

CASEY, M., EBINGER, C., KEIR, D., GLOAGUEN, R. & MOHAMED, F. 2006. Extension by dyke instrusion and faulting in the Main Ethiopian Rift. *In*: YIRGU, G., EBINGER, C.J. & MAGUIRE, P.K.H. (eds) *The Afar Volcanic Province within the East African Rift System*. Geological Society, London, Special Publications, **259**, 143–163.

CASSIGNOL, C. & GILLOT, P.-Y. 1982. Range and effectiveness of unspiked potassium–argon dating. *In*: ODIN, G.S. (ed.) *Numerical Dating in Stratigraphy*. John Wiley, New York, 159–172.

CHOROWICZ, J., COLLET, B., BONAVIA, F.F. & KORME, T. 1994. Northwest to north–northwest extension direction in the Ethiopian Rift deduced from the orientation of extension structures and fault slip analysis. *Bulletin of the Geological Society of America*, **105**, 1560–1570.

COGNE, J.P. 2003. Palaeomac, a Macintosh application for treating palaeomagnetic data and making plate reconstructions. *Geochemistry, Geophysics Geosystem*, **4**(1), 1007, doi: 101029/2001GC000227.

COX, A. 1969. Confidence limits for the precision parameter K. *Geophysical Journal of the Royal Astronomical Society*, **18**, 545–549.

DI PAOLA, G.M. 1972. The Ethiopian Rift Valley (between 7° and 8°40″ lat. North). *Bulletin Volcanologique*, T.XXXVI-4, 1–64.

EBINGER, C.J. 1989. Geometric and kinematic development of border faults and accommodation zones, Kivu–Rusizi Rift, Africa. *Tectonics*, **8**, 117–133.

EBINGER, C. & CASEY, M. 2001. Continental breakup in magmatic provinces: An Ethiopian example. *Geology*, **29**, 527–530; DOI: 10.1130/0091-7613.

EBINGER, C.J., YEMANE, T., WOLDEGABRIEL, G., ARONSON, J.L. & WALTER, R.C. 1993. Late Eocene–Recent volcanism and faulting in the southern main Ethiopian rift. *Journal of the Geological Society London*, **150**, 99–108.

FISCHER, R.A. 1953. Dispersion on a sphere. *Proceedings of the Royal Society of London Series A*, **217**, 295–305.

GILLOT, P.-Y. & CORNETTE, Y. 1986. The Cassignol technique for potassium–argon dating, precision and accuracy: Examples from the late Pleistocene to recent volcanics from southern Italy. *Chemical Geology*, **59**, 205–222.

GOUIN, P. 1979. *Earthquake history of Ethiopia and the Horn of Africa*, report. International Development Research Centre, Ottawa, Ontario, Canada, 258pp.

HALLS, H.C. 1976. A least-squares method to find a remanence direction from converging remagnetization circles. *Geophysical Journal of the Royal Astronomical Society*, **45**, 297–304.

HALLS, H.C. 1978. The use of converging remagnetization circles in palaeomagnetism. *Physics of the Earth and Planetary Interiors*, **16**, 1–11.

HOFFMAN, C., COURTILLOT, V., FÉRAUD, G., ROCHETTE, P., YIRGU, G., KETEFO, E., PIK, R. 1997. Timing of the Ethiopian flood basalt event and implications for plume birth and global change. *Nature*, **389**, 838–841.

HUCHON, P. & KHANBARI, K. 2003. Rotation of the syn-rift stress field of the northern Gulf of Aden margin, Yemen. *Tectonophysics*, **364**, 147–166.

JESTIN, F., HUCHON, P., GAULIER, J.-M. 1994. The Somalia plate and the East African Rift System: Present-day kinematics. *Geophysical Journal International*, **116**, 637–654.

KEIR, D., EBINGER, C., STUART, G., DALY, E. & AYELE, A. 2006. Seismicity of the Northern Main Ethiopian Rift, *Journal of Geophysical Research*, in press.

KENDALL, J.-M., STUART, G., EBINGER, C., BASTOW, I. & KEIR, D. 2005. Magma-assisted rifting in Ethiopia. *Nature*, **433**, 146–148.

KERANEN, K., KLEMPERER, S., GLOAGUEN, R. & EAGLE Working Group. 2004. Imaging a proto-ridge axis in the Main Ethiopian Rift. *Geology*, **39**, 949–952.

KIDANE, T., CARLUT, J., COURTILLOT, V., GALLET, Y., QUIDELLEUR, X., GILLOT, P.-Y. & HAILE, T. 1999. Palaeomagnetic and geochronological identification of the Reunion subchron in Ethiopian Afar, *Journal of Geophysical Research*, **104**, B5,10405–10419.

KIDANE, T., COURTILLOT, V., *ET AL*. 2002. New palaeomagnetic and geochronologic results from Ethiopian Afar: Block rotations linked to rift overlap and propagation and determination of a ~2 Ma reference pole for stable Africa. *Journal of Geophysical Research*, **107**, doi:10.1029/JB000645.

KIRSCHVINK, J.L. 1980. The least squares line and plane and the analysis of palaeomagnetic data. *Geophysical Journal of the Royal Astronomical Society*, **62**, 699–718.

KORME, T., CHOROWICZ, J., COLLET, B. & BONAVIA, F.F. 1997. Volcanic vents rooted on extension fractures and their geodynamic implications in the Ethiopian Rift. *Journal of Volcanology and Geothermal Research*, **79**, 205–222.

LAHITTE, P., GILLOT, P.-Y., KIDANE, T., COURTILLOT, V., BEKELE, A. 2003. New age constraints on the timing of volcanism in central Afar, and implications on emplacement and rifting. *Journal of Geophysical Research*, **108**, B2 doi: 10.1029/2001JB001689.

LEMAUX II, J., GORDON, R.G., ROYER, J.-Y. 2002. Location of the Nubia-Somalia boundary along the Southwest Indian Ridge. *Geology*, **30**, 339–442.

MAHATSENTE, R., JENTZSCH, G. & JAHR, T. 1999. Crustal structure of the Main Ethiopian Rift from gravity data: 3-dimensional modeling. *Tectonophysics*, **313**, 363–382.

MANIGHETTI, I., TAPPONNIER, P., GILLOT, P.-Y., COURTILLOT, V., JACQUES, E., RUEGG, J.-C., KING, G. 1998. Propagation of rifting along the Arabia-Somalia plate boundary: into Afar. *Journal of Geophysical Research*, **103**, 4947–4974.

MCELHINNY, M.W., MCFADDEN, P.L. & MERRILL, R.T. 1996. The time-averaged palaeomagnetic field 0–5 Ma, *Journal of the Geophysical Research*, **101**, 25,007–25,027.

MCFADDEN, P.L. & MCELHINNY, M.W. 1988. The combined analysis of remagnetization circles and direct observations in palaeomagnetism. *Earth and Planetary Science Letters*, **87**, 152–160.

MCFADDEN, P.L. & MCELHINNY, M.W. 1990. Classification of the reversal test in palaeomagnetism. *Geophysical Journal International*, **103**, 725–729.

MOHR, P.A. 1962. The Ethiopian Rift System. *Bulletin of the Geophysical Observatory, Addis Ababa*, **5**, 33–62.

PRÉVOT, M. & GROMMÉ, S. 1975. Intensity of magnetization of subaerial and sub-marine basalts and its possible change with time. *Geophysical Journal of the Royal Astronomical Society*, **40**, 207–224.

STRECKER, M.R. & BOSWORTH, W. 1991. Quaternary stress-field change and rifting processes in the East African Gregory Rift. *EOS, Transactions of the American Geophysical Union*, **72**, 17–22.

WOLDEGABRIEL, G., ARONSON, J.L. & WALTER, R.C. 1990. Geology, geochronology and rift basin development in the central sector of the Main Ethiopian Rift. *Bulletin of the Geophysical Society of America*, **102**, 439–458.

# Vertical deformation in the Main Ethiopian Rift: levelling results in its northern part, 1995–2004

LAIKE M. ASFAW[1], HENOK BEYENE[2], AMDE MKONNEN[2] & TADESSE OLI[2]

[1]*Geophysical Observatory, Addis Ababa University, PO Box 1176, Addis Ababa, Ethiopia*

[2]*Ethiopian Mapping Authority, PO Box 597, Addis Ababa, Ethiopia*

**Abstract:** A levelling line consisting of 43 benchmarks was established between the towns of Wolenchiti and Metehara in the northern part of the Main Ethiopian Rift in July 1995 by the Geophysical Observatory of Addis Ababa University and the Ethiopian Mapping Authority. The measurement was repeated in November 2003 on 30 of the surviving benchmarks. In both epochs of measurement, the standard accuracy attained is $\pm 4\ \mathrm{mm}\sqrt{K}$ corresponding to a first order levelling, where $K$ is the inter-benchmark distance in kilometres. The line crosses the northern and southern parts of the Nazret (Boset–Kone) and Sabure (Fantale–Dofen) magmatic segments respectively where 80% of rift deformation is believed to be localized. In eight years, interval height differences ranging from +3 mm to −22 mm are found along the line with the maximum subsidence rate of 2.8 mm $a^{-1}$ corresponding to the Kone–Gariboldi volcanic complex in the northern part of the Nazret magmatic segment. On the other hand, at the eastern end of the line, despite the existence of large fissures and an expanding lake in the vicinity suggesting possible significant subsidence, on the contrary relatively small vertical deformation is found. The strong subsidence measured at the Kone–Gariboldi volcanic complex is interpreted to be due to remnant processes of subsidence at these volcanic centres following withdrawal of magma in the recent past. Regarding the eastern end of the line where Lake Beseka is located, the result is particularly important in verifying that the rapid expansion of the lake associated with elevation increase of 40 cm $a^{-1}$ could not be attributed to tectonic subsidence or uplift in the region.

The region of interest is the northern part of the East African rift system within Ethiopia, which is south of the triple junction where the three rift systems of the Gulf of Aden, Red Sea and East Africa meet (inset Fig. 1). Several workers, (e.g. Boccaletti *et al.* 1997; Ebinger & Casey 2001) have shown that the Main Ethiopian Rift is bounded by NE–SW trending Miocene (18 Ma) border faults along which rifting was initiated. The active deformation centre moved to the axial part of the rift characterized by left-stepping magmatic segments oriented in a NNE–SSW direction (Fig. 1) by 1.6 Ma. It is shown by geodetic (Bilham *et al.* 1999) and seismic (Keranen *et al.* 2004) experiments that the magmatic segments are the locus of 80% of the deformation and that they are associated with intrusion of high-density material (e.g. Cornwell *et al.* 2006). In the region itself, visual manifestations of rift extension occur in the form of normal faults, tensional fissures and volcanic activity (e.g. Pizzi *et al.* 2006).

Compared to the Nazret magmatic segment, which had a significant earthquake only in 1993, since the establishment of the WWSSN station (AAE) in 1963, the Sabure magmatic segment and the Kesem border faults had been relatively more active seismically during the same epoch and afterwards with several episodes of seismicity (Asfaw & Kebede 1981; Keir *et al.* 2004). In one case, tensional fissures on sediments were associated with a swarm of earthquakes north of the Fantale Mountain in 1981.

Measurement of rift extension at latitude 9° N was made between 1994 and 1999 (Pan *et al.* 2002). Accordingly, in the five-year interval, extension rate of 2.6 mm $a^{-1}$ was found for GPS sites at Addis Ababa and Metehara, a rift floor site. The location of the sites partly explains the difference with the rate of 5 mm $a^{-1}$ calculated from global plate kinematics (Chu & Gordon 1999). A discrepancy between global rate and measured rate has also been noted by Williams *et al.* (2004) where measurement of fissure widths, radiocarbon dating of the sediments and fission track dating of the welded tuff on which the fissures occur give 1 mm $a^{-1}$ for the widening rate for fissures south of the Fantale volcano.

In connection with this work, it is important to note the acute environmental problems in the region caused by the rapid expansion of Lake Beseka at the southern end of the Sabure magmatic segment. Since 1976, as geodetic benchmarks show, the elevation of the lake has changed from 941 m above the Blue Nile datum (1957–1961

*From*: YIRGU, G., EBINGER, C.J. & MAGUIRE, P.K.H. (eds) 2006. *The Afar Volcanic Province within the East African Rift System*. Geological Society, London, Special Publications, **259**, 185–190.
0305-8719/06/$15.00 

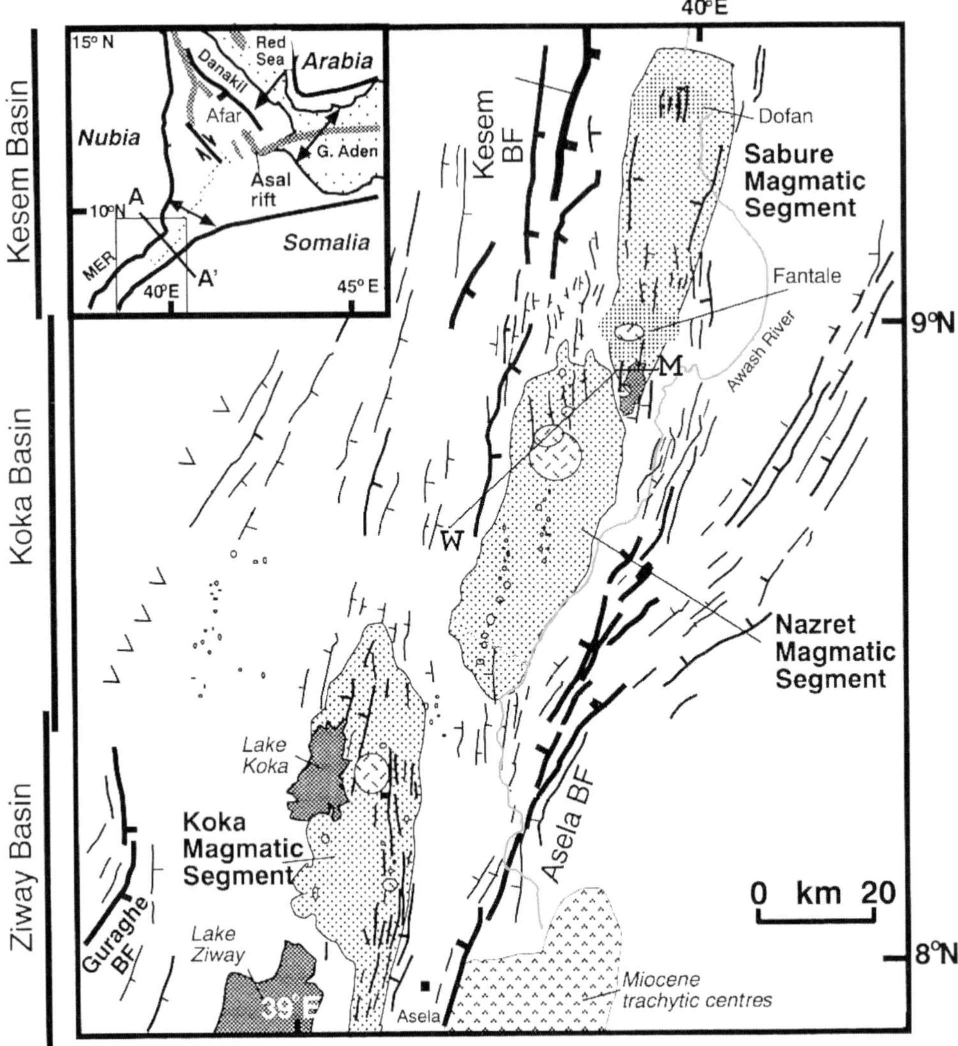


**Fig. 1.** The Main Ethiopian Rift showing border faults and magmatic segments. The levelling line crosses the northern part of the Nazret and southern part of Sabure magmatic segments. Based on Ebinger & Casey (2001) modified to show levelling cross-section W (Wolenchiti) to M (Metehara).

Ethiopian Geodetic Survey) to 951 m with a corresponding area increase from 30 $km^2$ to 45 $km^2$ (Berhanu 2001). The role of tectonic subsidence or uplift, if any, in the expansion of the lake has not been clear up to now.

## Levelling line and measurements

A levelling line was established between the towns of Welenchiti and Metehara by the Geophysical Observatory of Addis Ababa University and the Ethiopian Mapping Authority to measure the vertical deformation in the magmatic segments of Nazret and Sabure relative to the adjacent rift floor and to determine also if tectonic uplift or subsidence is the cause of the expansion of Lake Beseka at the southern end of the Sabure magmatic segment. The line is 77 km long and consists of 43 newly established benchmarks constructed from iron bolts driven into boulders, culverts and concrete monuments.

Initially, levelling started from a known benchmark named TBM34 (Fig. 2) located at Welenchiti

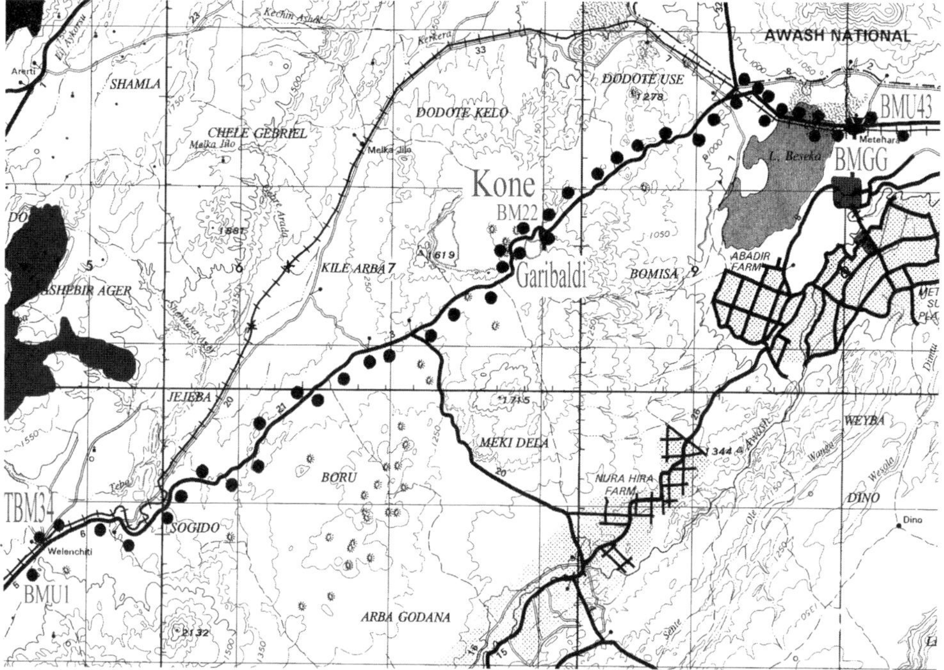


Fig. 2. Topographic map with elevation in metres showing the levelling line and location of benchmarks (•) starting west of Welenchiti town and terminating east of Metehara town.

railway station, and closed at a known benchmark BMGG which is found near Metehara town on the railway culvert.

The mean sea level heights were determined from August 14, 1995 to September 30, 1995. The height differences of the levelling line in forward and backward run were calculated in the field to check and achieve accuracy of $\pm 4$ mm $\sqrt{K}$ where $K$ is the distance in kilometres between the benchmarks (Table 1).

The re-measurement in 2003 was carried out on 30 of the benchmarks—13 being destroyed. All the time in determining the elevation of the benchmarks, direct differential or spirit levelling was performed and the operation was carried out in forward and backward run. In this case also, the result of the measurements obtained conform to accuracies of $\pm 4$ mm $\sqrt{K}$. Therefore in both epochs of measurement, 1995 and 2003, standard accuracy for first order levelling is attained.

## Measurement results

The result of the measurements corresponding to the deformation relative to the adjacent rift floor is shown in Table 1. From Table 1 and Figure 3b, the maximum vertical deformation value corresponding to subsidence is recorded for the Kone–Gariboldi volcanic complex. Where the previous benchmarks were destroyed, new ones were established in 2003 (Table 1). The newly established benchmarks are indicated with '(New) **A**' in Table 1 and Fig. 3b.

Figure 3a is a profile showing closure errors at the benchmarks and the corresponding allowable errors for first order levelling indicated with a ' + ' and ' − '. Figure 3b is a profile of elevation change at the benchmarks. The measurement errors for both 1995 and 2003 campaigns are within the allowable error of $\pm 4$ mm $\sqrt{K}$ for first order levelling.

In the 1995 survey, TBM34 with an elevation of 1455.8008 m was used as a reference. In the 2003 survey benchmark, TBM34 was found unreliable. As a result, BMU2, which was more secure than TBM34, was used as the reference. The elevation of BMU2 is 1440.8605 m.

## Discussion

Though vertical deformation played a significant role in the process of rift formation and development through the border faults, its contemporary role at the magmatic segments has not been of equally wide interest. As a result, a large proportion

**Table 1.** *Measurement results of 1995 and 2003. Height difference, elevation and elevation difference are in metres*

| Welenchiti–metehara level line 1995 & 2003 observation result | | | | | | | | | |
|---|---|---|---|---|---|---|---|---|---|
| Line | Ht diff of 2003 | Dist (km) | Adj | Adjusted ht diff 2003 | Adjusted elevation of 2003 | station name 2003 | Adjusted elevation of 2003 | station name 1995 | Change in elevation |
| *tbm34-bmu1* | 6.256 | | | | | *BMU1* | 1462.0794 | *BMU1* | |
| *tbm34* | | | | | | *TBM34* | 1455.8008 | *TBM34* | |
| *tbm34-bmu2* | −14.9577 | 2.492 | | | 1440.8605 | *BMU2* | 1440.86048 | *BMU2* | 0.0000 |
| *bmu2-bmu3* | −25.7139 | 1.894 | −0.0002 | −25.7141 | 1415.1464 | *BMU3* | 1415.15373 | *BMU3* | −0.0073 |
| *bmu3-bmu4A* | −48.1694 | 2.416 | −0.0002 | −48.1696 | 1366.9768 | *BMU4A* | 1366.44181 | *BMU4* | *NEW* |
| *bmu4A-bmu5A* | −59.8416 | 2.761 | −0.0002 | −59.8418 | 1307.1350 | *BMU5A* | 1306.86237 | *BMU5* | *NEW* |
| *bmu5A-bmu6* | −30.0386 | 1.898 | −0.0002 | −30.0388 | 1277.0962 | *BMU6* | 1277.10377 | *BMU6* | −0.0076 |
| *bmu6-bmu7* | −22.4670 | 1.886 | −0.0002 | −22.4672 | 1254.6290 | *BMU7* | 1254.63464 | *BMU7* | −0.0056 |
| *bmu7-bmu8* | −10.2330 | 1.903 | −0.0002 | −10.2332 | 1244.3959 | *BMU8* | 1244.39654 | *BMU8* | −0.0007 |
| *bmu8-bmu9A* | 25.4189 | 2.030 | −0.0002 | 25.4187 | 1269.8146 | *BMU9A* | 1269.21403 | *BMU9* | *NEW* |
| *bmu9A-bmu10* | −3.6825 | 1.952 | −0.0002 | −3.6827 | 1266.1319 | *BMU10* | 1266.13167 | *BMU10* | 0.0003 |
| *bm10-bmu11* | −39.1282 | 2.243 | −0.0002 | −39.1284 | 1227.0035 | *BMU11* | 1227.00374 | *BMU11* | −0.0002 |
| *bmu11-bmu12* | −16.8984 | 1.837 | −0.0002 | −16.8986 | 1210.1050 | *BMU12* | 1210.10531 | *BMU12* | −0.0003 |
| *bmu12-bmu13A* | 3.7412 | 1.996 | −0.0002 | 3.7410 | 1213.8460 | *BMU13A* | 1213.42847 | *BMU13* | *NEW* |
| *bmu13A-bmu14A* | 1.0247 | 1.922 | −0.0002 | 1.0245 | 1214.8705 | *BMU14A* | 1215.54632 | *BMU14* | *NEW* |
| *bmu14A-bmu15* | 36.5995 | 1.978 | −0.0002 | 36.5993 | 1251.4699 | *BMU15* | 1251.47661 | *BMU15* | −0.0067 |
| *bmu15-bmu16A* | 32.5776 | 2.302 | −0.0002 | 32.5774 | 1284.0473 | *BMU16A* | 1286.83299 | *BMU16* | *NEW* |
| *bmu16A-bmu17* | 52.7794 | 2.518 | −0.0002 | 52.7792 | 1336.8265 | *BMU17* | 1336.8416 | *BMU17* | −0.0151 |
| *bmu17-bmu18* | 28.0445 | 2.349 | −0.0002 | 28.0443 | 1364.8708 | *BMU18* | 1364.88679 | *BMU18* | −0.0160 |
| *bmu18-bmu19* | −18.6231 | 1.684 | −0.0001 | −18.6232 | 1346.2475 | *BMU19* | 1346.26808 | *BMU19* | −0.0206 |
| *bmu19-bmu20* | −28.7724 | 1.175 | −0.0001 | −28.7725 | 1317.4750 | *BMU20* | 1317.49401 | *BMU20* | −0.0190 |
| *bmu20-bmu21A* | 3.5920 | 1.826 | −0.0002 | 3.5918 | 1321.0669 | *BMU21A* | 1310.62979 | *BMU21* | *NEW* |
| *bmu21A-bmu22* | 12.6148 | 0.663 | −0.0001 | 12.6147 | 1333.6816 | *BMU22* | 1333.704 | *BMU22* | −0.0224 |
| *bmu22-bmu23* | −47.1409 | 1.318 | −0.0001 | −47.1410 | 1286.5406 | *BMU23* | 1286.55604 | *BMU23* | −0.0155 |
| *bmu23-bmu24* | −47.4894 | 1.673 | −0.0001 | −47.4895 | 1239.0510 | *BMU24* | 1239.06608 | *BMU24* | −0.0150 |
| *bmu24-bmu25* | −93.4828 | 2.062 | −0.0002 | −93.4830 | 1145.5681 | *BMU25* | 1145.58288 | *BMU25* | −0.0148 |
| *bmu25-bmu26* | −30.9874 | 1.794 | −0.0002 | −30.9876 | 1114.5805 | *BMU26* | 1114.59478 | *BMU26* | −0.0143 |
| *bmu26-bmu27A* | 17.2720 | 1.999 | −0.0002 | 17.2718 | 1131.8523 | *BMU27A* | 1131.51999 | *BMU27* | *NEW* |
| *bmu27A-bmu28A* | −16.4558 | 1.746 | −0.0002 | −16.4560 | 1115.3964 | *BMU28A* | 1121.54622 | *BMU28* | *NEW* |
| *bmu28A-bmu29* | −73.4254 | 2.038 | −0.0002 | −73.4256 | 1041.9708 | *BMU29* | 1041.98313 | *BMU29* | −0.0123 |
| *bmu29-bmu30A* | −17.1423 | 1.668 | −0.0001 | −17.1424 | 1024.8284 | *BMU30A* | 1024.18355 | *BMU30* | *NEW* |
| *bmu30A-bmu31* | −13.8681 | 1.676 | −0.0001 | −13.8682 | 1010.9601 | *BMU31* | 1010.96807 | *BMU31* | −0.0080 |
| *bmu31-bmu32* | −38.4711 | 2.245 | −0.0002 | −38.4713 | 972.4888 | *BMU32* | 972.498754 | *BMU32* | −0.0099 |
| *bmu32-bmu33* | −2.8296 | 0.830 | −0.0001 | −2.8297 | 969.6592 | *BMU33* | 969.668646 | *BMU33* | −0.0095 |
| *bmu33-bmu34* | −2.3102 | 0.491 | −0.0000 | −2.3102 | 967.3489 | *BMU34* | 967.357906 | *BMU34* | −0.0090 |
| *bmu34-bmu35* | −2.0557 | 0.474 | −0.0000 | −2.0557 | 965.2932 | *BMU35* | 965.301056 | *BMU35* | −0.0079 |
| *bmu35-bmu36* | −4.2722 | 0.520 | −0.0000 | −4.2722 | 961.0209 | *BMU36* | 961.027739 | *BMU36* | −0.0068 |
| *bmu36-bmu37A* | −3.9076 | 0.610 | −0.0001 | −3.9077 | 957.1133 | *BMU37A* | 955.727015 | *BMU37* | *NEW* |
| *bmu37A-bmu38* | 0.0007 | 1.621 | −0.0001 | 0.0006 | 957.1138 | *BMU38* | 957.120895 | *BMU38* | −0.0071 |
| *bmu38-bmu39* | 2.4574 | 0.613 | −0.0001 | 2.4573 | 959.5712 | *BMU39* | 959.574138 | *BMU39* | −0.0030 |
| *bmu39-bmu40A* | −1.7575 | 0.568 | −0.0000 | −1.7575 | 957.8136 | *BMU40A* | 956.11594 | *BMU40* | *NEW* |
| *bmu40A-bmu41A* | −3.9365 | 1.035 | −0.0001 | −3.9366 | 953.8770 | *BMU41A* | 952.853105 | *BMU41* | *NEW* |
| *bmu41A-Bm-gg* | 3.7854 | 0.760 | −0.0001 | −3.7853 | 957.6624 | *BM GG* | 957.6624 | *BM GG* | 0.0000 |
| *bmGG-bmu42* | −0.5390 | 1.108 | −0.0000 | −0.5390 | 957.1234 | *Bmu42* | 957.1225 | *Bmu42* | 0.0009 |
| *bmu42-bmu43* | −1.9477 | 1.325 | −0.0000 | −1.9477 | 955.1757 | *Bmu43* | 955.1726 | *Bmu43* | 0.0031 |

of recent geodetic campaigns in the region involved measurements of predominantly horizontal extension. It must be noted that GPS techniques do not compare in precision with the tedious levelling measurement where vertical deformation is of interest. The current work measures the vertical deformation across two magmatic segments relative to the adjacent rift floor.

Previous levelling measurements in a similar tectonic environment within the region were carried out in the Asal rift area at the western termination of the Gulf of Aden rift system. A levelling line crossing the Asal rift was measured in 1972 and remeasured in 1979 (Kasser *et al.* 1979) following a seismo-volcanic episode in 1978. Subsidence of up to 70 cm was found in the Asal floor while the flanks were uplifted by as much as 19 cm. On the other hand in Iceland, a sub-aerial part of the Mid-Atlantic Ridge, several levelling measurements were carried out at different sites (Tryggvason 1982) where vertical displacement of up to 7 mm on a fault in SW Iceland has been associated with earthquakes. At other sites, subsidence without extension has been observed.

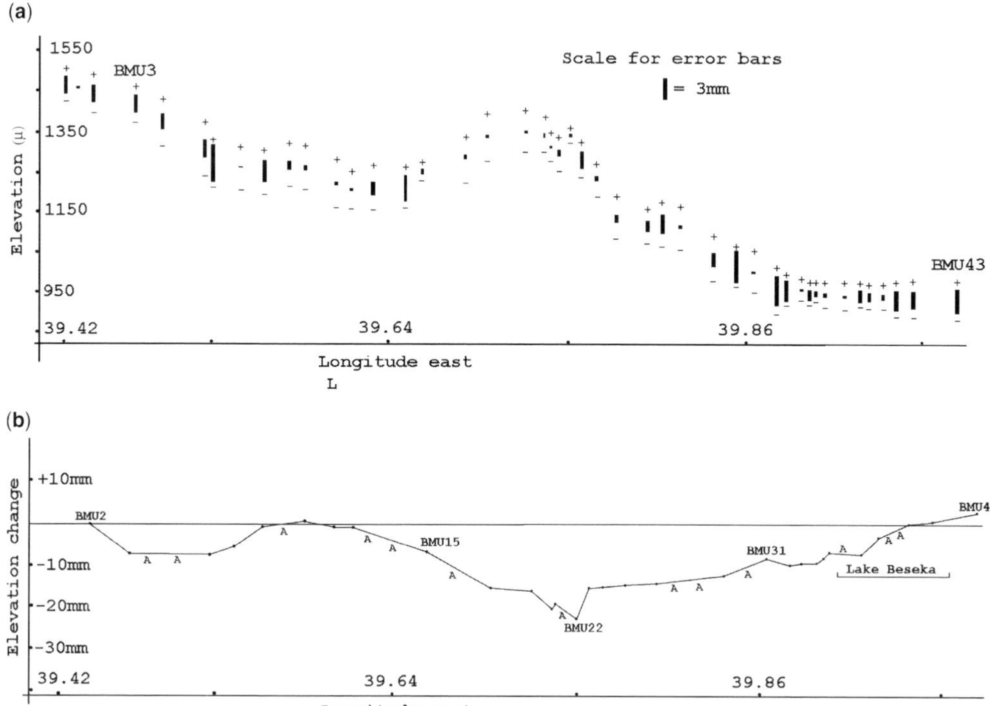


**Fig. 3.** Profiles: (**a**) closure error at benchmarks for the survey of 2003 indicated by bars according to scale shown. The corresponding (+) and (−) above and below the bars indicate the allowable error range for first order levelling having values of $\pm 4$ mm $\sqrt{K}$. The closure error for the 1995 survey is relatively low and is not shown here. Centres of bars correspond to elevations of benchmarks. (**b**) Elevation change profile, '**A**' stands for newly established benchmarks. A horizontal line segment shows the corresponding location of Lake Beseka.

From Table 1, the maximum vertical deformation of −22.4 mm corresponding to subsidence occurred at BMU22 near the Kone–Gariboldi volcanic complex (Figs 2 & 3b). The difference in elevation change in this part of the line (BMU15 to BMU31) is significant and merits an explanation. This is interpreted as a remnant effect of the subsidence that could be associated with withdrawal of magma at this volcanic complex in the recent past.

The Kone–Gariboldi volcanic complex has been described by Cole (1969) as a complex of at least three nested calderas. The basalt flow in the younger two calderas strongly resembles in appearance and freshness the historic flow at Fantale reported as having erupted in 1810 AD (Harris 1844).

On the other hand, as the results show, the observed subsidence attenuates towards Lake Beseka and reverses to small uplift east of the lake. The observed elevation increase of the lake amounting to 40 cm $a^{-1}$ (Berhanu 2001) cannot be explained in terms of tectonic-induced subsidence or uplift in the region. The amount of vertical deformation found for the lake segment (Fig. 3b) is between −6.8 mm and 0 mm in 8 years at the western and eastern sides of the lake respectively. This resolves the long-standing questions on the possible role of tectonics in the expansion of the lake.

The significance of this work lies in identifying the role of vertical deformation in the process of strain accommodation and localization as rifting proceeds at the magmatic segments in the axial part of the rift. Compared to the results of global plate kinematics and GPS surveys (Chu & Gordon 1999; Pan *et al.* 2002), the maximum vertical deformation rate in magmatic segments found in this study is of the same order of magnitude as that for horizontal extension.

The current work must be considered preliminary. In particular, the deformation measured is that of the magmatic segments relative to the adjacent rift floor. The next task should be to extend the line on both sides to stable points on the Nubian and Somalian plates so that the vertical deformation of the rift will be known relative to the 'rigid' plates.

The authors acknowledge funding by Addis Ababa University and partial support by the Ethiopian Mapping Authority. Our thanks go to C. Ebinger, R. Bendick and an anonymous referee for helpful comments on improving the paper.

## References

ASFAW, L.M. & KEBEDE, F. 1981. An earthquake swarm from the northern part of the Wonji Fault Belt and the state of seismicity of the rest of the Ethiopian Rift. *In*: WASSEF, A.M. (ed.) *Proceeding of the First International Conference on Crustal Movements in Africa*. UNECA, Addis Ababa, 86–103.

BERHANU, A. 2001. Geophysical investigation for lake level rise studies, northern part of Lake Beseka. M.Sc Thesis, Department of Geology and Geophysics, Addis Ababa University, 65 pp.

BILHAM, R., BENDICK, R., LARSON, K., MOHR, P., BRAUN, J., TESFAYE S. & ASFAW, L. 1999. Secular and tidal strain across the Main Ethiopian Rift. *Geophysical Research Letters*, **26**(18), 2789–2792.

BOCCALETTI, M., BONINI, M., MAZZUOLI, R., ABEBE, B., PICCARDI, L. & TORTORICI, L. 1998. Quaternary oblique extensional tectonics in the Ethiopian Rift (Horn of Africa). *Tectonophysics*, **287**, 97–116.

CHU, D. & GORDON, R. 1999. Evidence for motion between Nubian and Somalian plates along the southwest Indian ridge. *Nature*, **398**, 64–67.

COLE, J.W. 1969. Gariboldi volcanic complex, Ethiopia. *Bulletin Volcanologique*, **33**, 566–578.

CORNWELL, D.G., MACKENZIE, G., MAGUIRE, P.K.H., ENGLAND, R., ASFAW, L.M. & OLUMA, B. 2006. Northern Main Ethiopian Rift crustal structure from new high precision gravity data. *In*: YIRGU, G., EBINGER, C. & MAGUIRE, P.K.H. (eds) *The Afar Volcanic Province within the East African Rift System*. Geological Society, London, Special Publications, **259**, 307–321.

EBINGER, C.J. & CASEY, M. 2001. Continental breakup in magmatic provinces: An Ethiopian example. *Geological Society of America*, **29**(6), 527–530.

ETHIOPIAN GEODETIC SURVEY, 1957–1961. *Report on horizontal and vertical control surveys of the Blue Nile river basin*. US Department of Commerce Coast and Geodetic Survey, 563 pp.

HARRIS, W.C. 1844. *In*: *The Highlands of Ethiopia*, Vol. 3. Longman, Brown and Longman, London, Ch. 29 and 30.

KASSER, M., RUEGG, J.C., LEPINE, J.C. & TARANTOLA, A. 1979. Resultats des nouvelles measures geometriques sur le reseau de Djbouti implante en 1972 par l'Institut Geographique National. *Annales de Geophysique*, **35**(4), 171–176.

KERANEN, K., KLEMPERER, S.L., GLOAGUEN, R. & EAGLE GROUP, 2004. Three dimensional seismic imaging of a proto ridge axis in the Main Ethiopian Rift. *Geology*, **32**(11), 949–952.

KEIR, D., EBINGER, C., STUART, G., DALY, E. & AYELE, A. 2006. Strain accommodation by magmatism and faulting as rifting proceeds to breakup: Seismicity of the northern Ethiopian Rift. *Journal of Geophysical Research* (in press).

PAN, M., SJOBERG, L.E., ASFAW, L.M., ASENJO, E., ALEMU A. & HUNEGNAW, A. 2002. Analysis of the Ethiopian rift valley GPS campaigns in 1994 and 1999. *Journal of Geodynamics*, **33**, 333–343.

PIZZI, A., COLTORTI, M., ABEBE, B., DISPERATI, L., SACCHI, G. & SALVINI, R. 2006. The Wonji fault belt (Main Ethiopian Rift): Structural and geomorphological constraints and GPS monitoring, *In*: YIRGU, G., EBINGER, C. & MAGUIRE, P.K.H. (eds) *The Afar Volcanic Province within the East African Rift System*. Geological Society, London, Special Publications, **259**, 191–207.

TRYGGVASON, E. 1982. Recent ground deformation in continental and oceanic rift zones. *In*: PALMASON, G. (ed.) *Continental and Oceanic Rifts, AGU Geodynamics Series*, **8**, 17–29.

WILLIAMS, F.M., WILLIAMS, M.A.J. & AUMENTO, F. 2004. Tensional fissures and crustal extension in the northern part of the Main Ethiopian Rift. *Journal of African Earth Sciences*, **38**, 183–197.

# The Wonji fault belt (Main Ethiopian Rift): structural and geomorphological constraints and GPS monitoring

ALBERTO PIZZI[1], MAURO COLTORTI[2], BEKELE ABEBE[3], LEONARDO DISPERATI[4], GIORGIO SACCHI[2] & RICCARDO SALVINI[4]

[1]*Dipartimento di Scienze della Terra, Campus Universitario, Università di Chieti, Italy (e-mail: pizzi@unich.it)*

[2]*Dipartimento di Scienze della Terra, Università di Siena, Via di Laterina, 5-53100 Siena, Italy (e-mail: coltorti@unisi.it)*

[3]*Department of Earth Sciences, Addis Ababa University, Ethiopia (e-mail: bekelino@yahoo.com)*

[4]*Dipartimento di Scienze della Terra e Centro di Geotecnologie, Università di Siena, Via Vetri Vecchi, 34-52027 San Giovanni Valdarno (AR), Italy (e-mail: salvinir@unisi.it; disperati@unisi.it)*

**Abstract:** The Wonji Fault Belt (WFB), Main Ethiopian Rift, forms a network of faults oriented NNE–SSW with a Quaternary direction of extension oriented *c.* N95° E. Faults are spaced between 0.5 and 2 km, show a fresh steep scarp, recent activity and slip rates of up to 2.0 mm $a^{-1}$. This high value of deformation along the rift floor with respect to the plate separation rates suggests that most of the active strain could be accommodated by magma-induced faulting within the rift. However, the mountain front morphology associated with a displacement of 300–400 m since the Middle Pleistocene, tilted-blocks, brittle-seismic fault rock fabric and historical earthquakes with $M>6$ support a tectonic origin of the Asela boundary fault. Therefore, we propose a model that considers the possible coexistence of both magmatic deformation at the rift floor and brittle faulting at the rift margin. We also report the data relative to a GPS network installed in December 2004, along two transects across the WFB, between Asela and the Ziway Lake.

The Main Ethiopian Rift (MER) separates the African (Nubian) plate from the Somalian plate (Mohr 1967, 1987; Kazmin *et al.* 1980; Ebinger 1989, 2005; Jestin *et al.* 1994; Redfield *et al.* 2003; Wolfenden *et al.* 2004) (Fig. 1). The MER is bounded by discontinuous boundary fault systems striking between NNE–SSW in the south and NE–SW in the north. It is made by segments usually 15–25 km long, with a total displacement in correspondence of the main escarpment, west of Asela, of *c.* 400 metres (Abebe *et al.* 2005). At the latitude of Lake Ziway, *c.* 8° N, the Wonji fault belt (WFB) is located along the present eastern side of the rift (Mohr 1967, 1987; Boccaletti *et al.* 1999 and references therein).

The horizontal movements obtained with GPS measurements across the MER have been established in the order of between 3 to 6 mm $a^{-1}$ (Chu & Gordon 1999; Bilham *et al.* 1999; Walperdorf *et al.* 1999; Pan *et al.* 2002; Fernandez *et al.* 2004). The values obtained between the Nubian and Arabian plates, along the Red Sea and Somalian and Arabian plate, along the Gulf of Aden seem much higher (up to *c.* 18 mm $a^{-1}$; Izzeldin 1987; Redfield *et al.* 2003 and references within). Although the movements of the major plates have been investigated and presented in a series of articles, the behaviour of the local faults at plate margins has not been investigated in detail with GPS techniques. During December 2004, we installed a GPS network aimed to monitor movements along the WFB (Figs 1 & 2). This sector of the MER was chosen as the faults of the WFB are arranged in very closely spaced segments, and extremely high slip rates have been pointed out on some faults (Abebe *et al.* 1997, 2005; Boccaletti *et al.* 1999; Coltorti *et al.* 2002*c*). This paper describes work undertaken in 2004 and scientific problems that we wish to study in the future.

*From*: YIRGU, G., EBINGER, C.J. & MAGUIRE, P.K.H. (eds) 2006. *The Afar Volcanic Province within the East African Rift System*. Geological Society, London, Special Publications, **259**, 191–207.
0305-8719/06/$15.00 

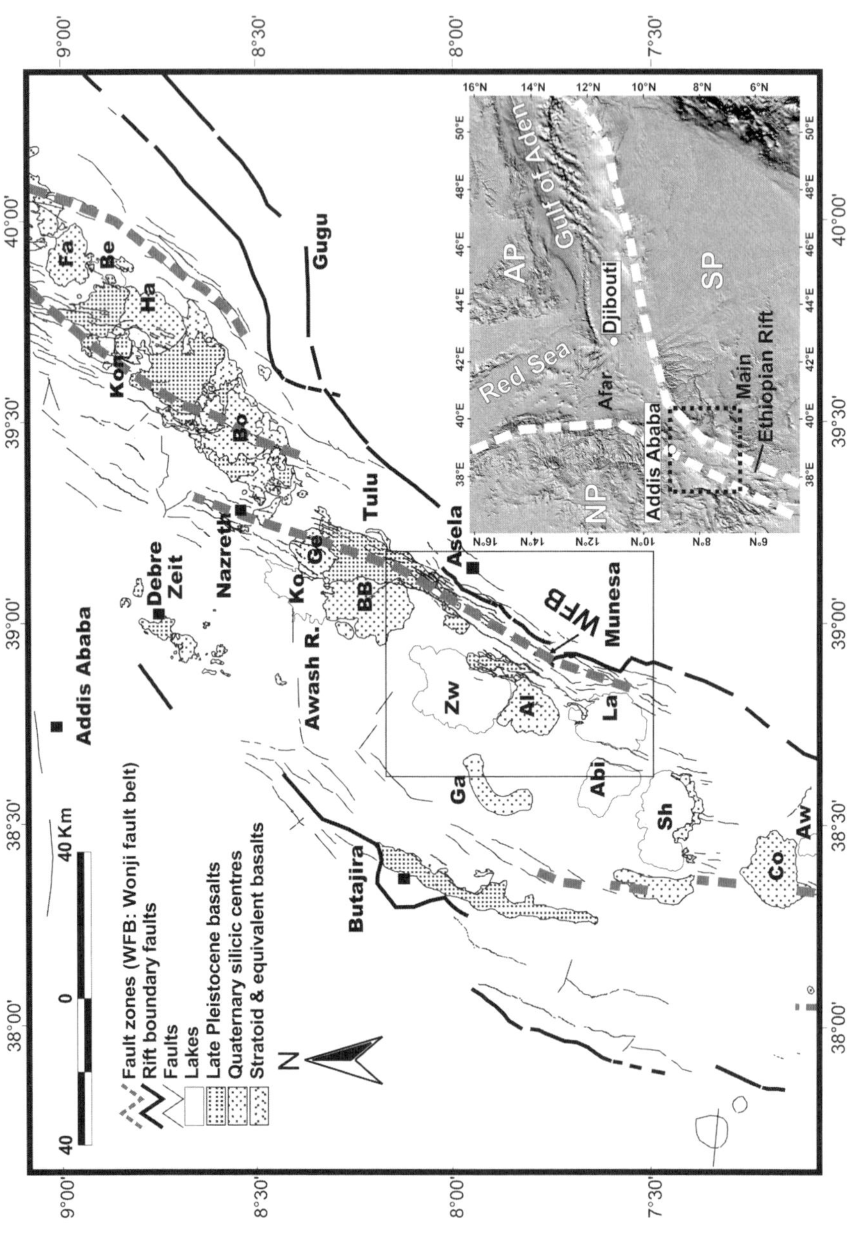


**Fig. 1.** Structural map of the northern sector of the Main Ethiopian Rift (MER) with major faults and location of study area (rectangle). Trend of fault zones includes the WFB that in the study area constitutes the eastern side of the MER Lakes. The inset map shows the plate tectonic configuration at the Afar triple junction. Artificially shaded topographic and bathymetric data (SRTM30 plus) with the light coming from the NW with a 60° inclination. Dashed line rectangle shows location of the structural map. Lakes: Aw, Awasa; Sh, Shala; Abi, Abijata; La, Langano; Zw, Zway; Ko, Koka; Be, Beseka. Volcanoes: Co, Corbetti; Ga, Gademota; Al, Aluto; BB, Bora Bericcio, Ge, Gedemsa; Bo, Bosetti; Ha, Hada; Kon, Kone; Fa, Fantale. NP, Nubian plate; SP, Somalian plate; AP, Arabian plate.

The new network was located along two transects that cross the main escarpment of the WFB to the north and south of Asela (Fig. 2). The transects have been chosen along two easily accessible roads that rise from Lake Ziway in the MER, to the top of the main escarpment at Asela.

In this paper, we present the results of an integrated geological and geomorphic investigation based on field survey data, aerial photographs and satellite imagery on the Late Quaternary geometry, kinematics and the slip-rates of the WFB. Pre-existing radiometric ages of the volcanic products, Pleistocene and Holocene lacustrine and fluvial deposits as well as geomorphic evidence provided age constraints on the recent deformation. In addition, we present the reference data of the GPS network that will be re-measured at the end of 2005.

## Geological background and timing constraints

The MER formed after the emplacement of the Ethiopian flood basalts (Trap), attributed to the

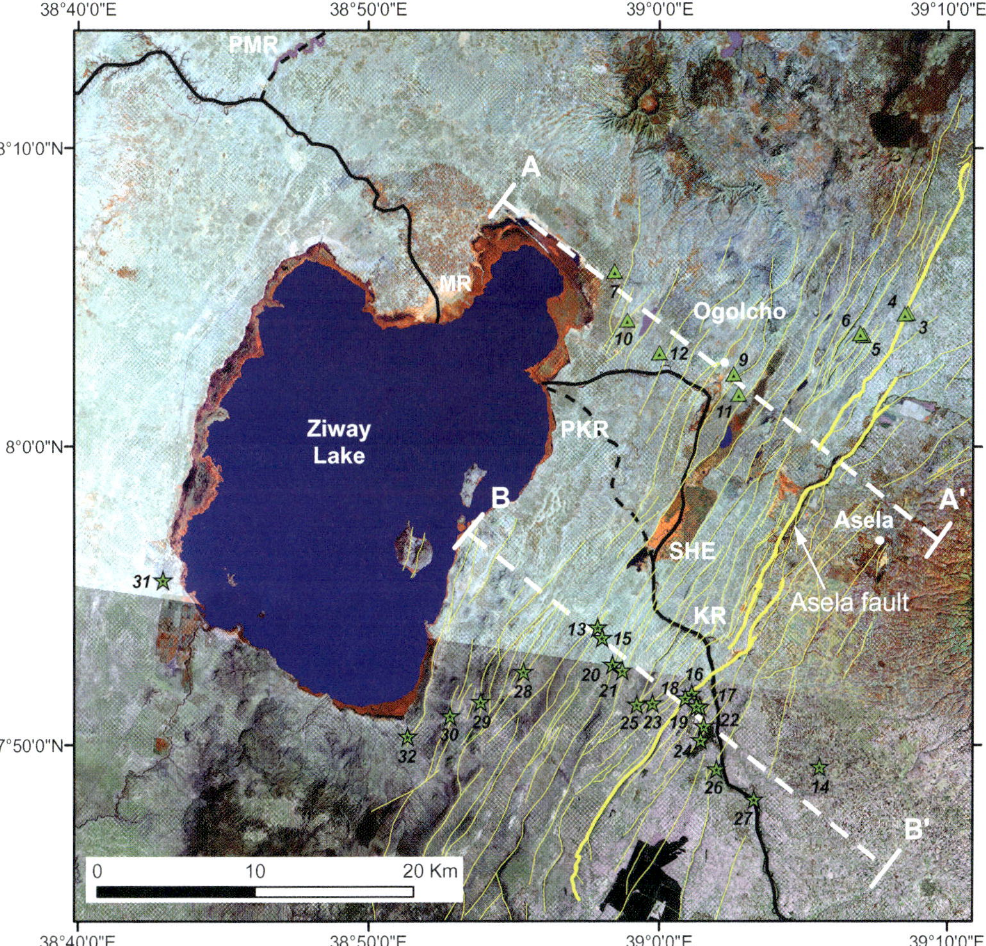


**Fig. 2.** Fault segments (thin yellow lines) constituting the WFB reported on the satellite image Landsat 7 ETM+ obtained with the overlapping of two scenes (168-055 dated 05.02.2000; 168-054 dated 05.12.2000); false colour composite 453 RGB; 30 m of spatial resolution. Green stars (southern transect), green triangles (northern transect) and related numeric labels (ID–# in Table 1) locate the GPS points established during the field survey in November–December 2004. Thick black line indicates the Katar River (KR) that in correspondence to the Shetemata Swamp (SHE) turns to the north and leaves a long-abandoned palaeocourse (thick black dots, PKR). Same symbols are used for the Meki River (MR) and the Palaeo-Meki (PMR). Thick yellow lines indicate the main escarpment along the eastern side of the WFB. The traces of sections A–A′ and B–B′ reported in Figure 3 are included.

timespan between 31 and 23 Ma (Chernet *et al.* 1998; Ayalew *et al.* 2002; Coulié *et al.* 2003; Redfield *et al.* 2003; Wolfenden *et al.* 2004; and references therein). Many authors hypothesized that these products were associated with the creation of a large topographic dome but no evidence has been reported. On the Highlands of Somalia, Yemen and Ethiopia, these Trap volcanic products lie over a planation surface that cuts: 1, alluvial plain and coastal marine sediments attributed to the Cretaceous–Eocene Amba Aradom formation in Ethiopia (Merla & Minucci 1938; Bosellini *et al.* 1997; Coltorti *et al.* 2006); 2, marine sandstones and conglomerates of the Medj–Zir Member in Yemen (Huchon *et al.* 1991). There are few constraints on the mechanism of uplift as data are scarce (Bonnefille *et al.* 1987; Hailemichael *et al.* 2002). Flood basalts do not fill tectonic depressions nor geomorphological features formed after uplift. We interpret this to show that uplift occurred after emplacement of the Trap basalts.

In the Afar depression, the oldest outcropping marine sediments have been attributed to the Early Miocene (Tiercelin *et al.* 1980) and are buried under the Lower Afar Stratoid Series (5.6–4.5 Ma; Chernet *et al.* 1998) and continental Plio-Pleistocene deposits (Renne *et al.* 1999). In the Adama basin, in the northwestern sector of the MER, the presence of tilted volcanic and volcanoclastic deposits (dated between *c.* 10 and 6 Ma) that are sealed by Pliocene deposits (*c.* 3–2 million years) led Wolfenden *et al.* (2004) to indicate that in this sector rifting started after 10 Ma. 30 km to the south, in the Lakes Region west of Asela, Quaternary extension is observed. On the eastern side of the MER, the bedrock is made up of aphyric basalts and basaltic agglomerates intercalated with fluvial deposits. A K–Ar age of 1.95 $\pm$ 0.3 Ma was obtained for the basalts along the main Asela escarpment (WoldeGabriel *et al.* 1990; Boccaletti *et al.* 1999; Abebe *et al.* 2005). These authors point out that basalts are overlain by pantelleritic and comenditic ignimbrites dated from 1.6 to 0.6 Ma. To the west of the main escarpment, the latter unit is recognized on the MER floor (west of Ogolcho) with a total vertical displacement of 500 m (Fig. 3). A series of fissural basaltic flows, cinder and spatter cones, locally displaced by the same set of NNE–SSW faults, were deposited on top of some horsts as well as inside many grabens. These volcanic products have been dated at 32–29 ka (WoldeGabriel *et al.* 1990; Abebe *et al.* 2005).

Lake sediments that crop out on the rift floor are used to study recent tectonic movements of the area. The Ziway, Langano, Abijata and Shalla Lakes are remnants of a much larger lake basin that reached its maximum size during the Late Pleistocene and Early Holocene (Grove *et al.* 1975; Gasse & Street 1978; Le Turdù *et al.* 1999; Benvenuti *et al.* 2002; Coltorti *et al.* 2002*a*). The lake evolution has been subdivided into four major phases: 1, megalake; 2, reduced lakes; 3, macrolake; and 4, separated lakes. Maximum expansion (megalake) was reached during the Interstadial Marine Isotope Stage (MIS) 3, between *c.* 45 and 30 ka (Le Turdù *et al.* 1999; Benvenuti *et al.* 2002; Coltorti *et al.* 2002*a*). During MIS 4, from *c.* 70 to 50 ka, the area was arid and aeolian sediments were deposited (Coltorti *et al.* 2002*a*). The Gademotta formation (Laury & Albritton 1975), made up of aeolian sediments and colluvial deposits, corresponds to this Stage but K/Ar dates obtained on two tephras (0.181 and 0.149 Ma) are unreliable. In fact, aeolian sediments containing similar Middle Stone Age tools, are covered without major unconformities by lacustrine deposits dated to the MIS 3 (Coltorti *et al.* 2002*a*) at the same stratigraphic position, in the Gademotta area and to the north of North Bay (Langano). During the Last Glacial Maximum (LGM), lakes largely dried up and evaporitic gypsum sediments were deposited inside small isolated ponds. The lake system expanded again during the Late Glacial Interstadials, particularly during Interstadial 1 (Dansgaard *et al.* 1993), and again during the Early Holocene. Grove *et al.* stated that the maximum Holocene lake level elevation, determined by a threshold generated by a palaeochannel of the Meki River (Fig. 2), is located a few metres above the present day thalweg that drained northwards to the Awash River. However, radiometric dates of soil and lacustrine deposits buried in the plain cut by this palaeochannel shows that the downcutting occurred during the Early Holocene, after 8 ka (Coltorti *et al.* 2002*a*). Before this time, the basin reached an elevation of 1720–1730 m, a similar elevation to the MIS 3. The Macrolake including Ziway, Langano, Abijata and Shalla Lakes formed after 5 ka and was soon followed by the present phase of separated lakes.

The displacement of the lake levels allows us to establish the minimum slip-rates of the faults in this sector. Along the WFB escarpments, and in particular slightly south of Asela, Holocene movements displaced a palaeochannel of the Katar River along a NNE–SSW oriented fault. This has caused formation of the Shetemata Swamp (SHE in Fig. 2) and diverted the new channel to the northern shore of the Ziway Lake (Coltorti *et al.* 2002*b*; Abebe *et al.* 2005). East of North Bay (Langano lake: see Fig. 1), a similarly oriented fault (Bole Fault) displaces a complex sequence of volcanic, fluvial and aeolian sediments deposited at the beginning of the Last Glaciation. In the hanging-wall

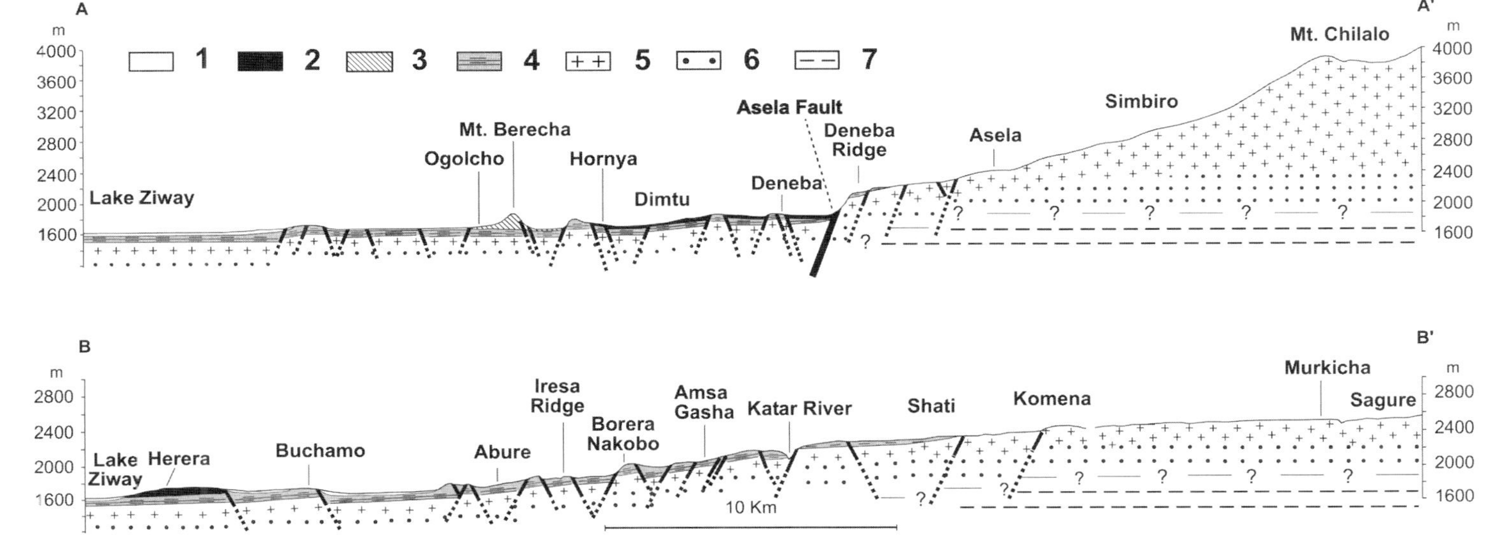


**Fig. 3.** Geological cross-section along the two transects west of Asela near to the traces of the GPS network. Location in Figure 1. These sediments do not crop out in the area but have been recognized in the eastern side of the Somalian plateau (Merla *et al.* 1973) and in a small horst on the western margin of the rift slightly south of the study area (Kella horst, Di Paola *et al.* 1993). 1, Lacustrine deposits; 2, Recent basaltic flows, scoria and lava cones (Late Quaternary); 3, Axial silicic centres (Quaternary volcanoes); 4, Ignimbrites dated 0.6 Ma; 5, Eastern margin volcanic products (Mt Chilalo ignimbrites and basalts: Pliocene–Early Quaternary?); 6, Trap (Tertiary); 7, Mesozoic sedimentary sequence.

block, Late Pleistocene and Holocene lacustrine sediments have been deposited. Therefore, very high slip rates, ranging from 0.5 mm $a^{-1}$ to 2.0 mm $a^{-1}$ have been suggested along these faults (Coltorti *et al.* 2002*c*; Abebe *et al.* 2005).

## Present-day kinematics

The zone of faulting in the MER seems to have narrowed from 60 km (since *c.* 12 Ma) to less than 20 km in the past 2 Ma (Ebinger & Casey 2001; Wolfenden *et al.* 2004). Within the MER, these 20 km-wide zones form right-stepping en echelon tectono-magmatic segments that correspond to the central and northern part of the WFB. The WFB is made up of closely spaced en echelon Quaternary faults and aligned cones oriented about N20°–30° E. These faults are slightly oblique to the NE–SW trend of the supposed Miocene border faults that bound the MER to the north of Asela (Wolfenden *et al.* 2004). Changes in the direction of extension and the role played by oblique/strike–slip kinematics have been outlined by several authors (e.g. Chorowicz *et al.* 1994; Boccaletti *et al.* 1999; Acocella & Korme, 2002; Korme *et al.* 2004), but the Late Quaternary kinematics of the northern MER is still debated. A dextral component of shear along the WFB, consistent with a *c.* NW–SE active extension, has been pointed out by Chorowicz *et al.* (1994) and recently by Acocella & Korme (2002, and reference therein) using field data. A left lateral component, consistent with a near E–W direction of Quaternary extension, has also been proposed (Mohr 1967; Gibson 1969; Bosworth *et al.* 1992; Coblentz & Sandiford 1994; Jestin *et al.* 1994; Abbate *et al.* 1995; Bonini *et al.* 1997; Boccaletti *et al.* 1999; Foster & Jackson 1998; Bilham *et al.* 1999; Chu & Gordon 1999; Ayele 2000; Wolfenden *et al.* 2004). There is also no consensus on what proportion of the strain is accommodated by magmatism and faulting in the MER–Afar rift zones. The tectonic history of the MER is matched by an intense volcanic activity up to historical times, characterized by a typical bimodal composition (e.g. WoldeGabriel *et al.* 1990). This bimodal composition gradually evolved to basalt-dominated volcanism which could indicate a period of increased extension (Boccaletti *et al.* 1999). Corti *et al.* (2003) and Mazzarini *et al.* (2004) propose that the development of the en echelon WFB segments favoured the uprising of melts and volcanic activity along the rift axis. On the other hand, active faulting along the WFB induced by dyking passing into normal faulting near the surface has been suggested on the basis of quantitative fault morphology and geophysical studies (Keranen *et al.* 2004; Ebinger 2005; Kendall *et al.* 2005; Kurz *et al.* 2005; Casey *et al.* 2006).

Our preliminary studies confirmed that Late Quaternary deformation between the eastern shores of Ziway and Langano Lakes and the margin of the eastern Ethiopian plateau was mainly accommodated by closely spaced NNE–SSW, and subordinate NE–SW and north–south trending faults (Figs 1 & 2).

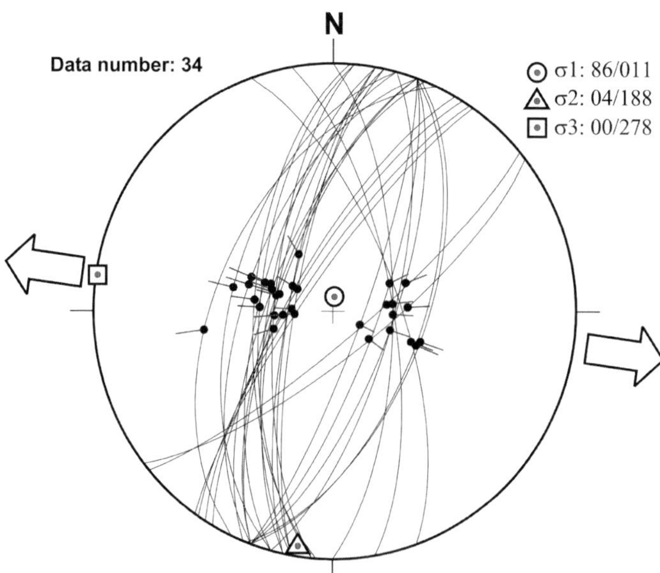


**Fig. 4.** Stereographic projection of the fault planes and associated slip vectors measured in the study area and palaeostress analysis (TENSOR program by D. Delvaux, 1993). See text for explanation.

These faults delimit a set of horsts, grabens and half-grabens with a typical spacing of 0.5–2 km (Fig. 3). The fact that geological and geomorphological fault displacements are similar suggests a very limited reworking by erosional processes and therefore recent activity of the fault system. Fault planes, both synthetic (west-dipping) and antithetic, show a maximum length of 20–30 km and are characterized by high-angle geometry with a mean dip of 65°–75°. In the cases where measurements of fault plane striations were possible, they consistently show normal/oblique kinematics (Fig. 4). Palaeostress analysis of fault-slip data affecting Quaternary terrain (i.e. fault plane striations and conjugate shear planes) revealed a sub-horizontal $\sigma 3$ axis oriented *c.* N95° E, which is consistent with the *c.* E–W Quaternary direction of extension. In particular, kinematic indicators observed along N30°–40° E oriented fault planes (which probably represent reactivated segments of pre-existing boundary faults along the eastern rift margin) showed a slight component of sinistral movements with pitch-angles $\geq$60°–70°, as already suggested by some authors (e.g. Boccaletti *et al.* 1999). This oblique component, however, is consistent with the evaluated N95° E axis of Late Quaternary extensional deformation as well as focal mechanisms interpretation (e.g. Ayele 2000).

## Mechanism of active faulting

Although much geological and geophysical evidence suggests a volcano-tectonic nature of deformation (Ebinger 2005; Kendall *et al.* 2005; Kurz *et al.* 2005; Casey *et al.* 2006), some doubts remain about the possible co-existence of magmatism and tectonic faulting occurring in a brittle-seismic extensional regime at the present stage of incipient break-up.

In this framework, the fault delimiting the main escarpment west of Asela (Figs 1–3) shows some atypical features for a magmatically induced normal fault. First, the Asela master fault (Fig. 2), one of the more external structures along the eastern side of the WFB, shows a clear mountain front morphology (Figs 5 & 6) with up to 300–

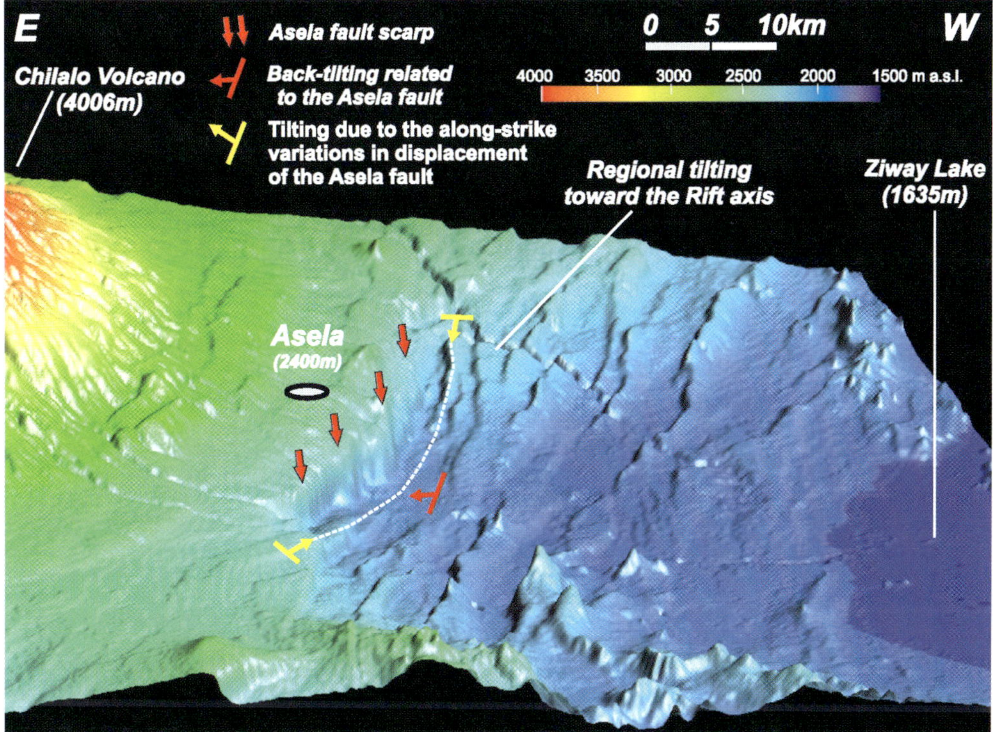


**Fig. 5.** Digital elevation model (SRTM data, at resolution of 3 arc sec.) of the area between eastern Lake Ziway and margin of the eastern Ethiopian plateau (*c.* 5× vertical exaggeration, Sun direction from the WNW and Sun angle = 45°). A clear mountain front morphology with up to 200–300 m of displacement characterizes the *c.* 20 km-long Asela fault scarp. Note back tilting associated with the downthrown block of the Asela fault whereas the regional tilting pattern of the faulted blocks is towards the rift axis. Yellow arrows indicate tilting due to along-strike variations in displacement of the Asela fault which affects a faulted block already tilted towards the rift axis (white dashed line points out hangingwall downthrown profile) (see text).

**Fig. 6.** Panoramic view of the central northern sector of the Asela fault. The fault scarp shows 'fresh' morphotectonic features suggesting its recent activity (see text). Photo is taken from Point 6 in Figure 2.

400 m of geological and topographical displacement delimiting the Ethiopian plateau for a total length of about 20 km.

Conversely, magmatically induced normal faults commonly present a low morphology, bilateral symmetry, offsets of a few tens of metres and a length up to 10 km (Mastin & Pollard 1988; Hackett *et al.* 1996). The downthrown block of the Asela fault is tilted back towards the rift border (i.e. towards the east) in the opposite direction to the riftward tilting pattern of the faulted blocks observed more to the west (Fig. 5). In particular, Figure 5 shows a west-tilted block within the hanging wall of the Asela fault deflected as a consequence of the along-strike variations in displacement of the fault (white dashed line in Fig. 5) and the back tilting (toward the rift border) at the central sector of the fault where displacement is maximum. The observed geometry differs from the tilted-block morphology of dyke-induced faults described by Gloaguen & Casey (2002). In particular, the Asela fault plane dips 75° W and has an orientation ranging from N10° E to N35° E observed both at the outcrop-scale (Fig. 7a, b) and at the map scale; such a pattern was probably controlled by the interaction of the newly formed N10° E faults with the pre-existing *c.* NE–SW oriented structures along the eastern rift margin. Kinematic indicators (Fig. 7c) measured along the major fault surface, as well as the orientation of minor conjugate fault planes within the fault zone, indicated a principal dip-slip normal movement providing an extension axis oriented E(SE)–W(NW), consistent with those evaluated for the Late Quaternary deformation in this sector.

Large landslides and coarse debris prevented the observation of dateable Quaternary deposits along the fault plane. Very well-preserved fault scarps and suspended valleys with waterfalls show very recent activity of the Asela fault (Figs 6 & 7a). The fault rock associated with the Asela fault is composed of cemented fault breccias *c.* 4 m thick with a typical grain size fining toward the slip plane which is affected by minor conjugate synthetic and antithetic structures (Fig. 7b). This fault rock fabric represents the typical fault zone in the seismogenic regime (Sibson 1977; Wise *et al.* 1984) and the ratio between the Asela fault zone

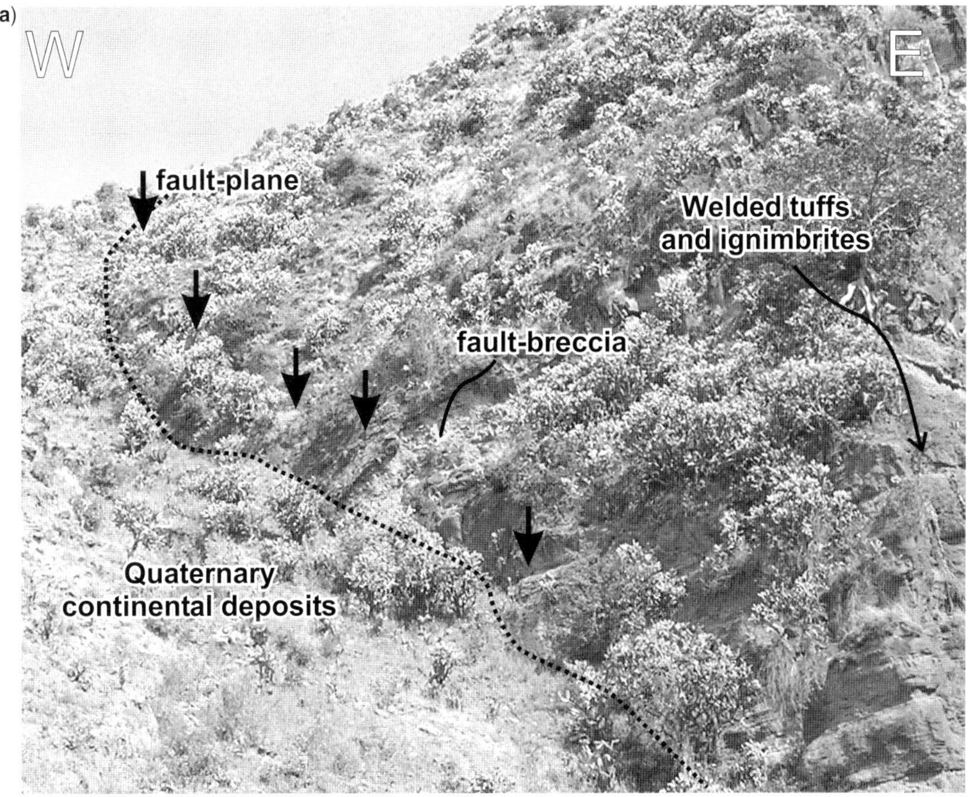


**Fig. 7.** (**a**) Panoramic view of the Asela fault scarp observed close to the Bolkesa waterfall.

(b)

(c)

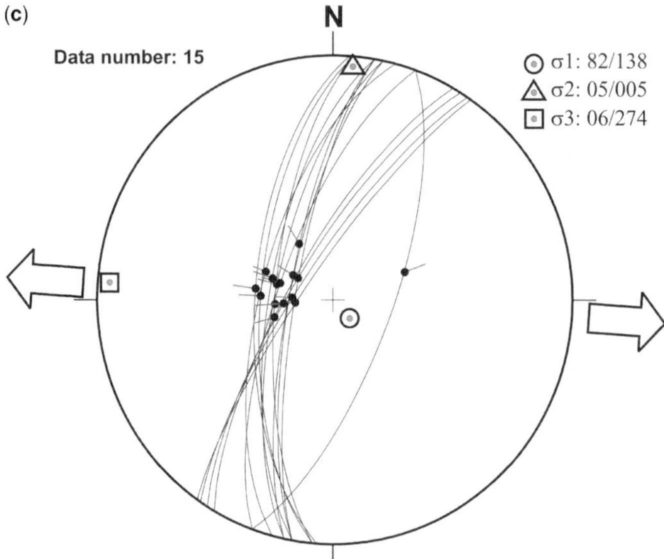


**Fig. 7.** *Continued.* (**b**) Details of the *c.* 4 m-thick cemented fault breccias associated with the Asela fault-rock observed at the Bolkesa waterfall. The fault displaces the ignimbrites (Asela unit) overlying the basal lava flows and is buried by coarse slope debris and alluvial fan deposits. The fault plane dips 75° W with an orientation ranging from N10° E to N35° E thick dotted-line: Asela fault slip-plane; thin dashed line: conjugate shear planes within the Asela fault zone; dotted–dashed line: bedding of the ignimbrites at the footwall block of the fault). Person at bottom left of the photo for scale.
(**c**) Stereographic projection and palaeostress analysis (TENSOR program by D. Delvaux, 1993) of the fault planes and kinematic indicators observed at the Bolkesa waterfall site (see text).

thickness (*c*. 4 m) and displacement (*c*. 400 m) is consistent with the range of value of 0.01–0.1 proposed for brittle fault zone growth through many rupture episodes (e.g. Robertson 1983; Scholz 1987). A brittle seismic origin of the Asela fault is also supported by the records of historical earthquakes in the region (e.g. the 1906 Langano and 1960 Awasa events) which have probably exceeded (magnitude) M6 (Gouin 1979; Ayele & Kulhanek 2000*a*; Kebede & van Eck 1997), whereas dyke-induced earthquakes in rift zones rarely exceed M5.5 (Hackett *et al.* 1996).

Fault rock fabrics and seismicity, however, cannot be unequivocally considered as a proof for the present-day tectonic activity of the Asela fault since numerous studies have shown that dyke injection can induce brittle failure and seismicity on pre-existing faults (e.g. Rubin & Gillard 1998) and because the 1906 and 1960 seismic events are very poorly constrained as they occurred prior to comprehensive instrumental recording in this region. Hofstetter & Beyth (2003) showed that there have been fewer than 20 earthquakes of $M > 4$ in the MER since 1960 and, based on the seismic moment release deficit, suggested that at least 50% of strain in the MER is accommodated aseismically. Following these authors and Keir *et al.* (2005, and reference therein), we believe that most of the strain in the MER can be accommodated aseismically by magma-induced faulting through the activation of closely spaced fault segments along the rift floor, whereas a brittle seismic behaviour could be still acting at the rift margin (Fig. 8). The magma-induced faulting hypothesis is consistent with the very high value of Late Quaternary slip rates (up to 2.0 mm $a^{-1}$) evaluated for some of the densely distributed faults between the Asela fault and Ziway Lake (Abebe *et al.* 2005). In fact, these faults rarely show evidence for brittle fault zones and slip planes, even though they have comparable length and similar fault-block lithology with respect to the Asela fault, and can hence be attributed to magmatic process. Tectonic active faulting localized at the rift margin is probably controlled at depth (9–14 km) by *c*. NE–SW pre-existing basement structures as suggested by geological and seismological evidence (e.g. Foster & Jackson 1998; Ayele 2000; Korme *et al.* 2004).

## The GPS network

GPS investigations within the MER have been carried out by Bilham *et al.* (1999) and Pan *et al.* (2002) (Fig. 9). The results are applicable to a short period only but point out the major problems that need to be solved to understand the present-day MER kinematics. In particular, a SE-directed horizontal movement of the Somalian plateau relative to the Ethiopian plateau has been ascertained in the order of 2.3–3.3 mm $a^{-1}$ from point measurements of the network E2 located out of the MER (Fig. 10 redrawn from Pan *et al.* 2002). Moreover, measurements that take the Addis Ababa station as a reference point, compared to those farther east,

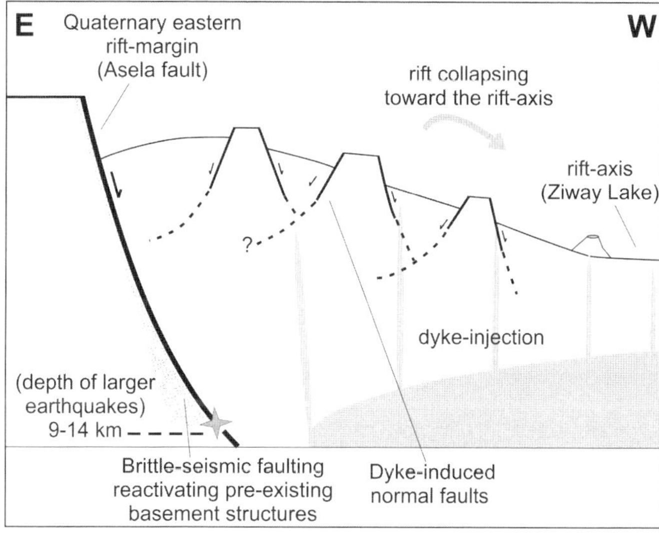


**Fig. 8.** Sketch (not to scale) of the proposed model showing the possible coexistence of brittle seismogenic faulting at the eastern rift margin (e.g. Asela fault) and dyke-induced normal faults along the rift floor (exaggerated topography).

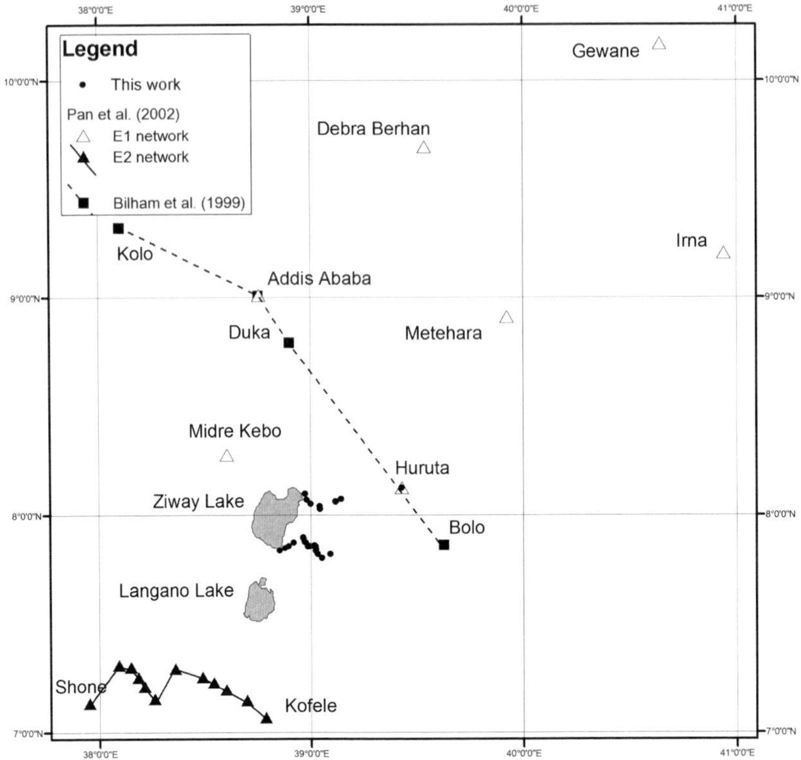


**Fig. 9.** The GPS network realized in the rift area by Bilham *et al.* (1999), Pan *et al.* (2002) and in this work. The position of the Bilham network and E2 transect of Pan *et al.* (2002) is approximate because of the uncertainties of the resolution of some of their data.

suggest that the eastern part of the Ethiopian plateau may be affected by active extension. The results obtained along a transect normal to the MER, located approximately 100 km to the south of the study area (transect E2 in Pan *et al.* 2002; Fig. 10), show that, on the western side, survey points are moving to the southeast at twice the rates relative to the station BM24 located on the African plate (6.1–6.7 mm $a^{-1}$ vs. 3.3 mm $a^{-1}$) while in the eastern side points move to the southwest faster (4.7–9.3 mm $a^{-1}$ vs. 2.3–3.1 mm $a^{-1}$) than the Somalian plate displacement rate. Hence, higher rates are recognized inside the MER than outside (rates of plates separation) and displacement vectors on the Ethiopian and Somalia plateau point toward the rift axis.

The occurrence of blocks tilted 20°–30° toward the rift axis has been used by Coltorti *et al.* (2002*b*) to suggest the existence of a series of sliding surfaces that could be connected to a collapse of the rift margin. In this model many incompetent layers (i.e. the clayey part of the Amba Aradom formation or the Mesozoic shales) may act as a décollement surface and could account for the above GPS data. These formations have been observed east of the eastern side of the Rift (Merla *et al.* 1973) as well as along the western rift border, in the Kella horst (Di Paola *et al.* 1993).

With regard to vertical movements, repeatability of the results from Pan *et al.* (2002) are much worse than horizontal co-ordinates, so the magnitudes of vertical movements are similar or smaller than the measurement accuracy threshold. A new GPS campaign along the network E2 (Pan *et al.* 2002) about 11 years after the first campaign could probably provide useful data of vertical movements.

The new GPS network (Fig. 2) was set up in order to try to solve those uncertainties left by and/or arising from the old GPS measurements. Namely, the new GPS network was addressed to monitor ground deformation within the MER, instead of providing results useful at the plate scale. Assuming that the main ground deformation is accommodated in this region through normal faults associated to the WFB, the network was arranged taking into account the fault pattern characterizing this sector of the MER. Hence, by monitoring the network,

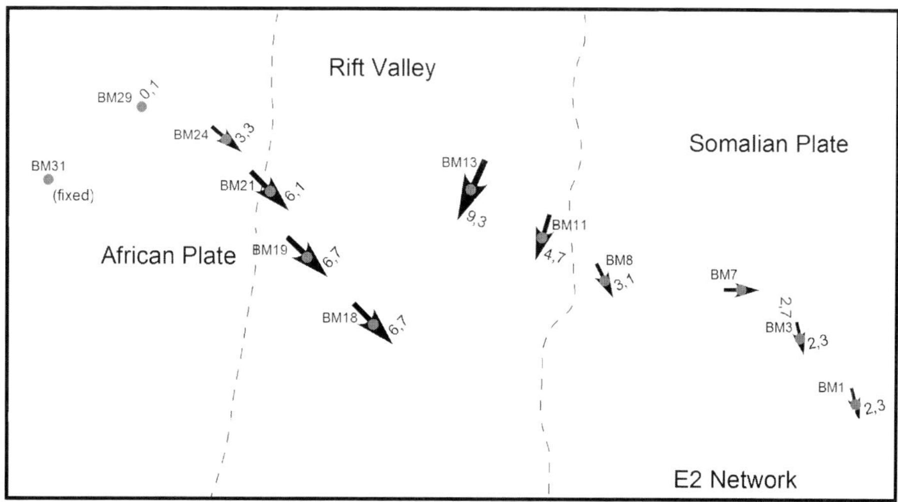


**Fig. 10.** Displacement rates (mm $a^{-1}$) of GPS stations of E2 network between 1994 and 1999 (redrawn from Pan *et al.* 1999). Arrow length is scaled to displacement rates and arrow direction shows direction of displacement. See location in Figure 9.

quantitative information about slip rates and kinematics of the WFB will be acquired.

The GPS network is made up of two transects crossing the WFB as represented in Fig. 2. The southern transect consists of 20 survey points, while the northern transect consists of nine (see Table 1 for geographic coordinates of survey points).

Survey points were positioned using survey marks made up of brass expansion anchors and pins fixed through epoxy resin set within holes drilled in competent bedrock. They were located in the immediate vicinity of main faults previously recognised within the WFB by means of aerial photograph and satellite image interpretation and fieldwork. Where possible, couples of survey points were positioned on both the hangingwall and footwall of faults in order to monitor relative displacements between the base and the top of steep escarpments. While the procedure was practical and quick on top of the main horsts, location of suitable survey points within the grabens was harder because of the occurrence of soft sediments and weathering products.

GPS measurements were performed in static arrangement by means of three Leica SR 530 receivers. The reference point used during the survey campaign was located near Ziway Lake. Other points of the network were occupied simultaneously during different measurement sessions in order to estimate repeatability and internal precision. Data were recorded at 5 s intervals for three to four hours per session. Baselines connected to the reference point

**Table 1.** *Geographic coordinates (WGS 84) of the GPS points established during the field survey in December 2004. Numeric labels reporting the ID-# item of Table 1 are represented in Figure 2.*

| ID-# | Longitude | Latitude |
|---|---|---|
| 3 | 39° 08′ 36″ E | 8° 04′ 28″ N |
| 4 | 36° 08′ 30″ E | 8° 04′ 26″ N |
| 5 | 39° 07′ 03″ E | 8° 03′ 43″ N |
| 6 | 39° 06′ 56″ E | 8° 03′ 47″ N |
| 7 | 38° 58′ 26″ E | 8° 05′ 52″ N |
| 9 | 39° 02′ 32″ E | 8° 02′ 26″ N |
| 10 | 38° 58′ 52″ E | 8° 04′ 14″ N |
| 11 | 39° 02′ 43″ E | 8° 01′ 45″ N |
| 12 | 39° 00′ 00″ E | 8° 03′ 08″ N |
| 13 | 38° 57′ 52″ E | 7° 53′ 56″ N |
| 14 | 39° 05′ 30″ E | 7° 49′ 16″ N |
| 15 | 38° 58′ 01″ E | 7° 53′ 35″ N |
| 16 | 39° 01′ 07″ E | 7° 51′ 44″ N |
| 17 | 39° 01′ 24″ E | 7° 51′ 18″ N |
| 18 | 39° 00′ 54″ E | 7° 51′ 34″ N |
| 19 | 39° 01′ 17″ E | 7° 51′ 19″ N |
| 20 | 38° 58′ 22″ E | 7° 52′ 42″ N |
| 21 | 38° 58′ 42″ E | 7° 52′ 30″ N |
| 22 | 39° 01′ 32″ E | 7° 50′ 39″ N |
| 23 | 38° 59′ 45″ E | 7° 51′ 25″ N |
| 24 | 39° 01′ 25″ E | 7° 50′ 10″ N |
| 25 | 38° 59′ 13″ E | 7° 51′ 22″ N |
| 26 | 39° 01′ 58″ E | 7° 49′ 12″ N |
| 27 | 39° 03′ 14″ E | 7° 48′ 11″ N |
| 28 | 38° 55′ 18″ E | 7° 52′ 28″ N |
| 29 | 38° 53′ 50″ E | 7° 51′ 27″ N |
| 30 | 38° 52′ 48″ E | 7° 50′ 59″ N |
| 31 | 38° 42′ 56″ E | 7° 55′ 31″ N |
| 32 | 38° 51′ 20″ E | 7° 50′ 17″ N |

range in distance between 18 km to 50 km, while other baselines are generally shorter than 5 km.

In order to allow an integration with existing GPS data, the network was also connected to the GPS network described in Pan *et al.* (2002) by occupying the Addis Ababa and Furi stations that are located to the west of the western border of the MER.

## Conclusions

The integrated geological and geomorphological investigations allow us to characterize the WFB in the eastern side of the Lake Region. The WFB forms a close network of faults, spaced between 0.5 and 2 km, with a very fresh steep fault scarp, evidence of recent activity and very high slip rates up to 2.0 mm $a^{-1}$. Structural analysis along Quaternary fault segments of the WFB showed normal/oblique kinematics and revealed a Late Quaternary direction of extension oriented *c*. N95° E. Based on the very high values of Late Quaternary slip rates and geological–structural evidence along the WFB, we suggest that most of the active strain could be accommodated aseismically in the MER floor by magma-induced faulting, in agreement with Hofstetter & Beyth (2003), Kendall *et al.* (2005) and Keir *et al.* (2006).

However, some of the features observed along the rift margin (i.e. Asela fault) such as: (i) clear mountain front morphology; (ii) displacement since the beginning of middle Pleistocene of 300–400 m; (iii) fault length of *c*. 20 km; (iv) tilted-block morphology; (v) brittle–seismic fault rock fabric; and possibly the occurrence of historical earthquakes in the region with $M>6$ support a tectonic origin of this boundary fault. Therefore, we suggest that the present-day coexistence of both magmatic deformation at the rift floor and brittle faulting at the rift margin cannot be excluded.

To explain the anomalous values of horizontal movements recognized by Pan *et al.* (2002) in the rift floor with respect to its margins, we also suggested the possible existence of shallow detachment levels, within the pre-rift sequence (Coltorti *et al.* 2002*c*), that may decouple the uppermost layers of the MER.

During December 2004, we installed a GPS network along two transects across the main faults of the WFB, between Asela and Ziway Lake. The network was connected to the existing GPS network of Pan *et al.* (2002) and would represent a powerful tool for the evaluation of ground displacement rates across the dense set of faults that separate the bottom of the MER from the Somalia plateau. Accurate monitoring of the movements will provide valuable data to understand extensional dynamics such as fault geometry and the important phenomenon of block tilting within the MER. The data will help to discriminate between 'deep-rooted' brittle faulting and shallower magma-induced deformation in the upper crust of the MER and to decipher how the transition between tectonic to magma-induced faulting is taking place in this zone of incipient break-up.

Financial supports from Siena University (PAR) 2003, responsible Prof. Mauro Coltorti and Ministry of University and Research (PRIN) 2003, Chieti Unit, responsible Prof. Alberto Pizzi. Many thanks to Derek Keir and Bernard Le Gall for providing useful suggestions during their revisions.

## References

Abbate, E., Passerini, P. & Zan, L. 1995. Strike-slip faults in a rift area: a transect in the Afar Triangle, East Africa. *Tectonophysics*, **241**, 67–97.

Abebe, B., Boccaletti, M., Bonini, M., Mazzuoli, R., Piccardi, L., Tortorici, L. & Trua, T. 1997. *Geological Map of Asela Region*. S.E.L.C.A., Florence, Italy.

Abebe, B., Coltorti, M. & Pizzi, A. 2005. Rates of Late Quaternary deformation along the Wonji fault belt in the Lakes Region, Main Ethiopian Rift. *Rendiconti Società Geologica Italiana*, **1**, 41–43.

Acocella, V. & Korme, T. 2002. Holocene extension direction along the Main Ethiopian Rift, Africa. *Terra Nova*, **14**, 191–197.

Ayele, A. 2000. Normal left-oblique fault mechanisms as an indication of sinistral deformation between the Nubia and Somalia plates in the Main Ethiopian Rift. *Journal of African Earth Sciences*, **31**, 359–367.

Ayele, A. & Kulhanek, O. 2000. Reassessment of source parameters for three major earthquakes in the East African rift system from historical seismograms and bulletins. *Annali di Geofisica*, **43**, 81–94.

Ayalew, D., Barbey, P., Marty, B., Reisberg, L., Yirgu, G. & Pik, R. 2002. Source, genesis and timing of giant ignimbrite deposits associated with Ethiopian continental flood basalts. *Geochimica et Cosmochimica Acta*, **66**, 1429–1448.

Benvenuti, M., Carnicelli, S., *et al.* 2002. The Ziway–Shala Lake Basin (Main Ethiopian Rift, Ethiopia): a revision of basin evolution with special reference to the Late Quaternary. *Journal of African Earth Sciences*, **35**, 247–269.

Bigazzi, B., Bonadonna, F.P., Di Paola, G.M. & Giuliani, A. 1993. K–Ar and fission-track ages of the last volcano tectonic phase in the Ethiopian Rift Valley (Tullu Moyè area). *Geology and mineral resources of Somalia and surrounding regions*, **113**, 311–322.

Bilham, R., Bendick, R., Larson, K., Braun, J., Tesfaye, S., Mohr, P. & Asfaw, L. 1999.

Secular and tidal strain across the Ethiopian Rift. *Geophysical Research Letters*, **27**, 2789–2793.

BOCCALETTI, M., MAZZUOLI, R., BONINI, M., TRUA, T. & ABEBE, B. 1999. Plio-Quaternary volcanotectonic activity in the northern sector of the Main Ethiopian Rift: relationships with oblique rifting. *Journal of African Earth Science*, **29**, 679–698.

BONINI, M., SOURIOT, T., BOCCALETTI, M. & BRUN, J.P. 1997. Successive orthogonal and oblique extension episodes in a rift zone: Laboratory experiments with application to the Ethiopian Rift. *Tectonics*, **16**, 347–362.

BONNEFILLE, R., VINCENS, A. & BUCHET, G. 1987. Palaeolimnology, stratigraphy and paleoenvironment of a Pliocene hominid site (2.9–3.3 m.y.) at Hadar, Ethiopia. *Palaeogeography, Palaeoclimatology, Palaeoecology*, **3**(60), 249–281.

BOSELLINI, A., RUSSO, A., FANTOZZI, P.L., ASSEFA, G. & SOLOMON, T. 1997. The Mesozoic succession of the Mekele outlier (Tigré Province, Ethiopia). *Memorie di Scienze Geologische*, **9**, 95–116.

BOSWORTH, W., STRECKER, M.R. & BLISNIUK, P.M. 1992. Integration of East African paleostress and present-day stress data: implications for continental stress field dynamics. *Journal of Geophysical Research*, **97**, 11851–11865.

CASEY, M., EBINGER, C., KEIR, D., GLOAGUEN, R. & MOHAMED, F. 2006. Strain accomodation in transitional rifts: extension by magma intrusion and faulting in Ethiopian rift magmatic segments. *In*: YIRGU, G., EBINGER, C.J. & MAGUIRE, P.K.H. (eds) *The Afar Volcanic Province within the East African Rift System*. Geological Society, London, Special Publications, **259**, 143–163.

CHERNET, T., HART, W.K., ARONSON, J.L. & WALTER, R.C. 1998. New age constraints on the timing of volcanism and tectonism in the northern Main Ethiopia Rift–southern Afar transition zone (Ethiopia). *Journal of Volcanology and Geothermal Research*, **80**, 267–280.

CHOROWICZ, J., COLLET, B., BONAVIA, F.F. & KORME, T. 1994. Northwest to north–northwest extension direction in the Ethiopian Rift deduced from the orientation of extension structures and fault-slip analysis. *Bulletin of the Geological Society of America*, **150**, 1560–1570.

CHU, D. & GORDON, R.G. 1999. Evidence for motion between Nubia and Somalia along the Southwest Indian Ridge. *Nature*, **398**, 64–67.

COBLENTZ, D.D. & SANDIFORD, M. 1994. Tectonic stresses in the African plate: constraints on the ambient lithospheric stress state. *Geology*, **22**, 831–834.

COLTORTI, M., CORBO, L. & SACCHI, G. 2002*a*. New evidence for the Late Pleistocene and Holocene climatic changes in the Lake Region. Proceedings of the Symposium: IAG International Geomorphologist Association. *In*: DRAMIS, F. (ed.) '*Climate Changes, Active Tectonics and Related Geomorpologic Effects in High Mountain Belts and Plateaux*', Addis Ababa, 9–10.12.2002, 30–35.

COLTORTI, M., PIZZI, A., CORBO, L. & SACCHI, G. 2002*b*. The gravitational collapse along the margin of the Ethiopian plateau near Asela (Ethiopia). Proceedings of the Symposium: IAG International Geomorphologist Association. *In*: DRAMIS, F. (ed.) '*Climate Changes, Active Tectonics and Related Geomorpologic Effects in High Mountain Belts and Plateaux*', Addis Ababa, 9–10.12.2002, 10–14.

COLTORTI, M., PIZZI, A., CORBO, L. & SACCHI, G. 2002*c*. Fault activity, river captures and delta-growth on the eastern side of the Ziway lake (Ethiopia). Proceedings of the Symposium: IAG International Geomorphologist Association. *In*: DRAMIS, F. (ed.) '*Climate Changes, Active Tectonics and Related Geomorpologic Effects in High Mountain Belts and Plateaux*', Addis Ababa, 9–10.12.2002, 15–19.

COLTORTI, M., DRAMIS, F. & OLLIER, C.D. 2006. Planation surfaces in Northern Ethiopia. *Geomorphology*, (in press).

CORTI, G., BOVINI, M., INNOCENTI, F., MANETTI, P., MAZZARINI, F. & ABEBE, T. 2003. Volcano-tectonic evolution of the main ethiopian rift. *Geophysical Research Abstracts*, **5**, 09405.

COULIÉ, E., QUIDELLEUR, X., GILLOT, P.Y., COURTILLOT, V., LEFÈVRE, J.C. & CHIESA, S. 2003. Comparative K–Ar and Ar/Ar dating of Ethiopian and Yemenite Oligocene volcanism: implications for timing and duration of the Ethiopian Traps. *Earth and Planetary Science Letters*, **206**, 477–492.

DANSGAARD, W., JOHNSEN, S.J., *ET AL.* 1993. Evidence for general instability of past climate from a 250 kyr ice-core record. *Nature*, **364**, 218–220.

DELVAUX, D. 1993. Tensor program for paleostress reconstruction: examples from the East African and the Baikal Rift Zones. Abstract supplement 1 to *Terranova*, **5**, 216.

DI PAOLA, G.M., BERHE, S.M. & ARNO, V. 1993. The Kella Horst: its origin and significance in crustal attenuation and magmatic processes in the Ethiopian rift valley. *In*: *Geology and Mineral Resources of Somalia and Surrounding Regions*, Istituto Agronomico Oltremare, Firenze, Relazioni e Monografie, **113**, 323–338.

EBINGER, C.J. 1989. Tectonic development of the western branch of the east African rift system. *Geological Society of America Bulletin*, **101**, 885–903.

EBINGER, C.J. 2005. Continental breakup: The East African perspective. *Astronomy and Geophysics*, **46**, 2.16–2.21.

EBINGER, C.J. & CASEY, M. 2001. Continental breakup in magmatic provinces: An Ethiopian example. *Geology*, **29**, 527–530.

FERNANDEZ, R.M.S., AMBROSIUS, B.A.C., NOOMEN, R., BASTOS, L., COMBRINCK, L., MIRANDA, J.M. & SPAKMAN, W. 2004. Angular velocities of Nubia and Somalia from continuous GPS data: implications on present-day relative kinematics. *Earth and Planetary Science Letters*, **222**, 197–208.

FOSTER, A.N. & JACKSON, J.A. 1998. Source parameters of large African earthquakes: implications for crustal rheology and regional kinematics. *Geophysics Journal International*, **134**, 422–448.

GASSE, F. & STREET, F.A. 1978. Late Quaternary lake-level fluctuations and environments of the northern Rift Valley and Afar region (Ethiopia and Djibouti). *Palaeogeography, Palaeoclimatology, Palaeoecology*, **24**, 279–325.

GIBSON, I. 1969. The structure and volcanic geology of an axial portion of the Main Ethiopian Rift. *Tectonophysics*, **8**, 561–565.

GROVE, A.T., STREET, F.A. & GOUDIE, A. 1975. Former lake levels and climatic change in the Rift Valley of southern Ethiopia. *The Geographical Journal*, **141**, 177–202.

GLOAGUEN, R. & CASEY, M. 2002. Dyke intrusion inferred from the morphology of induced faults in the Quaternary magmatic segments, Ethiopia. *AGU Fall Meeting Abstracts*, A1123.

GOUIN, P. 1979. *Earthquake History of Ethiopia and the Horn of Africa*. International Development Research Centre, Ottawa, Ontario.

HACKETT, W.R., JACKSON, S.M. & SMITH, R.P. 1996. Paleoseismology of volcanic environments. *In*: MCCALPIN, J.P. (ed.) *Paleoseismology*. International Geophysics Series, **62**, 147–181.

HAILEMICHAEL, M., ARONSON, J., SAVIN, S., TEVESZ, M. & CARTER, J. 2002. $\delta^{18}O$ in mollusk from Pliocene Lake Hadar and modern Ethiopian Lakes: implications history of the Ethiopian monsoon. *Palaeogeography, Palaeoclimatology, Palaeoecology*, **186**, 81–99.

HOFSTETTER, A. & BEYTH, M. 2003. The Afar depression: interpretation of the 1960–2000 earthquakes. *Geophysical Journal International*, **155**, 2, 715–732.

HUCHON, P., JESTIN, F., CANTAGREL, J.M., GAULIER, J.M., AL KHIRBASH, S. & GAFANEH, A. 1991. Extensional deformations in Yemen since Oligocene and the Africa–Arabia–Somalia triple junction. *Annales Tectonicae*, **2**, 141–162.

JESTIN, F., HUCHON, P. & GAULIER, J.M. 1994. The Somalia plate and the East African Rift system: present-day kinematics. *Geophysical Journal International*, **116**, 637–654.

KAZMIN, V., BERHE, S.F., NICOLETTI, M. & PETRUCCIANI, C. 1980. Evolution of the Northern part of the Ethiopian rift. *Atti Convegni Lincei*, **47**: 275–292, Rome.

KEBEDE, F. & VAN ECK, T. 1997. Probabilistic seismic hazard assessment for the Horn of Africa based on seismotectonic regionalisation. *Tectonophysics*, **270**, 221–237.

KEIR, D., EBINGER, C., STUART, G., DALY, E. & AYELE, A. 2006. Strain accommodation by magmatism and faulting as rifting proceeds to break-up: Seismicity of the northern Ethiopian Rift. *Journal of Geophysical Research* (in press).

KENDALL, J.M., STUART, G., EBINGER, C., BASTOW, I. & KEIR, D. 2005. Magma-assisted rifting in Ethiopia. *Nature*, **433**, 146–148.

KERANEN, L., KLEMPERER, S., GLOAGUEN, R. & EAGLE WORKING GROUP 2004. Imaging a proto-ridge axis in the Main Ethiopian rift. *Geology*, **39**, 949–952.

KORME, T., ACOCELLA, V. & ABEBE, B. 2004. The role of pre-existing structures in the origin, propagation and architecture of faults in the Main Ethiopian rift. *Gondwana Research*, **7**, 467–479.

KURZ, T., GLOAGUEN, R., EBINGER, C., CASEY, M. & ABEBE, B. 2005. Deformation distribution and type in the Main Ethiopian Rift (MER); a remote sensing study. *Journal of African Earth Sciences* (in press).

IZZELDIN, A.Y. 1987. Seismicity, gravity and and magnetic surveys in the central part of the Red Sea: their interpretation and implication for the structure and evolution of the Red Sea. *Tectonophysics*, **143**, 269–306.

LAURY, R.L. & ALBRITTON, C.C. 1975. Geology of Middle Stone Age Archaeological Sites in the Main Ethiopian Rift Valley. *Geological Society of America Bulletin*, **86**, 999–1011.

LE TURDU, C., HERCELIN, J.-J. ET AL. 1999. The Ziway–Shala lake basin system, Main Ethiopian Rift: influence of volcanism, tectonics, and climatic forcing on basin formation and sedimentation. *Palaeogeography, Palaeoclimatology, Palaeoecology*, **150**, 135–177.

MASTIN, L. & POLLARD, D. 1988. Surface deformation and shallow dike intrusion processes at Inyo Craters, Long Valley, California. *Journal of Geophysical Research*, **93**, 13,221–13,235.

MAZZARINI, F., CORTI, G., MANETTI, P. & INNOCENTI, F. 2004. Strain rate and bimodal volcanism in the continental rift: Debre Zeyt volcanic field, northern MER, Ethiopia. *Journal of African Earth Sciences*, **39**, 415–420.

MERLA, G., ABBATEE, E., AZZAROLI, A., BRUNI, P., CANUTI, P., FAZZUOLI, M., SAGRI, M. & TACCONI, P. 1973. A geological map of Ethiopia and Somalia (1: 2 000 000 scale) and comments with map of major landforms. *CNR, University of Florence*, Centro Stampa Firenze, Italy, 95pp.

MERLA, G. & MINUCCI, E. 1938. Missione geologica nel Tigrai-1. La serie dei terreni, *Rendiconti Reale Accademia d'Italia, Centro Studi per l'Africa Orientale Italiana*, Rome, **3**, 362pp.

MOHR, P.A. 1967. The Ethiopian Rift System. *Bulletin of Geophysical Observatory*, Addis Ababa, **11**, 1–65.

MOHR, P. 1987. Patterns of faulting in the Ethiopian Rift Valley. *Tectonophysics*, **143**, 169–179.

PAN, M., SJÖBERG, L.E., ASENJO, E., ALEMU A. & ASFAW, L.M. 2002. An analysis of the Ethiopian Rift Valley GPS campaigns in 1994 and 1999. *Journal of Geodynamics*, **33**, 333–343.

REDFIELD, T.F., WHEELER, W.H. & OFTEN, M. 2003. A kinematic model for the development of the Afar Depression and its paleogeographic implications. *Earth and Planetary Science Letters*, **216**, 383–398.

RENNE, P.R., WOLDEGABRIEL, G., HART, W.K., HEIKEN, G. & WHITE, T. 1999. Chronostratigraphy of the Miocene–Pliocene sagantole formation, Middle Awash Valley, Afar Rift, Ethiopia. *Geological Society of America Bulletin*, **111**, 869–885.

ROBERTSON, E.C. 1983. Relationship of fault-displacement to gouge and breccia thickness.

*Transactions of the American Institute of Mining Engineering*, **35**, 1426–1432.

RUBIN, A.M. & GILLARD, D. 1998. Dike–induced earthquakes: Theoretical considerations. *Journal of Geophysical Research*, **103**, 10,017–10,030.

SCHOLZ, C.H. 1987. Wear and gouge formation in brittle faulting. *Geology*, **15**, 493–495.

SIBSON, R.H. 1977. Fault rocks and fault mechanisms. *Journal of the Geological Society of London*, **133**, 191–213.

TIERCELIN, J.J., TAIEB, M. & FAURE, H. 1980. Continental sedimentary basins and volcano-tectonic evolution of the Afar Rift. *In*: CARRELLI, A. (ed.) Geodynamic evolution of the Afro-Arabian Rift system. *Atti Convegno Accademia Nazionale dei Lincei*, **47**, 491–504.

WALPERDORF, A., VIGNY, C., RUEGG, J.C., HUCHON, P., ASFAW, L.M. & AL KIRBASH, A. 1999. 5 years of GPS observations of the Afar triple junction area. *Geodynamics*, **28**, 225–236.

WISE, D.U., DUNN, D.E., ENGELDER, J.T., GEISER, P.A., HATCHER, R.D., KISH, S.A., ODOM, A.L. & SCHAMEL, S. 1984. Fault-related rocks: suggestions for terminology. *Geology*, **12**, 391–394.

WOLDEGABRIEL, G., ARONSON, J. & WALTER, R. 1990. Tectonic development of the Main Ethiopian rift. *Geological Society of America Bulletin*, **102**, 439–458.

WOLFENDEN, E., EBINGER, C., YIRGU, G., DEINO, A. & AYALEW, D. 2004. Evolution of the northern Main Ethiopian rift: Birth of a triple junction. *Earth and Planetary Science Letters*, **224**, 213–228.

# Dynamics of prolonged continental extension in magmatic rifts: the Turkana Rift case study (North Kenya)

WILLIAM VETEL[1] & BERNARD LE GALL

*Institut Universitaire Européen de la Mer (IUEM), UMR 6538 UBO/CNRS, 4 Place Nicolas, Copernic, 29280 Plouzané, France*

[1]*Now at: Estacion Regional del Noveste, Universidad Autonoma de Mexico, 83000 Hermosillo, Mexico (e-mail: wvetel@geologia.unam.mx)*

**Abstract:** The Turkana magmatic rift (Northern Kenya) initiated at 45 Ma as one of the nucleation zones of rifting in the East African Rift. It forms an anomalously broad-rifted zone (*c.* 200 km) striking with a north–south trend and lying within a NW–SE topographic depression, floored on both sides of the Turkana area by Cretaceous rifts in the Sudan and Anza plains. From a compilation of available data, combined with newly acquired remote sensing and DEM dataset, we propose a five-stage tectono-magmatic model for the Turkana rift evolution (45–23 Ma; 23–15 Ma; 15–6 Ma; 6–2.6 Ma and 2.6 Ma–Present). The corresponding 'restored' maps clearly show the changing spatial distribution of magmatism and fault/basin network with time, hence supplying some clues about dynamics of continental extension. First-order basement-rooted transverse faults zones are identified and their influence on nucleation and propagation of strain is demonstrated, whereas the role of magmatic 'soft-spots' as concentrating strain is minimized. Blocking of deformation as well as rift jump and lateral transfer of strain are discussed in relation to various types of fault interaction (dip direction, strikes and acute/obtuse angle of the intersecting faults). The causal links between rift nucleation 'cells' and inherited transverse weakness zones in the Turkana rift might also exist elsewhere along the eastern branch of the East African Rift, hence suggesting a complex and discontinuous mode of rift propagation.

The structural style of continental rifts depends on a combination of thermal, mechanical and kinematic parameters (strain rate, initial conditions, mechanical instabilities) that control the deformation of the stretched lithosphere. When a number of parameters interact concomitantly, such as long duration of rifting, strong structural inheritance, volcanic activity and changing stress field conditions, increasing structural complexities of rift domains may occur, as is the case for the *c.* 45 Ma-old Turkana magmatic rift (northern Kenya) in the central part of the eastern branch of the East African Rift. The *c.* 16 000 km$^2$ study area (*c.* 35–39° E long. and *c.* 2–6° N lat.) encompasses the Turkana depression and the external parts of the more poorly constrained Ethiopian and Kenyan plume-related domes (Fig. 1). The north–south-trending fault/basin structures of the Cenozoic Turkana rift cut at high angle the NW–SE axis of the Anza–Sudan Cretaceous rift system that outlines the Turkana depression as a whole (Figs 1 & 2). Its anomalous width (*c.* 200 km-wide in east–west direction), its subdued morphology, with an average elevation of 400–900 m which contrasts with the typical graben-like structure of the Suguta axial valley to the south, and its reduced crustal thickness (20 km) have long been recognized (KRISP Working Group 1991; Prodehl *et al.* 1994). However, many of these structural characteristics remain largely misunderstood and a complete rift model integrating both shallow and deep structures in the timespan of *c.* 45 Ma has not been so far established. The only accurate kinematic model applied to the Turkana rift is principally focused upon Oligo-Miocene fault/basins developed during the onset of rifting in the western part of the rifted zone, i.e. west of modern Lake Turkana (Morley *et al.* 1992, 1999*a*; Hendrie *et al.* 1994), and in the Ethiopian rift to the north (Ebinger *et al.* 2000).

The aim of this paper is to get a structural picture of the long-lived Turkana rift, encompassing the seismically imaged sedimentary basins, as well as the poorly investigated basement and magmatic domains. Our renewed structural interpretation of surface geology, mainly from remote sensing data, provides some clues about the dynamic processes that governed the Turkana rift evolution since Eocene–Oligocene time. The five-stage kinematic evolutionary model established for the 45 Ma-long rift history (45–23 Ma, 23–15 Ma, 15–6 Ma, 6–2.6 Ma and 2.6 Ma–Present) emphasizes the interplay between strain and magmatism. The dimension (*c.* 100 km) of individual structures

*From*: YIRGU, G., EBINGER, C.J. & MAGUIRE, P.K.H. (eds) 2006. *The Afar Volcanic Province within the East African Rift System*. Geological Society, London, Special Publications, **259**, 209–233.
0305-8719/06/$15.00 

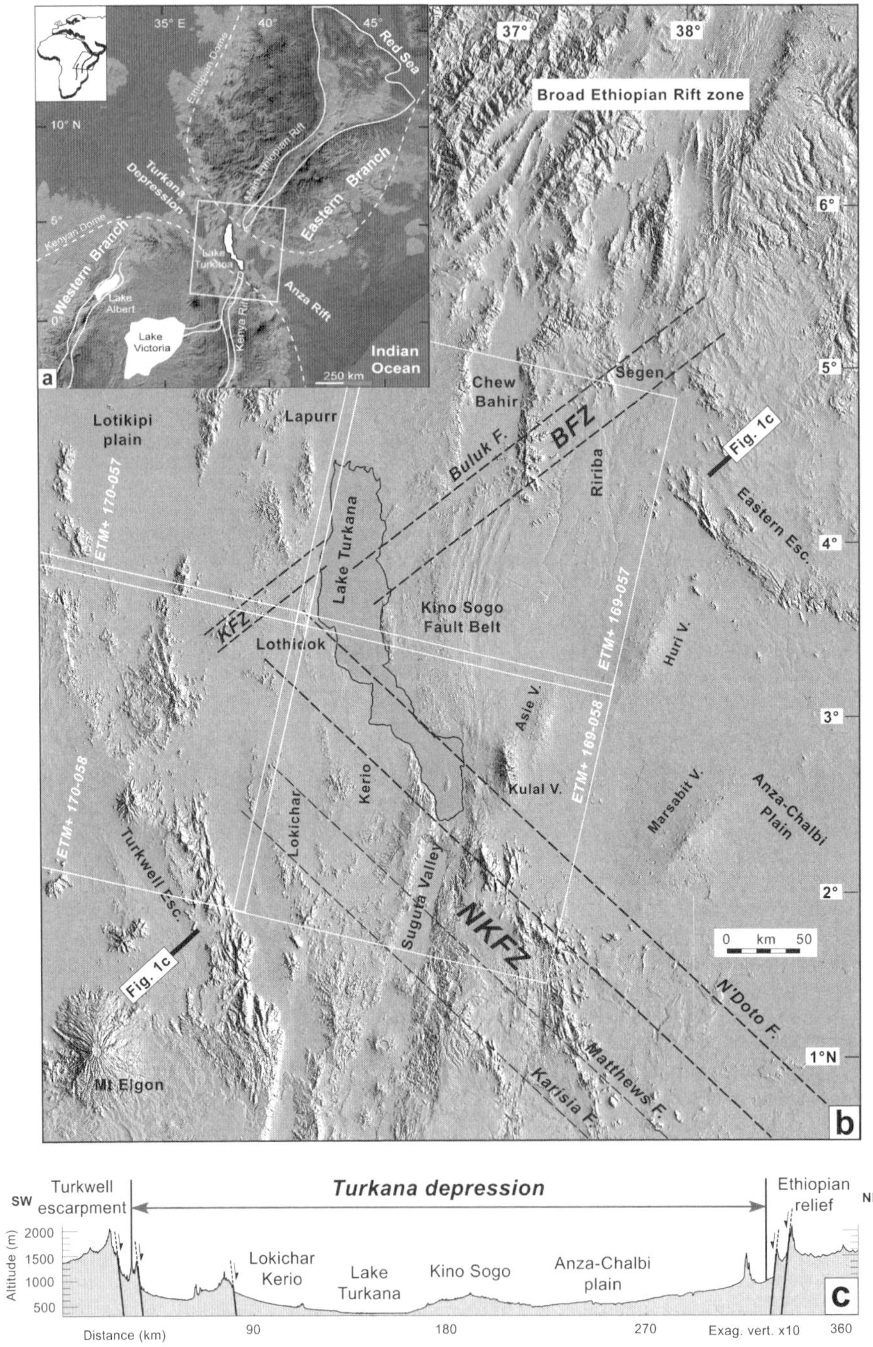

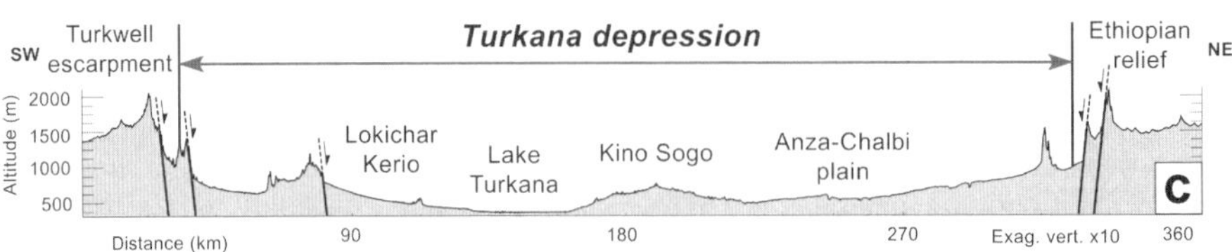


**Fig. 1.** Topographical framework of the Turkana rifted zone. (**a**) General topography of northeastern Africa (Gtopo 30 data). The north–south-trending Turkana rift (white square) is located within the transverse Turkana depression flanked by the Ethiopian and Kenyan plume-related domes (dashed lines). (**b**) Shuttle Radar Topography Mission DEM showing the relatively low relief of the Turkana rift. White squares correspond to the 4 Landsat ETM+ satellite images used in this work. Dashed lines show the trace of transverse fault zones in the Buluk (BFZ), Kataboi (KFZ) and N'Doto-Karisia (NKFZ) areas. Principal location names used in this paper are also indicated. (**c**) Topographic profile across the Turkana depression. Location on Figure 1b.

investigated here supplies an adequate resolution for discussing the influence of structural inheritance (with emphasis on large-scale transverse-faulted zones trending at N50° E and N140° E), distribution of magma and stress field on the style of rifting. From our tectono-magmatic model, it is clear that the 45 Ma evolution of the Turkana rift is more complex than a simple and linear propagation of strain and magmatism as was inferred from earlier reconnaissance studies (Hendrie *et al.* 1994; Morley 1994; Morley *et al.* 1999*a*).

## Dataset and methods

Our specific objective is to determine the Cenozoic tectono-magmatic evolution of the broad Turkana rift zone from *c.* 45 Ma (Late Eocene) to Present. For that purpose, an exhaustive analysis of available seismic imagery and geological maps has been performed and augmented by the structural interpretation of Landsat Enhanced Thematic Mapper Plus (ETM+) images and digital elevation models (Fig. 1). PROBE (offshore) and AMOCO (onshore) seismic reflection profiles across the syn-rift basins of the Lake Turkana area *sensu lato* have been partially reinterpreted from previous published works (Dunkelman *et al.* 1989; Morley *et al.* 1992, 1999*a*) (Fig. 2b, c). Four 1 : 250 000 geological maps (1° × 1°) (Charsley 1987; Ochieng *et al.* 1988; Wilkinson 1988; Dunkley *et al.* 1993) help us to correlate surface geology with structures identified from satellite imagery interpretation. Four Landsat ETM+ images (169-057, 27/01/2000; 169-058, 21/10/1999; 170-057, 12/07/2000; 170-058, 18/01/2000), with a ground resolution of 30 m (15 m for panchromatic band) (Girard & Girard 1999), have been processed by using Geomatica 9.2 software in order to enhance the trace of surface rift fault patterns. The recently available Shuttle Radar Topography Mission (SRTM) topographic data supply supportive evidence for recent fault scarps, hence leading us to elaborate a new and more constrained neotectonic model for the Turkana rift (Fig. 1). The five 'restored' tectono-magmatic maps applied to the successive rift stages are completed by two additional pre-rift structural maps of the underlying Proterozoic basement and the Anza Cretaceous rift, in order to further discuss the problem of structural inheritance.

The accuracy of our evolutionary rift model is directly dependent on the validity of the published radiometric dating (a few tens of K–Ar ages) acquired during the 1980s, principally on the main syn-rift volcanic complexes (e.g. Zanettin *et al.* 1983; McDougall & Watkins 1988; Ochieng *et al.* 1988; Wilkinson 1988; Morley *et al.* 1992; Dunkley *et al.* 1993; McDougall & Feibel 1999 for the Turkana area, and Davidson & Rex 1980; Ebinger *et al.* 2000 for the Chew Bahir/north Omo area). Other limitations of our work are related to the poorly imaged shallow structures on seismic profiles, and the uncertainties about the strain rate, the cumulative extension and the applied stress-field conditions for each time period. Consequently, the five retrograde tectonic maps proposed in the study are not true 'restored' maps and the orientation of the corresponding principal stress axes is deduced indirectly, in most cases, from 2D map fault arrangements and volcanic lineament azimuths. Despite those limitations, new insights about fault propagation and strain migration are discussed with respect to magmatic distribution and structural inheritance throughout the broad Turkana rift zone.

## Geological framework

### *Cenozoic basin and magmatic setting*

Since the acquisition of seismic reflection profiles in the late 1980s and early 1990s, the rifted domain centred over Lake Turkana is known to be a *c.* 200 × 100 km extensional province comprising three parallel belts of north–south half-grabens (Dunkelman *et al.* 1989; Morley *et al.* 1999*a*) (Fig. 2). Stratigraphic calibration from two boreholes (Loperot, 2950 m and Eliye Springs, 2964 m) indicates a progressive northwards and eastwards younging of the depocentres from the 6–7 km-deep Lokichar half-graben which is one of the oldest basins (*c.* 35 Ma) of the East African Rift as a whole (Morley 1999*c*; Morley *et al.* 1999*a*) (Fig. 2b, c). An array of half-grabens with alternating polarity extends to the east, involving the North Kerio, Turkwell, Turkana and Lake basins (Fig. 2a).

The rifting activity has been accompanied since *c.* 45 Ma by dominantly mafic volcanism, emplaced preferentially along interbasin structural highs forming the Lapurr, Lothidok, Kino Sogo and Napedet lava plateaus (Zanettin *et al.* 1983; McDougall & Watkins 1988; Ochieng *et al.* 1988; Wilkinson 1988; Morley *et al.* 1992; Dunkley *et al.* 1993; McDougall & Feibel 1999) (Fig. 2a). The available radiometric dataset suggests the global easterly shift of the magmatic activity from Oligo-Miocene (Turkana domain) to Present (shield volcanoes in the Anza–Chalbi plain) though three discrete magmatic centres persist in the Lake Turkana basins (volcanic islands). A quite similar lateral migration of strain during recent time is also indicated by: (1) the $<3$ Ma dense fault grid across the recent Kino Sogo volcanics between the Chew Bahir and Suguta grabens (Morley *et al.* 1999*a*; Vétel *et al.* 2005); and (2) the *c.* 1 Ma Ririba extensional branch further east (Figs 1 & 2a).

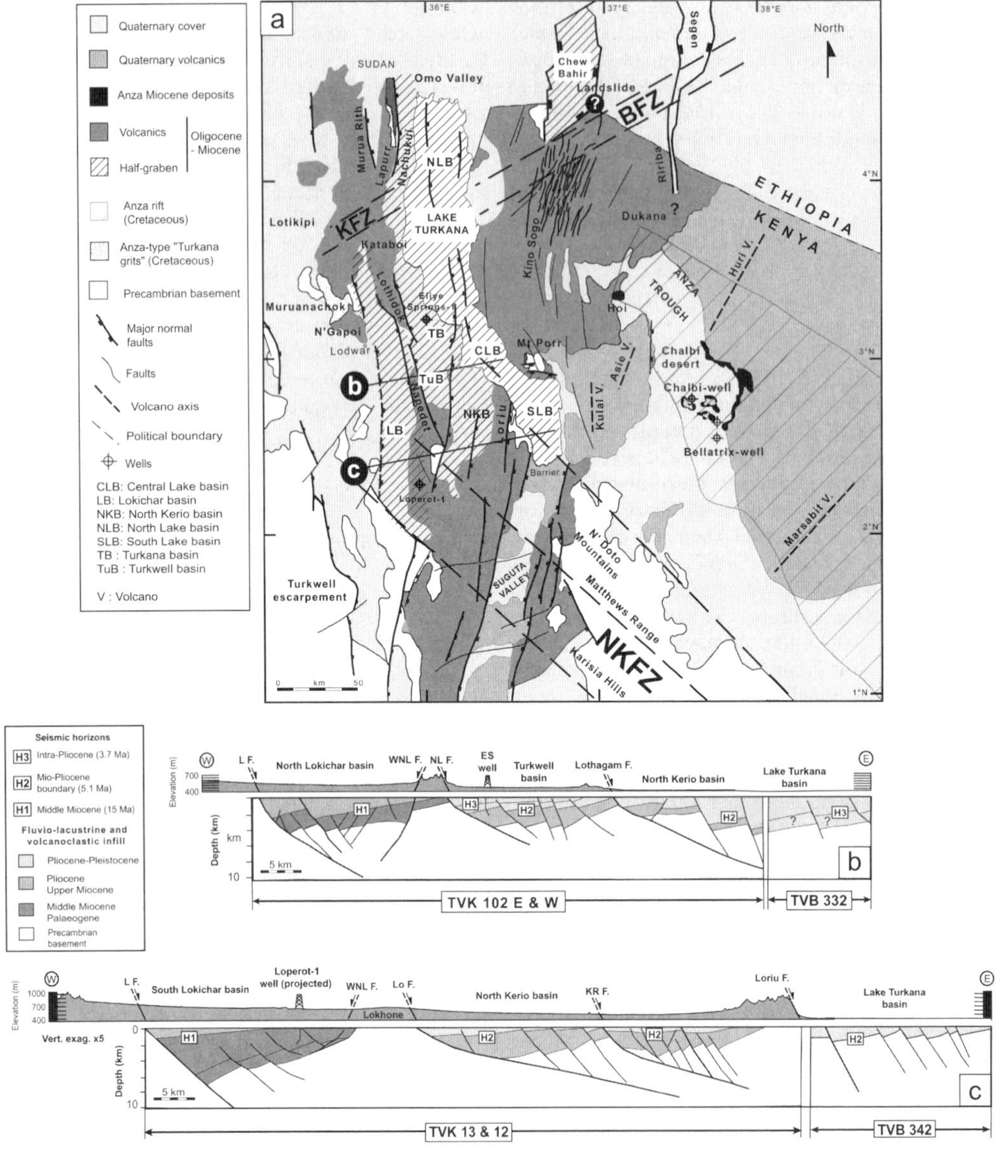


**Fig. 2.** Geological framework and deep geometry of the Turkana rift. (**a**) Structural map showing a series of north–south Oligo-Pliocene half-grabens. Eocene–Miocene volcanism is confined to the west whereas Plio-Quaternary lavas are restricted to the east. (**b**) and (**c**) Geological sections across the Turkana syn-rift basins from seismic reflection data (Amoco TVK and Probe TVB profiles, modified from Morley *et al.* 1999*a*). Location on Figure 2a.

Bulk extension recorded by the Turkana rifted zone since Oligocene times (*c.* 35 Ma) is estimated at 30–40 km by Hendrie *et al.* (1994) inducing a $\beta$ value of about 1.55–1.65. The fairly detailed image of the crust beneath the Turkana rift (KRISP Working Group 1991) shows the step-like morphology of the Moho along the rift axis, passing southwards from a thinned crust (*c.* 20 km-thick) beneath Lake Turkana into a normal thickness (*c.* 35 km) in Central Kenya (Mechie *et al.* 1994; Prodehl *et al.* 1994).

## Transverse fault zones

On the basis of interpreted satellite and DEM imagery, a number of first-order oblique fault zones are identified in the study area as prominent topographic scarps, intrabasin disrupted zones and magmatic lineaments (Figs 1 & 2a). Two main trends are recognized at N50° E, along the so-called Buluk and Kataboi fault zones (BFZ and KFZ in the text), and at N140–160° E, along the so-called N'Doto–Karisia fault zone (NKFZ).

### *The Buluk fault zone*

The BFZ expresses itself as a *c.* 30 km-wide faulted corridor between the Chew Bahir graben and the highly faulted Kino Sogo volcanic plateau (Fig. 3). Its N50° E internal fault structures cut through fractured lavas as young as 3.3 Ma (Hackman *et al.* 1990; Vétel *et al.* 2005). It is bounded to the north by the so-called Buluk fault that dies out to the NE in the Quaternary cover of the Chew Bahir trough (Fig. 3c). The BFZ is also outlined by numerous circular intrusions emplaced from Miocene (21–14 Ma-old Jibisa complex) to Quaternary in the Chew Bahir graben (WoldeGabriel & Aronson 1987; Hackman *et al.* 1990; Ebinger *et al.* 2000). The dominant dip-slip extensional component of the faults within the BFZ is suggested by the morphology of the fault scarp limiting the Kino Sogo uplifted fault block to the north (Fig. 3b–d). This fault geometry is consistent with the attitude of the normal fault imaged by seismic reflection data in 5–3.7 Ma old strata of the North Lake Turkana basin, close to the inferred tip zone of the Buluk fault (Fig. 3a, e).

In this offshore fault attenuation zone, the BFZ displays an underlapping convergent geometry (Morley 1999*b*) with the KFZ discussed below. The BFZ may also extend 70 km further to the NE across the Ririba extensional area where it correlates with a sharp NE–SW fault-like boundary running across the Segen–Ririba linked faulted basins (Ebinger *et al.* 2000) (Figs 1 & 2a).

### *The Kataboi fault zone*

The KFZ separates two distinct magmatic terrains in the uplifted footwall block of the Lapurr–Lothidok fault system, i.e. the Lapurr–Murua Rith complex (Eocene–Miocene) to the north and the Lothidok–N'Gapoi complex (Miocene) to the south (Fig. 4b). The KFZ is a *c.* 20 km-wide deformed zone that comprises a network of *c.* N50° E faults facing to the SE and truncating the inferred <15 Ma-old rhyolitic intrusions (Zanettin *et al.* 1983) enclosed within the KFZ (Fig. 4b, c). Some of these long-lived structures extend with a constant NE–SW strike across the Turkana lacustrine plain where they post-date Plio-Pleistocene Hominid sediments of the Nachukui Formation (Harris *et al.* 1988; Roche *et al.* 1999). Other fault structures in the KFZ swing towards a north–south direction and branch into the Lapurr extensional master fault system to the northeast. To the south, the KFZ fault network dissects and post-dates the dense rift-parallel fault array in the southern Lothidok complex (Fig. 4b–d). The present-day downthrown position of this southern volcanic block (Fig. 4e) is indicative of dominantly southeasterly-directed extensional displacement along the KFZ. Smaller-scale dip-slip faults observed in the Topernawi area confirm the importance of extensional tectonics through the BFZ (Fig. 4d).

### *The N'Doto-Karisia fault zone*

The NKFZ is the most important oblique fault zone recognized in the Turkana area (Le Gall *et al.* 2005) (Fig. 5). It forms a *c.* 110 × 250 km faulted corridor bounded by a system of NW–SE structures. To the north, the N'Doto fault *sensu stricto* is outlined by a number of discontinuous structures that differ in terms of age and tectonic style along-strike. They consist of the faulted extremity of the wedge-shaped Lothidok Miocene lava plateau (NW), and the prominent morphological scarp between the N'Doto basement relief and the Anza–Chalbi Plain (SE) (Fig. 5). To the south, the Karisia bounding fault is marked on both sides of the Suguta Valley by basement scarps (SE) and the oblique termination of the South Lokichar half-graben (NW) (Figs 1b & 5c). The existence of NW–SE transverse structures within the NKFZ, and their reactivation as dextral faults, are documented by Riedel fault network in <3.7 Ma strata of the North Kerio basin (see Le Gall *et al.* 2005 for further details) (Fig. 5d). The earlier activity of the NKFZ during Cretaceous rifting is suggested further SE in the Anza rift by its coincidence with the trace of the Cretaceous–Miocene Finan Gos border fault (see Fig. 7a) and its location over more than 400 km between a non-volcanic (NE) and a highly intruded (SW) sub-basin.

## Pre-rift geology

### *Proterozoic basement*

Pre-Cretaceous crystalline basement rocks in the Turkana rift are principally exposed SW of the NKFZ and NE of the BFZ (Fig. 6). Their present-day elevated structural position mainly results from footwall block uplift along master Cenozoic extensional faults in the Hamer Range, Lapurr, Lokichar, Lokhone, Loriu and N'Doto areas. Basement rocks typically consist of highly deformed granite gneisses of the *c.* 600 Ma Proterozoic Mozambique belt (Davidson *et al.* 1983; Shackleton

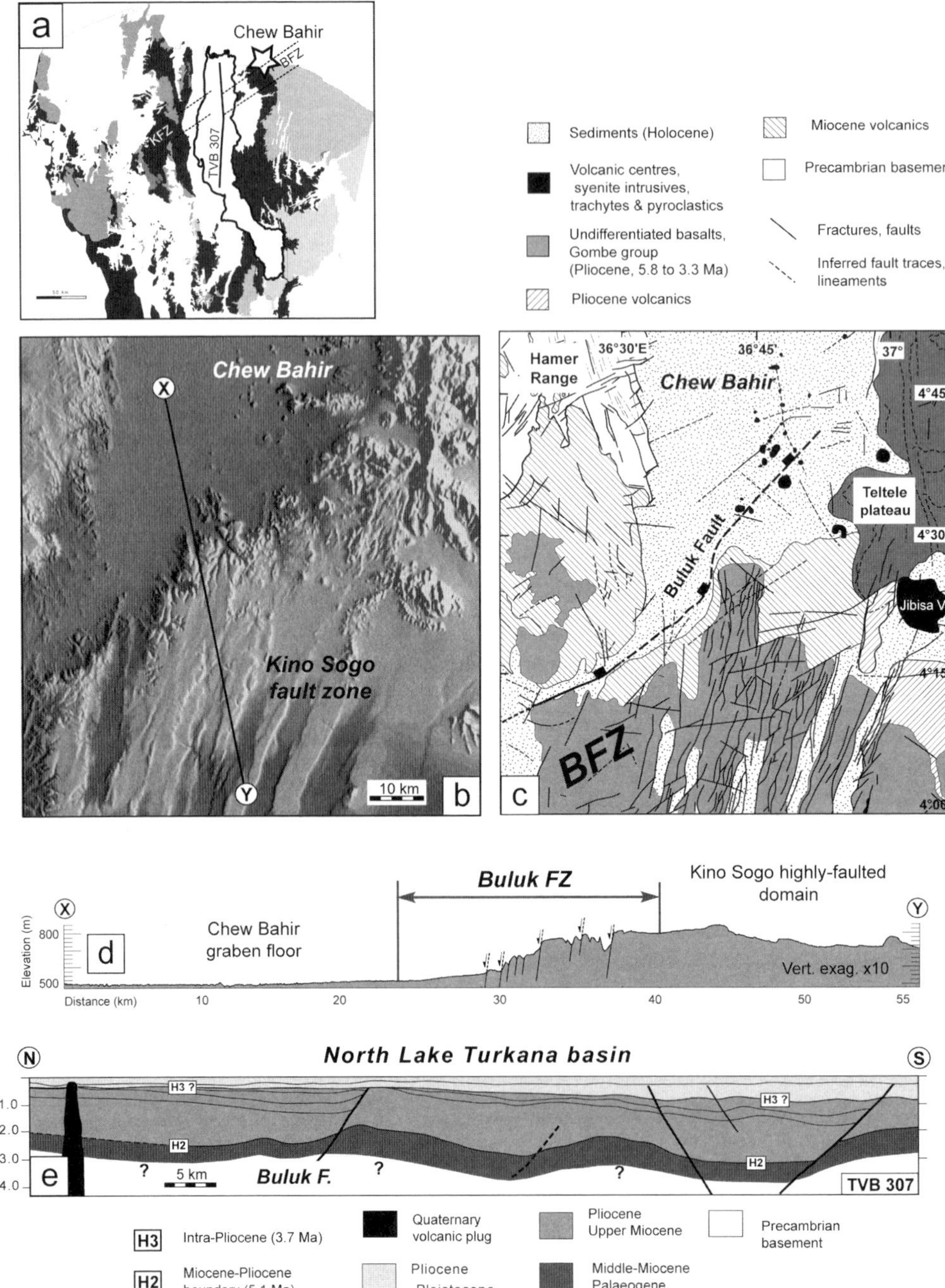


**Fig. 3.** Structural evidence for the Buluk Fault Zone (BFZ). (**a**) Location map of the N50° E BFZ. (**b**) DEM (SRTM data) showing the trace of the BFZ between the Chew Bahir trough and the Kino Sogo plateau. (**c**) Structural interpretation of Figure 3b. The trace of the Buluk fault is extrapolated from the sharp NE-trended limit of the Pliocene volcanic plateau to the SW up to the Chew Bahir intrabasinal intrusions to the NE. (**d**) Topographic section across the BFZ illustrating NW downthrown displacement along the BFZ (location on Fig. 3b). (**e**) Probe TVB 307 seismic line showing the western extremity of the Buluk fault across North Lake Turkana Pliocene basin (modified from Dunkelman *et al.* 1989). Location on Figure 3a.

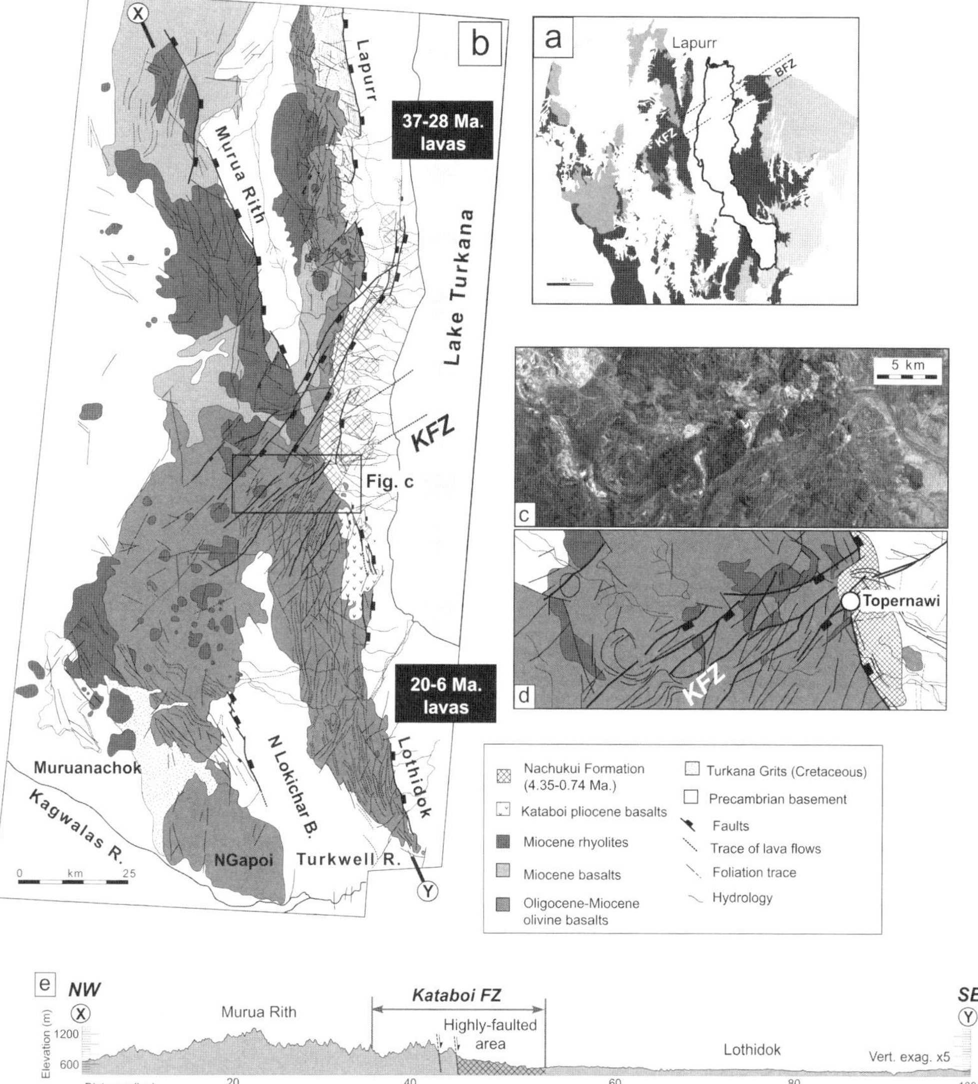


**Fig. 4.** Structural evidence for the Kataboi Fault Zone (KFZ). (**a**) Location of the KFZ in the SW continuation of the BFZ. (**b**) Structural map of the Lapurr–Lothidok volcanic area from Landsat ETM+ interpretation. The KFZ fault/fractures network separates the Eocene–Oligocene Lapurr domain from the highly faulted Miocene Lothidok domain. KFZ structures also affect the Pliocene/Recent lacustrine sediments to the east. (**c**) and (**d**) Detail view from Landsat ETM+ (scene 170-057) and structural interpretation illustrating the emplacement of Miocene rhyolite intrusions along N50° E faults of the KFZ. (**e**) Topographic section from Murua Rith to Lothidok showing the SE fault downthrown along the KFZ. Location on Figure 4b.

1993). The basement structural grain is dominated by ductile foliation planes, locally involved in coaxial folding, but extending generally with a steep attitude at an almost north–south constant azimuth. The ductile fabrics are cut by an intricate network of fault/fractures trending at north–south, N50–70° E and N140° E (Fig. 6). These structures are inhomogeneously distributed with a higher proportion of N50° E structures and local ENE–WSW Cretaceous dykes (dated at ~95 Ma; Ebinger *et al.* 2000) in the Hamer Range, i.e. in the vicinity of the similarly trending BFZ, and a high density of N140° E faults on the western side of the NKFZ. Though few age constraints exist

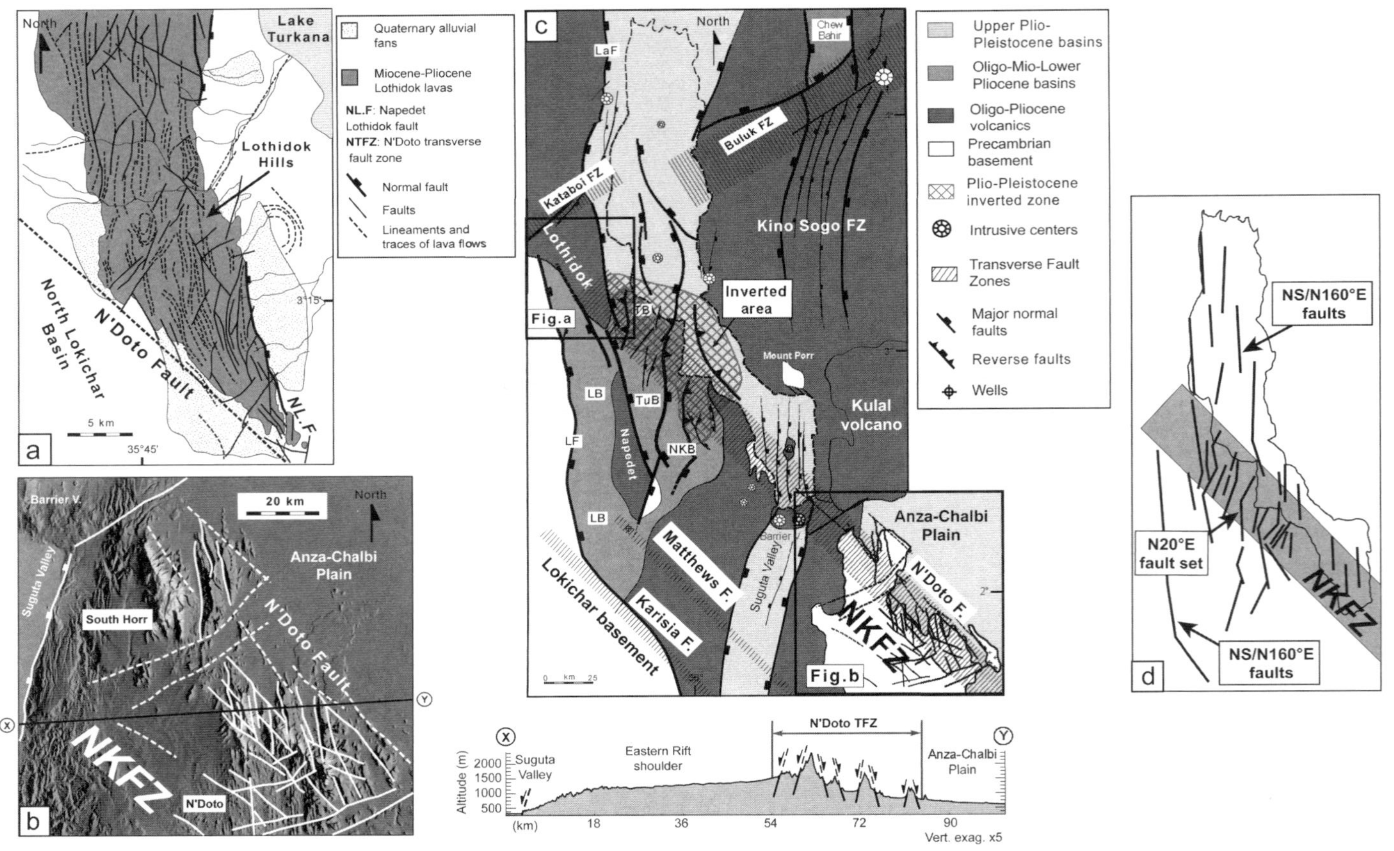


**Fig. 5.** Structural evidence for the N'Doto–Karisia fault zone (NKFZ). (**a**) Structural interpretation of the Landsat ETM+ (scene 170-058) in the southern extremity of the Lothidok Miocene plateau. Location in Figure 5c. The faulted origin of this boundary is evidenced by the sharp southern termination of the volcanics and the cartographic virgation of faults and lava flows. (**b**) Detail view of DEM (SRTM data) showing the transverse faulted boundary of the N'Doto basement relief, SE of Lake Turkana. Location in Figure 5c. The NKFZ is expressed by highly fractured relief with a maximum topographic elevation of 2000 m in the cross-section X-Y drawn on map 5b. (**c**) Synthetic structural map of the Turkana rift showing the trace of the NKFZ. (**d**) Structural sketch illustrating: (1) the variation of fault direction from north–south to N20° E; and (2) the increasing density of faults within the N140° E NKFZ.

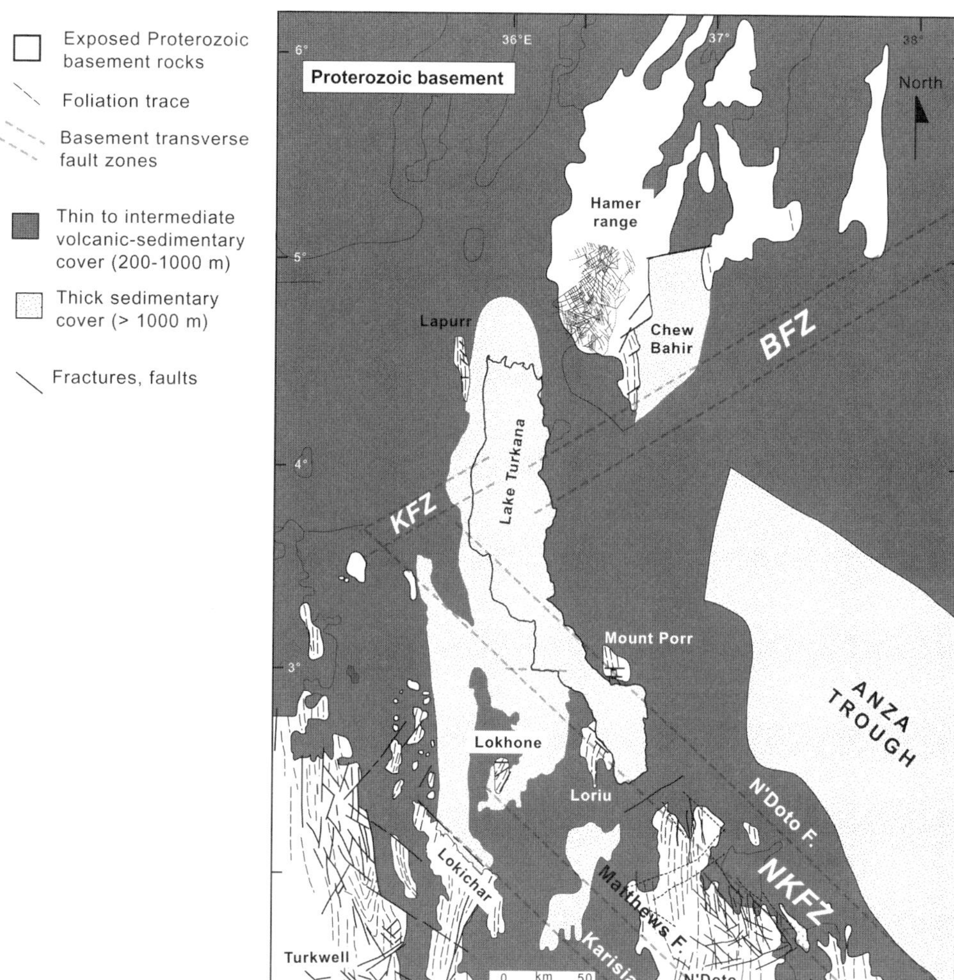


**Fig. 6.** Structural sketch map illustrating: (1) the present-day spatial distribution of basement massifs; (2) the trace of regional-scale Proterozoic structures; and (3) the location of deeply-buried basement beneath major syn-rift basins.

for these brittle structures, N50° E faults seem to have guided mafic and felsic dyke swarms, probably during the Cretaceous (Ebinger *et al.* 2000), but most of the transverse fault zones are presumed to have initiated as Proterozoic features by comparison with similar basement trends in other parts of the East African Rift (McConnell 1972; Smith & Mosley 1993; Coussement 1995; Morley 1999*a*).

## *Cretaceous Anza rift*

In the study area, the non-volcanic sub-basin of the N140° E Anza rift forms an asymmetrical graben structure on seismic reflection data (Morley *et al.* 1999*b*) (Fig. 7). Between the Lagh Bogal and Finan Gos bounding faults, highly faulted depocentres as thick as 7 km involve dominantly terrigenous series Neocomian–Palaeogene in age (Bosworth & Morley 1994; Morley *et al.* 1999*b*) (Fig. 7b). The onset of extensional faulting is followed by a thermal sag phase of basin subsidence during Palaeogene time (Morley *et al.* 1999*b*). The 60 km bulk extension estimated by Dindi (1994) from cumulative fault heave measurements on seismic profiles might have caused significant crust/lithosphere attenuation beneath the Anza rift; however, available gravimetric and refraction data do not further constrain the deep structure of the Anza rift (Prodehl *et al.* 1994). The prolongation of the Anza rift to the NW beneath the Lake

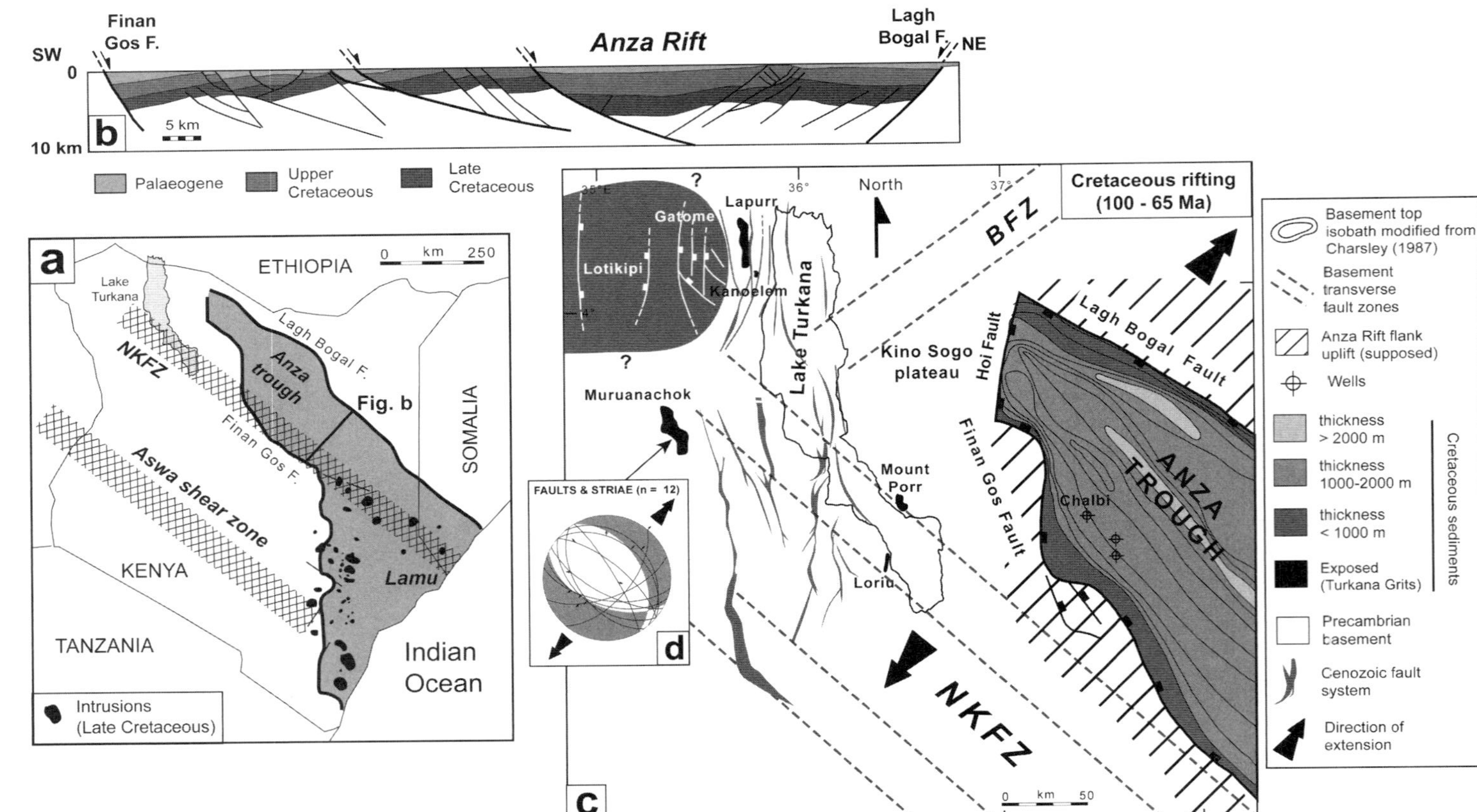


**Fig. 7.** Structural framework of the Anza Cretaceous rift. (**a**) General map showing: (1) spatial distribution of (Late Cretaceous?) magmatic intrusions (modified from Reeves *et al.* 1987); and (2) structural subdivision of the Anza basin along the NKFZ. (**b**) NE–SW structural cross-section with two main inward-facing border faults controlling the 8 km-deep basins (modified from Morley *et al.* 1999b). Location in Figure 7a. (**c**) Structural map of the northwestern termination of the Anza graben in the vicinity of Lake Turkana. Isobath lines of the basement top (established from geological cross-sections of Charsley 1987), suggest the NW faulted termination of the Anza trough along the N10° E Hoi fault. Origin of discrete Cretaceous (Turkana Grits) outcrops on the western Turkana area is discussed in text. Traces of modern Lake Turkana and major transverse fault zones (BFZ, NKFZ, KFZ) are drawn as geographical and structural references (idem for following 'restored' maps). (**d**) Faults and striae diagram ($n = 12$) about deformed 'Turkana Grits' in the Muruanachok area indicates NE–SW direction of extension.

Turkana area is a long-standing debate (Bosworth 1992; Hendrie *et al.* 1994). It is here suggested that the high-gradient isobath pattern of the pre--Cretaceous basement top at the northern extremity of the graben (Charsley 1987) (Fig. 7c) should indicate a syn-rift faulted boundary along the Hoi fault rather than a flexural margin extending to the NW beneath the Lake Turkana Cenozoic basins. This interpretation is supported by the absence of Cretaceous and Lower Tertiary sequences between the basement and thin (*c.* 200 m) Oligo-Pliocene volcanics in the Kino Sogo plateau to the NW (Fig. 7c) (Wilkinson 1988; Haileab *et al.* 2004; Vétel *et al.* 2005). It is thus assumed that the Anza trough was initially disconnected from the Cretaceous rifts preserved on the western side of the Turkana region as (1) scattered outcrops of quartz-bearing grits (Turkana Grits), presumably Cretaceous in age (Arambourg 1943; Williamson & Savage 1986; Morley *et al.* 1992); and (2) poorly constrained (nature and age) syn-rift series imaged on reflection seismic lines (TVK 4 and 7; Wescott *et al.* 1999) in the Lotikipi–Gatome area (Fig. 7c). Fault plane analysis in the (undated) Turkana Grits of the Muruanachok Hills (Fig. 7d) indicates a NE–SW extension, compatible with the Cretaceous palaeostress field inferred to have controlled the N140° E-trending Anza rift (Morley *et al.* 1999*b*). On the other hand, the connection between these poorly constrained 'Turkana' Cretaceous basins and the Sudan rift to the NW is still undetermined owing to a large data gap (e.g. Bosworth 1992; Ebinger & Ibrahim 1994).

## Rift stages (45 Ma to Present)

The 45 Ma Cenozoic rift history is marked in the Turkana sector by a succession of magmatic pulses and fault/basin activity which evolve in both time and space as discussed in the following sections on the basis of the 'restored' maps of Figs 8–12.

### *Palaeogene–Lower Miocene (45–23 Ma)*

Cenozoic rifting starts at *c.* 45 Ma in the Turkana area, contemporaneously with emplacement of widespread volcanism in SW Ethiopia (Ebinger *et al.* 2000) (Fig. 8). The earliest manifestation of magmatism in the Turkana area occurred as early as 45–35 Ma with extrusion of dominantly basaltic lava flows: (1) in the Lotikipi–Lapurr basinal area (up to 3 km thick) (Walsh & Dodson 1969; Bellieni *et al.* 1981; Zanettin *et al.* 1983; Morley *et al.* 1999*a*; Wescott *et al.* 1999) and (2) in the Nabwal, Balesa Koromto (proto-Kino Sogo), and Emuruabwin areas as north–south-elongated complexes (Dunkley *et al.* 1993; Morley *et al.* 1999*a*). Due to the poor resolution of the offshore TVB reflection seismic lines, it is difficult to assert whether the Oligo-Miocene volcanics of the proto-Kino Sogo and Lapurr uplifted fault blocks link laterally beneath Pliocene series in Lake Turkana basins. The absence of initial flood basaltic series in the Anza area might result from still-preserved Late Cretaceous rift flank uplift relief in Palaeogene times (Fig. 8a).

The first occurrence of strain is observed at *c.* 35 Ma, i.e. syn- to post-magmatic activity. This statement implies that the wedge-shaped series underneath the Oligo-Miocene volcanics on the TVK 4–7 seismic lines across the Lotikipi–Gatome faulted area (Wescott *et al.* 1999) are Cretaceous syn-rift deposits (Fig. 7). The fact that the corresponding syn-depositional faults occur with a steep attitude on the TVK 7 north–south seismic line (Wescott *et al.* 1999) confirms their high obliquity relative to the north–south striking Cenozoic basin, and thus suggests a Cretaceous origin.

Initial deformation in the Lokichar area resulted in two contrasted extensional settings comprising: (1) the North Lokichar half-graben where modest extension is recorded along the westerly-dipping West Lothidok Fault (Fig. 8a, c) and; (2) the South Lokichar half-graben where extension is expressed by a *c.* 6 km-thick lozenge-shaped depocentre (40 × 20 km), bounded by the north–south easterly-dipping Lokichar master fault and part of the N140° E Karisia fault (Fig. 8a, d). Evidence for syn-magmatic extension is also documented in the proto-Kino Sogo volcanic domain by rapid lava thickness variations (from 1200 m to 100 m) along the so-called Allia Bay westerly-directed fault (Vétel *et al.* 2005) (Fig. 8e). The east–west direction of extension governing the first increment of rifting in the Turkana area is mainly deduced from the dominant north–south strike of the main syn-depositional fault network. A similar result is obtained in the Kanoelem area from a swarm of north–south-trending dykes which are likely to have fed the Lapurr Oligocene lavas (Fig. 8a, b).

### *Lower Miocene (23–15 Ma)*

The effusive products of a second magmatic pulse (18–11 Ma) overlie with no major changes the broad Ethiopian rifted zone to the north (Stewart & Rogers 1996; Ebinger & Sleep 1998; George *et al.* 1998; Ebinger *et al.* 2000) (Fig. 9a). Conversely, the distribution of magmatism changed markedly in the Turkana area resulting in the concentration of the volcanic activity in the Lothidok–Kino Sogo (LKS) province at the intersection between the BFZ, NKFZ and Hoi faults. The lack of Lower Miocene sedimentary deposits

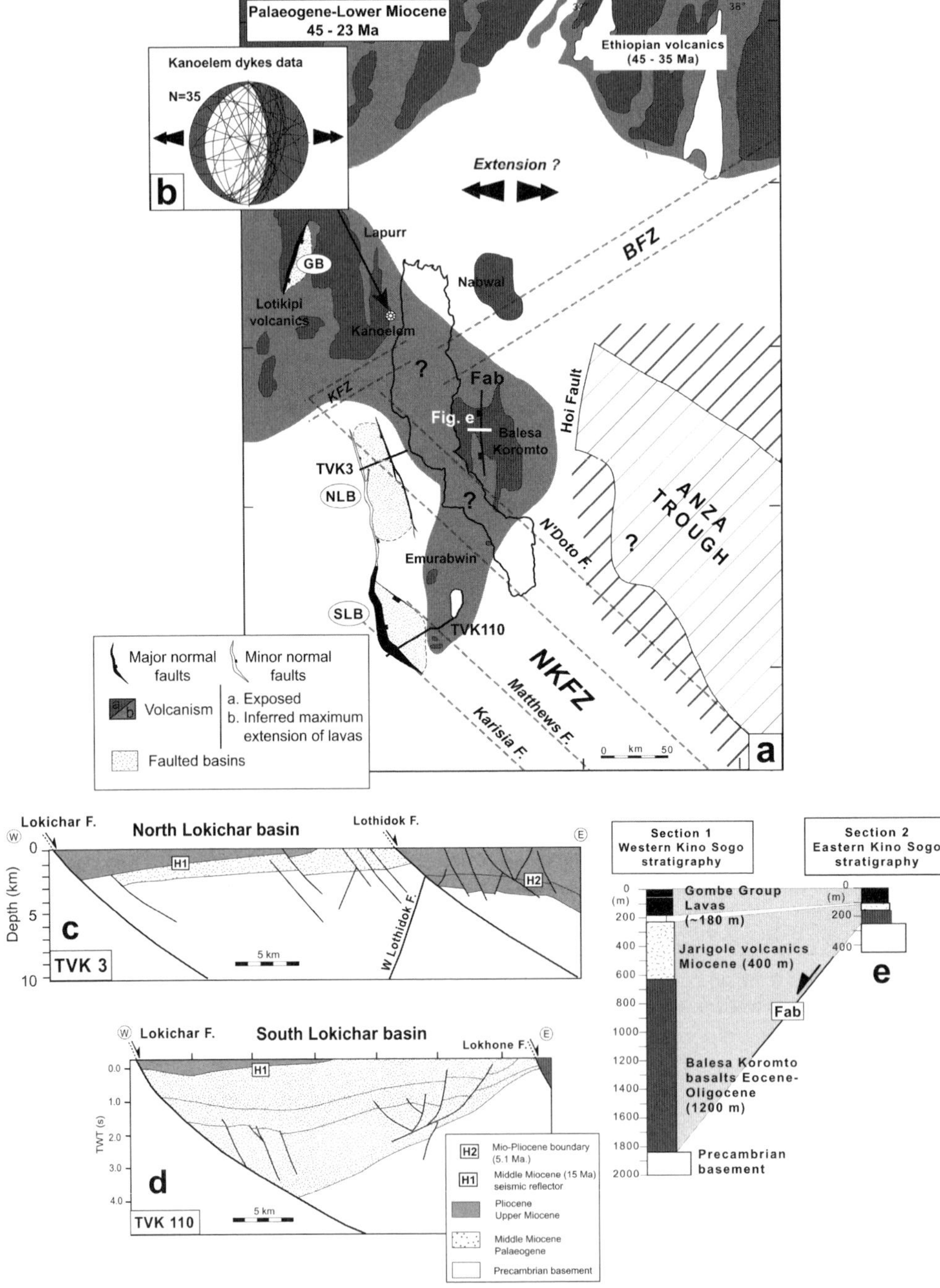


**Fig. 8.** Palaeogene–Lower Miocene (45–23 Ma) event in the Turkana area. (**a**) 'Restored' structural map showing magmatic and basinal domains. Bounding faults are drawn from the isopach map of Morley *et al.* 1999*a*. Contours of volcanic domains are smooth, compared to fault/basin configuration, because of the limitations (dating, ...) mentioned in the text. GB, Gatome basin; NLB, North Lokichar basin; SLB, South Lokichar basin. Same caption for following maps. (**b**) East–west extension deduced from dyke directions ($n = 35$) in the Kanoelem area. (**c**) and (**d**) Interpreted seismic lines TVK 3 and 110 through the North and South Lokichar half-graben. (**e**) Rapid thickness variations of Oligocene lavas (Balesa Koromto) in relation to normal faulting (Allia Bay fault, Fab).

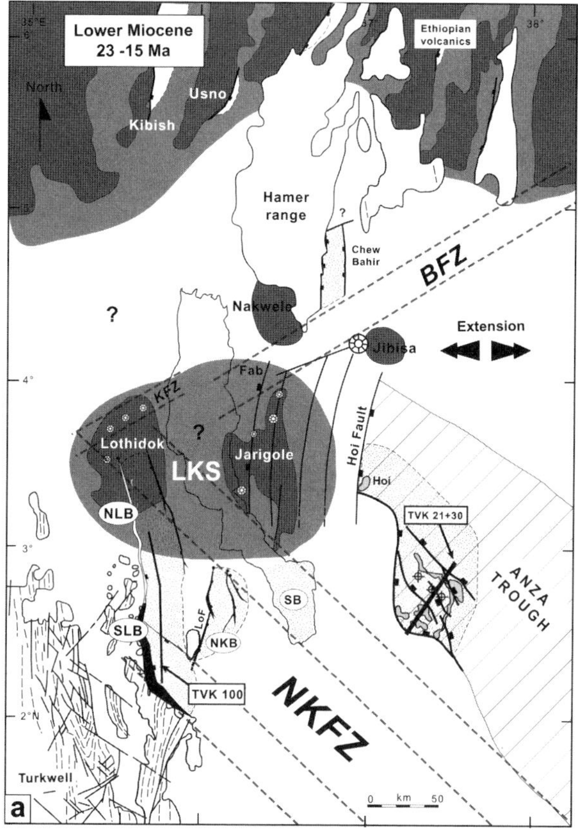


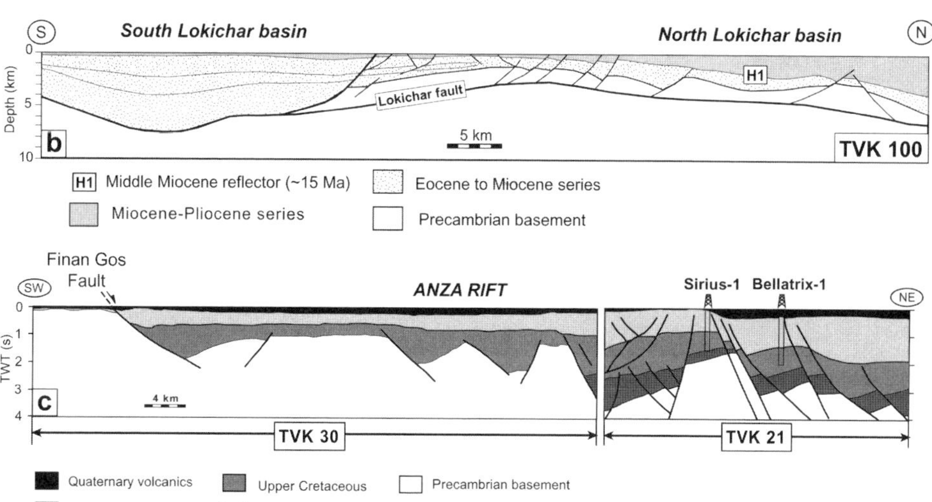


**Fig. 9.** Lower Miocene (23–15 Ma) event in the Turkana area. (**a**) Occurrence of magmatism in the LKS central domain surrounded by the Lokichar–Kerio basin (SW), the NW reactivated part of the Anza rift (SE) and the newly formed Chew Bahir trough (N). NKB, north Kerio basin; SB, south Lake Turkana basin. (**b**) Interpreted TVK 100 along-strike seismic line through the Lokichar basin (modified after Morley 1999*c*). Location on Figure 9a. (**c**) Composite geological section from interpreted TVK 30 and 21 seismic lines through the Anza reactivated rift basin (modified after Morley *et al.* 1999*b*).

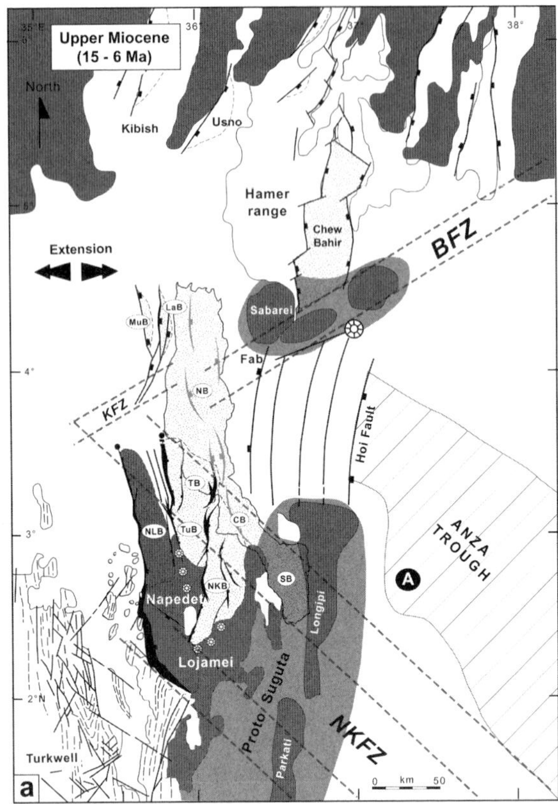


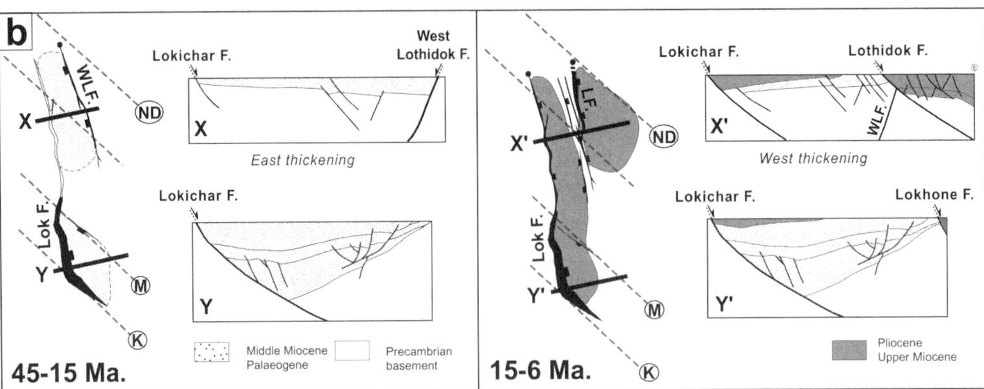


**Fig. 10.** Upper Miocene (15–6 Ma) event in the Turkana area. (**a**) Note: (1) the abandonment of magmatic activity in the LKS central domain; (2) the southerly migration of magmatism over the Lokichar–Kerio domain; (3) the emplacement of intrusions along part of the BFZ to the north; and (4) the locus of deformation through Lake Turkana basins. A, abandoned basin (Anza); CB, Central Basin; LAB, Lapurr Basin, MuB, Murua Rith Basin NB, North Basin; TB, Turkwell Basin. (**b**) Sketch sections and maps showing the timing (45–15 and 15–6 Ma) and spatial evolution of main border faults in the Lokichar–Lothidok basinal area.

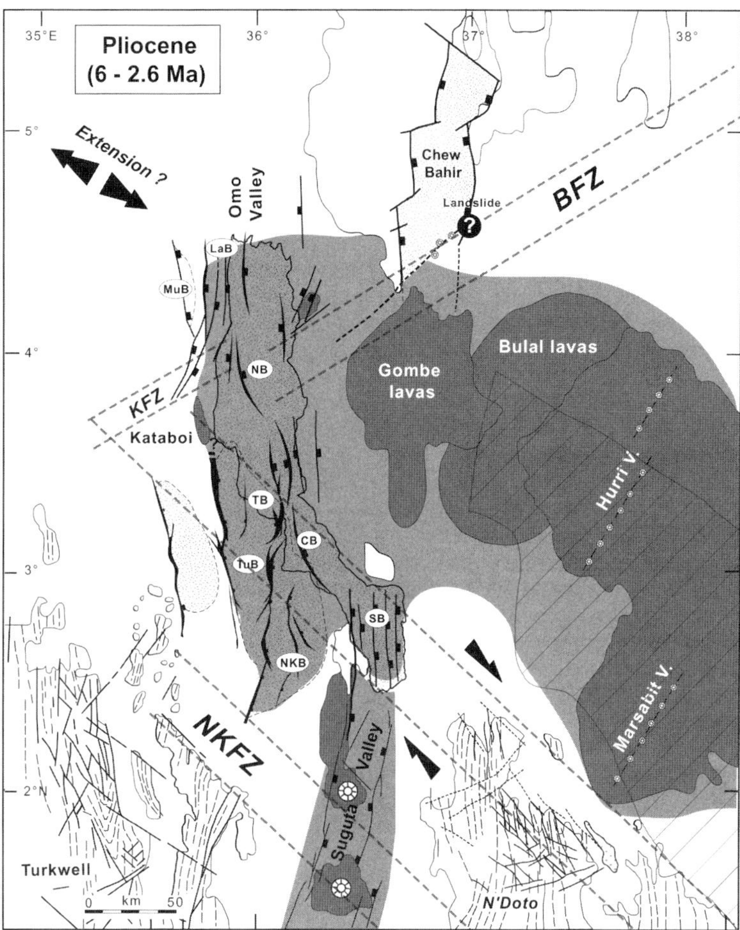


**Fig. 11.** Pliocene (6–2.6 Ma) event in the Turkana area. Note: (1) the emplacement of huge volumes of fissural-type lavas to the east over the Anza domain; and (2) the ongoing concentration of strain along the Lake Turkana and Suguta axes.

in this elliptical-shaped volcanic area is probably due to its thermal-induced domal topography (see below). At this stage, a number of intrusions were emplaced along rift-parallel faults (Jarigole intrusive centres) and N50° E transverse structures of the KFZ, BFZ and Chew Bahir area (21–14 Ma Jibisa complex; Hackman *et al.* 1990; Ebinger *et al.* 2000) (Fig. 9a).

Coeval with the 23–15 Ma volcanism, extensional strain largely developed over a *c.* 250 km-wide zone encompassing downthrown sedimentary basins around the LKS volcanic dome which in turn seems to have recorded very little (if any) deformation. This rift stage is dominated by the easterly migration of extension from the initial, but still active, Lokichar strained domain (Fig. 9a, b). Two small-sized half-grabens (20 × 30 km) with a dominant western polarity, initiated to the east in the North Kerio area, following N10–20° E basement ductile fabrics. A third discrete basin probably formed further east (South Lake Turkana basin) at the extremity of a wedge-shaped extended zone narrowing to the east within the NKFZ. Evidence for extensional fault reactivation is also documented at this stage outside the NKFZ, along Cretaceous faults bounding the northwestern termination of the Anza rift (Fig. 9a, c). Stratigraphic correlation between seismic data and surface outcrops leads us to regard the 'Turkana Grits' exposed in the Chalbi Plain as Lower Miocene series deposited in a 150 × 50 km, *c.* 2 km-deep faulted basin (Bellatrix well; Morley *et al.* 1999*b*), bounded to the west by the Finan Gos and Hoi rejuvenated faults (Fig. 9a, c).

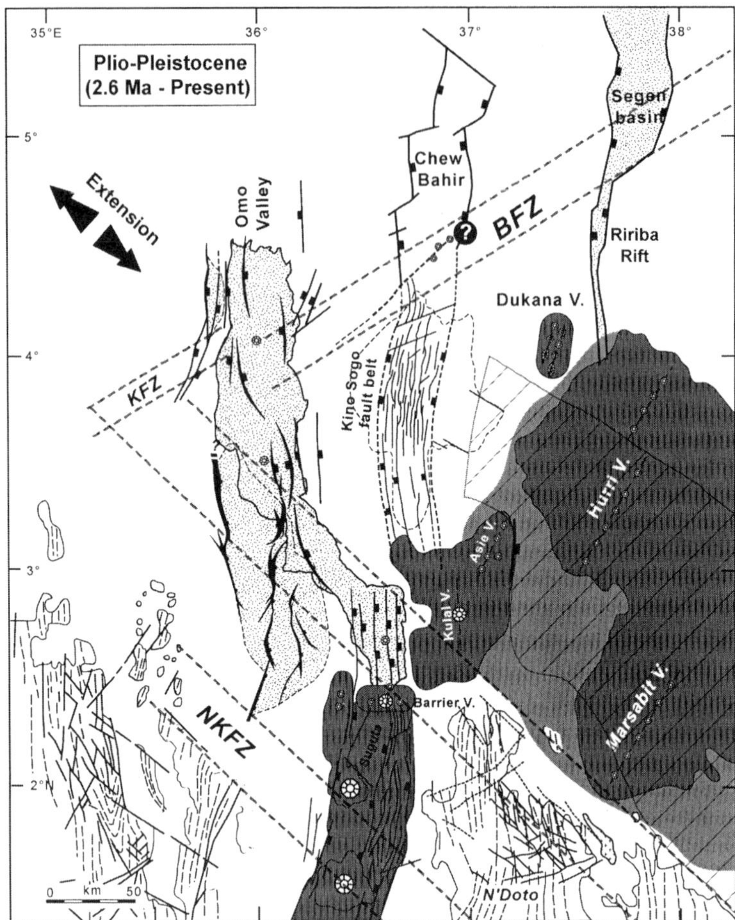


**Fig. 12.** Plio-Pleistocene (2.6 Ma–Present) event in the Turkana area. Note: (1) the cessation of magmatism (excepted a few isolated volcanic centres) over the Lake Turkana basinal area; and (2) the formation of the Kino Sogo and Ririba–Segen north–south-elongated fault belts.

In addition, the resulting east–west anomalously wide rifted zone is also significantly enlarged to the north in response to the opening of the isolated Chew Bahir graben, north of the BFZ (WoldeGabriel & Aronson 1987; Ebinger *et al.* 2000) (Fig. 9a). The surface fault pattern of this large trough clearly reflects the influence of N50° E (Buluk-type) trends along its southern oblique extremity and its dextrally offset western border fault (Vétel *et al.* 2005). Though the Chew Bahir deep geometry is unknown, interpretation of gravity records and projection of stratal dips further north over a set of north–south–N20° E small half-grabens (Kibish, Usno, Mago basin network; Ebinger & Ibrahim 1994; Ebinger *et al.* 2000) indicates depth values of 1 to 2 km. Those basins initiate over a *c.* 2–300 km-wide zone in southern Ethiopia, in close connection with the 18–11 Ma volcanic event (see above). With respect to the previous rift structures, most of the extensional faults developed at this stage with a constant north–south–N20° E trend, probably in response to a speculative-enough assumption that strike was perpendicular to extension.

## *Upper Miocene (15–6 Ma)*

Major changes occurred in the overall tectono-magmatic organization of the Turkana rift during the Upper Miocene (Fig. 10). Excluding the ongoing volcanic activity along the BFZ (Sabarei complex), magmatism continued to migrate southwards from the LKS central domain (which became inactive) into the Lokichar–Kerio area where lava flow extrusion and sedimentary deposits are spatially associated for the first time. The inferred feeding centres of volcanics, such as the Napedet and Lojamei dyke swarms, dated at 13–17 Ma (K–Ar) in the present work, preferentially occurred along north–south interbasinal structural highs (Fig. 10a). Further east, the N10° E Longipi–Parkati magmatic axis outlined the trace of the proto-Suguta trough.

Fault/basin development is dominated by: (1) the abandonment or the drastic reduction of activity of the most external basins (South Lokichar and northwest Anza); (2) the reorganization of rift fault pattern in the northern part of the previously deformed zone where a pair of east-facing half-grabens (North Lokichar/Lothidok) developed on top of earlier westerly-directed faults (Fig. 10b); and (3) the distribution of extension north of the KFZ/BFZ through the Chew Bahir and the newly formed Lapurr–Murua Rith basins. These latter are assumed to extend eastwards through modern Lake Turkana (Wescott *et al.* 1999) and it is suggested here that they might connect southwards via the poorly constrained Central Lake Turkana area into the southern basinal domain. To the north, the spatial link between the Chew Bahir trough and small basins in the south Ethiopian broad zone was also possibly established at this stage (Ebinger *et al.* 2000). The overall fault/basin evolution results in: (1) focusing of strain in the median part of the extended zone; (2) the north–south elongation of the fault/basin pattern, still compatible with an east–west extension for this time period; thanks to (3) the propagation of strain outside the NKFZ toward the north.

### *Pliocene (6–2.6 Ma)*

The 'restored' structural maps applied to the Pliocene (6–2.6 Ma) and Recent (2.6 Ma–Actual, see below) periods are better constrained because they benefit from both good resolution seismic images and accurate digital elevation models (Figs 11 & 12).

After a 6 Ma quiescent phase, a modest but widely distributed Pliocene magmatism is emplaced in the Turkana basins forming: (1) a *c.* 30 m-thick interlayered lava sequence (H2 on Fig. 2) dated at 5.1 $\pm$ 0.2 Ma (Eliye Springs borehole) (Morley *et al.* 1992; Shell 1992); and (2) the 4.41–5.05 Ma Kataboi basalts (K–Ar ages, present work) at the western extremity of the magmatic province (Fig. 11). The main locus of magmatism occurred to the south, in the Suguta axial trough, and to the east, over the NW–SE Cretaceous Anza rift where a huge volume of fissure-type lavas, a few tens of metres thick, spread over about 6000 km$^2$ (Hackman *et al.* 1990) (Fig. 11). This eastern volcanic province encompassed the Gombe Group (5.79–3.28 Ma; Wilkinson 1988; Haileab *et al.* 2004) in the Kino Sogo plateau to the west (see Fig. 12).

The focusing of strain, inferred to have initiated during the Upper Miocene over the north–south elongated Lake Turkana area (see above), is confirmed for the Pliocene stage from seismically imaged sedimentary infills (<5 Ma) in a linked system of small half-grabens with alternating polarities (Dunkelman *et al.* 1989). To the south, the graben-like Suguta trough nucleated along the earlier N10° E volcanic axis (Bosworth & Maurin 1993). At this stage, structural complexities locally recorded by extensional fault networks in the Kerio and Turkwell basins (increasing fault density and changing strikes) are assigned to the dextral reactivation of the NKFZ at depth under changing stress field conditions (permutation $\sigma 1/\sigma 2$ and clockwise rotation (*c.* 20°) of the principal stress axes) (Le Gall *et al.* 2005). Rotation of extension from east–west to N110° E during the Pliocene is also likely to account for the opening of the N30° E tensile crack/fracture networks that might have guided the ascent of magma beneath the Marsabit and Huri fissure-type shield volcanoes over the Anza province (Figs 11 & 12).

### *Upper Pliocene–Present (<2.6 Ma)*

During the last rifting stage, a new N10–20° E magmatic axis formed in the northern prolongation of the Suguta axial graben. It is outlined by the Dukana (0.9 Ma), Asie (2.7 Ma) and Kulal (1.7 Ma) volcanoes (Charsley 1987; Dunkley *et al.* 1993) that are partly faulted (Fig. 12), hence suggesting the ongoing lateral shift of strain eastwards into the previous Anza rift. The inferred broadening of the extended zone at this stage is confirmed by the nucleation of a new (poorly known) rift branch in the Ririba area, to the NE (Figs 1b & 12). This atypical extensional structure (100 × 10 km) dies out southwards into the *c.* 1 Ma Huri lavas and it gives way northwards, via the BFZ, into the wider Segen half-graben (Ebinger *et al.* 1993). The emplacement of the Ririba–Segen rift branch, 70 km east of the Chew Bahir trough, reinforced the north–south connection between the Turkana and south Ethiopian extensional provinces. In the main central deformed zone, development of faulted basins is accompanied by three additional processes that are: (1) the abandonment of external basins, initiated during the preceding stage and confirmed about the Lokichar basin; (2) the build-up of a dense north–south extensional grid fault across the Kino Sogo volcanic plateau (Fig. 12), e.g. between the Chew Bahir and the Suguta–South Lake Turkana basins, that tends to form a coalescent and highly segmented branch, 40 km-wide, east of the Lake Turkana fault/basin pattern (Vétel *et al.* 2005); and (3) partial inversion of the Kerio and Turkwell basins in relation to strike-slip tectonics along reactivated NKFZ structures under still-rotating stress-field ($\sigma$3 at N130° E) (Le Gall *et al.* 2005).

## Controlling factors on the Turkana rift evolution

The five rift stages described above for the last *c.* 45 Ma illustrate the complex tectono-magmatic history of the Turkana area. Although it is difficult to discriminate the respective role of structural inheritance, magmatism and stress-field variations, some new insights about dynamics of crustal extension are discussed below by comparison with concepts recently applied on extensional systems from natural examples and analogue modelling (e.g. Morley 1994; Morley *et al.* 1999*a*; Corti *et al.* 2003; Ziegler & Cloetingh 2004).

### *Shallow/surface processes*

The pre-rift arrangement of the Turkana area is dominated by both Proterozoic basement fabrics and Cretaceous rift structures (Fig. 13a). The consistent north–south-striking foliation trace in the basement is cut by three first-order transverse fault zones (NKFZ, BFZ and KFZ) that intersect at high angle (*c.* 90°) and are bisected by the inferred east–west extension that prevailed during Cenozoic rifting until *c.* 6 Ma. The sequential rift development depicted in Figure 13a to 13e highlights the key role played by the N50–70° E and N140° E inherited discontinuities on rift fault kinematics. The resulting various types of fault interaction (FIT) are shown in Figure 13f. Strain/magmatism relationships are discussed in the following section.

### *Role of transverse discontinuities*

The N50° E and N140° E fault networks contrast with regards to their nature, geometry and influence on rift fault development as illustrated on the five-stage kinematic model of Figure 13. The *c.* 110 km-wide NKFZ forms a long-lived and mechanically weak zone which partly controlled Cretaceous rift trends to the SE (Anza), and which was later the main locus of Cenozoic strain. By contrast, the N50° E fault structures extend as narrower faulted corridors (BFZ–KFZ) acting as transfer zones between the southwestern Turkana basinal domain and a three-arm rift system that connects northwards into the south Ethiopian broad rift zone (Fig. 13e).

During initial strain (45–23 Ma; Fig. 13a), the main basinal area in the Lokichar sector is strictly confined within the NKFZ, in response to an east–west extension. Distribution, timing and 3D-architecture of the oldest depocentres in this area (Morley *et al.* 1992) indicate two main types of fault interaction (Fig. 13f). The convergent and overlapping fault pattern displayed by the Lokichar and West Lothidok master faults (Fig. 13a) suggests the arrest of fault-tip propagation along the N140° E transverse faults, probably because of the acute angle (*c.* 50°) and the opposite sense of dip of the two intersecting fault networks ($FIT_1$ on Fig. 13f). In contrast, the obtuse angle (140°) and the similarly facing direction of the two cross-cutting fault systems to the south (Karisia) are more favourable for extension ($FIT_2$ on Fig. 13f). In these conditions, dimension of the lozenge-shaped main depocentre (South Lokichar) are directly controlled by the transverse fault spacing (*c.* 30–50 km) in the NKFZ.

At stage 23–15 Ma (Fig. 13b), in response to inhibition of master fault propagation, extension shifted eastwards throughout the eastern part of the NKFZ weakness zone, still recording east–west tensile stress. The arrest of the northerly fault/basin propagation outside the NKFZ results, in map-view, in a triangle-shaped extended zone involving regularly spaced (*c.* 20 km) basins (North Kerio and possibly South Lake Turkana) that decrease in size eastwards (Fig. 13b). The difficulty of strain in propagating northwards across the volcanic, and probably less brittle, LKS domain is expressed by: (1) a minor and more diffuse fault/fracture network throughout the LKS; (2) the transfer of extension to the NE along the BFZ into the newly formed Chew Bahir graben; and (3) the *c.* 50 km jump of strain eastwards into reactivated parts of the previous Anza rift. The combined effect of rift jump and lateral offset of strain leads to an anomalously wide rifted zone (Morley *et al.* 1992; Ebinger *et al.* 2000), *c.* 250 km in east–west direction parallel to the inferred extension, with a length/width ratio of about 1.2. The strike and dip direction of the north–south and N50° E intersecting faults bounding the Chew Bahir graben define a symmetrical pattern with respect to the Lokichar fault system, and are thus either favourable or unfavourable for basin opening ($FIT_3$ and $FIT_4$, respectively on Fig. 13f).

Time interval 15–6 Ma (Fig. 13c) is dominated by cessation of magmatic activity over the LKS where a major fault/basin rearrangement took place, coeval with opening of new basins still under east–west extension, along the oblique central segment and northern extremity of modern Lake Turkana. The coalescence of previous faults via newly formed structures leads to a more elongated and continuous

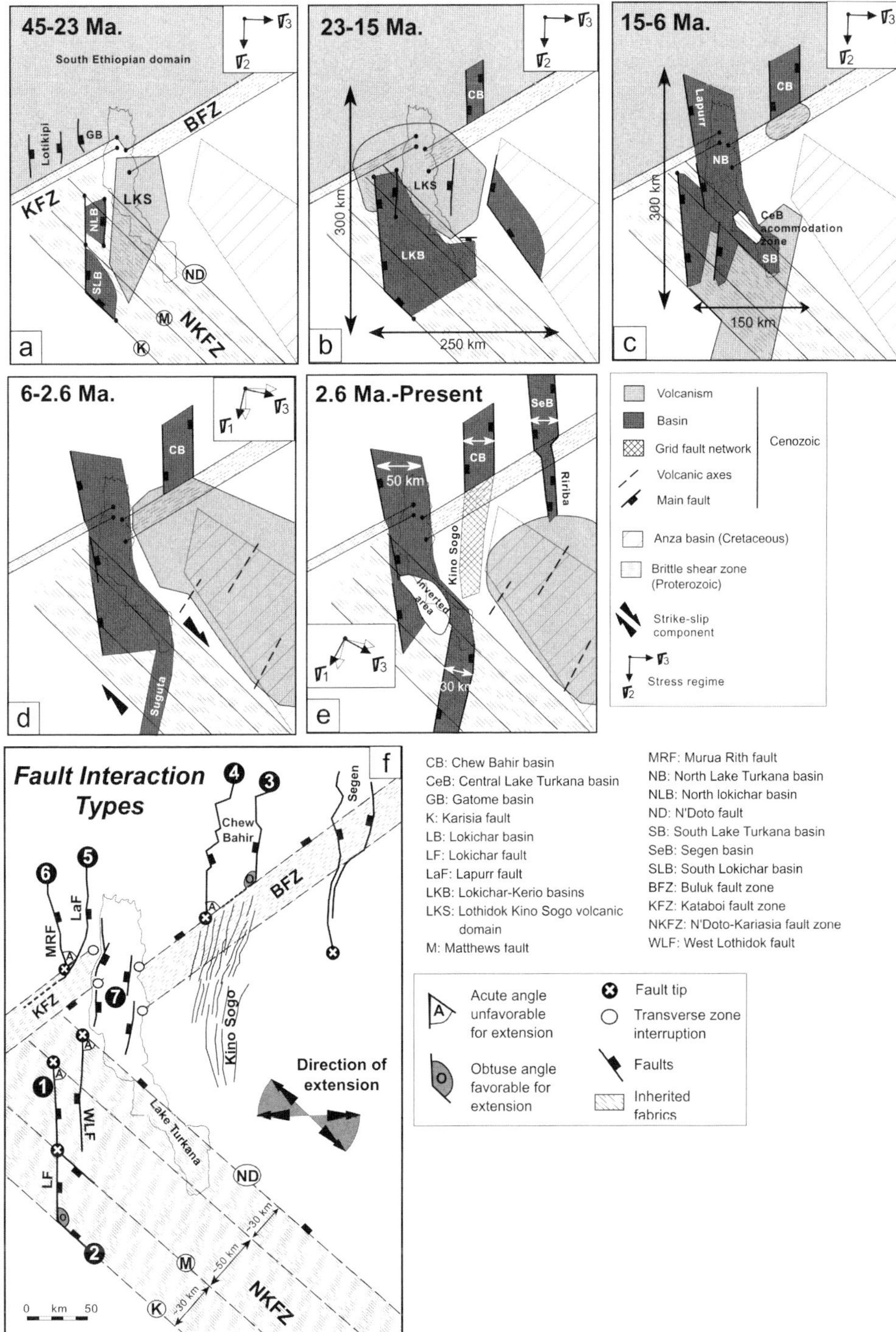


**Fig. 13.** Cartoons synthesizing the 45 Ma rifting history of the Turkana area. (**a**, **b**, **c**, **d**, **e**) Structural sketches elaborated from the restored maps of Figures 8–12. See text for discussion. (**f**) Various types of fault interaction (FIT) between north–south and transverse fault structures. The numbers refer to citation order of structures in the text.

rift pattern with a higher length/width ratio of *c.* 2. At this stage, a number of N50° E fault structures of the BFZ are likely to have acted as vertical pathways for ascending magmas.

The structural interaction between the N50° E and north–south rift-parallel fault networks leads to various FIT that are well documented in the Lapurr area ($FIT_{5-7}$ on Fig. 13f). $FIT_5$ corresponds to the tendency for the north–south Lapurr master fault to curve into alignment with the N50° E KFZ to the south. The decrease of extensional displacement from the Lapurr to the KFZ fault segment is consistent with their respective orientation with regards to the east–west extension, if assuming their synchronous activity. $FIT_6$ concerns the Murua Rith border fault which limits to the west a half-graben-like depression narrowing to the south towards the KFZ. This cartographic fault configuration indicates, in addition to the southerly attenuation of the relief in the western uplifted block, that propagation of the north–south fault system was restricted by N50° E discontinuities to the south. A quite similar fault attenuation process is likely to have occurred, with an opposite sense, along the North Lokichar symmetrical depressed zone, south of the KFZ (see Fig. 4b). The role of the KFZ–BFZ volcanic/fault lineament as a mechanical barrier for rift fault propagation is indirectly deduced at stage 15–6 Ma from the propagation of rift-parallel faults northwards, beyond the KFZ–BFZ disconnected segments, through their underlapping approaching zone, i.e. in the stepover region which is known to be a low-strength area (Morley 1999*b*) ($FIT_7$ on Fig. 13).

During the two last rifting stages (<6 Ma) (Fig. 13d, e), important, but spatially restricted, structural complexity occurred in the oldest and more extended North Kerio, Turkana and Turkwell basins where local inversion processes took place in relation to the dextral reactivation of N140° E fault structures (Le Gall *et al.* 2005) under changing stress field conditions (Bosworth *et al.* 1992). Concomitantly, the tendency of strain to migrate eastwards over the newly formed Kino Sogo fault grid (Vétel *et al.* 2005) and into the nascent Segen–Ririba branch contributes to a stronger rift fault/basin linkage (Chew Bahir/Kino Sogo/Suguta axis) and to a better-established connection of the Turkana rift with the adjoining south Ethiopian and central Kenyan rift patterns (Ebinger *et al.* 2000).

## *Strain/magmatism relation*

The overall Eocene–Present lava series covering the *c.* 300 km-wide zone in the Turkana rift is generally regarded as resulting from the easterly shift of magmatism with time (Morley *et al.* 1999*a*). However, a more intricate time/space magmatic distribution is evidenced on the 'restored' maps presented above when considering both lower-scale structures (*c.* 100 km) and shorter time intervals (*c.* 10 Ma). In fact, an east–west but also a north–south migration of magmatic activity coexist, resulting in complex relationships between strain and magmatism with time.

In the LKS, the *c.* 35–3 Ma volcanic activity is interrupted in the interval 15–6 Ma. At this stage, volcanism starts at *c.* 12 Ma, for the first time, further south in the Napedet–Longipi–Suguta domain (Morley *et al.* 1992), and continues until Recent time with intrabasinal mafic centres intruding the median part of the rift zone (volcanic islands of modern Lake Turkana). A system of fissure-type shield volcanoes is emplaced during the Plio-Quaternary in an off-axis position to the east, over part of the Anza rift (Hackman *et al.* 1990). The most fruitful data about strain/magmatism relationships are provided by the evolving fault and magmatic arrangement in central LKS volcanic province during the 23–15 and 15–6 Ma time periods (Fig. 14).

During stage 23–15 Ma, the distribution of the main downthrown faulted basins (Lokichar–Kerio, Chew Bahir and part of Anza) at the periphery of the LKS volcanic domain, as well as the systematically outward vergence of the corresponding master faults (Hoi, West Lothidok, South Chew Bahir and possibly Mount Porr), both suggest that extensional structures formed under the combined effect of east–west applied extension and gravity-driven collapse on the flank of the uprising and thermal-induced LKS volcanic dome (Fig. 14a, b). Extension in the LKS itself should have been accomplished through intense dyke emplacement at depth whereas brittle failure at shallow level probably occurred in response to regional east–west extension and bending stress that gives rise to a network of extrados-type fault/fractures (Fig. 14b). Quantifying extension in the LKS volcanic faulted domain is not an easy task and that makes it difficult to assess whether the changing style of deformation from the sedimentary downthrown basins (south) to the uprising volcanic dome (north) is also accompanied by strain rate variations.

During stage 15–6 Ma, the southerly migration of magmatism from the LKS into the Lokichar–Kerio area occurs synchronously with downthrown block-faulting throughout the deflated median part of the previous LKS magmatic dome (Fig. 14b). There, earlier development of suitably oriented (north–south) 'extrados' fractures (see above) might have further helped the northerly migration of faulting beyond the NKFZ. The resulting antagonist spatial distribution of faulting versus volcanism leads us to envisage that the longitudinal propagation of faults is made easier throughout a

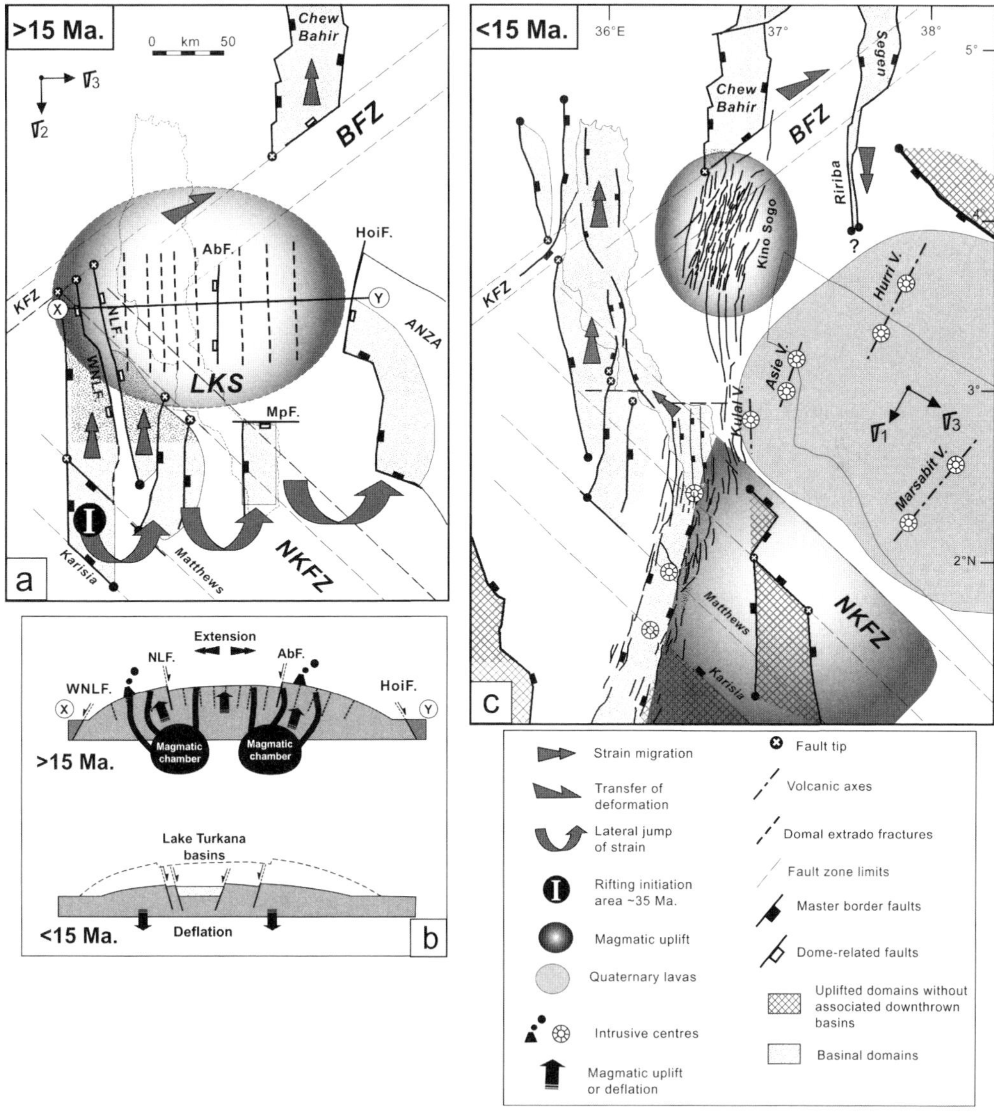


**Fig. 14.** Fault/magmatism interaction model. (**a**) Before 15 Ma, strain migrates from the South Lokichar (nucleation) area around the LKS magmatic dome. (**b**) 2-phase evolutionary model explaining the development of extensional faulting over the Lothidok–Kino Sogo magmatic dome in terms of reactivated bending-stress fractures. Trace of cross-section on Figure 14a. (**c**) The Kino Sogo fault belt disrupting the axial part of the Kino Sogo volcanic dome after 15 Ma might be assigned to processes shown on Figure 14b. The origin of off-axis fissure-type lavas over the Anza domain is discussed in the text.

crustal domain getting cooler and more brittle during an intervolcanic stage. This assessment is reinforced about the volcanics of the Kino Sogo fault belt which benefits from tighter timing constraints (Vétel *et al.* 2005) (Fig. 14c). This statement is in contradiction with classical concepts that inversely predict the more deformable state of crustal material weakened by a thermal event (Callot *et al.* 2001, 2002; Corti *et al.* 2003). From our evolutionary kinematic model, it is therefore suggested that within a heterogeneous domain, involving a weak faulted corridor (NKFZ) and a softened crustal province (LKS sector, similar to the soft-spot of Callot *et al.* 2001), and undergoing extension, strain preferentially focuses in the brittle pre-faulted areas and then migrates laterally along thermal-induced fractures in the adjoining palaeo-volcanic domes.

Another type of strain/magmatism relationship is documented about the fissure-type mafic shield volcanoes emplaced during Plio-Quaternary over the Anza province (Fig. 14c). Whatever the location of the magmatic chamber at depth beneath either the off-axis Anza area or the uplifted area flanking the axial rift trough to the east, the ascent of magma through the Anza upper crust might have exploited (inherited or newly formed?) NNE fissure swarms that opened during changing stress field conditions under a *c.* N30° E-directed shortening axis (Le Gall *et al.* 2005).

## Discussion and conclusions

From surface (remote sensing) and subsurface reflection seismic dataset in the Turkana rift, it is here emphasized that Cenozoic rifting in northern Kenya preferentially nucleates over an unstretched crust/lithosphere separating the Anza and Sudan NW–SE Cretaceous rifted (and probably strengthened) domains (Fig. 15). The origin of the mechanisms responsible for this rift location still remains to be defined. The 45 Ma tectono-magmatic development of the Turkana Cenozoic rift segment is discussed in this study on the basis of five 'restored' structural maps for the periods 45–23 Ma, 23–15 Ma, 15–6 Ma, 6–2.6 Ma, 2.6 Ma–Present. One of the most distinctive conclusions of this work is to assess that the first increment of extension within an heterogeneous continental crust, involving a pre-existing mechanical weakness zone (N140° E) and a syn-rift magmatic province, preferentially locates throughout the brittle and pre-fractured terrains, in contradiction with recent modelling that inversely suggests the locus of deformation in the magmatic 'soft spot' (Callot *et al.* 2001, 2002). In the deformed areas, the location and final geometry of the resulting basin/fault networks are directly controlled by the dimensions of the inherited transverse fault zones and the angular and dip direction of intersecting fault structures. Various types of fault interaction are defined in terms of locking, branching or propagation of extensional structures and their associated depocentres.

Outside the nucleation transverse zone, propagation of fault/basin network is highly controlled by additional parameters dealing mainly with the spatial distribution of volcanic provinces. That is documented about two magmatic domes in the central part of the rift which record extensional strain after the cessation of volcanic activity, i.e.

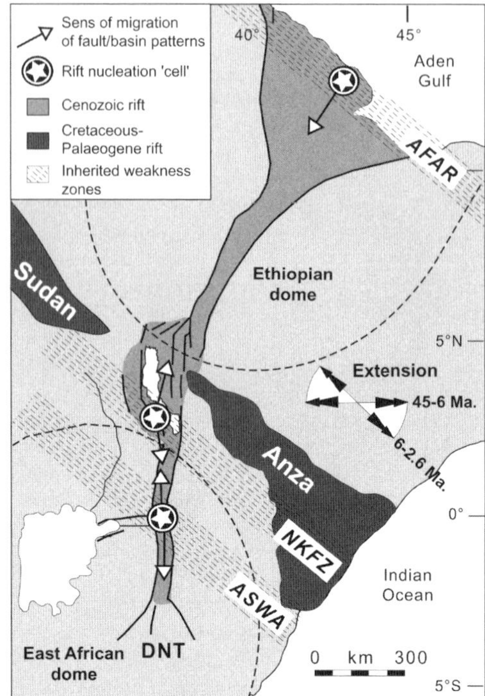


**Fig. 15.** Model of rift propagation along the eastern branch of the East African Rift. Extension is inferred to have: (1) initiated within discrete nucleation 'cells' located within first-order transverse inherited weakness zones (Afar? NKFZ and ASWA); and (2) propagated longitudinally (north and south) to form a most continuous rift branch.

when thermally induced topography is deflated. At this stage, previous inner volcanic dome structures, such as syn-magmatic and/or extrados fracture networks, are likely to have guided the rift-parallel fault pathway.

In response to marked changes in the timing and spatial distribution of both magmatism and brittle weakness zones during *c.* 45 Ma, the overall 2D-arrangement of the Turkana rift evolves from: (1) an anomalously broad deformed zone (250 × 200 km) resulting from rift jump and strain transfer (stage 23–15 Ma); to (2) a more typical elongated rift fault/basin pattern (*c.* 50 × 200 km) with strain focusing along either one (15–6 Ma), two (6–2.6 Ma) or three (2.6 Ma–Present) parallel branches that connect into the adjoining Ethiopian and Kenyan rift troughs.

At a greater scale, considering the age and location of the oldest sedimentary depocentres and/or eruptive centres along the eastern branch of the East African Rift leads us to investigate the mode of rift propagation from the Afar to the North Tanzanian rifts (Fig. 15). The deeply buried fault-bounded depocentres imaged by seismic (Mugisha *et al.* 1997) and magneto-telluric records (Hautot *et al.* 2000) in the Baringo/Bogoria area, *c.* 250 km south of the Turkana rift, are >21 Ma-old basins closely linked to the ASWA first-order N140° E basement fault zone (Chorowicz & Mukonki 1980; Coussement 1995) (Fig. 15). Strong structural similarities within the Turkana rift emphasize the role of inherited transverse discontinuities on initiation of strain and syn-rift basins. We therefore come to the conclusion that: (1) instead of propagating progressively southwards from the Afar/Aden ocean/continent transitional area to the rift tip zone in the North Tanzanian Divergence, extension nucleated more or less simultaneously, in the time interval *c.* 35–20 Ma, within discrete initial 'cells' (Turkana, Baringo); and (2) the locus of initial extension is highly controlled by the presence of major transverse brittle weakness zones. However, the mode of fault/basin linkage between nucleation zones and its possible relationships with syn-rift magmatism younging roughly to the south (e.g. Chapman *et al.* 1978) still remain to be clarified.

By contrast with the Afar and Central Kenya plume-related updomed rifts, the Turkana transverse faulted zone is inferred to have evolved as a long-lived downwarping domain where only local-scale domes (*c.* 100 km in diameter) existed in relation to individual magmatic centres. Indeed, though being volcanically active since the Oligocene, there is no strong supportive evidence for regional domal uplift in the Turkana depression that would indicate the presence of a Neogene thermal plume. Explaining the origin of its subdued rift topography implies a consideration of various parameters, about which consensus does not yet exist, and which are: (1) the existence of stationary mantle rising plume(s) beneath East Africa (McDonald 1994; Stewart & Rogers 1996; George *et al.* 1998; Rogers *et al.* 2000; George & Rogers 2002), (2) the absolute plate motion of Africa since Eocene times (Bonavia *et al.* 1995; Gripp & Gordon, 2002; Morley 2002; Calais *et al.* 2006); and (3) the interference between plume(s) and the East Africa heterogeneous lithosphere involving Cretaceous and Cenozoic rift systems.

Among the possible models that might account for the origin of the Turkana depression, the following hypotheses can be attempted: (1) lithospheric strengthening of the Anza–Sudan rift, by assimilation and cooling of mantle-derived material to the lower crust (Kusznir & Park 1987; Ziegler & Cloetingh 2004) at the end of Cretaceous–Palaeogene rifting, that prevents any bending of the weaker and normal-thickness lithosphere in the Turkana intervening zone; and (2) lateral channelling of Cenozoic (plume-induced?) melts at the base of the Anza–Sudan Cretaceous stretched lithosphere ('thin-spot model' of Thompson & Gibson 1991) that should have resulted in a drastic reduction of magma production and a decrease of thermally induced uplift in the Turkana rift zone.

The authors wish to thank Christopher Morley, Tanya Furman and Cynthia Ebinger for their helpful and constructive comments which improved the paper. We also thank Cynthia Ebinger for integrating our Kenya paper in this special issue devoted to the Ethiopian rift. This publication is contribution no. 964 of the IUEM, European Institute for Marine Studies (Brest, France).

## References

ARAMBOURG, C. 1943. Les Formations pretertiaires de la bordure occidentale du Lac Rudolphe (Afrique Orientale). *Comptes Rendus de l'Académie des Sciences de Paris*, **197**, 1663–1665.

BELLIENI, G.E., VICENTIN, J., ZANETTIN, B. & PICIRILLO, E.M. 1981. Oligocene transitional tholeiitic magmatism in northern Turkana (Kenya): comparison with coeval Ethiopian volcanism. *Bulletin Volcanologique*, **44**, 411–427.

BONAVIA, F., CHOROWICZ, J. & COLLET, B. 1995. Have wet and dry Precambrian crust largely governed Cenozoic intraplate magmatism from Arabia to east Africa? *Geophysical Research Letters*, **22**, 2337–2340.

BOSWORTH, W. & MAURIN, A. 1993. Structure, geochronology and tectonic significance of the northern Suguta Valley (Gregory Rift), Kenya. *Journal of the Geological Society of London*, **150**, 751–762.

BOSWORTH, W. & MORLEY, C.K. 1994. Structural and stratigraphic evolution of the Anza rift, Kenya. *Tectonophysics*, **236**, 93–115.

BOSWORTH, W., STRECKER, M.R. & BLISNIUK, P.M. 1992. Integration of East African paleostress and present-day stress data: implications for continental stress field dynamics. *Journal of Geophysical Research*, **97**, B8, 11851–11865.

CALAIS, E., EBINGER, C.J., HARTNADY, C. & NOCQUET, J.M. 2006. Kinematics of the East African Rift from GPS and earthquake slip vector data. *In*: YIRGU, G., EBINGER, C.J. & MAGUIRE, P.K.H. (eds) *The Afar Volcanic Province within the East African Rift System*. Geological Society, London, Special Publications, **259**, 9–22.

CALLOT, J.P., GRIGNE, C., GEOFFROY, L. & BRUN, J.P. 2001. Development of volcanic margins: two-dimensional laboratory models. *Tectonics*, **20**, 148–159.

CALLOT, J.P., GEOFFROY, L. & BRUN, J.P. 2002. Development of volcanic margins: three-dimensional laboratory models. *Tectonics*, **21**, 1052 (doi: 10.1029/2001TC901019).

CHAPMAN, G.R., LIPPARD, S. & MARTIN, J.E. 1978. The stratigraphy and structure of the Kamasia range, Kenya rift valley. *Journal of the Geological Society of London*, **135**, 265–281.

CHARSLEY, T.J. 1987. Geology of North Horr area. Ministry of Environment and Natural Resources, Mines and Geology Department, Republic of Kenya, 40pp.

CHOROWICZ, J. & MUKONKI, M.B. 1980. *Linéaments anciens, zones transformantes récentes et géotectoniques des fosses dans l'est Africain, d'après la télédétection et la microtectonique*. Museum Royal Africa Central, Tervuren, Belgium. Department of Geology and Mineralogy Annual Report, 143–146.

CORTI, G., BONINI, M., CONTICELLI, S., INNOCENTI, F., MANETTI, P. & SOKOUTIS, D. 2003. Analogue modelling of continental extension: a review focused on the relations between the patterns of deformation and the presence of magma. *Earth Science Reviews*, **63**, 169–247.

COUSSEMENT, C. 1995. *Structures transverses et extension intracontinentale. Le rôle des zones de failles d'Assoua et Tanganyika–Rukwa–Malawi dans la cinématique néogène du système de Rift Est-Africain*. PhD Thesis, Université de Bretagne Occidentale, Brest, 222pp.

DAVIDSON, A. (compiler) 1983. The Omo River Project, reconnaissance geology and geochemistry of parts of Illubabor, Kefa, Gemu Gofa, and Sidamo: Ministry of Mines and Energy, *Ethiopian Institute Geological Surveys Bulletin*, **2**, 1–89.

DAVIDSON, A. & REX, D.C. 1980. Age of volcanism and rifting in southern Ethiopia. *Nature*, **283**, 657–658.

DINDI, E.W. 1994. Crustal structure of the Anza graben from gravity and magnetic investigations. *Tectonophysics*, **236**, 359–371.

DUNKELMAN, T.J., ROSENDAHL, B.R. & KARSON, J.A. 1989. Structure and stratigraphy of the Turkana rift from seismic reflection data. *Journal of African Earth Sciences*, **8**, 489–510.

DUNKLEY, P.N., SMITH, M., ALLEN, D.J. & DARLING, W.G. 1993. The geothermal activity and geology of the northern sector of the Kenya Rift Valley. From *Research report SC/93/1*, British Geological Survey for Kenyan Ministry of Energy, 183pp.

EBINGER, C.J. & IBRAHIM, A. 1994. Multiple episodes of rifting in Central and East Africa: A re-evaluation of gravity data. *Geologisches Rundschau*, **83**, 689–702.

EBINGER, C.J. & SLEEP, N. 1998. Cenozoic magmatism throughout East Africa resulting from impact of a single plume. *Nature*, **395**, 788.

EBINGER, C.J., YEMANE, T., WOLDEGABRIEL, G., ARONSON, J.L. & WALTER, R.C. 1993. Late-Eocene-Recent volcanism and faulting in the southern Main Ethiopian Rift. *Journal of the Geological Society of London*, **150**, 99–108.

EBINGER, C.J., YEMANE, T., HARDING, D.J., TESFAYE, S., KELLEY, S. & REX, D.C. 2000. Rift deflection, migration and propagation: Linkage of the Ethiopian and Eastern rifts, Africa. *Geological Society of America Bulletin*, **112**, 163–176.

GEORGE, R.M. & ROGERS, N.W. 2002. Plume dynamics beneath the African plate inferred from the geochemistry of the Tertiary basalts of southern Ethiopia. *Contributions to Mineralogy and Petrology*, **144**, 286–304.

GEORGE, R., ROGERS, N. & KELLEY, S. 1998. Earliest magmatism in Ethiopia: Evidence for two mantle plumes in one flood basalt province. *Geology*, **26**, 10, 923–926.

GIRARD, C.M. & GIRARD, M.C. 1999. *Traitement des données de Télédétection*. Editions Dunod, Paris 528pp.

GRIPP, A.E. & GORDON, R.G. 2002. Young track of hotspots and current plate velocities. *Geophysical Journal International*, **150**, 321–361.

HACKMAN, B.D., CHARSLEY, T.J., KEY, R.M. & WILKINSON, A.F. 1990. The development of the East African Rift system in north–central Kenya. *Tectonophysics*, **184**, 189–211.

HAILEAB, B., BROWN, F.H., MCDOUGALL, I. & GATHOGO, P.N. 2004. Gombe Group basalts and initiation of Pliocene deposition in the Turkana depression, northern Kenya and southern Ethiopia. *Geological Magazine*, **141**, 41–53.

HARRIS, J.M., BROWN, F.H. & LEAKEY, M.G. 1988. Stratigraphy and paleontology of Pliocene and Pleistocene localities west of Lake Turkana, Kenya. *Contributions in Science* **399**, 1–128.

HAUTOT, S., TARITS, P., WHALER, K., LE GALL, B., TIERCELIN, J.J. & LE TURDU, C. 2000. Deep structure of the Baringo Rift Basin (central Kenya) from three-dimensional magnetotelluric imaging: Implications for rift evolution. *Journal of Geophysical Research*, **105**, B10, 23493–23518.

HENDRIE, D.B., KUSZNIR, N.J. MORLEY, C.K. & EBINGER, C.J. 1994. Cenozoic extension in northern Kenya: a quantitative model of rift basin development in the Turkana region. *In*: PRODEHL, C., KELLER, G.R. & KHAN, M.A. (eds) Crustal and Upper Mantle Structure of the Kenya Rift. *Tectonophysics*, **236**, 409–438.

KRISP Working Group, 1991. The Kenyan Rift: pure shear extension above a mantle plume. *Nature*, **345**, 223–227.

KUSZNIR, N.J. & PARK, R.G. 1987. The extensional strength of the continental lithosphere: its dependence on geothermal gradient and crustal thickness. *In*: COWARD, M.P., DEWEY, J.F. & HANCOCK, P.L. (eds) *Continental Extensional Tectonics*. Geological Society, London, Special Publications, **28**, 35–52.

LE GALL, B., VETEL, W. & MORLEY, C.K. 2005. Inversion tectonics during continental rifting: The Turkana rifted zone, Northern Kenya. *Tectonics*, **24**, TC2002, doi:10.1029/2004TC001637.

MCCONNELL, R.B. 1972. Geological development of the rift system of eastern Africa. *Geological Society of America Bulletin*, **83**, 2549–2572.

MCDONALD, R. 1994. Petrological evidence regarding the evolution of the Kenya Rift Valley. *Tectonophysics*, **236**, 373–390.

MCDOUGALL, I. & WATKINS, R.T. 1988. Potassium–argon ages of volcanic rocks from northeast of Lake Turkana, northern Kenya. *Geological Magazine*, **125**, 15–23.

MCDOUGALL, I. & FEIBEL, C.S. 1999. Numerical age control for the Miocene–Pliocene succession at Lothagam, a hominoid-bearing sequence in the northern Kenya Rift. *Journal of Geological Society, London*, **156**, 731–745.

MECHIE, J., KELLER, G.R., PRODEHL, C., GACIRI, S., BRAILE, L.W., MOONEY, W.D., GAJEWSKI, D. & SANDMEIER, K.J. 1994. Crustal structure beneath the Kenya Rift from axial profile data. *In*: PRODEHL, C., KELLER, G.R. & KHAN, M.A. (eds) Crustal and Upper Mantle Structure of the Kenya Rift. *Tectonophysics*, **236**, 179–199.

MORLEY, C.K. 1994. Interaction of deep and shallow processes in the evolution of the Kenya rift. *Tectonophysics*, **236**, 81–91.

MORLEY, C.K. 1999*a*. Influence of pre-existing fabrics on rift structure. *In*: MORLEY, C.K. (ed.) *Geoscience of Rift Systems—Evolution of East Africa*. AAPG Studies in Geology, **44**, 151–160.

MORLEY, C.K. 1999*b*. Aspect of Transfer Zone Geometry and Evolution in East African Rifts. *In*: MORLEY, C.K. (ed.) *Geoscience of Rift Systems-Evolution of East Africa*. AAPG Studies in Geology, **44**, 161–171.

MORLEY, C.K. 1999*c*. Marked along-strike variations in dip of the normal faults—the Lokichar fault, N. Kenya rift: a possible cause for metamorphic core complexes. *Journal of Structural Geology*, **21**, 479–492.

MORLEY, C.K. 2002. Tectonic settings of continental extensional provinces and their impact on sedimentation and hydrocarbon prospectivity. *SEPM, Special Publication*, **73**, 25–56.

MORLEY, C.K., WESCOTT, W.A., STONE, D.M., HARPER, R.M., WIGGER, S.T. & KARANJA, F.M. 1992. Tectonic evolution of the northern Kenyan Rift. *Journal of the Geological Society, London*, **149**, 333–348.

MORLEY, C.K, WESCOTT, W.A., HARPER, R.M., WIGGER, S.T., DAY, R.A. & KARANJA, F.M. 1999*a*. Geology and Geophysics of the Western Turkana Basins, Kenya. *In*: MORLEY, C.K. (ed.) *Geoscience of Rift Systems—Evolution of East Africa, AAPG Studies in Geology*, **44**, 19–54.

MORLEY, C.K., DAY, R.A., *ET AL*. 1999*b*. Geology and Geophysics of the Anza Graben. *In*: MORLEY, C.K. (ed.) *Geoscience of Rift Systems—Evolution of East Africa*. AAPG Studies in Geology, **44**, 67–90.

MUGISHA, F., EBINGER, C.J., STRECKER, M. & POPE, D. 1997. Two-stage rifting in the Kenya Rift: Implications for half-graben models. *Tectonophysics*, 278, 61–81.

OCHIENG, J.O., WILKINSON, A.F., KAGASI, J. & KIMONO, S. 1988. Geology of the Loiyangalani area. *Ministry of Environment and Natural Resources*, Mines and Geological Department, Republic of Kenya, 53pp.

PRODEHL, C., JACOB, B., THYBO, H., DINDI, E. & STANGL, R. 1994. Crustal structure on the northeastern flank of the Kenya Rift. *In*: PRODEHL, C., KELLER, G.R. & KHAN, M.A. (eds) Crustal and Upper Mantle Structure of the Kenya Rift. *Tectonophysics*, **236**, 271–290.

REEVES, C.V., KARANJA, F.M. & MCLEOD, I.N. 1987. Geophysical evidence for a failed Jurassic rift and triple junction in Kenya. *Earth and Planetary Science Letters*, **81**, 299–311.

ROCHE, H., DELAGNES, A., BRUGAL, J.P., FEIBEL, C., KIBUNJIAS, M., MOURREL, V. & TEXIER, P.J. 1999. Early hominid stone tool production and technical skill 2.34 Myr ago in West Turkana, Kenya. *Nature*, **399**, 57–60.

ROGERS, N., MCDONALD, R., FITTON, J.G., GEORGE, R., SMITH, M. & BARREIRO, B. 2000. Two mantle plumes beneath the East African Rift System: Sr, Nd and Pb isotope evidence from Kenya Rift basalts. *Earth and Planetary Science Letters*, **176**, 387–400.

SHACKELTON, R.M. 1993. Tectonics of the Mozambique Belt in East Africa. *In*: RICHARD, H.M., ALABASTER, T., HARRIS, N.B.W. & NEARY, C.R. (eds) *Magmatic Processes and Plate Tectonics*, Geological Society, London, Special Publications, **76**, 345–362.

SHELL Report, Well Resume Eliye Springs-1, 1992. National Oil Corporation of Kenya, Nairobi, 40pp.

SMITH, M. & MOSLEY, P. 1993. Crustal heterogeneity and basement influence on the development of the Kenya Rift, East Africa. *Tectonics*, **12**, 591–606.

STEWART, K. & ROGERS, N. 1996. Mantle plumes and lithosphere contributions to basalts from southern Ethiopia. *Earth and Planetary Science Letters*, **139**, 195–211.

THOMPSON, R.N. & GIBSON, S.A. 1991. Subcontinental mantle plumes, hotspots and pre-existing thinspots. *Journal of the Geological Society, London*, **148**, 973–977.

VETEL, W., LE GALL, B. & WALSH, J.J. 2005. Geometry and growth of an inner rift fault pattern: the Kino Sogo Fault Belt, Turkana Rift (North Kenya). *Journal of Structural Geology*, **27**, 2204–2222.

WALSH, J. & DODSON, R.G. 1969. Geology of the Northern Turkana. Report 82, Geological Survey of Kenya, 48pp.

WESCOTT, W.A., WIGGER, S.T., STONE, D.M. & MORLEY, C.K. 1999. Geology and geophysics of the Lotikipi plain. *In*: MORLEY, C.K. (ed.) *Geoscience of Rift Systems—Evolution of East Africa*. AAPG Studies in Geology, **44**, 55–65.

WILKINSON, A.F. 1988. Geology of the Allia Bay area. Report 109, *Ministry of Environment and Natural Resources*, Mines and Geological Department, Republic of Kenya, 54pp.

WILLIAMSON, P.G. & SAVAGE, R.J.G. 1986. Early rift sedimentation in the Turkana basin, northern Kenya. *In*: FROSTICK, L.E. *ET AL*. (eds) *Sedimentation in the African Rifts*. Geological Society of London, Special Publications, **25**, 267–283.

WOLDEGABRIEL, G.W. & ARONSON, J.L. 1987. Chow Bahir rift: A 'failed' rift in southern Ethiopia. *Geology*, **15**, 430–433.

ZANETTIN, B., VISENTIN, J., BELLIENI, G., PICCIRILLO, E.M. & RITA, F. 1983. Le volcanisme du bassin Nord-Turkana (Kenya) : âge, succession et évolution structurale. *Bulletin des Centres de Recherches Exploration Production Elf-Aquitaine*, **7**, 249–255.

ZIEGLER, P.A. & CLOETINGH, S. 2004. Dynamic processes controlling evolution of rifted basins. *Earth-Science Reviews*, **64**, 1–50.

# Part 4: Rifting in the Afar volcanic province: Geophysical studies of crustal structure and processes

This section describes geophysical studies in the northern Main Ethiopian Rift and the adjoining Ethiopian plateau, and in eastern Afar. The first paper involves a detailed analysis of the crust beneath Djibouti, while the remaining four papers present new results from the EAGLE project (Maguire *et al.* 2003) concentrating on variations in crustal structure beneath the rift in its transitional state between the Main Ethiopian Rift and south-western Afar. The implications of the results are discussed in relation to the influence of magmatic processes on strain localization as rifting proceeds to sea-floor spreading.

**Dugda *et al.*** present a detailed analysis of crustal structure beneath GEOSCOPE station ATD in Djibouti using $H-\kappa$ stacking of 48 high-quality receiver functions derived from 10 years of recorded data. They provide estimates of Moho depth, mean crustal velocity and crustal Poisson's ratio. A second analysis involving the joint inversion of receiver functions and surface wave group velocities shows how shear wave velocities vary with depth in the crust beneath ATD. While their results are different from some previous studies, this is considered due to the definition of Moho velocity. The structure identified beneath ATD is similar to that elsewhere beneath central and eastern Afar. The principal conclusion is that the high average $V_p$ and Poisson's ratio indicate a crust of mafic composition, inconsistent with models invoking the stretching of pre-existing crust. The preferred model involves the igneous crust being generated as the lithosphere extended, by the intrusion of mantle-derived magmas, the implications being that igneous rock is emplaced at the late syn-rift stage, extension being accommodated by the addition of large volumes of intrusive rock, mafic dykes, sills and underplate. This reinforces the transitional rift model of Ebinger & Casey (2001), who suggested that in such rifts, strain is accommodated locally within magmatic centres rather than along rift border faults.

**Stuart *et al.*** using the broadband passive array data from the EAGLE project, also determine receiver functions to derive crustal thickness and average $V_p/V_s$ for the northern Main Ethiopian Rift (NMER) and its flanking plateau. Small variations in crustal thickness on the western plateau correlate with the off-rift mantle low-velocity structure identified by Bastow *et al.* (2005). Beneath the eastern plateau, the crust is of similar thickness. From the $V_p/V_s$ results, the crust beneath the western plateau on the Nubian plate is more mafic than that beneath the Somalian plate to the south-east of the rift, which could result from more magmatic intrusion or different pre-rift crustal compositions. This is reflected in the results of the MT study by Whaler & Hautot (2006), who show that there is a fundamental difference in crustal resistivity beneath the eastern and western plateaus flanking the rift valley. There are very high $V_p/V_s$ ratios (>2.0) beneath the Quaternary volcanic chains on the flanks of rift, indicative of partial melt within the crust. Such $V_p/V_s$ ratios are also identified at the northeastern end of the EAGLE survey within the rift towards the Afar depression. This corroborates other geophysical evidence (e.g. Dugda *et al.* 2006) for increased magmatic activity as continental rifting evolves to oceanic spreading in Afar. Crustal thinning increases from south to north along the axis of the rift, consistent with the results of the EAGLE-controlled source survey (Maguire *et al.* 2006). These results reinforce Ebinger & Casey's (2001) model for the development of a transitional rift, itself consistent with the magma-assisted rifting hypothesis of Buck (2004).

**Maguire *et al.*** integrate the results of the EAGLE-controlled source survey. Their principal results show that the upper crust beneath the rift is intruded by cooled gabbroic intrusions, modifying the upper crustal velocity to *c.* 6.5 km s$^{-1}$ from *c.* 6.1 km s$^{-1}$ beneath the plateaus. These intrusions are arranged en echelon along the axis of the rift beneath the overlying Quaternary volcanic segments (e.g. Casey *et al.* 2006). They identify a high velocity (*c.* 7.4 km s$^{-1}$) 'underplated/intruded' layer at the base of the crust beneath the northwestern flank of the rift, suggesting it was emplaced during the Oligocene flood basalt episode, possibly modified by more recent melt addition. The body lies directly above the low-velocity mantle anomaly identified

*From*: YIRGU, G., EBINGER, C.J. & MAGUIRE, P.K.H. (eds) 2006. *The Afar Volcanic Province within the East African Rift System*. Geological Society, London, Special Publications, **259**, 235–237.
0305-8719/06/$15.00 

by Bastow *et al.* (2005). Along the axis of the rift, the crust thins from about 40 to about 26 km from SW to NE beneath Afar. This, together with the variation in crustal velocities and layer thicknesses along the EAGLE axial profile, suggests that continental crust to the south develops into Iceland-type igneous crust at the northern end of the profile, consistent with the results of Stuart *et al.* (2006). An intriguing observation of a near-continuous mantle reflector at a depth of ~15–25 km below the base of the crust, also seen beneath the Kenya (Eastern) rift profiles (Prodehl *et al.* 1994, Fuchs *et al.* 1997) is discussed. It may be an important observation in relation to the origin and development of magmatic continental rifts and subsequent volcanic margins.

**Whaler *et al.*** report the results of a magnetotelluric (MT) survey occupying sites along the central portion of the EAGLE cross-rift wide-angle seismic profile. Results show that the shallow structure correlates well with the Quaternary to Jurassic age rocks, with a highly conducting lens beneath the Boset volcano at <1 km depth possibly representing a magma body. Beneath the Boset magmatic segment, conductive material extending to at least mid-crustal depths is consistent with the results of the seismic (Keranen *et al.* 2004; Mackenzie *et al.* 2005) and gravity modelling (Cornwell *et al.* 2006) in suggesting the presence of a mafic intrusion at depth. An intriguing extremely conductive body is identified in the lower crust beneath the northern flank of the rift above the 'underplate/intrusive' lower crustal layer identified by Mackenzie *et al.* (2005) and above the anomalous low-velocity mantle zone identified by Bastow *et al.* (2005). It is suggested to be a more recent and/or more extensive and/or hotter mafic intrusion than beneath the Boset magmatic segment, that may be connected to the conductive zone beneath the rift, and postulated as a region of partial melt. The inferred geoelectric strike direction which, within the rift, matches the orientation of the magmatic segments and on the plateau is parallel to the border faults, mirrors the change in orientation of the fast split shearwave (Kendall *et al.* 2005; Keir *et al.* 2005). This together with the identification of the high conductivity, probable magmatic intrusions beneath the profile, again is consistent with the model of magma-assisted rifting (Buck 2004).

Finally, Cornwell *et al.* analyse new gravity data to produce a 2D crustal and upper mantle density model across the NMER along the EAGLE cross-rift wide-angle seismic and MT profile. An initial model was produced by converting the EAGLE cross-rift wide-angle seismic model velocities (Mackenzie *et al.* 2005) into densities. The final model was obtained by varying the model parameters within limits defined by the seismic model errors and the testing of the robustness of the velocity–density relationship used. Upper crustal high density bodies are introduced beneath the Boset magmatic segment and a further 12 km wide body beneath the northwestern margin of the rift. Both are coincident with seismic velocity and MT anomalies, and interpreted in terms of gabbroic intrusions at depth. A shallow high-density body beneath the axial Boset volcano is interpreted as either a dyke zone feeding the magmatic segment from basaltic lavas at depth, or representing a body of magma derived by partial fractionation at depth or melting of existing crust and sediments due to increased temperature caused by the mafic intrusive material below. The study provides supporting evidence for the underplate layer beneath the northwestern rift flank (Mackenzie *et al.* 2005; Maguire *et al.* 2006). Analysis of the Bouguer anomaly in the light of a long-wavelength regional anomaly derived along the profile, shows that relatively low-density upper mantle must exist beneath the northwestern plateau and the rift, consistent with the presence of low-velocity mantle material extending to a depth of at least 300 km identified by Bastow *et al.* (2005). The change in crustal and mantle properties consistent with other EAGLE seismic and MT results occurring at the southeast margin of the rift marks the extent of lithospheric modification by rifting processes. Pre-existing changes in lithospheric properties across the rift may occur, as suggested by Whaler & Hautot (2006), but cannot be isolated in this density model.

## References


BASTOW, I.D., STUART, G.W., KENDALL, J.-M. & EBINGER, C.J. 2005. Upper mantle seismic structure in a region of incipient continental break-up: northern Ethiopian rift. *Geophysical Journal International*, **162**, 479–493 doi:10.1111/j.1365-246x.2005.02666.x.

BUCK, W.R. 2004. Consequences of asthenospheric variability on continental rifting. *In*: KARNER, G.D., TAYLOR, B., DRISCOLL, N.W. & KOHLSTEDT, D.L. (eds) *Rheology and Deformation of the Lithosphere at Continental Margins*. Columbia University Press, New York, pp. 1–30.

EBINGER, C.J. & CASEY, M. 2001. Continental breakup in magmatic provinces: An Ethiopian example. *Geology*, **29**, 527–530.

KEIR, D., KENDALL, J.-M. & EBINGER, C. 2005. Variations in late syn-rift melt alignment inferred from shear-wave splitting in crustal earthquakes beneath the Ethiopian rift. *Geophysical Research Letters*, L23308, doi: 10.1029/2005GL024150.

KENDALL, J.-M., STUART, G., EBINGER, C., BASTOW, I. & KEIR, D. 2005. Magma-assisted rifting in Ethiopia. *Nature*, **433**, 146–148.

KERANEN, K., KLEMPERER, S.L., GLOAGUEN, R. & the EAGLE Working Group, 2004. Imaging a proto-ridge axis in the Main Ethiopian Rift, *Geology*, **32**, 949–952.

MACKENZIE, G.D., THYBO, H. & MAGUIRE, P.K.H. 2005. Crustal velocity structure across the Main Ethiopian Rift: Results from two-dimensional wide-angle seismic modelling. *Geophysical Journal International*, **162**, 994–1006 doi: 10.1111/j.1365-246X.2005. 02710.x.

MAGUIRE, P.K.H., EBINGER, C.J. *ET AL*. 2003. Geophysics project in Ethiopia studies continental breakup. *EOS Transactions American Geophysical Union*, **84**, 342–343.

# New constraints on crustal structure in eastern Afar from the analysis of receiver functions and surface wave dispersion in Djibouti

MULUGETA T. DUGDA & ANDREW A. NYBLADE

*Department of Geosciences, Penn State University, University Park, PA USA 16802 (e-mail: mulugeta@geosc.psu.edu; andy@geosc.psu.edu)*

**Abstract:** Crustal structure beneath the GEOSCOPE station ATD in Djibouti has been investigated using $H$–$\kappa$ stacking of receiver functions and a joint inversion of receiver functions and surface wave group velocities. We obtain consistent results from the two methods. The crust is characterized by a Moho depth of $23 \pm 1.5$ km, a Poisson's ratio of $0.31 \pm 0.02$, and a mean $V_p$ of *c.* 6.2 km s$^{-1}$ but *c.* 6.9–7.0 km s$^{-1}$ below a 2–5 km-thick low-velocity layer at the surface. Some previous studies of crustal structure for Djibouti placed the Moho at 8 to 10 km depth, and we attribute this difference to how the Moho is defined (an increase of $V_p$ to 7.4 km s$^{-1}$ in this study vs. 6.9 km s$^{-1}$ in previous studies). The crustal structure we obtained for ATD is similar to crustal structure in many other parts of central and eastern Afar. The high Poisson's ratio and $V_p$ throughout most of the crust indicate a mafic composition and are not consistent with models invoking crustal formation by stretching of pre-existing Precambrian crust. Instead, we suggest that the crust in Afar consists predominantly of new igneous rock emplaced during the late syn-rift stage where extension is accommodated within magmatic segments by dyking. Sill formation and underplating probably accompany the dyking to produce the new and largely mafic crust.

Afar is a tectonically active region in between continental rifting and oceanic rifting where the Red Sea oceanic ridge, the Gulf of Aden oceanic ridge, and the continental East African Rift System (EARS) meet in a rift–rift–rift triple junction (Fig. 1). Understanding the nature and origin of the crust in Afar is important because Afar is one of the few places where it is possible to study the development of magmatic segmentation during rifting, the formation of volcanic rifted margins, and more generally, how continental rifts evolve into oceanic rifts.

There are a number of conflicting views about the thickness and composition of the crust in Afar, and how it formed. For example, some studies have reported estimates of crustal thickness between 14 and 26 km, and from these estimates inferred that the crust may be more continental than oceanic in nature (Makris & Ginzburg 1987; Berkhemer *et al.* 1975). Other studies have reported estimates of crustal thickness between 8 and 10 km (Ruegg 1975; Sandvol *et al.* 1998), suggesting an oceanic origin for the crust, and yet others have argued for completely new igneous crust that is much thicker than typical oceanic crust based on such observations as the velocity structure of the crust and Poisson's ratio (Mohr 1989; Dugda *et al.* 2005).

In this paper, crustal structure is imaged beneath the GEOSCOPE station ATD in eastern Afar (Djibouti) using receiver functions and surface wave dispersion measurements. We report new estimates of Moho depth, Poisson's ratio and shear velocity structure, and combine these estimates with seismic images of crustal structure from other parts of Afar to re-examine the nature and origin of the crust in Afar. This study differs from previous studies of receiver functions for station ATD in that 10 years of data have been used, enabling us to obtain high-quality receiver function stacks and to investigate azimuthal variations in structure beneath the station. In addition, Rayleigh wave group velocities have been used to constrain crustal shear wave velocities.

## Tectonic and geodynamic setting

The seismic station ATD is located on the south side of the Gulf of Tadjoura in Djibouti (Figs 1 and 2). The Gulf of Tadjoura is the western extension of the Gulf of Aden ridge, and represents the penetration of the ridge into Afar, where it joins with the East African and Red Sea rifts (Courtillot 1980; Cochran 1981; Manighetti *et al.* 1997; Courtillot *et al.* 1999). Station ATD is located on 10–14 Ma rhyolites, and away from the main zones of rifting in the Gulf of Tadjoura (Fig. 2) (Manighetti *et al.* 1997, 1998).

The tectonic and geodynamic setting of Afar has been studied extensively (e.g. Hayward & Ebinger

*From*: YIRGU, G., EBINGER, C.J. & MAGUIRE, P.K.H. (eds) 2006. *The Afar Volcanic Province within the East African Rift System*. Geological Society, London, Special Publications, **259**, 239–251.
0305-8719/06/$15.00 

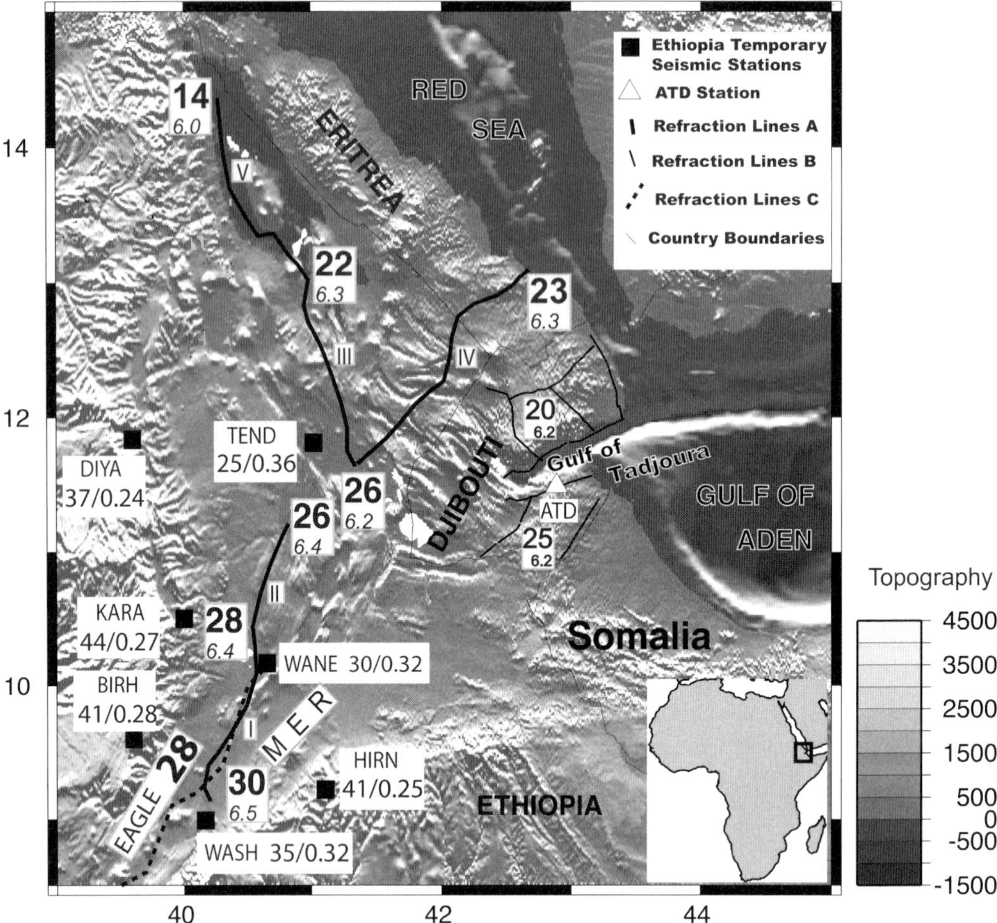


**Fig. 1.** Shaded relief map of the Afar region showing location of GEOSCOPE station ATD (white triangle), political boundaries (thin solid lines), temporary broadband seismic stations operated between 2000 and 2002 (black squares), and seismic refraction lines (A, Berkhemer, 1975, solid bold lines labelled I to V; B, Ruegg, 1975, medium thickness lines in Djibouti; C, EAGLE, Maguire *et al.* 2003, bold dashed line). Station names are shown in the white boxes next to seismic stations, Moho depth (in km), and crustal Poisson's ratio. Estimates of Moho depth (in km) and mean crustal $V_p$ (in km $s^{-1}$) are shown in the white boxes next to seismic refraction lines, assuming the Moho is where P-wave velocities increase to $\geq$7.4 km $s^{-1}$.

1996; Manighetti *et al.* 1997, 1998; Acton *et al.* 2000; Ebinger & Casey 2001; Tesfaye *et al.* 2003; Audin *et al.* 2004; Wolfenden *et al.* 2005). The Red Sea and Gulf of Aden rifts began forming in the Oligocene when Arabia first separated from Africa. The western Gulf of Aden initially opened between 15 and 10 Ma and entered Afar at *c.* 5 Ma (Courtillot *et al.* 1999; Hofmann *et al.* 1997). The eruption of flood basalts in Ethiopia and Yemen occurred at about 31 Ma, concurrent with or immediately prior to the opening of the Red Sea and Gulf of Aden (d'Acremont *et al.* 2005; Wolfenden *et al.* 2005; Ukstins *et al.* 2002; Hofmann *et al.* 1997). The opening of the Main Ethiopian Rift (MER) to form the third arm of the Afar triple junction commenced around 11 Ma (Wolfenden *et al.* 2005; Chernet *et al.* 1998).

Rift models for extension in rheologically layered continental lithosphere, as found in Afar, can be grouped into two categories: (1) those invoking mechanical stretching, where strain is accommodated by large offset faults in a brittle upper crust and by ductile deformation in the lower crust; and (2) those invoking extension caused by dyke intrusion within magmatic segments (e.g. Ebinger & Casey 2001; Buck 2004; Ebinger

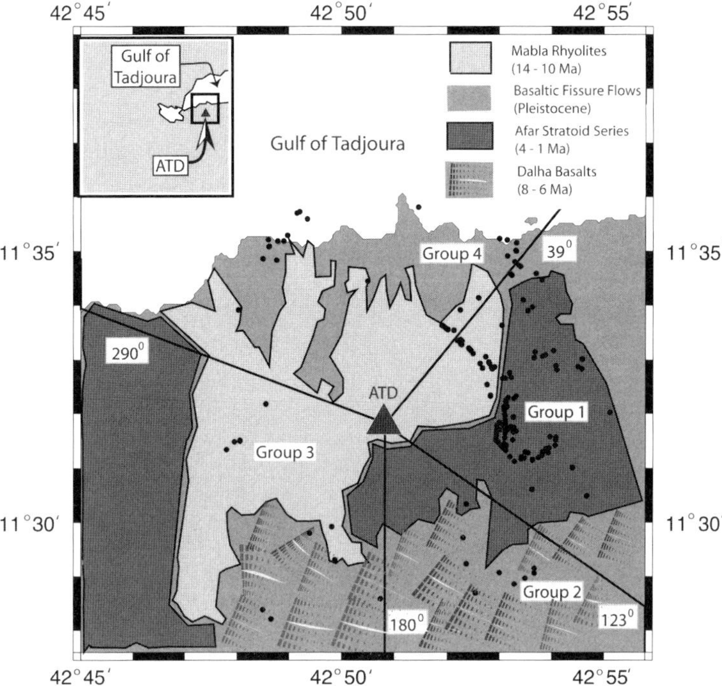


**Fig. 2.** Sketch map showing the geology around station ATD (solid triangle) (after Varet 1975), the distribution of Ps conversion points at a depth of 25 km (dots), and the four regions used to group receiver functions.

2005). The latter proposal has been supported for the MER and Afar by geodetic data indicating that magmatic segments accommodate >80% of the extension (Bilham *et al.* 1999), and by gravity and morphotectonic observations (Hayward & Ebinger 1996; Manighetti *et al.* 1998). A dyke intrusion model is also supported for the Afar and MER by recent shear-wave splitting studies using seismic data from the Ethiopian Afar Geoscientific Lithospheric Experiment (EAGLE) and permanent seismic stations in the region (Ayele *et al.* 2004; Kendall *et al.* 2005).

## Previous studies of crustal structure in Afar

Detailed information about crustal P-wave velocity structure in Djibouti comes from the deep seismic sounding experiment of Ruegg (1975) (Fig. 1). To the south of the Gulf of Tadjoura, Ruegg (1975) reported a velocity structure consisting of five layers with P-wave velocities, from top to bottom, of 4.0 km $s^{-1}$, 6.4 km $s^{-1}$, 6.9 km $s^{-1}$, 7.1 km $s^{-1}$, and 7.4 km $s^{-1}$, and layer thicknesses of 3.6 km, 5.6 km, 4.6 km, 11 km and 10 km, respectively. Ruegg (1975) interpreted the Moho to be at about 10–11 km depth between layers with velocities of 6.4 km $s^{-1}$ and 6.9 km $s^{-1}$ (i.e. velocities of about 6.9 km $s^{-1}$ and higher were considered to be indicative of mantle rock). Ruegg (1975) found similar structure to the north of the Gulf of Tadjoura, with increase in velocity from 7.1 km $s^{-1}$ to 7.4 km $s^{-1}$ occurring at a somewhat shallower depth of about 20 km.

Crustal structure has been investigated in other parts of the Afar using seismic refraction, surface wave dispersion, gravity and other geophysical data. Figure 1 summarizes Moho depths ($H$), crustal $V_p$ values, and crustal Poisson's ratios reported in previous studies. Much of the information available about crustal structure in Afar comes from seismic refraction surveys conducted in the mid-1970s (Berkhemer *et al.* 1975) (lines I–V, Fig. 1). Makris & Ginzburg (1987), revising the previous interpretation of Berkhemer *et al.* (1975), reported Moho depths along refraction lines I and II of 30 km in the south and 26 km in the north. For profiles III and V, they found crustal thickness variations of 26 to 14 km, with a change in the middle of profile V from about 26 km to about 20 km. For refraction line IV, they reported that crustal thickness thins from 26 to 23 km toward the Red Sea coast. The values of P-wave structure reported by Ruegg (1975) for the Gulf of Tadjoura region is similar to the structure reported by Makris & Ginzburg (1987) for other parts of the Afar, but the layer thicknesses are somewhat different, and also the interpretation of what velocity indicates mantle rock (i.e. how the Moho is defined). Makris & Ginzburg (1987) interpret P-wave velocities as high as 7.1–7.2 km $s^{-1}$ as crustal velocities, and they concluded that the Moho is marked by an increase in velocity to 7.5 km $s^{-1}$.

Receiver functions from station ATD were modelled for crustal structure by Sandvol *et al.* (1998). They applied a grid search technique to model a receiver function stack comprised of 11 receiver functions from a limited range in backazimuth (1 event came from the west and 10 came from the east). They obtained a Moho depth of 8 km, suggesting oceanic-like crust beneath Djibouti, similar to the interpretation of crustal structure by Ruegg (1975). More recently, Dugda *et al.* (2005) analysed receiver functions from a temporary broadband station at Tendaho (TEND, Fig. 1) in central Afar, obtaining a Moho depth of $25 \pm 3$ km, in good agreement with estimates of crustal thickness along refraction lines II, III and IV (26 km, Fig. 1) by Makris & Ginsburg (1987).

In contrast to Moho depth, there is more uniformity in the crustal Poisson's ratios reported for the Afar. Ruegg (1975) reported a high Poisson's ratio of 0.28 to 0.33 for the Gulf of Tadjoura region, and in their global study of the continental crust, Zandt & Ammon (1995) obtained a crustal Poisson's ratio for station ATD of $0.29 \pm 0.02$. From an analysis of surface waves crossing Afar, Searle (1975) reported that Poisson's ratio increases from 0.25 at the surface to 0.29 at the deepest portion of the crust. Dugda *et al.* (2005) reported a high crustal Poisson's ratio of 0.36 for Tendaho (TEND, Fig. 1).

## New estimates of crustal structure for Djibouti

Crustal structure has been examined in this study using data from the GEOSCOPE station ATD and two complementary methods: (1) the $H$–$\kappa$ receiver function stacking method; and (2) a joint inversion of receiver functions and Rayleigh wave group velocities. The first method provides estimates of Moho depth and crustal Poisson's ratio, while the second method provides information about how shear wave velocities vary with depth in the crust.

### *Crustal structure from* H–κ *analysis of receiver functions*

Receiver functions have been modelled for several decades using a variety of methods (e.g. Langston 1979; Taylor & Owens 1984). For this study, we have used the $H$–$\kappa$ stacking technique ($H$ = Moho depth and $\kappa = V_p/V_s$) of Zhu & Kanamori (2000) because it provides robust estimates of crustal thickness and Poisson's ratio. It is well known that H and κ trade off strongly (Ammon *et al.* 1990; Zandt *et al.* 1995), and in an effort to reduce the ambiguity introduced by this trade-off, Zhu & Kanamori (2000) incorporated the later arriving crustal reverberations PpPs and PpSs + PsPs in a stacking procedure whereby the stacking itself transforms the time–domain receiver functions directly to objective function values in $H$–$\kappa$ parameter space. The objective function for the stacking is

$$s(H,\kappa) = \sum_{j=1}^{N} w_1 r_j(t_1) + w_2 r_j(t_2) - w_3 r_j(t_3) \quad (1)$$

where $w_1$, $w_2$, $w_3$ are weights, $r_j(t_i)$, $i = 1, 2, 3$ are the receiver function amplitude values at the predicted arrival times $t_1$, $t_2$, and $t_3$ of the Ps, PpPs, and PsPs + PpSs phases for the jth receiver function, and $N$ is the number of receiver functions used. The $H$–$\kappa$ stacking algorithm is based on the premise that the weighted stack sum of the receiver function amplitudes should attain its maximum value when $H$ and κ attain their correct values for a particular crust. By performing a grid search through $H$ and κ parameter space, the $H$ and κ values corresponding to the maximum value of the objective function can be determined (Zhu & Kanamori 2000). The $H$–$\kappa$ method provides a better estimate of Moho depth and $V_p/V_s$ ratio than a simple stack method because it uses the correct ray parameter in the stacking of each receiver function.

Events used in this study come from distances of 30°–100° and have magnitudes greater than 5.5. Most of the events are from the east (the Indonesian and Western Pacific subduction zones), but 43 out of the 183 events come from other azimuths. For computing the receiver functions, a time-domain iterative deconvolution method (Ligorria & Ammon 1999) was used, and to evaluate the quality of the receiver functions, a least-squares misfit criterion was applied. The misfit criteria provides a measure of the closeness of the receiver functions to an ideal case, and it is calculated by using the difference between the radial component and the convolution of the vertical component with the already determined radial receiver function. Receiver functions with a misfit of 10% and less were used in our analysis.

The receiver functions were filtered with a Gaussian pulse width of 1.6. Both radial and tangential receiver functions were examined for evidence of lateral heterogeneity in the crust and for dipping structure. Events with large amplitude tangential receiver functions were not used.

In applying the $H-\kappa$ technique, it is necessary to select weights $w_1$, $w_2$, and $w_3$, and a value for $V_p$. More weight is typically given to the phase that is most easily picked. Given a range of plausible values for average crustal $V_p$ values for rifted continental crust (5.8–6.8 km $s^{-1}$; e.g. Fuchs *et al.* 1997; Prodehl *et al.* 1994 and references therein), crustal thickness can vary by almost 4 km while the $V_p/V_s$ ratio can change by 0.02, as shown in Table 1. Thus, when estimating errors for the $H-\kappa$ method, the uncertainty in mean crustal velocity, as well as the sensitivity of the results to variations in the weights ($w_1$, $w_2$, and $w_3$), need to be considered.

**Table 1.** *Results from* $H-\kappa$ *stacking of receiver functions for the different groups shown in Figure 2 for a range of plausible mean crustal* $V_p$

| Group | $V_p$ | Moho depth (km) | $V_p/V_s$ | Poisson's ratio ($\sigma$) |
|---|---|---|---|---|
| 1 | **5.8** | 20.7 | 1.84 | 0.29 |
| | **6.0** | 21.4 | 1.84 | 0.29 |
| | **6.2** | 22.2 | 1.84 | 0.29 |
| | **6.4** | 23.1 | 1.83 | 0.29 |
| | **6.6** | 23.8 | 1.83 | 0.29 |
| | **6.8** | 24.7 | 1.82 | 0.29 |
| 2 | **5.8** | 20.8 | 1.97 | 0.33 |
| | **6.0** | 21.6 | 1.97 | 0.33 |
| | **6.2** | 22.4 | 1.96 | 0.32 |
| | **6.4** | 23.3 | 1.95 | 0.32 |
| | **6.6** | 24.1 | 1.95 | 0.32 |
| | **6.8** | 25.0 | 1.94 | 0.32 |
| 3 | **5.8** | 21.8 | 1.92 | 0.31 |
| | **6.0** | 22.7 | 1.91 | 0.31 |
| | **6.2** | 23.4 | 1.91 | 0.31 |
| | **6.4** | 24.2 | 1.91 | 0.31 |
| | **6.6** | 25.1 | 1.90 | 0.31 |
| | **6.8** | 25.9 | 1.90 | 0.31 |
| 4 | **5.8** | 21.8 | 1.87 | 0.30 |
| | **6.0** | 22.5 | 1.87 | 0.30 |
| | **6.2** | 23.4 | 1.86 | 0.30 |
| | **6.4** | 24.2 | 1.86 | 0.30 |
| | **6.6** | 25.1 | 1.85 | 0.29 |
| | **6.8** | 25.9 | 1.85 | 0.29 |
| All | **5.8** | 21.1 | 1.89 | 0.31 |
| | **6.0** | 21.9 | 1.89 | 0.31 |
| | **6.2** | 22.6 | 1.89 | 0.31 |
| | **6.4** | 23.5 | 1.88 | 0.30 |
| | **6.6** | 24.4 | 1.87 | 0.30 |
| | **6.8** | 25.2 | 1.87 | 0.30 |

We used the $H-\kappa$ stacking together with a bootstrap algorithm and normally distributed values of $V_p$ and phase weights to simultaneously find the best values of $H$ and $\kappa$, as well as the errors associated with these values. We began by incorporating uncertainty in mean crustal velocity into error estimates for $H$ and $\kappa$ by specifying a normal distribution of $V_p$ values so that 95% of the values selected fell between 5.9 and 6.5 km $s^{-1}$, with a mean value of 6.2 km $s^{-1}$, which is the mean crustal velocity in the refraction line near ATD (Fig. 1). For the weights ($w_1$, $w_2$ and $w_3$), we also used a normal distribution such that the sum of the weights add up to 1.00 but 95% of the values for $w_1$ fall between 0.55 and 0.65 with a mean of 0.6, for $w_2$ they fall between 0.25 and 0.35 with a mean value of 0.3, and for $w_3$ they vary between 0.05 and 0.15 with a mean value of 0.1.

Once values for $V_p$ and the weights were selected, we then used the bootstrap algorithm of Efron & Tibshirani (1991), together with the $H-\kappa$ stacking, to estimate $H$ and $\kappa$ with statistical error bounds. While performing the $H-\kappa$ stacking, the contribution of each of the receiver functions to the determination of $H$ and $\kappa$ was also weighted based on the least squares misfit value of the receiver functions. The procedure of selecting $V_p$ and weights from the distribution described above and then performing the $H-\kappa$ stacking with bootstrapping was repeated 200 times. After each time, new average values of $H$ and $\kappa$ and their uncertainties were computed. It was found that after repeating the procedure 50–60 times (out of 200), the error values for $H$ and $\kappa$ stabilized.

$H-\kappa$ stacking was performed on only the highest quality receiver functions (48 receiver functions spanning 10 years of data from 1993 to 2002). To examine azimuthal variation in crustal structure, the receiver functions were split into four groups from different azimuths (Fig. 2), and the stacking was performed on the receiver functions within each group. The groups were based on the clustering of the events with backazimuth, except for group 1, which we chose to be similar to the range of backazimuths represented in the receiver functions used by Sandvol *et al.* (1998). The result for group 1 is shown in Figure 3, and the results for all the groups, as well as for an $H-\kappa$ stack using all 48 receiver functions, are summarized in Table 2. The results are consistent between the various groups and give little indication of azimuthal variation in crustal structure.

For comparison with the results of Sandvol *et al.* (1998), we also computed stacks of the receiver functions by simply averaging them together (Fig. 4). Many more receiver functions were used

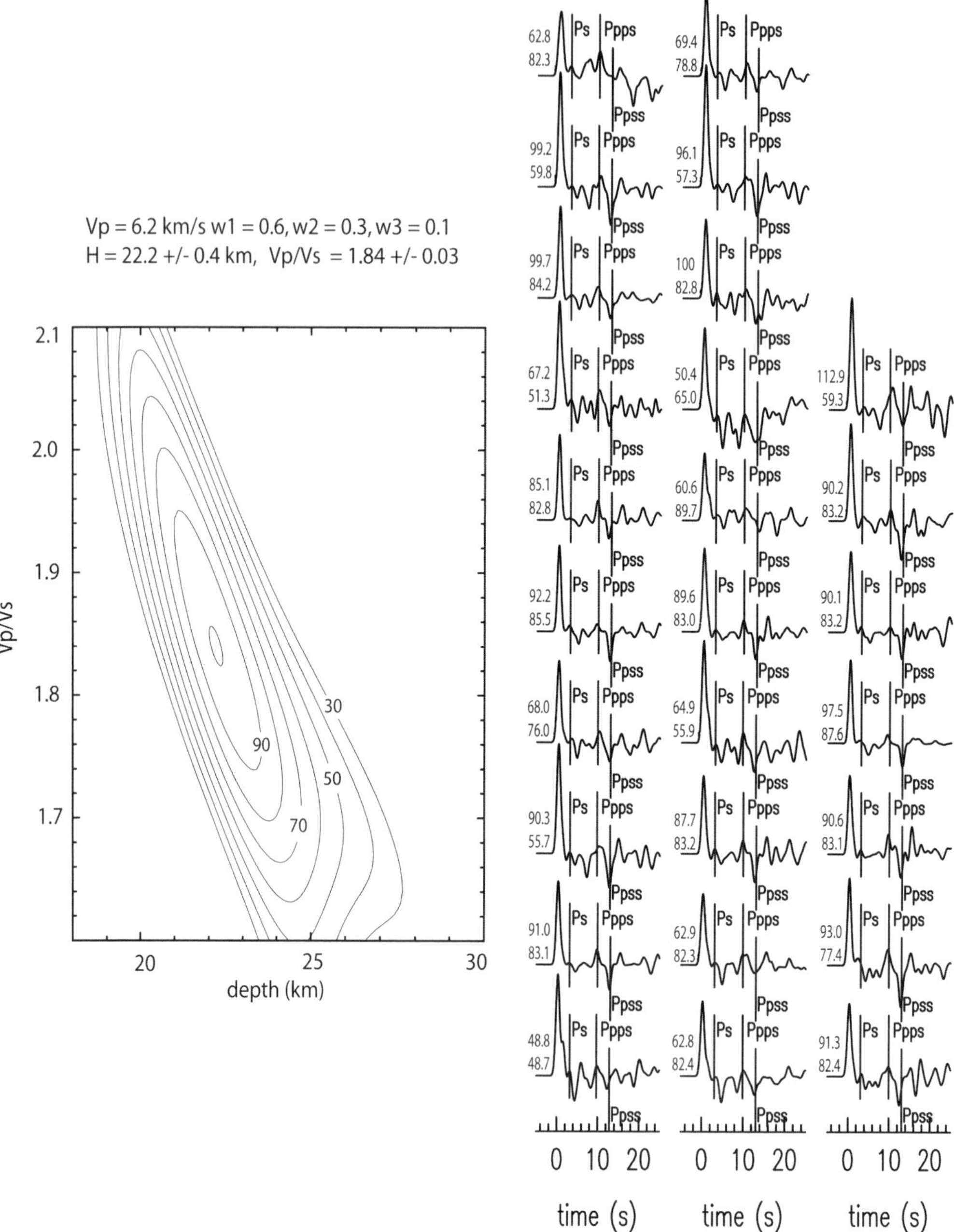


**Fig. 3.** $H-\kappa$ stack of receiver functions in group 1 for a mean crustal $V_p$ of 6.2 km $s^{-1}$. To the left of each receiver function, the top number gives the event azimuth and the bottom number gives the event distance in degrees. Contours map out percentage values of the objective function given in the text.

**Table 2.** H–κ *stacking results for the groups of receiver functions shown in Figure 2 using a mean crustal* $V_p$ *of 6.2 km s*$^{-1}$ *and the error estimation procedure described in the text*

| Group | Moho depth (km) | Depth uncertainties (+/−) | $V_p/V_s$ | $V_p/V_s$ uncertainties (+/−) | Poisson's ratio ($\sigma$) | No. events |
|---|---|---|---|---|---|---|
| 1 | 23 | 1.2 | 1.85 | 0.03 | 0.29 | 27 |
| 2 | 23 | 1.7 | 1.95 | 0.05 | 0.32 | 4 |
| 3 | 23 | 2.0 | 1.92 | 0.09 | 0.31 | 9 |
| 4 | 24 | 2.4 | 1.86 | 0.10 | 0.30 | 8 |
| All | 23 | 1.5 | 1.90 | 0.04 | 0.31 | 48 |

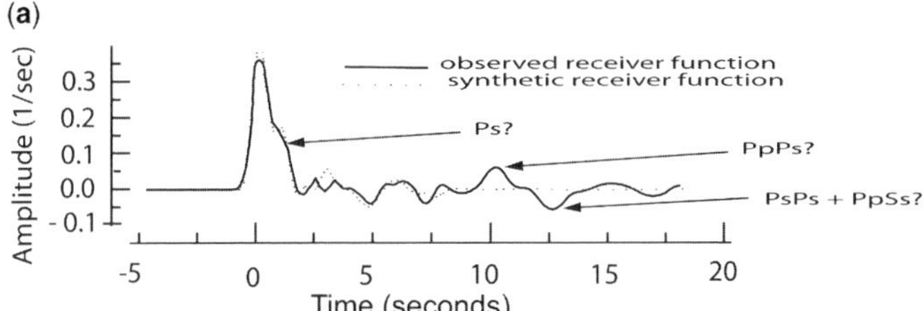


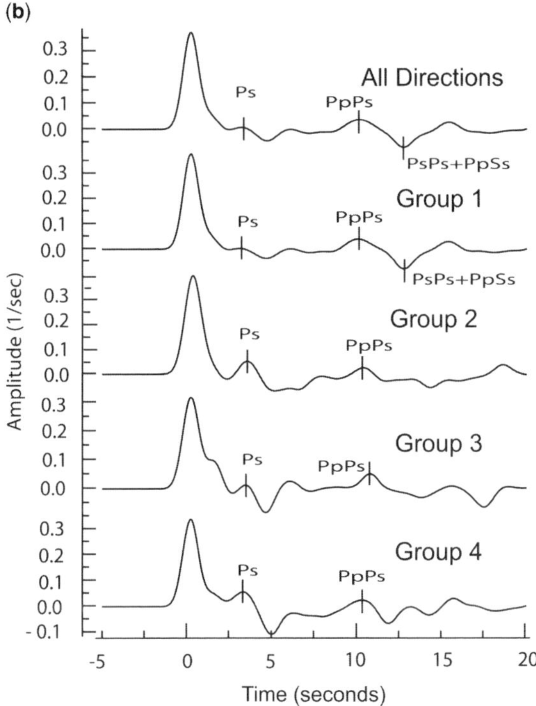


**Fig. 4.** (**a**) Receiver function stack and synthetic for a crustal thickness of 8 km for station ATD (redrawn from Sandvol *et al.* 1998). Crustal reverberations not interpreted by Sandvol *et al.* (1998) are labelled along with the phase they interpreted to be the Moho Ps conversion. (**b**) Simple stacks of receiver functions from this study for the different groupings shown in Figure 2.

for this (183 total), as compared to the $H$–κ stacking. The Ps conversion points at 25 km depth for all 183 events are shown on Figure 2.

### *Crustal structure from joint inversion of receiver functions and surface wave dispersion measurements*

Another approach to addressing the non-uniqueness inherent in interpreting receiver functions (besides $H$–κ stacking), is to invert jointly the receiver functions with observations that constrain crustal velocities, such as surface wave dispersion measurements. Joint inversions of receiver functions and surface wave dispersion measurements have been used by many authors to obtain improved models of crustal structure (e.g. Last *et al.* 1997; Ozalaybey *et al.* 1997; Du & Foulger 1999; Julia *et al.* 2000, 2005). An advantage of this method compared to $H$–κ stacking is that the inversion produces a model of shear wave velocities in the crust (in addition to Moho depth), and therefore details of crustal structure can be examined.

We used the method developed by Julia *et al.* (2000) to jointly invert the receiver functions from station ATD and surface wave group velocities. In the inversion, we used three groups of receiver functions each corresponding to a range of ray parameters from 0.04 to 0.049 (with an average value of 0.044), from 0.05 to 0.059 (with an average value of 0.056), and 0.060 to 0.069 (with an average value of 0.065) (Fig. 5). In addition, for each grouping of receiver functions, we computed and stacked two sets of receiver functions that have overlapping frequency bands: a low-frequency band of $f \leq 0.5$ Hz and a high frequency band of $f \leq 1.25$ Hz. By inverting receiver function stacks over a range of ray parameter and frequency, details of crustal structure can often be imaged, such as sharp versus gradational seismic discontinuities (Julia *et al.* 2005).

Rayleigh wave group velocities between 10 to 40 sec period were used in the inversion. The dispersion measurements come from Benoit (2005), who conducted a surface wave tomography study of eastern Africa by adding to the dispersion measurements of Pasyanos *et al.* (2001) new measurements made with data from the 2000–2002 Ethiopia broadband seismic experiment (Nyblade & Langston 2002).

The inversion was performed using two different starting velocity models to determine how sensitive the inversion results are to the starting model. In the first case (Fig. 5a), the starting model had a gradational velocity structure above 35 km over a half space. In the second case (Fig. 5b), the starting model was based on Ruegg's (1975) P-wave velocity profile converted to an S-wave model using a Poisson's ratio of 0.31.

The results from the inversions using the different starting models are nearly identical (Fig. 5), and therefore the inversion results do not appear to be influenced significantly by the starting model. The velocity models (Fig. 5) show major discontinuities at depths of about 23 km and 25 km, with a change of shear-wave velocity from about 3.75 km $s^{-1}$ to 4.2 km $s^{-1}$, which we interpret to be the Moho.

To estimate the uncertainties in our model results, we followed the approach of Julia *et al.* (2005) and repeatedly performed inversions using a range of weighting parameters, constraints and Poisson's ratio. Similar to the results of Julia *et al.* (2000, 2003, 2005), by repeating the inversions for many combinations of model parameters and data, we found the uncertainties in the shear wave velocities to be about 0.1 km $s^{-1}$ and uncertainties in the depth of discontinuities to be about 2–3 km.

## Comparison of results

The results of the $H$–κ stacking show little variation in crustal thickness or Poisson's ratio with backazimuth. The crustal thickness is $23 \pm 1.5$ km within each grouping and Poisson's ratio ranges from 0.29 to 0.32 (Tables 1 and 2). Ps conversion points shown in Fig. 2 illustrate that the receiver functions do not sample crust under the Gulf of Aden ridge. The results of the joint inversion are similar, indicating a Moho at 23 to 25 km depth.

Our results from both analyses ($H$–κ stacking and joint inversion) are consistent with the seismic velocity structure given in Ruegg (1975) showing a velocity discontinuity at *c.* 24–25 km depth from 7.1 km $s^{-1}$ to 7.4 km $s^{-1}$. By selecting the P to s conversion on the receiver functions at about 3 s after the P arrival as coming from the Moho, we are favouring an interpretation of crustal structure that identifies rocks with velocities as high as 7.1 km $s^{-1}$ as crustal rock, similar to the interpretation of Makris & Ginzburg (1987) in other parts of the Afar. The Poisson's ratio of $0.31 \pm 0.02$ we obtained is consistent with the estimate from Ruegg (1975) of 0.28 to 0.33, as well as from Zandt & Ammon (1995) of $0.29 \pm 0.02$. In addition, the shear velocity structure of the crust that we obtained from our joint inversion is remarkably similar to Ruegg's (1975) model at all depths.

As reviewed earlier, Sandvol *et al.* (1998) reported a crustal thickness estimate for ATD of 8 km from analysing receiver functions. In their analysis, they used a Poisson's ratio of 0.25 and

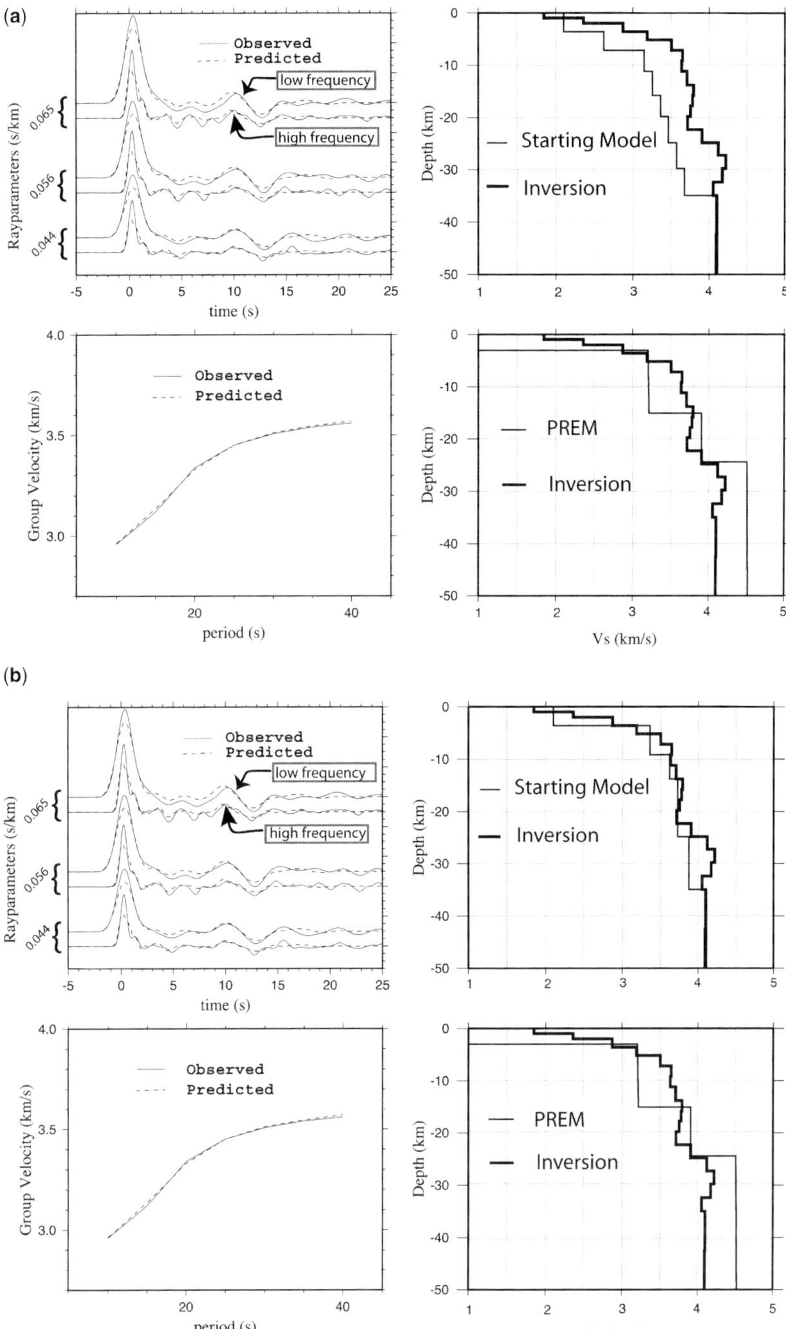


**Fig. 5.** Crustal shear wave velocity structure beneath station ATD from the joint inversion of receiver functions and Rayleigh wave group velocities for two different starting models: (**a**) a gradational velocity structure over a half space; and (**b**) a model based on Ruegg's (1975) P-wave velocity profile converted to an S-wave profile using a Poisson's ratio of 0.31. For both (a) and (b), upper left panel displays predicted and observed receiver functions for three different ray parameters, plus high- and low-frequency band receiver functions in each ray parameter; lower left panel shows predicted and observed surface wave group velocities; top right panel shows the starting model for the inversion and resulting model; bottom right panel displays the joint inversion result and the PREM (Preliminary Reference Earth Model; Dziewonski & Anderson 1981) model for comparison.

receiver functions from 11 events, 1 from the west and the rest from the east (backazimuths of *c.* 39° and *c.* 123°). Thus, the stack of the receiver functions they modelled primarily reflected structure beneath the area of group 1 in Fig. 2.

In comparison to Sandvol *et al.* (1998), our analysis of receiver functions is more comprehensive. We stacked many more receiver functions with good signal to noise ratios, and to check for heterogeneous structure, the receiver functions were examined in four groupings from different backazimuths, as mentioned previously. The stacked receiver functions for group 1 (Fig. 4b) show a Moho Ps phase that is much stronger than in the stack from Sandvol *et al.* (1998) (Fig. 4a). We interpret the Ps phase picked by Sandvol *et al.* (1998) to represent a shallower discontinuity, perhaps the discontinuity at about 4 km or 10 km depth seen on the refraction line from Ruegg (1975). The Ps phase that we picked as the Moho Ps conversion is not as clear on the stack used by Sandvol *et al.* (1998) as in our stack (Fig. 4b), but it is nonetheless apparent (Fig. 4a). In addition, the two reverberation phases picked by our $H-\kappa$ stacking algorithm can be seen in the stack used by Sandvol *et al.* (1998) at *c.* 10 and *c.* 12.5 seconds (Fig. 4a). The synthetic receiver function obtained by Sandvol *et al.* (1998) for their preferred crustal model does not fit the arrivals of the receiver function at those times (Fig. 4a).

## Discussion

As described earlier, the Moho in the Afar has been interpreted either as an increase in velocity from about 6.4 km s$^{-1}$ to 6.9 km s$^{-1}$ or as an increase in velocity from about 7.1 km s$^{-1}$ to 7.4 or 7.5 km s$^{-1}$. We favour the latter interpretation because a Moho at *c.* 23–25 km depth produces a Ps conversion that arrives at the time of the clearest Ps conversion on the receiver function stack for each grouping (Fig. 4b). The timing of the crustal reverberations is also well matched. In addition, the joint inversion yields a velocity structure for the crust with a clear velocity discontinuity at depths of 23–25 km (Fig. 5). Our preferred interpretation from the $H-\kappa$ stacking (Table 2) is based on a mean crustal $V_p$ of 6.2 km s$^{-1}$. Results summarized in Table 1 show that even for a mean crustal $V_p$ of 5.8 km s$^{-1}$ (i.e. similar to the mean crustal $V_p$ from Ruegg's model if the 7.1 km s$^{-1}$ layer is considered to be in the mantle), the Moho depth is still around 21 km.

To the north of the Gulf of Tadjoura, Ruegg's (1975) refraction profiles show an increase in velocity from 7.1 km s$^{-1}$ to 7.4 km s$^{-1}$ at about 20 km depth. Hence, the crust to the north of the Gulf of Tadjoura may be somewhat thinner than to the south, but there is little indication of thin (i.e. 8–10 km thick) crust across areas of Djibouti away from the main spreading centres.

By combining our model of crustal structure beneath station ATD with the estimates of crustal structure from Makris & Ginzburg (1987), and Ruegg's (1975) velocity models for other parts of Djibouti, it appears that crustal thickness and composition may be fairly uniform across many parts of central and eastern Afar (Fig. 1). Moho depths are between 20–25 km, mean $V_p$ is around 6.2 km s$^{-1}$ but about 6.9–7.0 km s$^{-1}$ below a 2–5 km-thick low-velocity layer at the surface, and Poisson's ratio is about 0.30 or higher.

What are the tectonic implications of this crustal structure for understanding the transition from continental rifting to sea-floor spreading? Mohr (1989) reviewed two plausible models for the nature of Afar crust which have different tectonic origins; (1) stretched (thinned) Precambrian continental crust modified by igneous intrusions; and (2) new igneous crust created by the addition of large volumes of mafic magma and lesser amounts of silicic magma capped by coeval flood lavas. Using estimates of crustal stretching, crustal structure and sea-floor spreading parameters for the Red Sea and Gulf of Aden basins, together with the geology of Afar indicating a region affected predominantly by fissure volcanism, Mohr (1989) argued in favour of Model 2.

Poisson's ratio is particularly diagnostic of crustal modification and was not commented on by Mohr (1989). Below the melting point of many rocks, mineralogy is the most important factor influencing Poisson's ratio (Christensen 1996), with the abundance of quartz and plagioclase feldspar having a dominant effect on the common igneous rocks. Granitic rocks have a Poisson's ratio of about 0.24, while intermediate composition rocks have values of around 0.27 and mafic rocks about 0.30 (Christensen 1996; Tarkov & Vavakin 1982).

Crustal Poisson's ratios in Afar can be used to further argue against Model 1. The Precambrian basement surrounding Afar is mostly Neoproterozoic Mozambique Belt. Dugda *et al.* (2005) have reported that average Mozambique Belt crust in eastern Africa is approximately 40 km thick, has a Poisson's ratio of about 0.25 and an average $V_p$ of 6.5 km s$^{-1}$. It is very difficult to take such a crust and create the crust in Afar by simply stretching it. The resulting thinned crust would not have a sufficiently high Poisson's ratio to account for the observed Poisson's ratio in Afar of *c.* 0.30, nor would it have a sufficiently high average $V_p$.

Model 2 could have an appropriately high Poisson's ratio to account for the observed high ratio found in Afar, but for this model to be viable there needs to be a reasonable explanation for how to generate the new igneous crust. One possibility is that the early Afar crust was initially oceanic in origin but was then modified by plume-generated melts from the mantle. Mohr (1978) suggested this possibility, noting that most of the anomalous crustal thickening would have occurred in the lower crustal layer (15–20 km thick in Afar compared to 4–5 km in oceanic crust), and that this amount of crustal thickening was consistent with the excessive magmatism found in Afar.

Another possibility, also proposed by Mohr (1989), is that the new igneous crust of Afar (Model 2) was generated by the intrusion of mantle-derived magmas breaking the crust. The formation of new igneous crust in this way is supported by Ebinger & Casey (2001), who suggested that in transitional rift settings extensional strain is accommodated locally within magmatic centres instead of along rift border faults. According to them, border faults (detachments) play an active controlling role in the continental break-up process during the early stages of rifting, but, in the late syn-rift stages, crustal extension results primarily from dykes intruding into the crust. Recent results from the Ethiopian Afar Geoscientific Lithosphere Experiment indicate that within the northern end of the main Ethiopian rift strain is indeed being accommodated within magmatic segments (Keir *et al.*, in press; Bendick *et al.* 2006). Consequently, the Afar crust could have been created in a similar way, with the seismic structure described above reflecting the product of extension via the addition of large volumes of intrusive rock, predominantly mafic in composition, as dykes, sills and underplate.

## Conclusions

A crustal thickness of $23 \pm 1.5$ km and a crustal Poisson's ratio of 0.29 to 0.33 have been obtained for station ATD in Djibouti (eastern Afar) from an $H-\kappa$ stacking analysis of receiver functions and a joint inversion of receiver functions and surface wave dispersion measurements. These results are consistent with the seismic velocity structure of the crust in Djibouti obtained from seismic refraction profiles (Ruegg 1975). By combining our results of crustal structure beneath station ATD with the estimates of crustal structure elsewhere in Afar, it appears that crustal thickness and composition may be fairly uniform across many parts of central and eastern Afar, with Moho depths between 20–25 km. The high Poisson's ratio and high $V_p$ throughout most of the crust indicates a mafic composition.

The high crustal Poisson's ratio and high mean crustal $V_p$ throughout much of Afar, as well as the crustal thickness, are not consistent with models invoking crustal formation through stretching of pre-existing Precambrian crust. Instead, we suggest that crust in Afar consists predominantly of new igneous rock emplaced as part of the extensional process. During late syn-rift stages in the main Ethiopian rift, extensional strain is accommodated within magmatic segments through dyke intrusion. In addition to dyking, sill formation and underplating associated with the magmatic centres probably combine to help form new igneous crust. The formation of the new igneous (mafic) crust in Afar could have also taken place through modification of oceanic crust that was subsequently altered by plume-derived magmas from the mantle.

We thank Chuck Ammon and Tanya Furman for helpful comments, and Cecile Doubre, Graham Stuart, Cindy Ebinger, Peter Maguire and Ian Bastow for constructive reviews. This research has been funded by the National Science Foundation (grants EAR 993093 and 0003424).

## References

ACTON, G.D., TESSEMA, A., JACKSON, M. & BILHAM, R. 2000. The tectonic and geomagnetic significance of paleomagnetic observations from volcanic rocks from central Afar, Africa. *Earth and Planetary Science Letters*, **180**, 225–241.

AMMON, C.J., RANDALL, G.E. & ZANDT, G. 1990. On the nonuniqueness of receiver function inversions. *Journal of Geophysical Research*, **95**, 15303–15318.

AUDIN, L., QUIDELLEUR, X., ET AL. 2004. Palaeomagnetism and K–Ar and $^{40}Ar/^{39}Ar$ ages in the Ali Sabieh area (Republic of Djibouti and Ethiopia), constraints on the mechanism of Aden Ridge propagation into southeastern Afar during the last 10 Myr. *Geophysical Journal International*, **158**, 327–345.

AYELE, A., STUART, G. & KENDALL, J.-M. 2004. Insights into rifting from shear wave splitting and receiver functions: an example from Ethiopia. *Geophysical Journal International*, **157**, 354–362.

BENDICK, R., BILHAM, R., ASFAW, L. & KLEMPERER, S.L. 2006. Distributed Nubia–Somalia relative motion and dyke intrusion in the Main Ethiopian Rift. *Geophysical Journal International* (in press).

BENOIT, M. 2005. The upper mantle seismic velocity structure beneath the Arabian Shield and East Africa. PhD thesis, Department of Geosciences, Pennsylvania State University.

BERKHEMER, H., BAIER, B., ET AL. 1975. Deep seismic soundings in the Afar region and on the highland of Ethiopia. *In*: PILGER, A. & ROSLER, A. (eds) *Afar Depression of Ethiopia*. E. Schweizerbart, Stuttgart, 89–107.

BILHAM, R., BENDICK, R., LARSON, K., MOHR, P., BRAUN, J., TESFAYE, S. & ASFAW, L. 1999. Secular and tidal strain across the main Ethiopian rift. *Geophysical Research Letters*, **26**, 2789–2792.

BUCK, W.R. 2004. Consequences of asthenospheric variability on continental rifting. *In*: KARNER, G.D., TAYLOR, B., DRISCOLL, N. & KOHLSTEDT, D. (eds) *Rheology and Deformation of the Lithosphere at Continental Margins*. Columbia University Press, 92–137.

CHERNET, T., HART, W.K., ARONSON, J.L. & WALTER, R.C. 1998. New age constraints on the timing of volcanism and tectonism in the northern Main Ethiopian Rift–southern Afar transition zone (Ethiopia). *Journal of Volcanology and Geothermal Research*, **80**, 267–280.

CHRISTENSEN, N.I. 1996. Poisson's ratio and crustal seismology. *Journal of Geophysical Research*, **101**, 3139–3156.

COCHRAN, J.R. 1981. The Gulf of Aden; structure and evolution of a young ocean basin and continental margin. *Journal of Geophysical Research*, B, **86**, 263–287.

COURTILLOT, V. 1980. Opening of the Gulf of Aden and Afar by progressive tearing. *Physics of the Earth and Planetary Interiors*, **21**, 343–350.

COURTILLOT, V., JAUPART, C., MANIGHETTI, I., TAPPONNIER, P. & BESSE, J. 1999. On causal links between flood basalts and continental breakup. *Earth and Planetary Science Letters*, **166**, 177–195.

D'ACREMONT, E., LEROY, S., BESLIER, M.-O., BELLAHSEN, N., FOURNIER, M., ROBIN, C., MAIA, M. & GENTE, P. 2005. Structure and evolution of the eastern Gulf of Aden conjugate margins from seismic reflection data. *Geophysical Journal International*, **160**, 869–890.

DZIEWONSKI, A.M. & ANDERSON, D.L. 1981. Preliminary reference Earth model. *Physics of the Earth and Planetary Interiors*, **25**, 297–356.

DU, Z.J. & FOULGER, G.R. 1999. The crustal structure beneath the northwest fjords, Iceland, from receiver functions and surface waves. *Geophysical Journal International*, **139**, 419–432.

DUGDA, M.T., NYBLADE, A.A., JULIA, J., LANGSTON, C.A., AMMON, C.J. & SIMIYU, S. 2005. Crustal structure in Ethiopia and Kenya from receiver function analysis: implications for rift development in eastern Africa. *Journal of Geophysical Research*, **110**, B01303, doi:10.1029/2004JB003065.

EBINGER, C. 2005. The Bullerwell Lecture: Continental break-up: The East African Perspective. *Astronomy & Geophysics*, **46**, 2.16–2.21, doi:10.1111/j.1468-4004.2005.46216.x.

EBINGER, C.J. & CASEY, M. 2001. Continental breakup in magmatic provinces: An Ethiopian example. *Geology*, **29**, 527–530.

EFRON, B. & TIBSHIRANI, R. 1991. Statistical data analysis in the computer age. *Science*, **253**, 390–395.

FUCHS, K., ALTHERR, R., MULLER, B. & PRODEHL, C. (eds) 1997. Structure and dynamic processes in the lithosphere of the Afro-Arabian Rift System, Special Issue, *Tectonophysics*, **278**.

HAYWARD, N. & EBINGER, C.J. 1996. Variations in the along-axis segmentation of the Afar rift system. *Tectonics*, **15**, 244–257.

HOFMANN, C., COURTILLOT, V., FERAUD, G., ROCHETTE, P., YIRGU, G., KETEFO, E. & PIK, R. 1997. Timing of the Ethiopian flood basalt event and implications for plume birth and global change. *Nature*, **389**, 838–841.

JULIA, J., AMMON, C.J., HERRMANN, R.B. & CORREIG, A.M. 2000. Joint inversion of receiver functions and surface wave dispersion observations. *Geophysical Journal International*, **143**, 99–112.

JULIA, J., AMMON, C.J. & HERRMANN, R.B. 2003. Lithospheric structure of the Arabian Shield from the joint inversion of receiver functions and surface-wave group velocities. *Tectonophysics*, **371**, 1–21.

JULIA, J., AMMON, C.J. & NYBLADE, A.A. 2005. Evidence for mafic lower crust in Tanzania, East Africa, from joint inversion of receiver functions and Rayleigh wave dispersion velocities. *Geophysical Journal International*, **162**, 555–569.

KEIR, D., EBINGER, C., STUART, G., DALY, E. & AYELE, A. 2006. Strain accommodation by magmatism and faulting as rifting proceeds to break-up: Seismicity of the northern Ethiopian Rift. *Journal of Geophysical Research* (in press).

KENDALL, J.-M., STUART, G.W., EBINGER, C.J., BASTOW, I.D. & KEIR, D. 2005. Magma-assisted rifting in Ethiopia. *Nature*, **7022**, 146–148.

LANGSTON, C.A. 1979. Structure under Mount Rainier, Washington, inferred from teleseismic body waves. *Journal of Geophysical Research*, **84**, 4749–4762.

LAST, R.J., NYBLADE, A.A., LANGSTON, C.A. & OWENS, T.J. 1997. Crustal structure of the East African plateau from receiver functions and Rayleigh wave phase velocities. *Journal of Geophysical Research*, **102**, 24 469–24 483.

LIGORRIA, J.P. & AMMON, C.J. 1999. Iterative deconvolution and receiver-function estimation. *Bulletin of the Seismological Society of America*, **89**, 1395–1400.

MAGUIRE, P.K.H., EBINGER, C.J. *ET AL.* 2003. Geophysical Project in Ethiopia Studies Continental Breakup. *EOS Transactions, AGU*, **84**, 337, 342–343.

MAKRIS, J. & GINZBURG, A. 1987. The Afar Depression: transition between continental rifting and sea floor spreading. *Tectonophysics*, **141**, 199–214.

MANIGHETTI, I., TAPPONNIER, P., COURTILLOT, V., GRUSZOW, S. & GILLOT, P.Y. 1997. Propagation of rifting along the Arabia–Somalia plate boundary: The gulfs of Aden and Tadjoura. *Journal of Geophysical Research*, **102**, 2681–2710.

MANIGHETTI, I., TAPPONNIER, P., *ET AL.* 1998. Propagation of rifting along the Arabia-Somalia plate boundary in to Afar. *Journal of Geophysical Research*, **103**, 4947–4974.

MOHR, P. 1978. Afar. *Annual Review of Earth and Planetary Sciences*, **6**, 145–172.

MOHR, P. 1989. Nature of the crust under Afar: new igneous, not thinned continental. *Tectonophysics*, **167**, 1–11.

NYBLADE, A.A. & LANGSTON, C.A. 2002. Broadband seismic experiments probe the East African rift. *EOS Transactions AGU*, **83**, 405–408.

OZALAYBEY, S., SAVAGE, M.K., SHEEHAN, A.F., LOUIE, J.N. & BRUNE, J.N. 1997. Shear-wave velocity structure in the northern Basin and Range Province from the combined analysis of receiver functions and surface waves. *Bulletin of the Seismological Society of America*, **87**, 183–99.

PASYANOS, M.E., WALTER, W.R. & HAZLER, S.E. 2001. A surface wave dispersion study of the Middle East and North Africa for monitoring the Comprehensive Nuclear-Test-Ban Treaty. *Pure and Applied Geophysics*, **158**, 1445–1474.

PRODEHL, C., KELLER, G.R. & KHAN, M.A. 1994. Crustal and upper mantle structure of the Kenya Rift. *Tectonophysics*, **236**, 1–483.

RUEGG, J.C. 1975. Main results about the crustal and upper mantle structure of the Djibouti region (T.F.A.I.). *In*: PILGER, A. & ROSLER, A. (eds) *Afar Depression of Ethiopia*. E. Schweizerbart, Stuttgart, 120–134.

SANDVOL, E., SEBER, D., CALVERT, A. & BARAZANGI, M. 1998. Grid search modeling of receiver functions; implications for crustal structure in the Middle East and North Africa. *Journal of Geophysical Research*, **103**, 26 899–26 917.

SEARLE, R.C. 1975. The dispersion of surface waves across southern Afar. *In*: PILGER, A. & ROSLER, A. (eds) *Afar Depression of Ethiopia*. E. Schweizerbart, Stuttgart, 113–120.

TARKOV, A.P. & VAVAKIN, V.V. 1982. Poisson's ratio behavior in crystalline rocks: application to the study of the Earth's interior. *Physics of the Earth and Planetary Interiors*, **29**, 24–29.

TAYLOR, S.R. & OWENS, T.J. 1984. Frequency-domain inversion of receiver functions for crustal structure. *Earthquake Notes*, **55**, 7–12.

TESFAYE, S., HARDING, D.J. & KUSKY, T.M. 2003. Early continental breakup boundary and migration of the Afar triple junction, Ethiopia. *Bulletin of the Geological Society of America*, **115**, 1053–1067.

UKSTINS, I.A., RENNE, P.R., WOLFENDEN, E., BAKER, J., AYALEW, D. & MENZIES, M. 2002. Matching conjugate volcanic rifted margins; $^{40}Ar/^{39}Ar$ chronostratigraphy of pre- and syn-rift bimodal flood volcanism in Ethiopia and Yemen. *Earth and Planetary Science Letters*, **198**, 289–306.

VARET, J. 1975. Carte geologique de l'Afar central et septentrional au 1/500 000. Geologic map of central and northern Afar, 1:500,000. *Reunion Annuelle des Sciences de la Terre*, **3**, 370.

WOLFENDEN, E., EBINGER, C., YIRGU, G., RENNE, P.R. & KELLEY, S.P. 2005. Evolution of a volcanic rifted margin: Southern Red Sea, Ethiopia. *Geological Society of America Bulletin*, **117**, 846–864, doi: 10.1130/B25516.1.

ZANDT, G. & AMMON, C.J. 1995. Poisson's ratio of Earth's crust. *Nature*, **374**, 152–155.

ZANDT, G., MYERS, S.C. & WALLACE, T.C. 1995. Crust and mantle structure across the Basin and Range-Colorado Plateau boundary at 37 degrees N latitude and implications for Cenozoic extensional mechanism. *Journal of Geophysical Research*, **100**, 10,529–10,548.

ZHU, L. & KANAMORI, H. 2000. Moho depth variation in southern California from teleseismic receiver functions. *Journal of Geophysical Research*, **105**, 2969–2980.

# Crustal structure of the northern Main Ethiopian Rift from receiver function studies

G.W. STUART[1], I.D. BASTOW[1] & C.J. EBINGER[2]

[1]*School of Earth and Environment, University of Leeds, Leeds, UK (e-mail: graham@earth.leeds.ac.uk)*

[2]*Department of Geology, Royal Holloway, University of London, Egham, UK*

**Abstract:** The northern Main Ethiopian Rift captures the crustal response to the transition from continental rifting in the East African rift to the south, to incipient seafloor spreading in the Afar depression to the north. The region has also undergone plume-related uplift and flood basalt volcanism. Receiver functions from the EAGLE broadband network have been used to determine crustal thickness and average $V_p/V_s$ for the northern Main Ethiopian Rift and its flanking plateaus.

On the flanks of the rift, the crust on the Somalian plate to the east is 38 to 40 km thick. On the western plateau, there is thicker crust to the NW (41–43 km) than to the SW (<40 km); the thinning taking place over an off-rift upper mantle low-velocity structure previously imaged by travel-time tomography. The crust is slightly more mafic ($V_p/V_s \sim 1.85$) on the western plateau on the Nubian Plate than on the Somalian Plate ($V_p/V_s \sim 1.80$). This could either be due to magmatic activity or different pre-rift crustal compositions. The Quaternary Butajira and Bishoftu volcanic chains, on the side of the rift, are characterized by thinned crust and a $V_p/V_s > 2.0$, indicative of partial melt within the crust.

Within the rift, the $V_p/V_s$ ratio increases to greater than 2.0 (Poisson's ratio, $\sigma > 0.33$) northwards towards the Afar depression. Such high values are indicative of partial melt in the crust and corroborate other geophysical evidence for increased magmatic activity as continental rifting evolves to oceanic spreading in Afar. Along the axis of the rift, crustal thickness varies from around 38 km in the south to 30 km in the north, with most of the change in Moho depth occurring just south of the Boset magmatic segment where the rift changes orientation. Segmentation of crustal structure both between the continental and transitional part of the rift and on the western plateau may be controlled by previous structural inheritances. Both the amount of crustal thinning and the mafic composition of the crust as shown by the observed $V_p/V_s$ ratio suggest that the magma-assisted rifting hypothesis is an appropriate model for this transitional rift.

The northern Main Ethiopian Rift (NMER) provides a unique expression of the crustal response to both the transition from the continental extension in the East African rift to incipient seafloor spreading of the Afar depression and a pre-rift episode of flood basalt magmatism (e.g. Ebinger & Casey 2001). The transition is marked at the surface by increased magmatism and the transfer of extensional strain accommodation from Mid-Miocene border faults to magma intrusion within about 20 km-wide Quaternary magmatic segments near the centre of the rift (Wolfenden *et al.* 2004). The rift transects the uplifted Ethiopia–Yemen plateau, which is thought to have developed above a Palaeogene mantle plume (e.g. Schilling 1973; Ebinger & Sleep 1998). The plateau is covered by up to 2 km of flood basalts and rhyolites, the majority of which erupted around 30 Ma, approximately coincident with the opening of the Red Sea and Gulf of Aden rifts (e.g. Hofmann *et al.* 1997; Wolfenden *et al.* 2004). Uplift of the Ethiopian plateau commenced between 20–30 Ma, soon after the major flood basalt eruptions (Pik *et al.* 2003). On top of the flood basalts are less voluminous syn-rift shield volcanoes that formed between 30–10 Ma (Mohr 1983). Most of the northern Quaternary volcanism in the Ethiopian rift has occurred around the central magmatic segments (Fig. 1), where about 80% of present-day extensional strain is concentrated (Bilham *et al.* 1999; Ebinger & Casey 2001).

Here we present a receiver function study of broadband seismic recordings from the northern Ethiopian rift and its surrounding plateau regions (Fig. 1) to determine how crustal structure has been modified by flood basalt magmatism and rifting processes. Receiver functions capture P- to S-wave conversions at velocity contrasts in the receiver crust and mantle recorded in the P-wave coda from distant teleseismic earthquakes. They are used here to provide estimates of crustal thickness and average $V_p/V_s$ ratio (Zandt & Ammon 1995), which can then be related to bulk crustal composition via Poisson's ratio (e.g. Christensen 1996; Chevrot & van der Hilst 2000).

*From*: YIRGU, G., EBINGER, C.J. & MAGUIRE, P.K.H. (eds) 2006. *The Afar Volcanic Province within the East African Rift System*. Geological Society, London, Special Publications, **259**, 253–267.
0305-8719/06/$15.00 

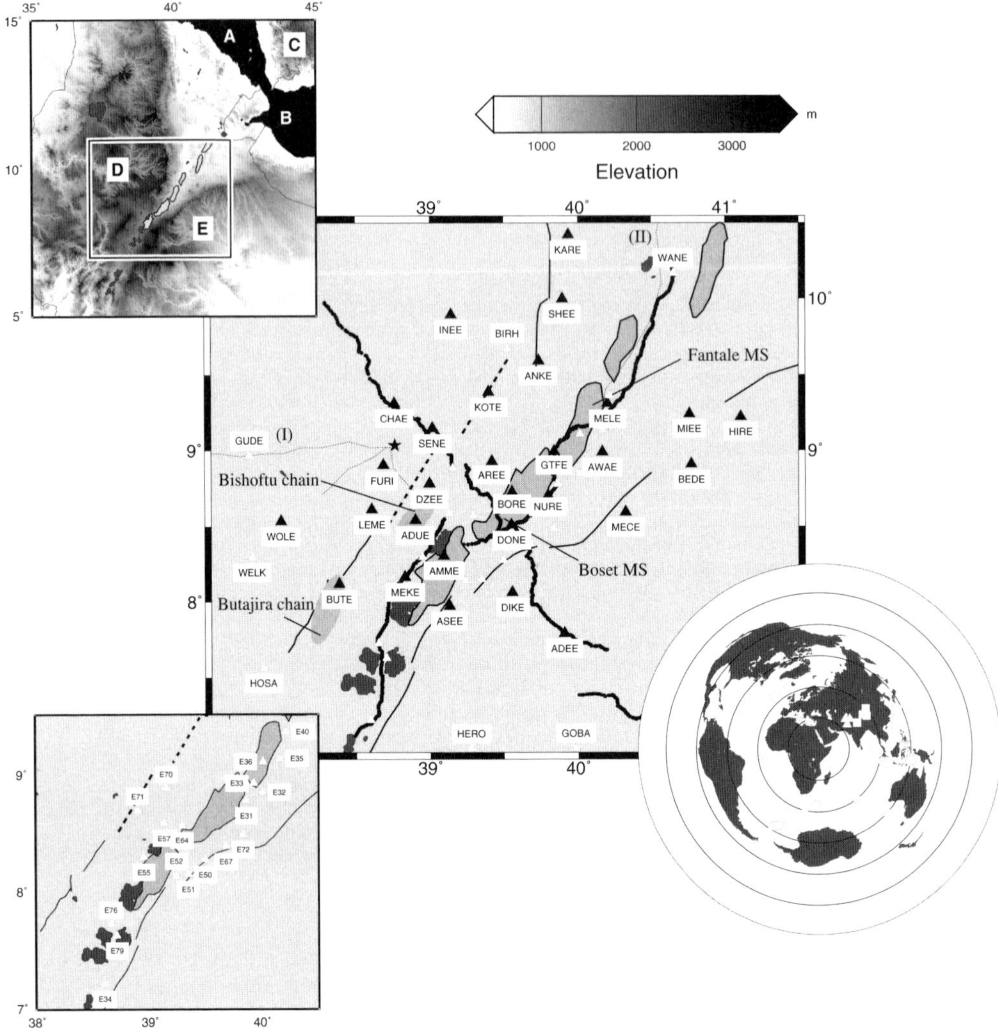


**Fig. 1.** Location of seismological stations. Filled triangles are EAGLE Phase I passive experiment stations and permanent station FURI. Open triangles are EAGLE Phase II passive experiment stations. Station AAE (Addis Ababa) is shown by a black star. Open triangles with names are from the Ethiopia 2000–2002 network (Nyblade & Langston 2002). Major Mid-Miocene border faults and Quaternary magmatic segments (MS) are shown in black lines. Dashed lines are faulted monoclines. Pecked lines follow the controlled source profile of Berckhemer *et al.* (1975); solid thick line shows location of EAGLE controlled-source experiment. Top inset shows regional topography. A, Red Sea; B, Gulf of Aden; C, Arabian Plate; D, Nubian Plate; E, Somalian Plate. Bottom right inset shows distribution of earthquakes used in the receiver function analysis. Different receiver function source regions are indicated by different symbols. Concentric circles on plot indicate 30° intervals from centre of EAGLE passive network. Bottom left inset shows names and locations of EAGLE Phase II stations. Mid-Miocene border faults and Quaternary magmatic segments are also shown.

Controlled source seismic refraction/wide-angle reflection surveys provide the most detailed 2D information about crustal structure in the region. Berckhemer *et al.* (1975) and Makris & Ginzburg (1987) report on interpretations of seismic profiles on the western plateau and axially along the northern part of the Main Ethiopian Rift and into the Afar depression (Fig. 1). On the western plateau, the Moho depth varies from 38 km to 44 km on their profile I, and it is underlain by normal mantle velocity (8.1 km $s^{-1}$). Within the rift, the crust thins from 30 km at the southern end of their profile II (Fig. 1) to 26 km in Afar and 20 km on the Red Sea coast. Makris & Ginzburg (1987) interpret sub-Moho velocities of 7.4–7.5 km $s^{-1}$ in the rift as high-temperature upper mantle. As part of the EAGLE (Ethiopia Afar Geoscientific Lithospheric Experiment) project, controlled source seismic wide-angle reflection/refraction profiles (Fig. 1) were carried out across and along the axis of the rift (Maguire *et al.* 2006).

Along the cross-rift profile, Mackenzie *et al.* (2005) show the crustal thickness to vary from approximately 50 km on the western plateau to 40 km beneath the eastern plateau on the Somalia plate. A *c.* 15 km-thick, reflective lower crustal high-velocity (7.4 km $s^{-1}$) layer, interpreted to be crustal underplate emplaced during the Oligocene plume activity and more recent extensional episodes, was modelled under the western plateau. The cross-rift profile shows that the crust thins to about 35 km in a 50 km-wide zone under the rift. Along the axis of the rift, the crust thins from 40 km in the south to 26 km in the north (Keller *et al.* 2004; Maguire *et al.* 2006). The major proportion of this thinning is interpreted to occur in a 70 km-wide zone centred around the Boset magmatic segment (Fig. 1). Low upper mantle P-wave velocities (*c.* 7.7 km $s^{-1}$) in the rift, observed from the controlled source experiments, suggest the presence of partial melt (Mackenzie *et al.* 2005). Further geophysical evidence for the presence of partial melt in the crust and upper mantle underlying the rift comes from surface waves (Knox *et al.* 1998), travel-time tomography (Bastow *et al.* 2005), shear wave splitting (Kendall *et al.* 2005) and magnetotellurics (Whaler & Hautot 2006).

Whilst controlled source experiments give detailed P-wave crustal structure along 2D profiles, receiver functions can more readily provide information of spatial variations. Dugda *et al.* (2005) analysed receiver functions from a distributed temporary broadband network in Ethiopia deployed between 2000 and 2002 (Nyblade & Langston 2002). Beneath the plateaus on either side of the rift, crustal thicknesses range between 33 and 44 km and Poisson's ratios vary from 0.23 to 0.28. Within the rift, the crust thins from 38 to 27 km and Poisson's ratios increase from 0.27 to 0.35. Dugda *et al.* (2005) conclude that the crust beneath the western plateau has not been greatly modified by Cenozoic rifting and magmatism; the crust beneath the rift, on the other hand, has been thinned and extensively modified by the addition of mafic material.

The present study analyses receiver functions from the EAGLE passive deployments of up to 79 broadband sensors (Maguire *et al.* 2003). The network was centred on the Boset magmatic segment and covered a region 250 × 350 km within the NMER and its flanks (Fig. 1). The EAGLE phase I passive network provides the bulk of our data. This consisted of 29 broadband seismometers (30 sec natural period), deployed with an average station spacing of about 40 km for 16 months from October 2001 until January 2003 (Fig. 1). This network was augmented for four months (October 2002–February 2003) by 50 Phase II broadband (30 sec natural period) seismometers for phase II, distributed with a station spacing of 10 km within the rift itself (Fig. 1). Five Phase I stations were in common with those of the receiver function study of Dugda *et al.* (2005). Dugda *et al.* (2005) results from stations not common with EAGLE observations are incorporated into the present study (Table 1). Our sampling of crustal variations is more detailed and more directly applicable to the rift itself than that of Dugda *et al.* (2005).

## Methodology

Receiver function analysis is well developed and is a technique for identifying P- to S-wave conversions and their multiple reverberations from interfaces in the crust and mantle within the P-wave coda of teleseismic waveforms (e.g. Langston 1979). Receiver functions are generated by deconvolving the vertical component of a P-wave seismogram from the corresponding radial and tangential components. In this study, receiver functions are computed using the source equalization approach (Ammon 1991).

The distribution of the recording stations and earthquakes studied are shown in Figure 1. We have analysed teleseismic recordings from earthquakes with mb >5.5 recorded on Phase I and Phase II broadband instruments of the EAGLE experiment (Maguire *et al.* 2003). 70 earthquakes (Fig. 1) were of sufficient quality to progress with receiver function analysis. Receiver function stacks from earthquakes of similar backazimuth and epicentral distance were computed to increase the signal to noise. Both radial and tangential receiver functions were examined for evidence of lateral heterogeneity and dipping structure. A clear arrival interpreted as the P- to S-wave conversion from the Moho (Pms), was found in a time window from 4.5 to 5.5 s after the P-wave on most receiver functions (Fig. 2). The amplitude of Pms was noticeably smaller for stations in the rift as compared with those on the flanks. Station DONE (Fig. 1), close to the Boset volcano, had a particularly poorly developed Moho conversion (see Fig. 4). Sedimentary reverberations also made the identification of Pms from stations on thick Quaternary sequences in the rift difficult (e.g. station MEKE, Fig. 1).

**Table 1.** *Table of receiver function results.*

| Station | $H$(km) | $V_p/V_s$ | $\sigma$ | $V_p$ (km s$^{-1}$) | Phases(s) | t1 (s) |
|---|---|---|---|---|---|---|
| ADEE | 37.6 ± 1.6 | 1.79 ± 0.01 | 0.27 | 6.25 | $P_s$, $P_p P_s$ | 5.0 |
| ADUE | 33.9 ± 1.4 | 1.89 ± 0.02 | 0.31 | 6.15 | $P_s$, $P_p P_s$ | 5.1 |
| AMME | 37.6 ± 1.5 | 1.95 ± 0.01 | 0.32 | 6.10 | $P_s$, $P_p P_s$ | 6.0 |
| ANKE | 37.6 ± 1.6 | 1.90 ± 0.01 | 0.31 | 6.25 | $P_s$, $P_p P_s$ | 5.7 |
| AREE | 33.6 ± 1.3 | 1.90 ± 0.02 | 0.31 | 6.15 | $P_s$, $P_p P_s$ | 4.9 |
| ASEE | (38.2) | 1.95 | 0.32 | 6.10 | $P_s$ | 5.5 |
| AWAE | (36.0) | 2.09 | 0.35 | 6.25 | $P_s$ | 5.2 |
| BEDE | 41.9 ± 1.7 | 1.79 ± 0.01 | 0.27 | 6.25 | $P_s$, $P_p P_s$ | 5.4 |
| BORE | 32.0 ± 1.3 | 1.90 ± 0.01 | 0.31 | 6.15 | $P_s$, $P_p P_s$ | 4.7 |
| BUTE | 32.0 ± 1.1 | 2.06 ± 0.01 | 0.35 | 6.10 | $P_s$, $P_p P_s$ | 5.7 |
| CHAE | 40.7 ± 1.6 | 1.85 ± 0.01 | 0.30 | 6.25 | $P_s$, $P_p P_s$ | 5.7 |
| DIKE | 42.6 ± 1.7 | 1.73 ± 0.01 | 0.25 | 6.25 | $P_s$, $P_p P_s$ | 5.1 |
| DZEE | (38.4) | 1.96 | 0.32 | 6.25 | $P_s$ | 5.6 |
| FURI* | (37.4) | 2.00 | 0.33 | 6.25 | $P_s$, $P_p P_s$ | 5.5 |
| GTFE | 31.4 ± 2.4 | 2.07 ± 0.05 | 0.35 | 6.25 | $P_s$, $P_p P_s$ | 5.5 |
| INEE | 40.7 ± 1.6 | 1.90 ± 0.02 | 0.31 | 6.25 | $P_s$, $P_p P_s$ | 6.1 |
| HIRE | (39.4) | 1.77 | 0.27 | 6.25 | $P_s$ | 4.9 |
| KARE | 43.8 ± 1.8 | 1.80 ± 0.01 | 0.28 | 6.25 | $P_s$, $P_p P_s$ | 5.8 |
| KOTE | 45.7 ± 1.5 | 1.79 ± 0.02 | 0.27 | 6.25 | $P_s$, $P_p P_s$ | 6.0 |
| LEME | 33.3 ± 1.3 | 2.00 ± 0.01 | 0.33 | 6.25 | $P_s$, $P_p P_s$ | 5.5 |
| MECE | 37.6 ± 1.7 | 1.80 ± 0.02 | 0.28 | 6.25 | $P_s$, $P_p P_s$ | 5.1 |
| MELE | (35.0) | 2.09 | 0.35 | 6.25 | $P_s$ | 4.9 |
| MIEE | (35.0) | 2.09 | 0.35 | 6.25 | $P_s$ | 4.8 |
| NURE | 32.6 ± 6.5 | 1.95 ± 0.13 | 0.32 | 6.15 | $P_s$, $P_p P_s$ | 5.2 |
| SENE | 41.9 ± 1.6 | 1.79 ± 0.01 | 0.27 | 6.25 | $P_s$, $P_p P_s$ | 5.5 |
| SHEE | (35.9) | 2.09 | 0.35 | 6.25 | $P_s$ | 5.2 |
| E31 | (35.4) | 2.00 | 0.33 | 6.15 | $P_s$ | 4.6 |
| E32 | (35.2) | 2.00 | 0.33 | 6.15 | $P_s$ | 4.7 |
| E33 | (35.5) | 2.00 | 0.33 | 6.15 | $P_s$ | 4.8 |
| E34 | (38.9) | 1.95 | 0.32 | 6.10 | $P_s$ | 6.0 |
| E35 | (34.3) | 2.09 | 0.35 | 6.25 | $P_s$ | 4.7 |
| E36 | (35.0) | 2.09 | 0.35 | 6.25 | $P_s$ | 4.9 |
| E40 | (35.0) | 2.09 | 0.35 | 6.25 | $P_s$ | 4.8 |
| E50 | (36.9) | 1.93 | 0.32 | 6.15 | $P_s$ | 4.9 |
| E51 | (38.6) | 1.93 | 0.32 | 6.15 | $P_s$ | 5.6 |
| E52 | (38.6) | 1.93 | 0.32 | 6.15 | $P_s$ | 5.8 |
| E55 | (38.3) | 1.95 | 0.32 | 6.10 | $P_s$ | 5.8 |
| E57 | (37.3) | 1.93 | 0.32 | 6.15 | $P_s$ | 5.2 |
| E64 | 33.6 ± 1.4 | 1.89 ± 0.02 | 0.31 | 6.15 | $P_s$, $P_p P_s$ | 4.9 |
| E67 | (37.0) | 1.93 | 0.32 | 6.15 | $P_s$ | 5.2 |
| E70 | 35.7 ± 5.5 | 1.90 ± 0.11 | 0.31 | 6.25 | $P_s$, $P_p P_s$ | 5.3 |
| E71 | 33.9 ± 7.3 | 2.00 ± 0.16 | 0.33 | 6.25 | $P_s$, $P_p P_s$ | 5.7 |
| E72 | (38.4) | 1.93 | 0.32 | 6.15 | $P_s$ | 5.4 |
| E76 | (40.2) | 1.95 | 0.32 | 6.10 | $P_s$ | 6.5 |
| E79 | (39.2) | 1.95 | 0.32 | 6.10 | $P_s$ | 6.0 |
| AAUb | 37.0 ± 2.7 | 1.92 ± 0.09 | 0.31 | 6.50 | $P_s$, $P_p P_s$ | 5.63 |
| ARBA | 30.0 ± 2.6 | 1.80 ± 0.18 | 0.28 | 6.50 | $P_s$, $P_p P_s$ | 5.43 |
| BELA | 38.0 ± 2.2 | 1.97 ± 0.05 | 0.33 | 6.50 | $P_s$, $P_p P_s$ | 5.87 |
| BIRH | 41.0 ± 3.6 | 1.84 ± 0.09 | 0.29 | 6.50 | $P_s$, $P_p P_s$ | 5.50 |
| DMRK | 41.0 ± 2.4 | 1.83 ± 0.05 | 0.29 | 6.50 | $P_s$, $P_p P_s$ | 5.43 |
| GOBA | 42.0 ± 2.6 | 1.75 ± 0.06 | 0.26 | 6.50 | $P_s$, $P_p P_s$ | 5.00 |
| GUDE | (36.5) | 1.86 ± 0.10 | 0.30 | 6.50 | $P_s$, $P_p P_s$ | 5.01 |
| HERO | 42.0 ± 2.2 | 1.83 ± 0.03 | 0.29 | 6.50 | $P_s$, $P_p P_s$ | 5.57 |
| HOSA | 37.0 ± 2.8 | 1.96 ± 0.06 | 0.32 | 6.50 | $P_s$, $P_p P_s$ | 5.66 |
| JIMA | 36.0 ± 1.2 | 1.85 ± 0.06 | 0.29 | 6.50 | $P_s$, $P_p P_s$ | 4.89 |
| NAZA | 27.0 ± 2.0 | 2.21 ± 0.06 | 0.37 | 6.50 | $P_s$, $P_p P_s$ | 5.19 |

*(continued)*

**Table 1.** *Continued*

| Station | $H$(km) | $V_p/V_s$ | $\sigma$ | $V_p$ (km s$^{-1}$) | Phases(s) | t1 (s) |
|---|---|---|---|---|---|---|
| NEKE | 34.0 ± 2.4 | 1.81 ± 0.06 | 0.27 | 6.50 | $P_s$, $P_p$ $P_s$ | 4.40 |
| WANE | 30.0 | 1.92 | 0.31 | 6.50 | $P_s$, $P_p$ $P_s$ | 4.40 |
| WELK | 33.0 ± 2.4 | 1.83 ± 0.07 | 0.29 | 6.50 | $P_s$, $P_p$ $P_s$ | 4.38 |

Ethiopia and Kenya broadband seismic experiment (Nyblade & Langston, 2002). Data are after Dugda *et al.* (2005).
H, crustal thickness; $\sigma$, Poisson's Ratio; $V_p$, average crustal P-wave velocity; t1, the observed time difference between direct P and the converted S-wave from the Moho (Pms) for the average slowness of the stack (distance 67°). Crustal thicknesses in brackets are assumed using values of $V_p/V_s$ from surrounding stations. Data for FURI have been derived from the receiver function results presented by Ayele *et al.* (2004).

With the receiver function stacks from earthquakes with similar source regions, the arrival times of the conversion from the Moho, Pms and its reverberant, PpPms, can be used to determine Moho depth and $V_p/V_s$ following the equations shown in Zandt *et al.* (1995). The results depend slightly on the mean crustal velocity assumed (Zandt & Ammon 1995). There is a trade-off between the determination of crustal thickness ($H$) and average crustal $V_p/V_s$ ratio (e.g. Ammon *et al.* 1990; Zandt *et al.* 1995). In an effort to reduce the ambiguity introduced by this trade-off, Zhu & Kanamori (2000) incorporated later arriving crustal reverberations from the Moho, namely PpPms and PsPms/PpSms, in a stacking procedure. The method performs a grid search through reasonable values of crustal thickness, $H$ and average crustal $V_p/V_s$, for the maximum value of the weighted sum of receiver function amplitudes at the predicted arrival times of the three phases. In this study, we search a range $20 \leq H \leq 50$ km and $1.5 \leq V_p/V_s \leq 2.3$. There is a simple direct relationship between $V_p/V_s$ and Poisson's ratio, $\sigma$, often used to characterize the bulk composition of the crust (Zandt & Ammon 1995):

$$\sigma = 0.5(1 - [(V_p/V_s)^2 - 1]^{-1})$$

We have used the crustal multiple, PpPms, for the determination of crustal thickness and the average $V_p/V_s$ ratio following the methodology of Zhu & Kanamori (2000), weighting Pms and the reverberation amplitudes in the ratio 6 : 4; there were few convincing examples of the phase PsPms/PpSms. Figure 2 shows examples of the determination of crustal parameters from receiver functions using the technique of Zhu & Kanamori (2000). Simple stacks of receiver functions for EAGLE Phase I stations (with phases Pms and Ppms marked at times derived from the appropriate crustal models) are shown in Figure 3. The peak identified as the PpPms phase on the noisy receiver function stack of station LEME might be questioned. However, peaks either side of that chosen give unreasonable $V_p/V_s$ ratios, which are incompatible with those found for the surrounding stations. For stations with poorly developed crustal multiples, either due to strong sedimentary layer reverberations or laterally inhomogeneous structure, we have measured the Pms arrival time and assumed a $V_p/V_s$ from a nearby station to determine crustal thickness. Dugda *et al.* (2005) investigate the effect of Pms and PpPms sampling different crustal structure for stations close to the rift flanks. They conclude that this problem reduces rather than increases the value of $V_p/V_s$ determined from the Moho reverberations.

Zhu & Kanamori (2000) point out that crustal thickness, $H$, and average $V_p/V_s$ determination are not very sensitive to variations in the average P-wave velocity for the crust, $V_p$. Bootstrapping sensitivity analysis shows that in Ethiopia, on average, $H$ can be determined to ±1.5 km and the average crustal $V_p/V_s$ ratio to ±0.04 for reasonable variations in $V_p$ (Dugda *et al.* 2005). To put it another way, for a change in assumed $V_p$ of 0.1 km s$^{-1}$ the crustal thickness changes by 0.6 km and the $V_p/V_s$ ratio by less than 0.01 (Zhu & Kanamori 2000). The standard errors in Table 1 have been computed by determining the sensitivity of the Zhu & Kanamori method to variations in $V_p$ and the weights of Pms and the PpPms reverberation in the computation of the stack. $V_p$ is varied between 5.9 and 6.6 km s$^{-1}$ at regular intervals of 0.1 km s$^{-1}$ and the amplitude weights of Pms to PpPms between 0.55–0.45 and 0.65–0.35, respectively, at regular intervals of 0.01. We have constrained the average P-wave velocity for the crust in the region of study from the recent EAGLE controlled-source experiments (Maguire *et al.* 2006; Mackenzie *et al.* 2005). Within the rift, three average crustal velocity regions are defined: in the south $V_p \sim 6.10$ km s$^{-1}$; in a central zone running from north of MEKE to south of AWAE, $V_p \sim 6.15$ km s$^{-1}$; and in the north, $V_p \sim$ 6.25 km s$^{-1}$ (Keller *et al.* 2004). The eastern plateau has a $V_p \sim 6.25$ km s$^{-1}$ whilst on the western plateau the crust above the underplate has a $V_p \sim 6.25$ km s$^{-1}$ (Mackenzie *et al.* 2005). These average crustal velocities are lower than the $V_p \sim 6.50$ km s$^{-1}$ assumed by Dugda *et al.* (2005). We have incorporated the Dugda *et al.* (2005)

**(a)** CHAE: $V_P$=6.25 kms$^{-1}$, $V_P/V_S$=1.85±0.01; $\sigma$=0.30; $H$=40.70±1.6 km

Western Plateau

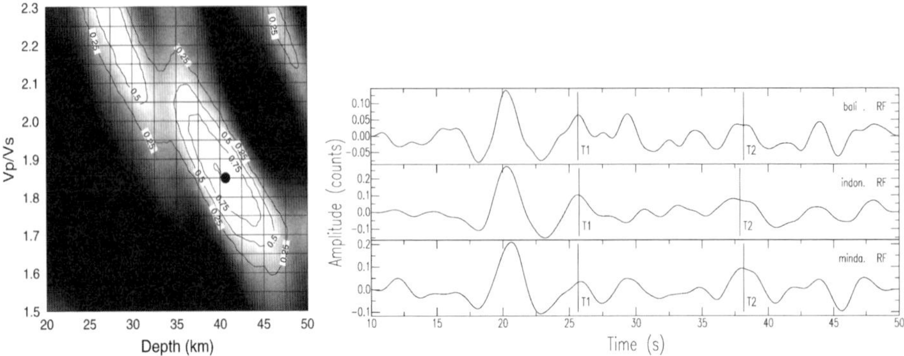


**(b)** AREE: $V_P$=6.15 kms$^{-1}$, $V_P/V_S$=1.90±0.02; $\sigma$=0.31; $H$=33.64±1.3 km

Rift Valley

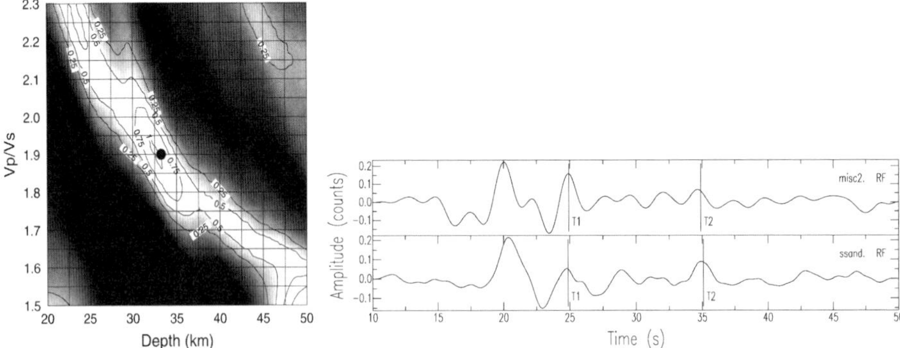


**(c)** ADEE: $V_P$=6.25 kms$^{-1}$, $V_P/V_S$=1.79±0.01; $\sigma$=0.27; $H$=37.60±1.6 km

Eastern Plateau

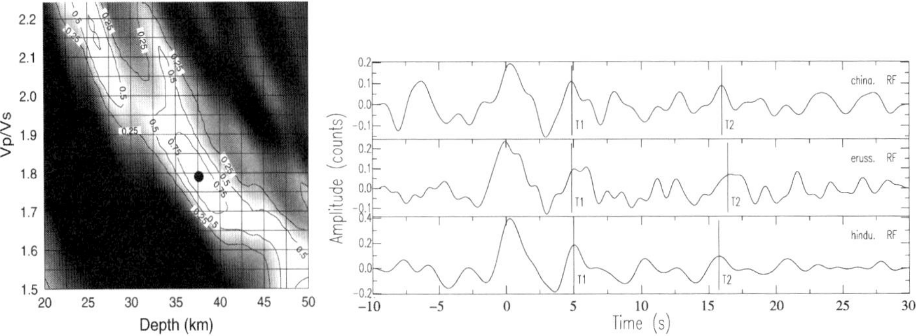


**Fig. 2.** Examples of crustal thickness ($H$) versus $V_p/V_s$ plots from the method of Zhu & Kanamori (2000) and receiver functions for: (**a**) western plateau station CHAE; (**b**) rift valley station AREE; and (**c**) eastern plateau station ADEE. On each plot, the arrival times of Pms ($T1$) and PpPms ($T2$), based on the crustal thickness and $V_p/V_s$ listed in Table 1, are shown. Black dot on contour plots mark maximum of the stack.

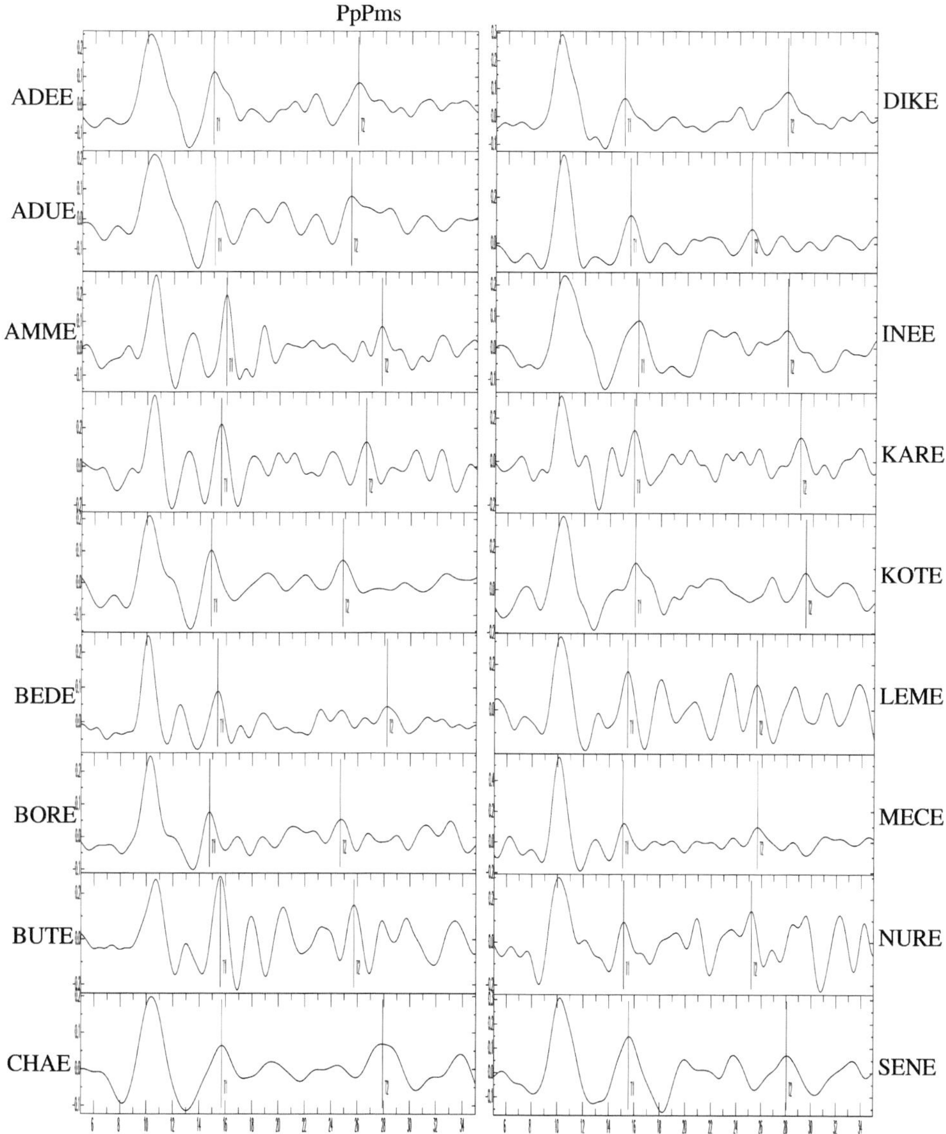


**Fig. 3.** Simple stacks of EAGLE Phase I receiver functions at which the Zhu & Kanamori (2000) stacking technique has been used. On each plot, arrival times of Pms and PpPms, based on crustal thickness and $V_p/V_s$ listed in Table 1, are shown.

results as published, except for GUDE, for which we have re-interpreted the arrival time of the crustal multiple to produce a crustal $V_p/V_s$ value more compatible with the surrounding stations.

## Results

Figure 4a plots the observed average time-difference between the direct P-wave and the converted S-wave from the Moho; the average distance to the earthquakes used is 67°. These times now need converting to Moho depth using estimated values for average crustal $V_p$ and $V_p/V_s$. Figure 4b shows the $V_p/V_s$ ratios determined from the technique of Zhu & Kanamori (2000) for stations where the phase PpPms was readily identifiable. Table 1 tabulates the results.

There are two areas within the rift where $V_p/V_s > 2.0$ ($\sigma > 0.33$), characteristic of partial melt in the crust (Watanabe 1993). First, there is

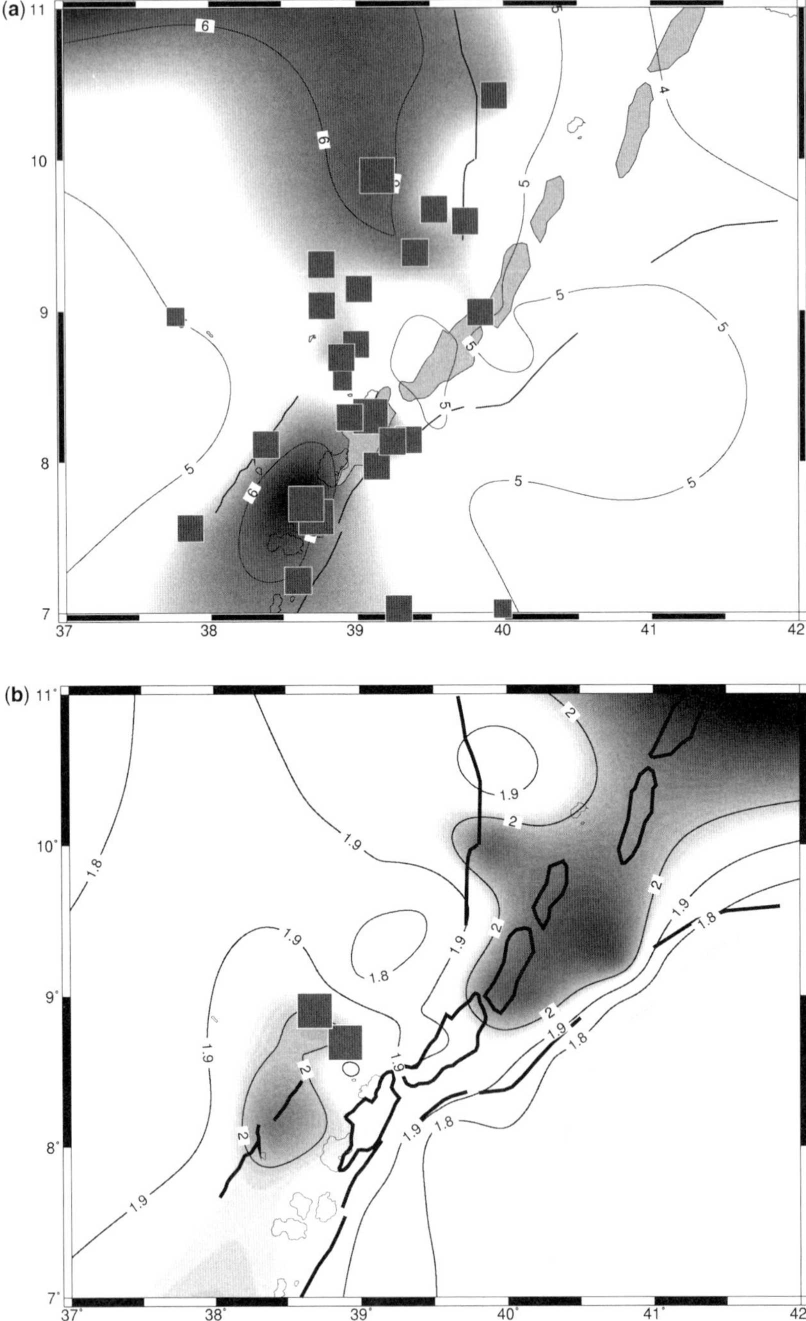


**Fig. 4.** Contour maps of (a) mean P–to–Pms time; (b) mean $V_p/V_s$ ratio; and (c) mean crustal thickness. Symbol size is proportional to the value of the parameter plotted. Shading is used to highlight regions with higher values.

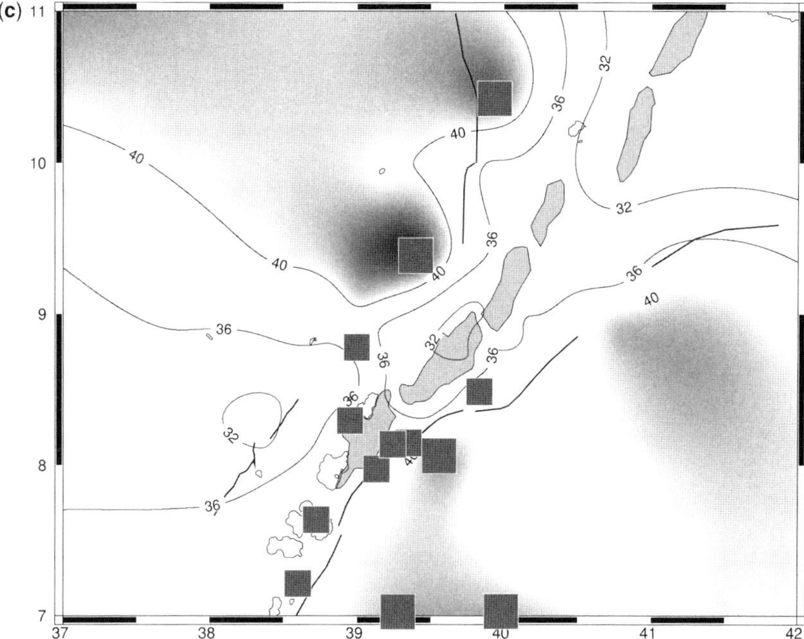


**Fig. 4.** *Continued.*

the region that runs north of the Fantale magmatic segment and into the incipient oceanic spreading region of Afar. Historic fissural basalt flows at Fantale and Kone volcanoes (near stations GTFE and DONE respectively, Fig. 1) as recently as 1810 AD indicate ongoing volcanism (Harris 1844). Secondly, there is a region around the Quaternary Butajira and Bishoftu volcanic chains on the western flank of the rift (Fig. 1). On the eastern plateau, $V_p/V_s$ is characteristically less than 1.8 ($\sigma < 0.26$). On the more-elevated western plateau, $V_p/V_s$ is greater than 1.8 ($\sigma > 0.28$).

Figure 4c shows the crustal thickness variation, determined from the receiver functions. There is a general correlation of Moho depth with elevation in line with local isostatic compensation. Our results show that within the rift the crust thins from around 38 km in the south to less than 30 km in the north, as the rift opens out into the incipient oceanic spreading of the Afar depression. These crustal thickness results are compatible with those determined by controlled source experiments (Berkhemer *et al.* 1975, Maguire *et al.* 2006). Our data suggest that the crustal thickness variation is not uniform along axis; the Moho rises sharply just south of the Boset magmatic segment, where the rift changes orientation from NNE–SSW to NE–SW, and the crust apparently thickens in a region north of the Fantale magmatic segment (Figs 1 & 4c). We are confident of the former observation; the more rapid change in crustal thickness is imaged, with improved spatial resolution, on the wide-angle reflection / refraction interpretation (Maguire *et al.* 2006) and in the gravity field (Mickus *et al.* 2004). However, our station distribution is such that the apparent thickening of the crust north of the Fantale magmatic segment in Fig. 4c is probably a sampling artefact, though magmatic underplate cannot be ruled out.

Thinner crust may lie to the NW and directly overlie the low-velocity upper mantle anomaly imaged by Bastow *et al.* (2005); Bouguer gravity profiles also suggest our station distribution does not sample the zone of maximum crustal thinning (Tiberi *et al.* 2005).

The crust on the Somalian plate to the SE of the rift is approximately 38 km thick in the north, increasing with elevation to the south to over 40 km. BEDE has a crustal thickness of 42 km, an exception to this pattern. On the western plateau, there is thicker crust to the north (41–45 km) than to the south (<40 km) as noted by Dugda *et al.* (2005). The boundary between the two regions occurs along an off-rift upper mantle low-velocity structure imaged in the seismic tomographic study of Bastow *et al.* (2005). In the NW part of the Ethiopian plateau, the crust thickens towards the uplifted rift flanks (Fig. 4c); to the SW, the region around the Butajira and Bishoftu volcanic chains is underlain by anomalously thin crust (<35 km) and is associated with high values of $V_p/V_s$. The receiver function crustal thickness estimates compare favourably with those estimated from controlled-source recordings on the western plateau along profile I of Berckhemer *et al.* (1975). The interpretation of the elevated western plateau section of the EAGLE across-rift controlled-source profile shows high-velocity (7.4 km $s^{-1}$) lower crustal underplate at depths of 33–48 km (Mackenzie *et al.* 2005). A comparison with our co-located receiver function results implies that what we have interpreted as the major Moho P-to-S conversion comes from the top of this underplate layer; even so, the controlled source interface is interpreted to be up to 5 km higher than the equivalent receiver function boundary.

Figure 5 shows an across-rift profile of stacked receiver functions from the EAGLE Phase I experiment. The variation in arrival time of Pms on the across-rift profile (Fig. 5) illustrates the crustal thinning taking place within the rift and the asymmetry in crustal thickness of the flanks. Station DONE, on the Boset magmatic segment, shows little Pms energy in a region thought to have melt in the lower crust from high conductivities at these depths (Whaler & Hautot 2006).

## Discussion

Average crustal $V_p/V_s$ or Poisson's ratio can be used to complement petrological studies in the investigation of the bulk composition of the crust (e.g. Chevrot & van der Hilst 2000). Laboratory experiments have shown that Poisson's ratio shows little variation with temperature and, for pressures greater than 100–200 MPa, it shows little dependence on pressure (Christensen 1996). Mineralogy is the most important factor influencing Poisson's ratio prior to the onset of melting. The relative abundance of quartz ($\sigma = 0.09$) and plagioclase feldspar ($\sigma = 0.30$) has a dominant effect (Christensen 1996): for felsic quartz-rich rocks, such as granite, $\sigma = 0.24$ ($V_p/V_s = 1.71$); for intermediate rocks, $\sigma = 0.27$ ($V_p/V_s = 1.78$); and for mafic plagioclase-rich rocks, such as gabbro, $\sigma = 0.30$ ($V_p/V_s = 1.87$). In continental settings, a Poisson's ratio above 0.30 is often associated with the presence of partial melt or serpentinite (Watanabe 1993); serpentinite is unlikely in this tectonic setting.

The high values of Poisson's ratio for stations in the rift imply that mafic crust underlies the NMER, with highest values beneath chains of Quaternary eruptive volcanic centres. Velocities of upper crustal layers beneath the rift are found to be 5–10% higher than outside the rift, which Keranen *et al.* (2004) and Mackenzie *et al.* (2005) interpret as resulting from mafic intrusions associated with the Quaternary magmatic segments. A decrease in

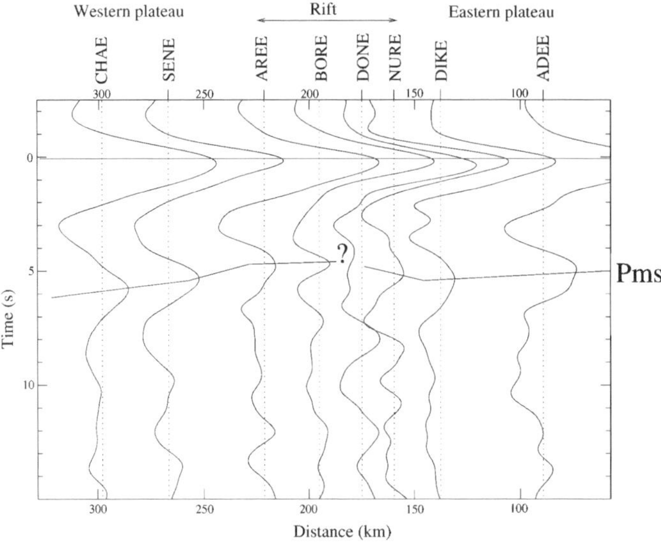


**Fig. 5.** Stacked receiver functions of EAGLE phase I stations for a cross-rift profile. Distances are from the south-eastern end of the EAGLE cross-rift controlled-source survey.

the proportion of upper crust relative to lower crust led Berckhemer *et al.* (1975) and Mohr (1989) to speculate that strain was accommodated by mechanical stretching of the brittle crust and by mafic intrusions and underplating in the lower crust. The axial EAGLE controlled-source line also shows discrepant upper and lower crustal stretching from the Boset magmatic segment northwards (Keller *et al.* 2004; Maguire *et al.* 2006). Our new and existing data provide insights into extensional processes beneath this transitional rift zone. Figure 4b shows two regions where $V_p/V_s$ is greater than 2.0 ($\sigma > 0.33$): the northern part of the rift as it enters the Afar depression, and beneath the Butajira and Bishoftu volcanic chains. The high $V_p/V_s$ ratios in these regions require partial melt in the crust that is consistent with xenolith evidence of non-fully consolidated dykes/veins that traverse the lithosphere beneath Bishoftu and Butajira (Rooney *et al.* 2005). Major element composition of Quaternary magmatic products erupted in these volcanic chains suggests the onset of melting in the depth range 53–88 km (Rooney *et al.* 2005). Melt fractionation is thought to take place at crustal depths.

Around the Boset magmatic segment, there is evidence of magma in the crust from our receiver functions and the magnetotelluric survey of Whaler & Hautot (2006). The stacked receiver function at DONE (Fig. 5) has a poorly developed Pms conversion, possibly attributable to attenuation of the converted S-wave; this is the region where Whaler & Hautot (2006) show high conductivity, inferred to be melt, down to lower crustal levels. Dugda *et al.* (2005) found the station NAZA at the southern end of the Boset segment to have a high average crustal Poisson's ratio of 0.35. Crustal anisotropy measurements of Keir *et al.* (2005) indicate approximately 3% melt within the upper crust beneath the Boset and Fantale magmatic segments.

An increase in volume of magmatism and decrease in depth of intrusions along the rift northwards to Afar is suggested from shear-wave splitting results (Ayele *et al.* 2004; Kendall *et al.* 2005), seismic tomography (Bastow *et al.* 2005) and geological considerations (Ebinger & Casey 2001). Segmentation of crustal structure both between the continental and transitional part of the rift and on the western plateau may be controlled by previous structural inheritances. Sub-Moho Pn velocities in the rift are low ($7.5–7.7$ km $s^{-1}$). These velocities are attributed to the presence of high-temperature mantle material with 3–5% partial melt (Mackenzie *et al.* 2005). The variability of the primary receiver function phase Pms (the Moho P to S conversion) and the deduced crustal thickness in the northern Ethiopian rift (Fig. 5) probably reflects different amounts of crustal extension and/or thickness of underplate and crustal intrusions within the rift. The composition of the ignimbrites, spanning the period 30 Ma to present, shows an increasing amount of crustal contamination with time, consistent with melting pre-rift basaltic underplate during progressive stretching and intrusion events (Ayalew & Yirgu 2003). All these factors suggest that the strain within the crust is accommodated by comparable amounts of stretching and localized magma intrusion beneath Quaternary magmatic segments (Ebinger & Casey 2001). Thus our results support the magma-assisted rifting models of Buck (2004), whereby magma intrusion to progressively shallower lithospheric levels allows breakup at smaller plate-driving forces.

The global average Poisson's ratio is $0.29 \pm 0.02$ ($V_p/V_s$ 1.84) for Precambrian shields and $0.27 \pm 0.03$ ($V_p/V_s$ 1.78) for Proterozoic platforms (Zandt & Ammon 1995). Chevrot & van der Hilst (2000) comment on the correlation between crustal thickness and Poisson's ratio for the Australian crust. In Proterozoic domains, $V_p/V_s$ increases with increasing crustal thickness. This is explained by an increase in thickness of mafic lower crust by underplating during tectonic cycles. In contrast, for Phanerozoic crust, Poisson's ratio decreases with increasing crustal thickness (Chevrot & van der Hilst 2000), which requires an increasing proportion of felsic crustal material, probably produced by crustal thickening during orogenesis. Dugda *et al.* (2005) pointed out that on the western Ethiopian plateau Poisson's ratio decreases with increasing crustal thickness as it does in the rift. This is in agreement with our additional results. They use this observation and similarity of the determined $V_p/V_s$ values (*c.* 1.8) to the global average for Precambrian crust to suggest that beneath the flood basalts there is Proterozoic crust largely unmodified by Cenozoic magmatic activity. Thus, our results differ from those of Chevrot & van der Hilst (2000). Increasing Poisson's ratio as this Proterozoic crust thins can be explained by mafic intrusions within the crust and extensional thinning of the brittle felsic upper crust (e.g. Berckhemer *et al.* 1975; Mohr 1989).

Crustal thickness estimates from receiver functions (Fig. 4c), seismic refraction (Maguire *et al.* 2006), gravity studies (Tiberi *et al.* 2005) and structural considerations (Ebinger & Casey 2001) suggest we can divide the crust along the axis of the NMER into three categories: a southern section typical of continental rifting; a northern oceanic spreading section in Afar; and a region in between, which is transitional in character. Elsewhere, typical continental rifts show crustal thinning of up to 10 km, e.g. East Africa (Prodehl *et al.* 1994), Baikal (Gao *et al.* 2004) and Rio Grande (Wilson *et al.* 2003). Notable from our results is the small amount of thinning (4–5 km) south of the Boset magmatic segment, where crustal thickness is greater than in the younger, less-evolved Kenya rift (Prodehl *et al.* 1994). We ascribe the small degree of stretching in the southern part of the rift to either smaller degrees of extension distributed across a broader region (Ebinger *et al.* 2000) and/or syn-rift magmatic underplating, which counters the mechanical stretching effects. Further north in the rift, the strain is localized to a 20 km-wide zone, thereby accounting for the larger degree of thinning imaged in the refraction and receiver function data.

Existing structural inheritances can play a role in rift propagation (Vauchez *et al.* 1997). Previous studies of the East African rift system have documented the controlling influence of pre-Cenozoic lithosphere thickness and rheology variations in the location of the rift, and the composition of magmatic products (e.g. Ebinger & Sleep 1998; Chesley *et al.* 1999; Nyblade & Brazier 2002). The NMER formed within Precambrian metamorphic crustal basement of Pan-African age. Pre-rift flood basalts mask the basement across most of the study region, making it impossible both to evaluate the continuity of Pan-African sutures exposed outside the flood basalt province and fault systems in the Mesozoic sedimentary strata beneath the flood basalts (Church 1991). From a few limited exposures in the plateau region, Precambrian terrain boundaries, commonly marked by ophiolites, strike sub-parallel to the rift in a N–NNE direction (Vail 1983; Abdelselam & Stern 1996) with NW- and ENE-striking shear zones (Purcell 1976). On the other hand, Korme *et al.* (2004) conclude that the length, pattern and segmentation of major active fault zones within the rift are controlled by pre-existing Mesozoic NW–SE-trending and Oligocene earth–west-trending faults. Mackenzie *et al.* (2005) find similar upper crustal velocities on either side of the rift and imply the upper crust is of similar origin. Whaler & Hautot (2006) show that there is a fundamental difference in crustal resistivity structure beneath the eastern and western plateau. Our receiver function study is equivocal on the subject; the differences in crustal thickness and average $V_p/V_s$ along the flanks of the rift are as large as across the rift; the across-rift crustal differences could be ascribed to either pre-rift differences in crustal structure across Precambrian tectonic domains, or local variations in degree of magmatic modification of the crust. A conclusion to be drawn from our receiver function results is that magmatic processes associated with rifting, as shown by gross crustal properties, such as Moho depth and average $V_p/V_s$, have fundamentally altered the crust in the rift. Pre-existing lithospheric heterogeneities add additional complexity to these patterns.

Dugda *et al.* (2005) argue that the Precambrian crust beneath the plateau has not been significantly modified by Cenozoic rifting and volcanism since crustal thickness and average $V_p/V_s$ is similar to the global average for Precambrian crust (Zandt & Ammon 1995) and close to that determined for similar Mozambique Belt crust in Tanzania and Kenya. Mackenzie *et al.* (2005), on the other hand, conclude from wide-angle reflection refraction data that the crust of the uplifted Ethiopian plateau has been underplated by up to 15 km of lower crustal high-velocity mafic material associated with the Oligocene flood basalts and Neogene rifting processes. Dugda *et al.* (2005) ascribed the observed crustal thickness variation between the northern portion of the western plateau (41 km) compared to the southern plateau (35 km), which is accompanied by little change in average Poisson's ratio, to variability in flood basalt thickness at the surface (1–2 km thicker in the north according to Pik *et al.* 2003) and/or differences in crustal underplating. We note that this change occurs across an off-rift upper mantle low-velocity feature (Bastow *et al.* 2005), which suggests a deeper lithospheric cause than surface basalts.

We have investigated whether the conversion from the bottom of the underplate on the western plateau, assumed to be the Moho by Mackenzie *et al.* (2005), has been overlooked in our receiver function analysis. Figure 6 shows the synthetic receiver function derived from the P-wave model of Mackenzie *et al.* (2005), with and without the underplate layer, compared with the observed stacked receiver function at station SENE (Fig. 1). This station is directly located on the seismic line over the region of thickest underplate. We have used our estimated average $V_p/V_s$ (1.8) for the station to derive the equivalent S-wave velocity–depth model. When the arrival times and amplitudes of the S-wave conversion from the EAGLE controlled-source model interfaces are compared to the observed receiver function (Fig. 6), it becomes clear that the synthetic waveform from the conversion at the top of the underplate is early and of reduced amplitude; the conversion from the bottom of the underplate is not seen on the observed receiver function. This implies either that the base of the underplate is gradational and the wide-angle reflection/refraction waveforms have imaged a boundary not resolved by the receiver function technique, or that the variability of the underplate and the spatial resolution of the two methods is such that they are sampling different regions of the lower crust.

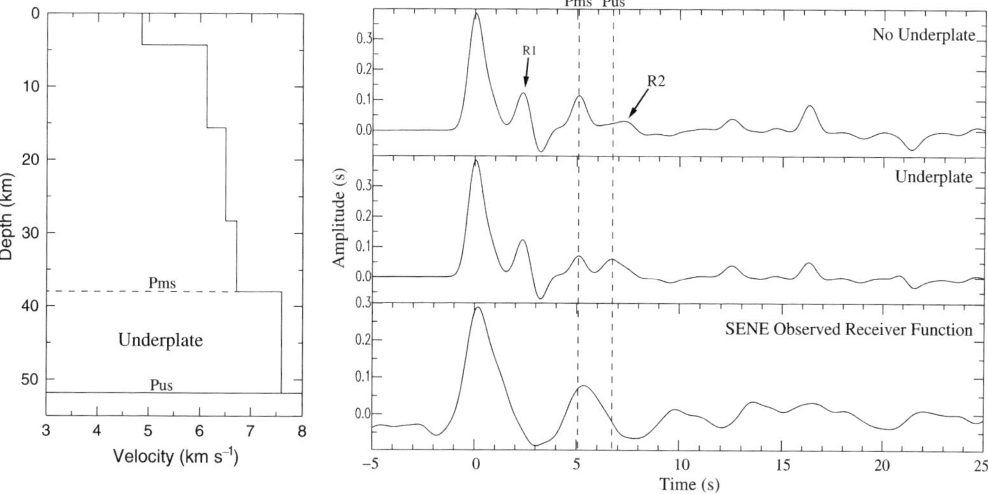


**Fig. 6.** Forward modelled synthetic receiver functions for EAGLE controlled-source crustal structure (Mackenzie *et al.* 2005), with and without an underplate layer, for station SENE compared with the observed receiver function. Pms is the P–to–S conversion from the receiver function Moho; Pus is the P–to–S conversion from the bottom of the underplate. R1 and R2 are the reverberations PpPms from the bottom of the first and second layer respectively. The P-wave velocity model used is also shown; $V_p/V_s$ is 1.80 in all layers.

## Conclusions

Receiver functions from the EAGLE broadband network have been used to determine crustal thickness and average $V_p/V_s$ for the NMER and its flanking plateaus. On the flanks of the rift, the crust on the Somalian plate to the east is 38–40 km thick. On the western plateau, there is thicker crust to the NW (41–43 km), than to the SW (<40 km); the thinning takes place over an off-rift upper mantle low-velocity structure previously imaged by travel-time tomography. The crust, on the western plateau, thickens towards the uplifted flanks of the NMER. The crust is slightly more mafic ($\sigma \sim 0.28$) on the Ethiopian plateau of the Nubian plate than on the Somalian plate ($\sigma \sim 0.26$). This could either be due to magmatic activity or different pre-rift crustal compositions. The Quaternary Butajira and Bishoftu volcanic chains, on the flanks of the rift, are marked by thinned crust and a $V_p/V_s > 2.0$, indicative of partial melt within the crust. Segmentation both between the continental and transitional part of the rift and on the western plateau may be controlled by previous structural inheritances, for example NW–SE Precambrian shear zones, reactivated by Mesozoic extension, which also affected Cenozoic rifting. The discrepancy in crustal thickness estimates on the western Ethiopian plateau between the EAGLE controlled-source experiment and receiver function analyses can partly be explained if the top of the underplate has been interpreted as the Moho in the receiver function studies. In this case, the base of the underplate should be gradational as there is little evidence in the receiver function waveforms for a sharp velocity boundary.

Within the rift, the $V_p/V_s$ ratio increases to greater than 2.0 (Poisson's ratio, $\sigma > 0.33$) northwards towards the Afar depression. Such high values are indicative of partial melt in the crust and corroborate other geophysical evidence for increased magmatic activity as continental rifting evolves to oceanic spreading in Afar. Along the axis of the rift, crustal thickness varies from around 38 km in the south to 30 km in the north, with most of the change in Moho depth occurring just south of the Boset magmatic segment where the rift changes orientation. Both the amount of crustal thinning and the basic composition of the crust as shown by the observed $V_p/V_s$ suggest that the magma-assisted rifting hypothesis of Buck (2004) is an appropriate break-up model for this transitional rift.

This work was supported by NERC grant NER/A/S/2000/01003 and 01002. The seismological equipment was loaned from the SEIS-UK equipment pool, with field assistance from Alex Brisbourne. We acknowledge the discussions and support from the EAGLE Working Group, in particular D. Keir, K.Whaler, T. Rooney, C. Tiberi and P. Maguire. IB was funded by a NERC scholarship NER/S/A/2001/06267.

## References


ABDELSALAM, M. & STERN, R. 1996. Sutures and shear zones in the Arabian–Nubian Shield, *Journal of African Earth Sciences*, **23**, 289–310.

AMMON, C.J. 1991. The isolation of receiver effects from teleseismic P waveforms. *Bulletin of the Seismological Society of America*, **81**, 2504–2510.

AMMON, C.J., RANDALL, G.E. & ZANDT, G. 1990. On the non-uniqueness of receiver function inversions. *Journal of Geophysical Research*, **95**, 15,303–15,318.

AYALEW, D. & YIRGU, G. 2003. Crustal contribution to the genesis of Ethiopian plateau rhyolitic ignimbrites: basalt and rhyolite geochemical provinciality. *Journal of the Geological Society, London*, **160**, 47–56.

AYELE, A., STUART, G. & KENDALL, J.-M. 2004. Insights into rifting from shear-wave splitting: an example from Ethiopia. *Geophysical Journal International*, **157**, 354–362.

BASTOW, I.D., STUART, G.W., KENDALL, J.-M. & EBINGER, C.J. 2005. Upper mantle seismic structure in a region of incipient continental break-up: northern Ethiopian rift. *Geophysical Journal International*, **162**, 479–493.

BERCKHEMER, H., BAIER, B., BARTELSON, H., BEHLE, A., BURKHARDT, H., GEBRANDE, H., MAKRIS, J., MENZEL, H., MILLER, H. & VEES, R. 1975. Deep seismic soundings in the Afar region and on the highland of Ethiopia. *In*: PILGER, A. & ROESLER, A. (eds) *Afar Between Continental and Oceanic Rifting*. Schweizerbart, Stuttgart, 89–107.

BILHAM, R., BENDICK, R., LARSON, K., MOHR, P., BRAUN, J., TESFAYE, S. & ASFAW, A. 1999. Secular and tidal strain across the Main Ethiopian Rift. *Geophysics Research Letters*, **26**, 2789–2792.

BUCK, R. 2004. Consequences of asthenospheric variability on continental rifting. *In*: KARNER, G., TAYLOR, B., DRISCOLL, N.W. & KOHLSTEDT, D.L. (eds) *Rheology and Deformation of the Lithosphere at Continental Margins*. Columbia University Press, New York, pp. 1–30.

CHESLEY, J.T., RUDNICK, R.L. & LEE, C.-T. 1999. Re–Os systematics of mantle xenoliths from the East African Rift; age, structure and history of the Tanzanian Craton. *Geochimica et Cosmochimica Acta*, **63**, 1203–1217.

CHEVROT, S. & VAN DER HILST, R.D. 2000. The Poisson's ratio of the Australian crust: geological and geophysical implications. *Earth and Planetary Science Letters*, **183**, 121–132.

CHRISTENSEN, N.I. 1996. Poisson's ratio and crustal seismology. *Journal of Geophysical Research*, **101**, 3139–3156.

CHURCH, W.R. 1991. Discussion of 'Ophiolites in northeast and East Africa: Implications for Proterozoic crustal growth', *Journal of the Geological Society, London*, **48**, 600–606.

DUGDA, M.T., NYBLADE, A.A., JULIA, J., LANGSTON, C.A., AMMON, C.J. & SIMIYU, S. 2005. Crustal structure in Ethiopia and Kenya from receiver function analysis: Implications for rift development in eastern Africa. *Journal of Geophysical Research*, **110**, doi 10.1029/2004JB003065.

EBINGER, C. & CASEY, M. 2001. Continental breakup in magmatic provinces: an Ethiopian example. *Geology*, **29**, 527–530.

EBINGER, C. & SLEEP, N. 1998. Cenozoic magmatism throughout E. Africa resulting from impact of a single plume. *Nature*, **395**, 788–791.

EBINGER, C., YEMANE, T., HARDING, D., TESFAYE, S., REX, D. & KELLEY, S. 2000. Rift deflection, migration and propagation: Linkage of the Ethiopian and Eastern rifts, Africa. *Geological Society of America Bulletin*, **102**, 163–176.

GAO, S.S., LIU, K.H. & CHEN, C. 2004. Significant crustal thinning beneath the Baikal rift zone: New constraints from receiver function analysis. *Geophysics Research Letters*, **31**, L210610, doi:10.1029/2004GL020813.

HARRIS, W.C. 1844. *The Highlands of Ethiopia*, Vol. 3, Longman, Brown and Longman, London.

HOFMANN, C., COURTILLOT, V., FERAUD, F. & ROCHETTE, P. 1997. Timing of the Ethiopian flood basalt event: Implications for plume birth and global change. *Nature*, **389**, 838–841.

KEIR, D., KENDALL, J.-M., EBINGER, C. & STUART, G. 2005. Variations in late syn-rift melt alignment inferred from shear-wave splitting in crustal earthquakes beneath the Ethiopian rift. *Geophysical Research Letters* **32**, L23308, doi: 10.1029/2055GLO24150.

KELLER, G.R., HARDER, S.H., O'REILLY, B., MICKUS, K., TADESSE, K., MAGUIRE, P.K.H. & EAGLE Working Group, 2004. A preliminary analysis of crustal structure variations along the Ethiopian Rift, 2004. *Proceedings of the International Conference on the East African Rift System*, June 20–24, Addis Ababa, 97–101.

KENDALL, J.-M., STUART, G., EBINGER, C., BASTOW, I. & KEIR, D. 2005. Magma-assisted rifting in Ethiopia. *Nature*, **433**, 146–148.

KERANEN, K., KLEMPERER, S.L., GLOAGUEN, R. & EAGLE Working Group, 2004. Three-dimensional seismic imaging of a protoridge axis in the Main Ethiopian rift. *Geology*, **32**, 949–952.

KNOX, R.B., NYBLADE, A.A. & LANGSTON, C. 1998. Upper mantle S velocities beneath Afar and western Saudi Arabia from Rayleigh wave dispersion. *Geophysical Research Letters*, **25**, 4233–4236.

KORME, T., ACOCELLA, V. & ABEBE, B. 2004. The role of pre-existing structures in the origin, propagation and architecture of faults in the Main Ethiopian Rift. *Gondwana Research*, **7**, 467–479.

LANGSTON, C. 1979. Structure under Mount Rainier, Washington, inferred from teleseismic body waves. *Journal of Geophysical Research*, **84**, 4749–4762.

MACKENZIE, G.D., THYBO, H. & MAGUIRE, P.K.H. 2005. Crustal velocity structure across the Main Ethiopian Rift: Results from 2-dimensional wide-angle seismic modelling. *Geophysical Journal International*, **162**, 996–1006.

MAGUIRE, P.K.H., EBINGER, C.J. ET AL. 2003. Geophysical project in Ethiopia studies continental break-up. *EOS, Transactions of the American Geophysical Union*, **84**, **337**, 342–343.

MAGUIRE, P.K.H., KELLER, G.R. *ET AL.* 2006. Crustal structure of the Northern Main Ethiopian Rift from the EAGLE controlled-source survey; a snapshot of incipient lithospheric break-up. *In*: YIRGU, G., EBINGER, C.J. & MAGUIRE, P.K.H. (eds) *The Afar Volcanic Province within the East African Rift System.* Geological Society, London, Special Publications, **259**, 269–291.

MAKRIS, J. & GINZBURG, A. 1987. The Afar Depression—Transition between continental rifting and sea-floor spreading. *Tectonophysics*, **141**, 199–214.

MICKUS, K., TADESSE, K., EBINGER, C., OLUMA, B. & ALEMU, A. 2004. Preliminary analysis of the gravity field of the Main Ethiopian Rift. *Proceedings of the International Conference on the East African Rift System*, June 20–24, Addis Ababa, 143–147.

MOHR, P. 1983. Ethiopian flood basalt province. *Nature*, **303**, 577–584.

MOHR, P. 1989. Nature of the crust under Afar — new igneous not thinned continental. *Tectonophysics*, **167**, 1–11.

NYBLADE, A.A. & LANGSTON, C.A. 2002. Broadband seismic experiments probe the East African rift. *EOS, Transactions of the American Geophysical Union*, **83**, 405, 408–409.

NYBLADE, A.A. & BRAZIER, R.A. 2002. Precambrian lithospheric controls on the development of the East African rift system. *Geology*, **30**, 755–758.

PIK, R., MARTY, B., CARIGAN, J. & LAVÉ, J. 2003. Stability of the Upper Nile drainage network (Ethiopia) deduced from (U–Th)/He thermochronometry: implications for uplift and erosion of the Afar plume dome. *Earth and Planetary Science Letters*, **215**, 73–88.

PRODEHL, C., KELLER, G.R. & KHAN, M.A. (eds) 1994. Crust and upper mantle structure of the Kenya rift. *Tectonophysics*, **236**, 483pp.

PURCELL, P.G. 1976. The Marda fault zone, Ethiopia. *Nature*, **261**, 569–571.

ROONEY, T.O., FURMAN, T., YIRGU, G. & AYALEW, D. 2005. Structure of the Ethiopian lithosphere: Evidence from mantle zenoliths. *Geochimica et Cosmochimica Acta*, **69**, 3889–3910.

SCHILLING, J-G. 1973. Afar mantle plume: Rare earth evidence. *Nature*, **242**, 2–5.

TIBERI, C., EBINGER, C., BALLU, V., STUART, G. & OLUMA, B. 2005. Inverse models of gravity data from the Red Sea–Aden–East African rifts triple junction zone. *Geophysical Journal International*, **163**, 775–787.

VAIL, J.R. 1983. Pan-African crustal accretion in north-east Africa. *Journal of African Earth Sciences*, **1**, 285–294.

VAUCHEZ, A., BARROUL, G. & TOMMASI, A. 1997. Why do continents break up parallel to ancient orogenic belts? *Terra Nova*, **9**, 62–66.

WATANABE, T. 1993. Effects of water and melt on seismic velocities and their application to characterization of seismic reflectors. *Geophysical Research Letters*, **20**, 2933–2936.

WHALER, K.A. & HAUTOT, S. 2006. The electrical resistivity structure of the crust beneath the northern Ethiopian Rift (EAGLE Phase III). *In*: YIRGU, G., EBINGER, C. & MAGUIRE, P.K.H. (eds) *The Afar Volcanic Province within the East African Rift System.* Geological Society, London, Special Publications, **259**, 293–305.

WILSON, D., ASTER, R. & the RISTRA Team, 2003. Imaging crust and upper mantle seismic structure in the southwestern United States using teleseismic receiver functions. *Leading Edge*, **22**, 232–237.

WOLFENDEN, E., EBINGER, C., YIRGU, G., DEINO, A. & AYALEW, D. 2004. Evolution of the northern Main Ethiopian rift: birth of a triple junction. *Earth and Planetary Science Letters*, **224**, 213–228.

ZANDT, G. & AMMON, C.J. 1995. Continental crust composition constrained by measurements of crustal Poisson's ratio. *Nature*, **374**, 152–154.

ZANDT, G., MYERS, S.C. & WALLACE, T.C. 1995. Crust and mantle structure across the Basin and Range – Colorado Plateau boundary at 37° N latitude and implications for Cenozoic extensional mechanism. *Journal of Geophysical Research*, **100**, 10 539–10 548.

ZHU, H. & KANAMORI, H. 2000. Moho depth variation in southern California from teleseismic receiver functions. *Journal of Geophysical Research*, **105**, 2969–2980.

# Crustal structure of the northern Main Ethiopian Rift from the EAGLE controlled-source survey; a snapshot of incipient lithospheric break-up

P.K.H. MAGUIRE[1], G.R. KELLER[2], S.L. KLEMPERER[3], G.D. MACKENZIE[1,*], K. KERANEN[3], S. HARDER[2], B. O'REILLY[4], H. THYBO[5], L. ASFAW[6], M.A. KHAN[1] AND M. AMHA[7]

[1]*Department of Geology, University of Leicester, Leicester LE1 7RH, UK (e-mail: pkm@le.ac.uk)*

[2]*Department of Geological Sciences, University of Texas at El Paso, El Paso, TX 79968, USA*

[3]*Department of Geophysics, Stanford University, Stanford, CA94305-2215, USA*

[4]*Dublin Institute of Advanced Studies, 5 Merrion Square, Dublin, Ireland*

[5]*Geological Institute, University of Copenhagen, Oster Voldgade 10, DK-1350, Copenhagen, Denmark*

[6]*Geophysical Observatory, Addis Ababa University, P.O. Box 1176, Addis Ababa, Ethiopia*

[7]*Ethiopian Science and Technology Commission, P.O. Box 2490, Addis Ababa, Ethiopia*

**Now at: Compagnie Générale de Géophysique, Vantage West, Great West Rd., Brentford, Middlesex, TW8 9GG, UK*

**Abstract:** The Ethiopia Afar Geoscientific Lithospheric Experiment (EAGLE) was undertaken to provide a snapshot of lithospheric break-up above a mantle upwelling at the transition between continental and oceanic rifting. The focus of the project was the northern Main Ethiopian Rift (NMER) cutting across the uplifted Ethiopian plateau comprising the Eocene–Oligocene Afar flood basalt province. A major component of EAGLE was a controlled-source seismic survey involving one rift-axial and one cross-rift *c*. 400 km profile, and a *c*. 100 km diameter 2D array to provide a 3D subsurface image beneath the profiles' intersection. The resulting seismic data are interpreted in terms of a crustal and sub-Moho P-wave seismic velocity model. We identify four main results: (1) the velocity within the mid- and upper crust varies from 6.1 km $s^{-1}$ beneath the rift flanks to 6.6 km $s^{-1}$ beneath overlying Quaternary axial magmatic segments, interpreted in terms of the presence of cooled gabbroic bodies arranged en echelon along the axis of the rift; (2) the existence of a high-velocity body ($V_p$ 7.4 km $s^{-1}$) in the lower crust beneath the northwestern rift flank, interpreted in terms of about 15 km-thick, mafic underplated/intruded layer at the base of the crust (we suggest this was emplaced during the eruption of Oligocene flood basalts and modified by more recent mafic melt during rifting); (3) the variation in crustal thickness along the NMER axis from *c*. 40 km in the SW to *c*. 26 km in the NE beneath Afar. This variation is interpreted in terms of the transition from near-continental rifting in the south to a crust in the north that could be almost entirely composed of mantle-derived mafic melt; and (4) the presence of a possibly continuous mantle reflector at a depth of about 15–25 km below the base of the crust beneath both linear profiles. We suggest this results from a compositional or structural boundary, its depth apparently correlated with the amount of extension.

The northern Main Ethiopian Rift (NMER) lies between the onshore extensions of the southern Red Sea and the Gulf of Aden rifts and the remainder of the East African rift system to the south (Fig. 1). It transects the Ethiopian plateau, a 1000 km-wide Palaeogene flood basalt province at 2500 m elevation (e.g. Mohr & Zanettin 1988), and is thus a prime locale to develop our understanding of the lithospheric processes resulting from continental break-up above what is widely believed to be a mantle plume (e.g. Montelli *et al.* 2004).

As continental rifting proceeds to sea-floor spreading, asthenospheric processes controlling magma supply should begin to dominate over lithospheric mechanical processes in the architecture of

*From*: YIRGU, G., EBINGER, C.J. & MAGUIRE, P.K.H. (eds) 2006. *The Afar Volcanic Province within the East African Rift System.* Geological Society, London, Special Publications, **259**, 269–291.
0305-8719/06/$15.00 

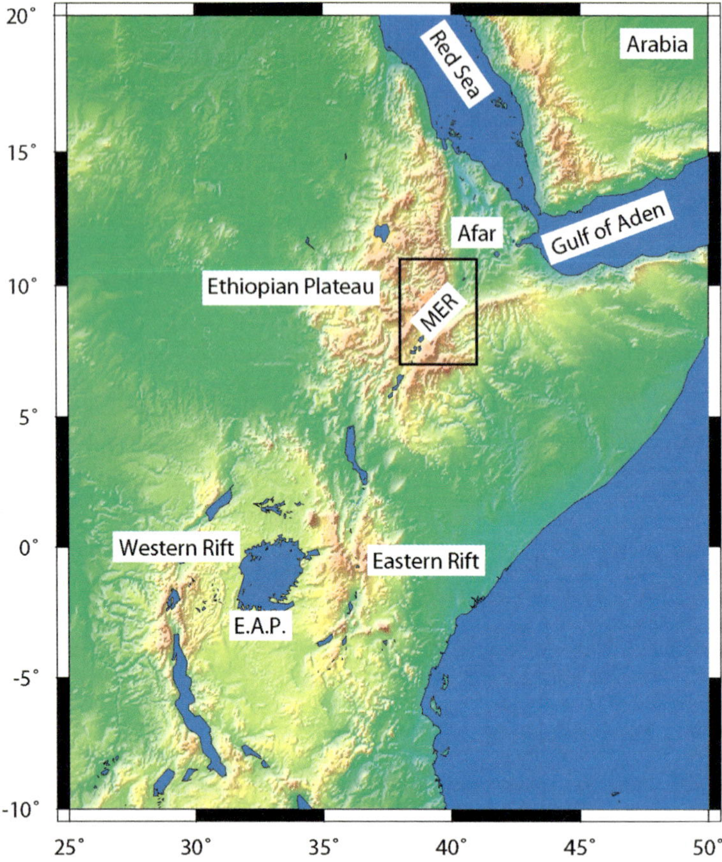


**Fig. 1.** Location map of the East African rift system including the Red Sea, Gulf of Aden, Main Ethiopian Rift (MER), the Eastern (Kenya) rift and the Western rift. Box section indicates region of Fig. 2. Ethiopian plateau is the area of high topography bisected by the MER. E.A.P., East African Plateau.

rifting, extensional strain within the crust being accommodated by axial dyking rather than faulting on rift basin flanking border faults. Ebinger & Casey (2001) proposed such a model for the transitional NMER at the southwestern corner of the Afar depression. Testing this model was a main objective of the Ethiopia Afar Geoscientific Lithospheric Experiment (EAGLE) (Maguire *et al.* 2003).

EAGLE was designed to examine crust and mantle processes occurring beneath the NMER and southern Afar, imaging 3D variations in crustal thickness and upper mantle structure, and characterizing the distribution of strain and magmatism across a typical transitional rift segment. It also involved examination of upper mantle flow associated with rifting above a mantle plume by mapping upper mantle anisotropy (e.g. Kendall *et al.* 2006), thereby providing a snapshot of the lithosphere immediately prior to separation.

A major component of EAGLE was a controlled-source wide-angle reflection/refraction experiment which has resulted in high-resolution 2D seismic velocity models both across and along the axis of the NMER and its transition into Afar, as well as a 3D model of the crustal structure at the intersection of the two 2D profiles. The experiment was designed to complement the EAGLE passive seismic (Bastow *et al.* 2005; Kendall *et al.* 2005, 2006; Stuart *et al.* 2006), local seismicity (Keir *et al.* 2006), magnetotelluric (Whaler & Hautot 2006) and geodetic (Bendick *et al.* 2006) studies. The controlled-source experiment has resulted in a model of absolute crustal and upper-mantle seismic P-wave velocities beneath the NMER providing:

(1) estimates of crustal thinning along and across this transitional sector of the rift, to relate to

present geodetic strain and the geodynamic processes resulting in lithospheric break-up;
(2) a model for the distribution of magmatism within the crust beneath the rift, and its relationship to the surface distribution of magmatic segments and rift architecture;
(3) information on the distribution of magmatic underplate and its contribution to surface uplift, with implications concerning support for the Ethiopian plateau; and
(4) the necessary control on starting models for local earthquake tomography, and the conversion of teleseismic receiver functions to depth.

We report here the first results and implications of this controlled-source seismic programme.

## Tectonic and geological setting

The collision between East and West Gondwana had a profound effect on the geology of East Africa. The late Proterozoic Mozambique belt, representing a Himalayan-scale orogeny, abuts the Archaean complexes of western Ethiopia and the Nyanza craton to the west of the Eastern (Kenya) rift (e.g. Stern 1994). In Ethiopia, the belt is transected by numerous predominantly north–south to NNE–SSW suture zones and associated ophiolitic slivers, and NW–SE trending shear zones (e.g. Abdelsalam & Stern 1996; Allen & Tadesse 2003). Mesozoic marine transgression over this region before the break-up of the Gondwana supercontinent resulted in a thick sedimentary sequence. This was followed in the early Cretaceous by a major tectonic uplift event probably related to rifting in the Sudan (Schull 1988, Bosellini *et al.* 2001). It was followed by deposition of up to 3 km of fluviatile sediments in the area of what is now the Ethiopian plateau and NMER (e.g. Tadesse *et al.* 2003).

The Ethiopian flood-basalt province comprises approximately 350 000 $km^3$ of Oligocene basalts and associated felsic volcanics generated between 31 and 29 Ma following the inferred impact of the Afar mantle plume on the base of the lithosphere. Wolfenden *et al.* (2004) summarize the tectono-magmatic evolution of the region. Extrusion of flood basalts post-dated extension in the Gulf of Aden at *c.* 35 Ma but predated that in the Red Sea at *c.* 28 Ma. Extension began between 18 and 15 Ma in southern Ethiopia, but only after 11 Ma in the NMER where it was associated with further basaltic volcanism (Chernet *et al.* 1998; Bonini *et al.* 2005). Sea-floor spreading in the Gulf of Aden propagated westward into the Afar depression since 16 Ma, and Red Sea sea-floor spreading commenced at *c.* 4 Ma (e.g. d'Acremont *et al.* 2005). Sea-floor spreading has yet to occur in the NMER.

Within the northern Main Ethiopian Rift, structural and stratigraphic patterns indicate a migration of extensional strain from 2.5 Ma to the present, from the marginal border faults to a narrow (20 km) wide-sub-axial zone of narrow, near north–south trending groups of aligned eruptive centres (Wolfenden *et al.* 2004). These in turn are cut by small offset faults and dykes and arranged in a right-stepping en echelon pattern (Boccaletti *et al.* 1999). Recent GPS measurements (Bilham *et al.* 1999) have shown that 80% of the extensional strain is concentrated in these magmatic segments (red-hatched zones in Fig. 2). It is these segments that Ebinger & Casey (2001) proposed as the locus of strain by active dyke injection. Most of the Quaternary volcanism has occurred within these magmatic segments, but other Pliocene–Quaternary volcanic centres are also located near to the NW rift margin (e.g. at Zikwala and Yerer volcanoes (see Fig. 6a)) and on the southeastern rift shoulder (e.g. Chilalo (see Fig. 6a)) demonstrating the spatial complexity of the crust/mantle magmatic system.

Within the rift valley, there are a series of asymmetric structural basins, bounded on one side by steep border faults (*c.* 60 km long and with >3 km throw). The basins contain Pleistocene volcanic, clastic and lacustrine strata that overlie the Miocene–Pliocene felsic and mafic sequences of the Kessem and Balchi Formations (Wolfenden *et al.* 2004). These extend into the Debre Zeit area around 8° 45 N. They are assumed to overlie Mesozoic sediments distributed widely over the study area.

## Previous geophysical studies

Recent teleseismic studies have shown that the upper mantle beneath Afar and the northern Main Ethiopian Rift is characterized by an elongate region of low wave-speed anomalies to depths of >400 km. The vertical and lateral extent of this anomaly suggests that it originates from a broad thermal upwelling connected to the so-called African superplume in the lower mantle beneath Southern Africa (Benoit *et al.* 2006). Bastow *et al.* (2005) imaged these low velocities as a 75 km-wide sheet extending to depths of around 300 km below the NMER and interpreted the feature as evidence for mantle upwelling. At depths of >100 km, north of 8.5° N this low-velocity zone broadens as the rift evolves towards the oceanic spreading centre in Afar, and appears to be connected to the deeper low-velocity structure.

Measurements of seismic anisotropy using regional surface waves demonstrate coherence between sub-lithospheric fast shear-wave orientations and

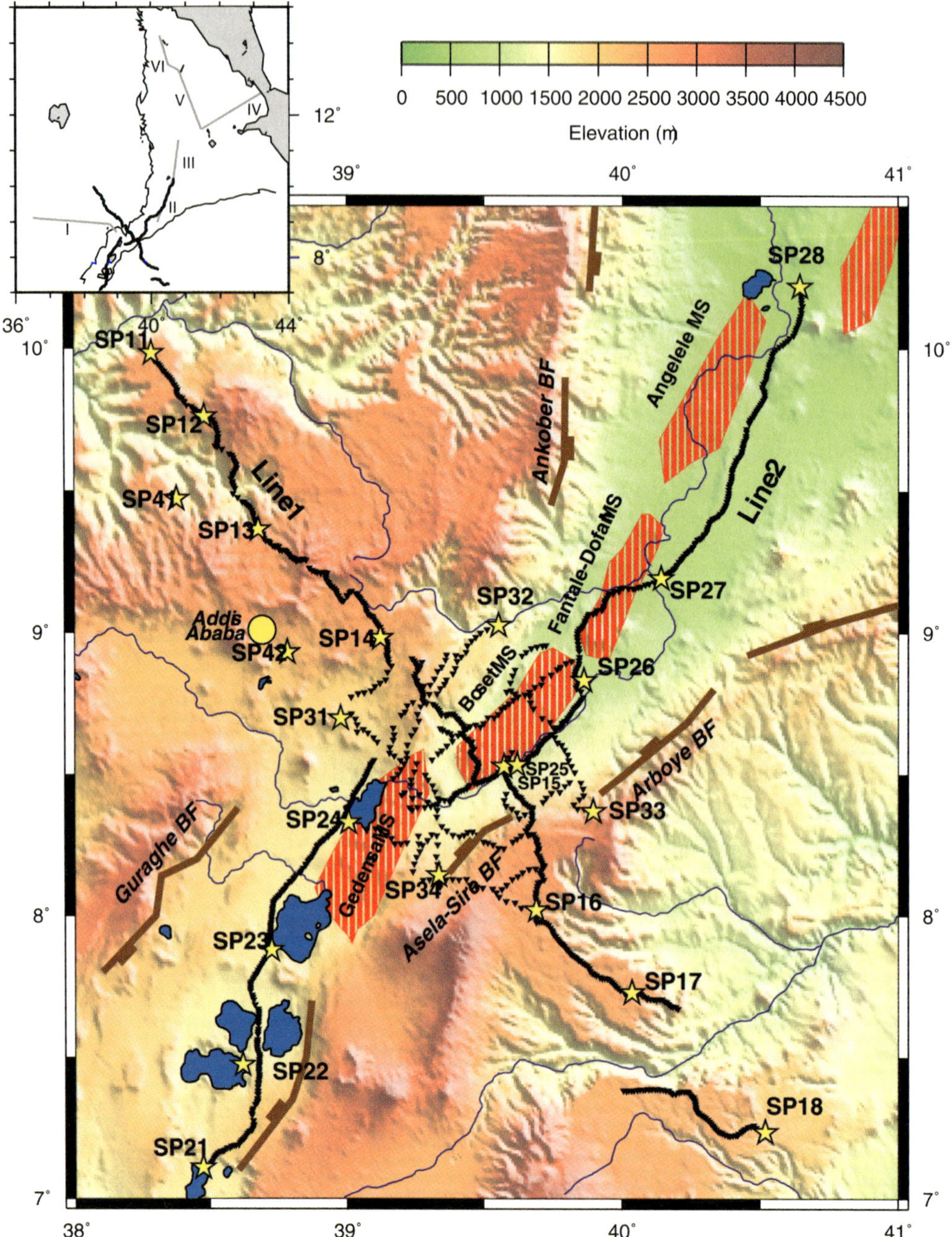


**Fig. 2.** Location map of the EAGLE controlled-source project. Yellow stars, shotpoints (e.g. SP11); black small inverted triangles, controlled source recorders; red hatched areas, magmatic segments (e.g. Boset MS) (after Wolfenden *et al.* 2004); brown lines – border faults (e.g. Arboye BF). Inset map identifies seismic refraction profiles I–VI of Berckhemer *et al.* (1975) in grey, and EAGLE profiles in black.

the trend of the African superplume rising toward the base of the lithosphere beneath Afar (Kendall *et al.* 2006). Above 150 km, the pattern of shear-wave anisotropy is more variable. Analysis of the splitting of teleseismic phase (SKS) and local shear-waves within the rift valley consistently shows rift-parallel directions (Kendall *et al.* 2005; Keir *et al.* 2005). Gashawbeza *et al.* (2004) and Kendall *et al.* (2005) show that the regional lithospheric anisotropy due to Proterozoic accretion of the Mozambique belt is only modified by extensional fabrics within a narrow zone less than 100 km wide beneath the geomorphic rift valley. Average fast shear-wave directions within the rift that are perpendicular to the modern WNW extension direction (Gashawbeza *et al.* 2004) are most likely due to an alignment of $<0.1\%$ melt fraction throughout the upper 70–90 km of the Earth (Kendall *et al.* 2005), parallel to the sheet-like upwelling in the upper mantle. There is a possible increase in the volume of melt from south to north along the rift (Gashawbeza *et al.* 2004). Kendall *et al.* (2006) also show that the anisotropy beneath the Ethiopian plateau is variable and may correspond to both pre-existing fabric and ongoing melt-migration processes.

Despite these significant results, knowledge of crustal structure is needed to evaluate shallow tectonic and magmatic processes in relation to those occurring in the lithospheric mantle and asthenosphere. The only previous crustal refraction data from Ethiopia are six profiles recorded in 1972 with lengths of 120 to 300 km. The recording utilized 15 analogue stations deployed at spacings of 5 to 7 km (Berckhemer *et al.* 1975). They have far lower resolution than our study which utilized over 1000 digital recorders spaced at 1 to 2.5 km. Of the 1972 data, Profile I trending east–west on the western plateau and Profile II coincident with the northern part of our Line 2 are of particular interest; the other profiles are further north in Afar (inset to Fig. 2). Profile I shows an upper crust of velocity 6.1 km $s^{-1}$ and thickness 15 to 23 km above a lower crust of velocity 6.65 km $s^{-1}$. The modelled Moho varied between 33 and 44 km depth (from west to east) above a normal mantle of velocity 8.0 km $s^{-1}$ (Makris & Ginzburg 1987).

Although the 1972 data are too sparse to constrain lateral variations in crustal velocities, Berckhemer *et al.* (1975) note that a high-density lower-crustal body beneath the rift escarpment is required to explain the gravity data. Profile II of Berckhemer *et al.* (1975) shows slightly lower upper-crustal velocities extending to only 8 km depth, above a faster lower crust (6.8 km $s^{-1}$). The highest velocity observed, 7.3 km $s^{-1}$, was presumed to represent partially molten upper mantle beneath the Moho at about 26 km depth. Analysis of the remaining profiles within the Afar region shows crustal thicknesses as low as 16 km (Berckhemer *et al.* 1975). Makris & Ginzburg (1987) provide a modest reinterpretation of the 1972 dataset, including recognition of an upper mantle reflector at *c.* 45 km depth (precise depth dependent on assumed upper-mantle velocity) in southern Afar extending 50 to 100 km north of our northernmost shotpoint SP28. All these older results—within the uncertainties inevitable with their limited recording equipment—are consistent with our newer data discussed below, though some of our geological interpretations differ. For example, Makris & Ginzburg (1987) considered the Afar crust to be continental (albeit stretched and intruded to varying degrees) based on the presence of at least some 6.1 km $s^{-1}$ 'granitic upper crust' material along each of their profiles, whereas our data show a more rapid increase to higher velocities in southern Afar, leading us to question the degree to which Precambrian basement survives in the axis of the northernmost NMER. Prodehl *et al.* (1997) provide a comprehensive overview and comparative crustal velocity columns of all prior seismic refraction results from the Afro-Arabian rift system, including those north and south of the region discussed in this paper.

Similarly broad-brush estimates of crustal thickness are offered by teleseismic studies and gravity inversions, which are quite compatible with our newer data within their anticipated analytical and interpretational uncertainties. Receiver-function studies (Hebert & Langston 1984; Ayele *et al.* 2004; Dugda *et al.* 2004; Stuart *et al.* 2006) yield estimates of crustal thickness of 27–38 km beneath the NMER, and 33–44 km beneath the eastern and western plateaus. Results from gravity studies (Mahatsente *et al.* 1999; Tiberi *et al.* 2005) suggest that the crustal thickness beneath the Main Ethiopian Rift axis decreases from about 32–33 km in the south to 24 km beneath the southern Afar depression, while thicker crust [*c.* 40 km (Tiberi *et al.* 2005); 38–51 km (Mahatsente *et al.* 1999)] is present beneath the plateau. Tiberi *et al.* (2005) suggest that magmatic underplating may exist beneath several collapsed caldera structures within the rift at around 8° N, 39° W, and Mahatsente *et al.* (1999) predict dense intrusions in both the middle and lower crust beneath the magmatic segments along the axis of the rift.

## EAGLE controlled-source survey

Two *c.* 400 km-long wide-angle reflection/refraction profiles centred on the Boset magmatic segment (Fig. 2), together with a *c.* 100 km-diameter 2D array spanning the rift at the intersection of the two profiles, were undertaken in January 2003. The NW–SE striking Line 1 was a cross-rift profile extending from the Blue Nile Gorge on the western plateau to the Bale Mountains on the eastern plateau (Fig. 2). The NE–SW striking Line 2 was an along-axis profile, crossing a number of basins from Lake Awassa in the south to Gewane in the north, and also the Gedemsa,

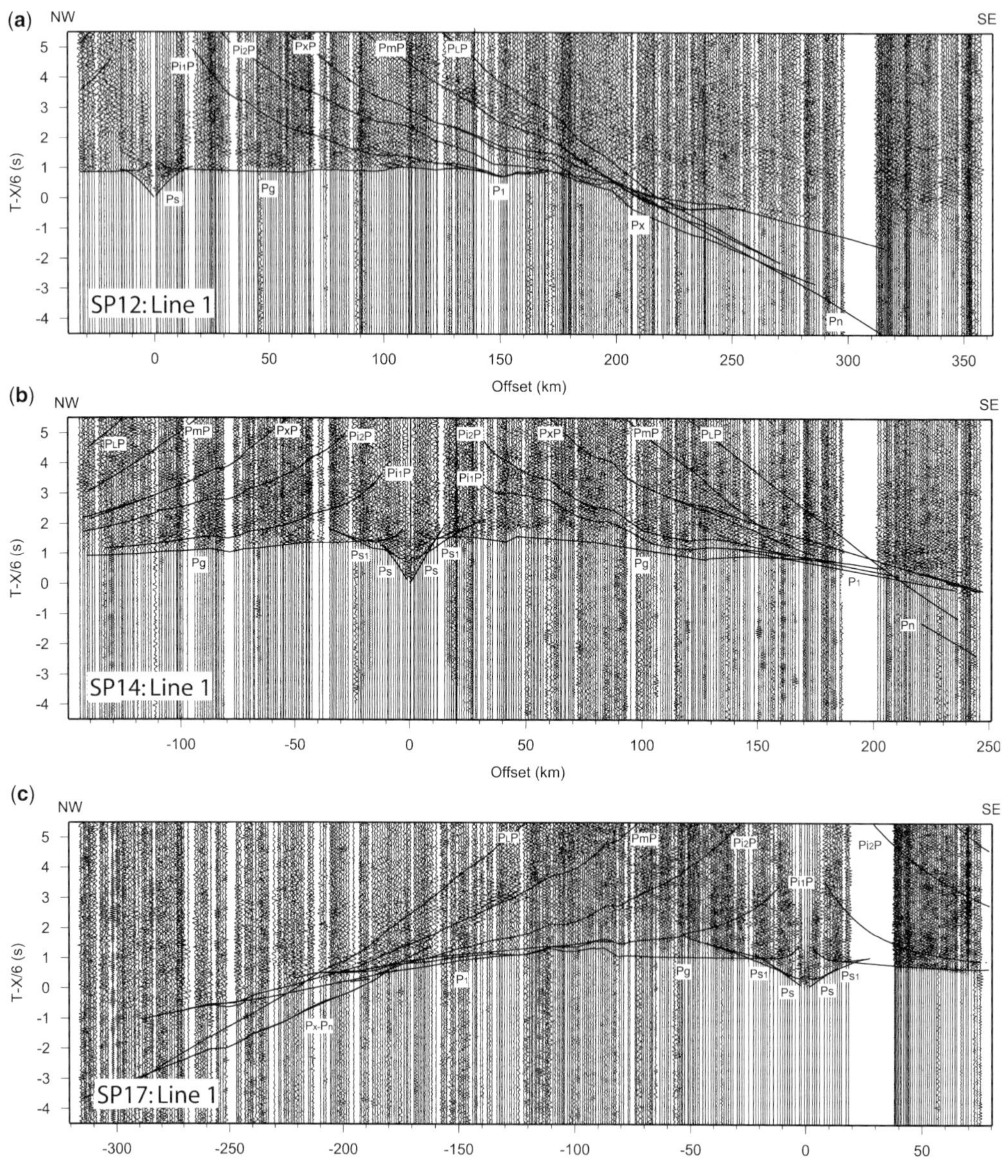


**Fig. 3.** Example seismic record sections from: (a) SP12 – Line 1; (b) SP14 – Line 1; (c) SP17 – Line 1; (d) SP25 – Line 2; (e) SP28 – Line 2. Sections are reduced at 6 km $s^{-1}$ with trace-normalized amplitudes and are bandpass-filtered from 4–16 Hz. Calculated travel-time curves from the velocity model shown in Fig. 5 are overlain. Phase labelling: Ps and $Ps_1$, sedimentary diving waves; Pg, crystalline basement diving wave; $P_1$, diving wave in mid–upper crust; $P_2$, diving wave in lower crust; Px, diving wave in high-velocity lowest crustal layer (HVLC); Pn, mantle diving wave; $Pi_1P$, reflection from mid–upper crust reflector; $Pi_2P$, reflection from top of lower crust; PxP, reflection from top of HVLC layer; PmP, Moho reflection; $P_LP$, mantle reflection.

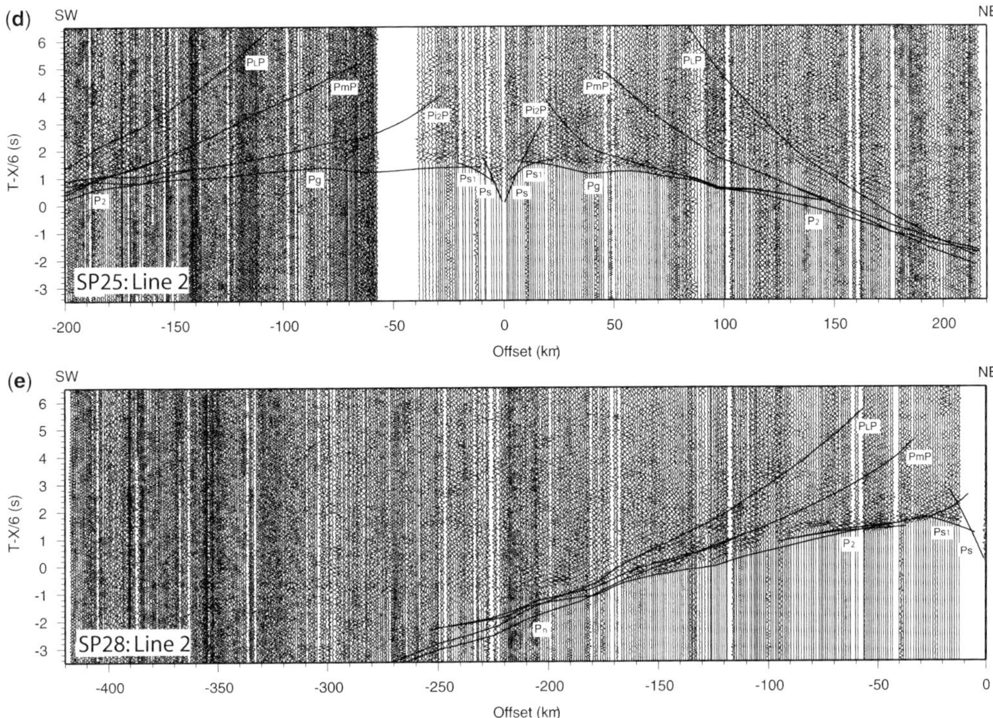


**Fig. 3.** *Continued.*

Boset, Fantale–Dofan and Angelele magmatic segments (Fig. 2). The profiles intersected above the Boset magmatic segment. The 2D array lay substantially within the NMER.

A total of 23 explosive charges ranging from 50 to 5750 kg with an average shot size of 1100 kg were detonated in boreholes up to 50 m deep, two lakes (Shala (SP22) and Arenguarde (SP31)) and two quarries (SP41, SP42) (Fig. 2). Shot spacing along the two profiles was a nominal 50 km, with a number of offline shots to extend the ray path azimuthal coverage into the 2D array. The receiver spacing was approximately 1 km along the two 400 km profiles, and 2.5 km within the 2D array, but only where accessible along the road network. Approximately 1000 Reftek single channel 'Texan' seismographs using 4.5 Hz geophones, together with 93 Guralp 6TD 3-component broadband recorders along Line 1, were distributed in one deployment in order to maximize the number of available seismograms. The EAGLE passive broadband arrays (see Stuart *et al.* 2006; Kendall *et al.* 2006) also recorded the 23 shots. Data were recorded with a sample rate of 100 sps for both instrument types. GPS timing was used throughout, with shot and instrument locations being measured using multiple GPS readings.

## Data

Overall data quality is good; best on Line 1 with substantial energy propagation along the entire length of the profile; variable on Line 2 with the southern half of the profile having to be deployed alongside a main highway, but good to the north; and good on the 2D array, although the larger station spacing on that array (*c.* 2.5 km) limits our ability to correlate phases following the first arrival. Representative examples of the seismic data, displayed in the form of distance versus reduced time record sections, are shown in Fig. 3 for shotpoints SP12, SP14 and SP17 on Line 1 and SP25 and SP28 on Line 2.

Seismic phases that have been identified include two from the sediment–volcanic layers close to and within the rift, Ps and $Ps_1$ (e.g. Fig. 3b–e), whereas only a single such phase is observed from seismic sections derived from shots away from the rift on the plateau (Fig. 3a). On only a few sections can reflections both from within and from the base of the sediment–volcanic layer be identified (e.g. Fig. 3b).

The upper crust has a velocity of *c.* 6.1 km s$^{-1}$ across the whole study area bar the northeastern end of Line 2, based on a consistent Pg arrival, the refracted wave through the crystalline upper crust. This phase is overtaken on a number of sections on Line 1 by $P_1$, the diving wave from a mid–upper crustal layer (e.g. Fig. 3a–c). This phase is difficult to distinguish from Pg, and is identified as it emerges from the reflection $Pi_1P$ originating from this mid-upper crustal layer boundary. On Line 1, $P_1$ and $Pi_1P$ are not seen from every shotpoint, the causative boundary being identified as an 'intermittent' horizon beneath the profile. These phases are not seen on Line 2. This is possibly due to poorer-quality recordings with lower signal-to-noise ratio, or to the reflector not being present beneath this profile. At extreme wide-angle distances, a strong crustal reflection (e.g. Fig. 3a–c) has been correlated as $Pi_2P$, the reflected phase from the top of a lower-crustal layer. This phase is also identified on Line 2 (Fig. 3d).

A generally strong PmP Moho reflection can be identified on all record sections. Mackenzie *et al.* (2005) describe a weak reflected phase, PxP, preceding the PmP phase on the northwestern side of the rift on Line 1. This is interpreted as originating from the top of a lower crustal high-velocity layer, which gives rise to a diving wave with an apparent velocity of 7.4 km s$^{-1}$(e.g. Fig. 3a). On Line 1, there is a weak, low-amplitude diving wave from below the Moho, Pn with an apparent velocity of 8.05 km s$^{-1}$, which can only be definitely identified emerging from beneath the eastern side of the rift (Fig. 3a,b). From SP17 (Fig. 3c), the diving wave beyond 180 km from the shotpoint is weak, and emerges from within the lower crustal layer identified beneath the northwestern side of the rift. Along the axis of the rift on Line 2, the highest velocity identified is 7.5 km s$^{-1}$, derived from a Pn diving wave in the mantle from the northeasternmost shotpoint, SP28 at Gewane (Fig. 3e). This Pn phase is equivalent to that identified by Berckhemer *et al.* (1975) from this region (Profile II: Fig. 2 inset). Additional processing is underway to attenuate noise, but as yet, no other shotpoint on Line 2 has produced clear Pn arrivals.

Mackenzie *et al.* (2005) describe a weak reflected phase, $P_LP$ from the upper mantle (Fig. 3a–c), which is identified from beneath both the northwestern and southeastern flanks of the rift, and beneath the rift itself on Line 1. An equivalent phase is seen on Line 2 (Fig. 3d,e) occurring at noticeably shorter offset beneath the northeastern end of Line 2 than elsewhere. Phase identification was confirmed through the checking of reciprocal times where possible, which proved especially useful for the weaker phases.

## Modelling and model description

For the data from Lines 1 and 2, our travel-time modelling used a combination of travel-time tomography and forward modelling and inversion with the RAYINVR code of Zelt & Smith (1992). The line lengths are beyond the limit for modelling using a flat Earth approximation, and a local Cartesian coordinate system was defined, and model offsets were calculated within this reference frame. The curvature of the Earth appears as a slight bulge on the resultant profile models (Figs 4 & 5). The local Cartesian coordinate system is centred on a point 4942 m below SP25, a point chosen to give all shot and receiver points positive elevations. Since SP28 has the lowest elevation, it has an elevation of zero in the local Cartesian coordinate system.

Following 2D first-arrival travel-time tomography to obtain preliminary velocities for raytracing, initial models were obtained through trial-and-error forward raytrace modelling to fit the picked travel times using a top-down (layer-stripping) approach. Damped least-squares inversion was then used to minimize the resultant travel-time residuals. For Line 1, the model was modified after qualitative comparison of the record sections with synthetic seismograms using the TRAMP code of Zelt & Forsyth (1994).

Interface depth and layer velocity errors were assessed via selective perturbation of the model parameters. For Line 1, the interface depth errors are $\pm 1$ km for intracrustal layers and $\pm 2$ km for the Moho, and $\pm 0.1$ km s$^{-1}$ for upper crustal velocities and $\pm 0.2$ km s$^{-1}$ for lower crustal velocities (Mackenzie *et al.* 2005). For Line 2, the equivalent errors have not yet been defined, but are likely to be somewhat greater than the Line 1 values. Model resolution for Line 1 was assessed from the number of rays traced for any solution, the RMS travel-time residual and the normalized $\chi^2$ value (Zelt & Forsyth 1994). It has not yet been fully assessed for Line 2. To provide an indication of the model resolution, diagrams of source-to-receiver rays are shown for both Line 1 and Line 2 models (Fig. 4a,b). Rays are only shown in Fig. 4 for source–receiver pairs for which the specific phases were visually identified.

Although elevation variations along the profiles are taken into account, the lateral distribution of the shots and recording stations about the best-fit line through each of the two profiles leads to velocity and depth errors in the final models resulting from the crooked line geometry. Each shot record section is analysed with the seismograms at their

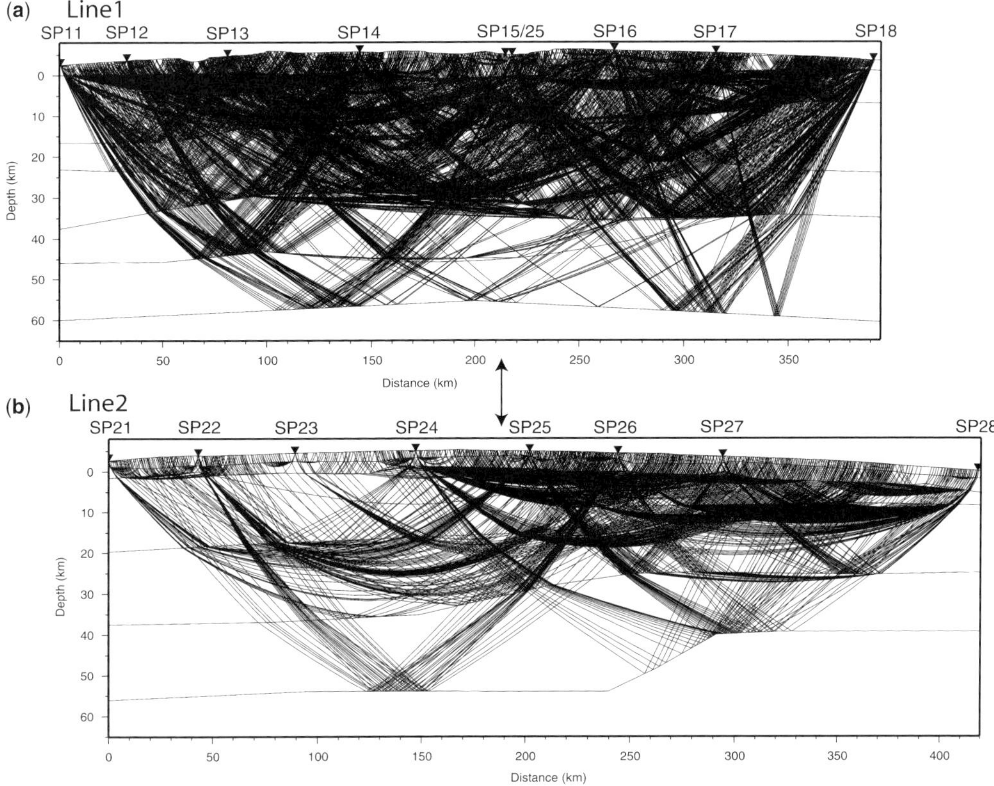


**Fig. 4.** Ray-coverage for Line 1 and Line 2 final models. Lines 1 and 2 are aligned at their intersection (double-headed arrow). RAYINVR (Zelt & Smith 1992) requires layer boundaries to be continuous across the model. Rays are traced from each shotpoint (e.g. SP11) to those recording stations providing phase picks used in the modelling. The low ray-coverage beneath the SW end of Line 2 reflects the poorer-quality data from this region (see text for comment). **(a)** Two-point ray-traced model for Line 1. **(b)** Two-point ray-traced model for Line 2.

true distance from the shot. However, the final 2D model has each shotpoint separated by its true distance from SP11 (for Line 1) and from SP21 (for Line 2). The errors introduced are most severe for large lateral offsets of near-shot stations with respect to the best-fit line through the profile. This situation is common in wide-angle crustal profiling and comment is made only where a noticeable error might be introduced.

The best estimate of the 'cross-over' point for the two profiles is at a distance of 214 km from SP11 and 189 km from SP21 (Fig. 5); the complex geometry is the reason the intersection point on the two lines seems discrepant with respect to SP 15/25. The 2D array data were interpreted using 3D first arrival travel-time tomography (see Keranen *et al.* 2004). Note that the Line 1 and Line 2 modelling procedure utilizes sharp boundaries in seismic velocity (Fig. 5) inferred to represent abrupt lithological changes, whereas the 2D array modelling procedure forces a smoothly varying velocity function (Fig. 6) which is computationally easier to handle in three dimensions.

## *Surface layers*

The surface layer velocities are best defined along the two crustal profiles (Fig. 5). There is a 2–5 km-thick layer of velocity *c.* 5.0 km $s^{-1}$ extending the whole length of Line 1. In the central part of the profile, it is covered by a layer of velocity *c.* 3.3 km $s^{-1}$, thickening into the rift. The Line 2 data show that these layers are at least semi-continuous along the rift valley. The total thickness of these layers, which from their velocities are interpreted as sediments and volcanics, appears marginally discrepant at the intersection of the two profiles, being 5.5 km on Line 1 and 4.2 km on Line 2, slightly outside the range of the estimated depth errors. This may be explained by the crooked-line geometry, which produced subsurface ray-paths that do not intersect beneath the

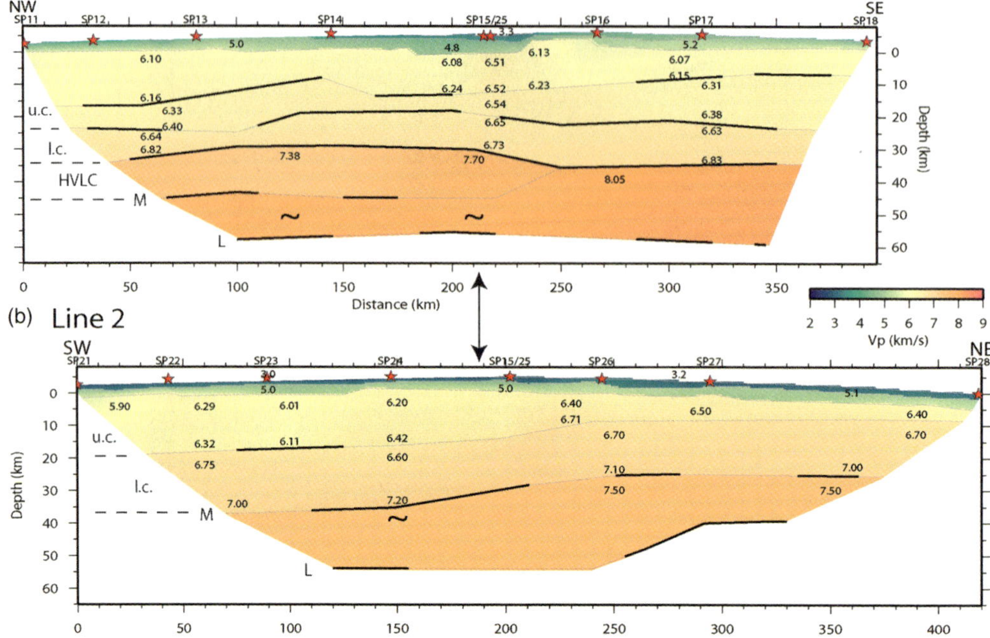


**Fig. 5.** Final ray-trace P-wave velocity models for **(a)** Line 1 and **(b)** Line 2. Model outlines indicate regions sampled by ray-paths (see Fig. 4 for exact coverage). P-wave velocities in km $s^{-1}$. '$\sim$' indicates there is no velocity information available. The depths to the reflectors beneath those regions identified by '$\sim$' have been estimated using laterally extrapolated known velocities within the same layer. Bold lines at layer boundaries are those regions sampled by reflected rays. u.c., upper crust; l.c., lower crust; HVLC, high-velocity lowest crust; M, Moho; L, lithospheric mantle reflector. Lines 1 and 2 are aligned at their intersection (double-headed arrow).

profile intersection at the surface. Given the topographic variations in the region where the lines intersect, the *c.* 50 km distance between shotpoints being of the order of over half the width of the rift valley, the presence of several volcanic centres, and what must be complex shallow subsurface structure that is beyond the resolution of our data, we feel that this tie is as good as one would expect.

## *The crust*

The crustal velocity models beneath Line 1 and Line 2 appear different, in terms of both the numbers of layers and the range of velocities identified at their intersection (Figs 5 & 8). Some differences may be resolved during ongoing modelling of Line 2: to date, the emphasis has been on arrivals that can be correlated along at least most of the profile. However, the strong three-dimensionality of the crust in the NMER (Fig. 6), and the refraction-profiling geometry that dictates that velocities beneath the profile intersection are obtained from quite different shotpoints along each profile, means that it would be a mistake to expect complete agreement. Nonetheless, we restrict our interpretations to the main crustal layers as discussed below.

A brief description of the crustal layering is as follows: beneath both controlled-source profiles, the crust can be divided into an upper crust with velocities of 6.1–6.4 km $s^{-1}$ above a lower crust of velocity *c.* 6.7 km $s^{-1}$ along Line 1, but which includes higher velocities (to 7.1 km $s^{-1}$) beneath Line 2. Beneath Line 1, the upper crust includes high velocities (*c.* 6.5 km $s^{-1}$) beneath a narrow 20–30 km-wide section beneath the rift valley; such velocities are also seen in the upper crust beneath the northern half of Line 2 (Fig. 5) and are resolved into discrete bodies in our 3D analysis (Fig. 6). Beneath Line 1, we identify an intra-upper crustal reflector that may, or may not be continuous across the rift. An anomalous high-velocity (7.4 km $s^{-1}$) lowermost crust has been identified beneath the *c.* 6.7 km $s^{-1}$ layer beneath the northwestern portion of the plateau on Line 1, terminating to the SE in the vicinity of the rift.

The principal observations are as follows:

(1) There is evidence of a high-velocity zone in the upper crust beneath the axis of the rift on

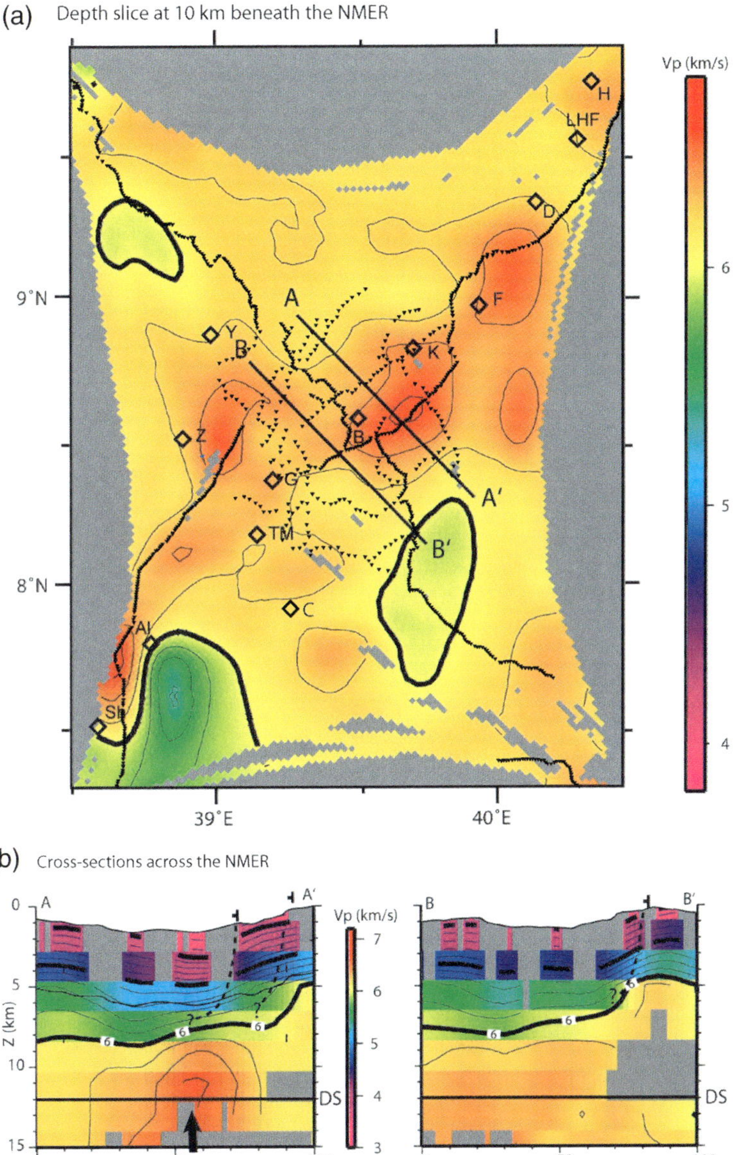


**Fig. 6.** Horizontal slice 10 km below the surface beneath the 2D array. The final model has been smoothed over a 6 × 6 × 2 km (xyz) spatial node and a 2 km velocity node distribution. Receiver positions (EAGLE Lines 1 & 2 and 2D array) shown by inverted triangles. Thick contour line marks 6.0 km s$^{-1}$ with minor contours at 0.2 km s$^{-1}$ intervals. High-velocity bodies (red) interpreted as solidified magmatic intrusions beneath the rift floor. Sections AA′ and BB′ displayed in Figure 6b. Diamonds identify volcanoes. Along the rift axis from south to north are: Sh, Shala, Al, Aluto, TM, Tullu Moje, G, Gedemsa, B, Boset, K, Kone, F, Fantale, D, Dofan, LHF, Liado Hayk Field, and H, Hertale. On the rift flanks are C, Chilalo, Y, Yerer, and Z, Zikwala (after Keranen *et al.* 2004 (reproduced with permission of the Geological Society of America)). (**b**) Rift-perpendicular cross-section AA′ extending through the Boset magmatic segment between Boset and Kone volcanoes. Model smoothing as for Figure 6a. The thick contour line at 6.0 km s$^{-1}$ is identified with other contours at 0.2 km s$^{-1}$ intervals. High-velocity ($V_p$) body (vertical arrow) beneath rift valley is interpreted as solidified mafic intrusion into Precambrian basement. Fault symbols and proposed subsurface continuation of faults (dashed lines) are marked. Depth of slice in Figure 6a is marked by horizontal line. Areas with no ray coverage are shown in grey. BB′ shows rift-perpendicular section between magmatic segments (after Keranen *et al.* 2004 (reproduced with permission of the Geological Society of America)).

Line 1, with velocities of *c.* 6.5 km $s^{-1}$. Similar velocities identified in the crustal model derived from the 2D array (Fig. 6) lie beneath the Boset magmatic segment. There is an equivalent velocity anomaly identified on Line 2, but slightly displaced to the NE from the cross-over of the two profiles. This would be expected from the 'wiggly' nature of Line 2, in particular in the vicinity of the Boset magmatic segment. The results from the 2D array (Keranen *et al.* 2004), which take full account of the nature of the subsurface and array geometry, definitely show that there is more than one such high-velocity zone and that these are segmented along the axis of the rift (e.g. as shown at 10 km depth below the surface in Fig. 6a).

(2) The upper crustal layer beneath Line 1 ($V_p$ 6.1–6.4 km $s^{-1}$) is about 28 km thick beneath the plateau on both sides of the rift valley (Fig. 5a). Immediately beneath the rift, and for about 40 km to the NW, the upper crust appears to thin to 23 km along Line 1. Beneath the rift along Line 2, the upper-crustal thickness decreases slowly northwards from SP22 (20 km) to SP 24 (18 km). It then thins dramatically across a 50 km-long zone in the region of the Boset magmatic segment to become only about 8 km thick beneath the northern part of Line 2 (Fig. 5b). There is a difference of only 4 km in the estimated depth to the base of the upper crust at the Line 1/Line 2 intersection, which we believe is acceptable given the rapid change in the depth of this layer near the intersection and the complexities present in this area.

(3) The upper crustal layer ($V_p$ 6.1–6.4 km $s^{-1}$) on Line 1 contains an intralayer reflection which may or may not be continuous along the profile. This reflector shallows from 20 km depth beneath the surface to the NW of the rift, to 12 km beneath the surface to the SE. Another intracrustal reflector is also modelled on the same profile beneath the rift axis. It is unclear whether these reflectors are continuous. No such reflector can be positively identified along the axis of the rift, but the lower data quality along Line 2 may preclude identification of a weak reflection resulting from such a boundary.

(4) The lower crust ($V_p$ 6.6–7.1 km $s^{-1}$) on Line 1 appears to change in thickness from 14 km beneath the eastern plateau to 9–12 km beneath the northwestern end of the profile where it overlies an anomalous layer of velocity 7.4 km $s^{-1}$ described below. There is no obvious change in thickness of the lower crustal layer immediately beneath the rift along Line 1, and its thickness is also relatively uniform at 16–20 km along the length of Line 2. While the velocity errors on Line 2 are not yet well defined, there may be an apparent correlation between the variation in velocity observed in the lower crust and that identified in the upper crust (Fig. 5b). The topmost lower crust velocity peaks at about 6.75 km $s^{-1}$ at 50 km from SP21, 6.9 km $s^{-1}$ at 215 km from SP21, and at 6.8 km $s^{-1}$at 300 km from SP21, each location being beneath a velocity 'high' in the upper crust.

## *Upper mantle and anomalous lower crust*

Beneath the southeastern flank of the rift, a normal mantle (P-wave velocity 8.05 km $s^{-1}$) underlies the lower crust at a depth of *c.* 40 km. Beneath the axis of the rift on Line 1, the velocity drops to 7.7 km $s^{-1}$, consistent with the along-axis velocity of 7.5 km $s^{-1}$ observed on Line 2 (as yet derived primarily from a diving phase from SP28 (Fig. 2)). The crust thins from about 40 km in the SW to 26 km in the NE beneath Line 2, with a consistent estimate of around 35 km beneath the intersection of the two profiles.

The most 'anomalous' layer in the derived model underlies the plateau NW of the rift on Line 1. Here, a 15 km-thick layer of P-wave velocity 7.4 km $s^{-1}$ lies at a depth of approximately 33 km (HVLC in Fig. 5a). It is defined by reflections from its top and base. There is no velocity estimate for the material underlying it, but it cannot be more than 8.0 km $s^{-1}$ otherwise it would have produced a first arrival diving wave, which is not observed. The depth (*c.* 48 km beneath the surface) to the base of this layer is significantly greater than the Moho depth beneath the eastern flank of the rift. The location of the layer underlies thinned lower crust beneath the northwestern flank of the rift (Fig. 5a).

There are distinct variations in the characteristics of the Moho reflections (PmP) identified on the cross-rift profile Line 1 (Fig. 7). From 90–130 km NW of SP18, PmP with bounce points beneath the southeastern margin of the rift is a simple phase with clear low-frequency content (Fig. 7c). Between offsets of 90–150 km from SP17, PmP with bounce points beneath the rift has a variable-amplitude, higher frequency, very reverberative character (Fig. 7b). Such reverberative characteristics are also identified from PmP reflections with bounce points beneath the northwestern margin of the rift between 110–150 km from SP12 (Fig. 7a). These reverberations occur behind the onset of the phase originating from the base of the 7.4 km $s^{-1}$

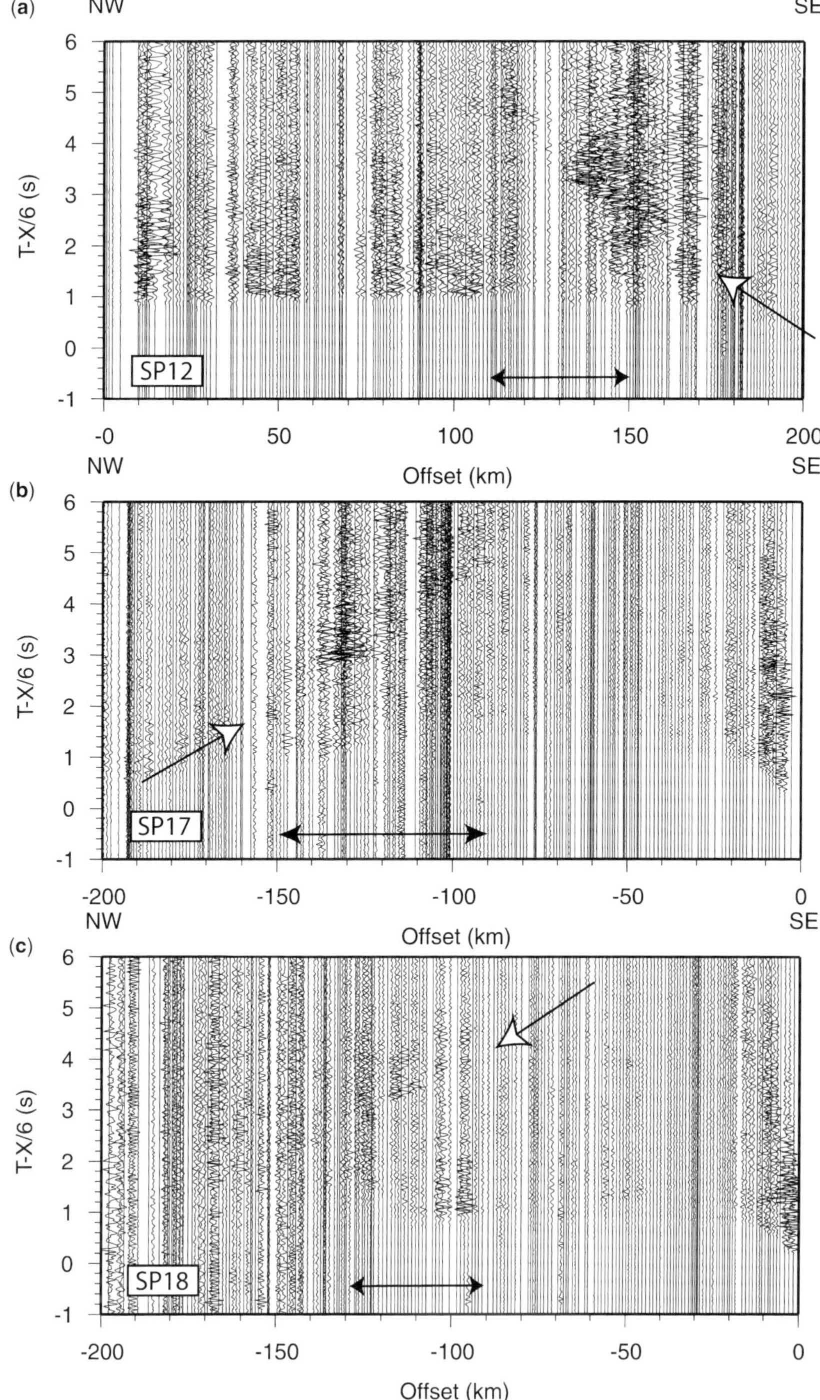


**Fig. 7.** Variability of PmP phase (identified by open-ended arrow) occurring between *c.* 2 to 5 seconds reduced time and offsets marked by arrows. Display parameters as for Fig. 3. **(a)** SP12 – Line 1 (strong reverberation between 110–150 km offset); **(b)** SP17 – Line 1 (strong reverberation between 90–150 km offset); **(c)** SP18 – Line 1 (non-reverberative phase between 90–130 km offset).

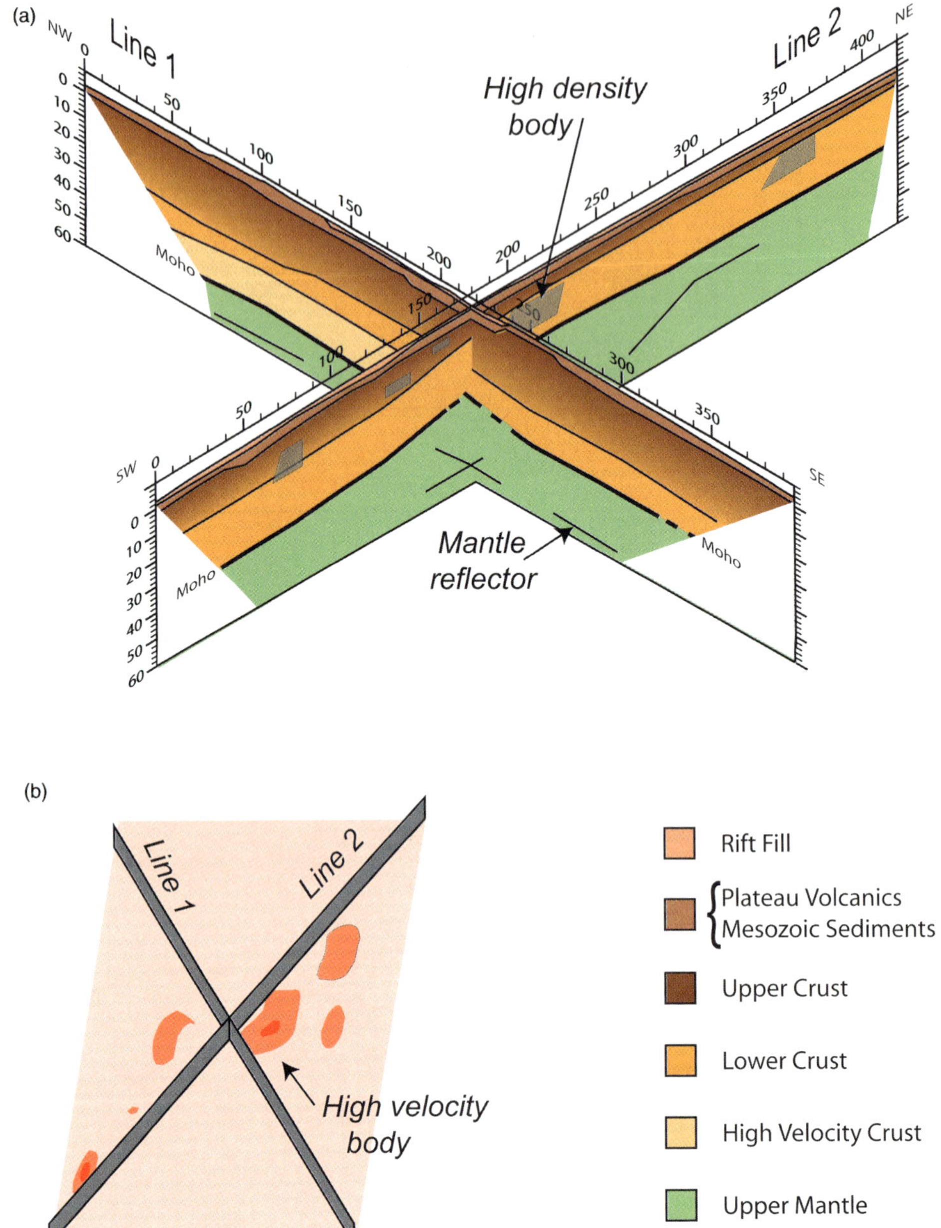


**Fig. 8.** Cartoon depicting two 3D views of the controlled source seismic results; A – from a southerly elevation; B – from a near vertical elevation. Lines 1 and 2 form a fence diagram. Red polygons are areas of seismic velocity higher than 7 km.s$^{-1}$ from Figure 6a projected onto a horizontal plane at 10 km depth. Grey polygons are mafic bodies with density $\sim 3000$ kg.m$^{-3}$ projected onto a vertical plane along the NE-SW axis that is Line 2..

layer, suggesting that it is this layer that is producing them.

On a number of the shotpoint record sections, there is also a reverberative reflection that occurs beneath the Moho. Although its depth cannot be defined precisely owing to the limited data constraining sub-Moho velocities, it can be identified not only beneath the eastern and western margins of the rift but also beneath the rift itself, where it may be slightly shallower than beneath the rift flanks (Fig. 5a). Below Line 2, where it is sampled beneath two regions, the same reflector is about 60 km beneath the southern part of the profile, and about 40 km beneath the northern part (Fig. 5b); thus its depth appears to correlate with crustal thickness.

## Interpretation

The layer thicknesses and velocities derived from the EAGLE controlled-source data confirm that the crust beneath the Main Ethiopian Rift is still predominantly continental in nature, with the principal modifications being: high-velocity intrusions segmented along the axis of the rift; alteration to a massively intruded, thinned crust in the southwest corner of Afar; and the presence of an anomalous 'underplate' layer beneath the plateau northwest of the rift. This crust overlies a mantle which appears normal beneath the southeastern part of Line 1 but modified by thermal processes elsewhere, and which includes a prominent reflector 10–20 km below the Moho along and adjacent to the rift.

### *Sediments and volcanics*

The near-surface sedimentary and volcanic layer of velocity *c.* 5.0 km $s^{-1}$ that is continuous along the length of both Line 1 and Line 2 most likely represents a combination of Mesozoic sediments and Oligocene flood basalts. In the central part of Line 1 and along Line 2, the lower-velocity overlying layer, thickening into the rift, correlates with the distribution of both the Quaternary sediments and the Pliocene–Miocene basalt and ignimbrite units associated with the early stages of extension in the NMER (Wolfenden *et al.* 2004). The thickening of these sedimentary–volcanic layers both on the cross-rift profile (Line 1) and beneath the 2D array (Fig. 6b) into the Arboye border fault (Fig. 2) of the Adama basin is consistent with structural studies (Wolfenden *et al.* 2004). This suggests that the varying thickness of the equivalent layers along Line 2 may result from this profile overlying other structural sub-basins along its length. The thickness of these layers along Line 2 only varies significantly in a few places, perhaps because the recording profile followed roads in physiographic lows that are also structural lows, so that a relatively uniform basement depth is detected. Alternatively, it may be because rift formation in the NMER has indeed created only limited along-axis basement structural variations, but this seems unlikely based on results from many other rifts. There is also significant basement topography to the SE of the rift on Line 1 near SP16 that does not correlate with surface topography and may result from Mesozoic extension (Korme *et al.* 2004)

### *Upper-crustal magmatic segmentation*

The P-wave velocity of *c.* 6.1 km $s^{-1}$ beneath the sedimentary–volcanic layer on Line 1 and the southern part of Line 2 is consistent with the presence of Precambrian crystalline basement. The high velocity of *c.* 6.5 km $s^{-1}$ beneath the axis of the rift on Line 1, highlighted in the 3D crustal model derived from the 2D array (Fig. 6) and lying beneath the Boset magmatic segment, suggests the presence of a gabbroic magmatic intrusion based on these velocities and the modelled density of 3000 kg·$m^{-3}$ (Cornwell *et al.* 2006; Mahatsente *et al.* 1999). Keranen *et al.* (2004) discuss the along-axis segmentation of these high-velocity zones (Fig. 6a) and their correlation with the positions of the Quaternary magmatic segments arranged en echelon along the rift. These results together with examination of the Line 2 upper-crustal layer show that such high-velocity zones can be identified in particular beneath the Lake Shala caldera, the Gedemsa magmatic segment, the northern end of the Boset and the southern end of the Fantale–Dofan magmatic segments. Interestingly, the Boset magmatic segment and those to the NE appear slightly offset to the SE of the high-velocity zones, while those magmatic segments to the SW of Boset appear offset to the NW of the high-velocity zones (Fig. 6a).

### *Along-axis crustal thinning*

The thinning of the upper crust beneath the northern part of Line 2 is consistent with previous results from the area (Berckhemer *et al.* 1975; Makris & Ginzburg 1987). We infer significant extension of the brittle upper crust, whereas the lower crust of relatively uniform thickness along the axis of the NMER is at first sight unaffected. Previous authors (e.g. Makris & Ginzburg 1987) have suggested that this lower crustal layer may have been significantly intruded by mafic material beneath southwestern Afar, so masking crustal stretching. The velocity of this layer (*c.* 6.9 km $s^{-1}$), though variable, does not change significantly between the southern and northern parts of Line 2. This might suggest there is little petrological difference between the highly extended region at the northeastern end of the profile in southwestern Afar and the less-extended region to the SE. However, significant elevation of the geotherm in southwestern Afar would result in a lowering of velocity, which would then suggest the presence of more mafic intrusive material in the lower crust beneath Afar than further to the SW.

## *Upper mantle beneath the NMER*

There is normal mantle of P-wave velocity 8.05 km $s^{-1}$ beneath the southeastern flank of the rift identified beneath Line 1. The sub-Moho Pn mantle diving wave velocity of 7.7 km $s^{-1}$ modelled beneath the rift axis on Line 1, and the equivalent velocity seen at the same boundary along the axis of the rift, is similar to that identified beneath the axis of the Eastern (Kenya) Rift (Mechie *et al.* 1994*a*). In Kenya, such a mantle velocity has been interpreted from the joint analysis of seismic velocities and xenolith compositions (Mechie *et al.* 1994*b*) to result from 3–5% partial melt trapped in the mantle. This would be consistent with Bastow *et al.*'s (2005) conclusions that upper mantle low-velocity zones beneath the NMER result from anomalously high temperatures and local areas of melting or melt-ponding. EAGLE shear-wave splitting data also indicates partial melt in the upper mantle and crust beneath the rift axis (Kendall *et al.* 2005; Keir *et al.* 2005).

The upper mantle reflector at a depth of about 60 km beneath the rift valley and its flanks, and identified at 40 km beneath southwestern Afar, is discussed separately below.

## *Anomalous lower crust*

If taken at face value, the anomalous lower crust ($V_p$ 7.4 km $s^{-1}$) beneath the northwestern part of Line 1 merges with a higher-velocity upper mantle layer under the rift valley region. However, we feel confident that this is an artefact of the modelling algorithm that requires interfaces to be continuous across the model. In the NW where it is part of the crust, this layer is characterized by a distinct reflection from its base, whereas there is no equivalent reflection from the *c.* 7.5–7.7 km $s^{-1}$ layer below the rift axis and above the mantle reflector at 60 km (which is also observed beneath the anomalous crustal layer). We therefore suggest this anomalous lower crustal layer terminates at or near the rift's western escarpment but could have extended across at least some of the rift valley region before the recent extension and magmatism. To the NW, this layer appears reverberative in the seismic sections (Fig. 7a), and forward modelling of Line 1 gravity data is also consistent with its presence (Cornwell *et al.* 2006). The lateral extent of this layer under the plateau may perhaps be limited, because Berckhemer *et al.* (1975) and Makris & Ginzburg (1987) did not identify a layer with *c.* 7.5 km $s^{-1}$ velocity beneath their Profile 1 (inset to Fig. 2), and their estimate of depth to 'normal mantle' beneath the plateau is equivalent to the base of our anomalous layer beneath Line 1.

However, one may question whether the 1972 refraction data truly have the resolution to recognize the high-velocity lower crust. Stuart *et al.* (2006) incorporating results from a previous survey (Dugda *et al.* 2004) use receiver functions to derive crustal thickness and average $V_p/V_s$ estimates over the NMER and its flanking plateaus. Their thickness estimates derived from both within the NMER and on its southern margin are consistent with those derived from the controlled-source study. However, those obtained from stations close to or coincident with the northwestern end of Line 1 on the plateau, have apparently identified the top rather than the base of the anomalous crustal layer. This discrepancy is discussed in detail in Stuart *et al.* (2006). It may in part be due to the different signal characteristics (wave-type, wave-length, spatial sampling, ray-path orientation) used by the receiver function method and wide-angle controlled-source analysis procedures. However, the complex lower crust/upper mantle structure in the vicinity of this anomalous lower crustal layer may also be a contributory factor. It is therefore apparently not possible at present to constrain the areal extent of this anomalous underplate layer.

Mackenzie *et al.* (2005) argue that the anomalous layer under the plateau NW of the rift valley is equivalent to a magmatic underplate layer that is characteristic of continental volcanic margins (e.g. Kelemen & Holbrook 1995; Menzies *et al.* 2002). The existence of the layer beneath the northwestern plateau, where the maximum thickness of flood basalts is observed, suggests a direct connection to the Oligocene volcanics. Such a link would imply uplift of the plateau associated with the emplacement of the flood basalts, consistent with the initiation of the Blue Nile canyon at 25–29 Ma (Pik *et al.* 2003) and with geochemical studies of the Oligocene plateau rhyolitic ignimbrites which originate from the fractional crystallization of $\geq$3 km of gabbroic underplate (Ayalew *et al.* 2002). The recent felsic magmatism may also have been generated from partial melting of existing gabbroic underplate followed by fractionation at upper crustal levels (Trua *et al.* 1999; Ayalew *et al.* 2006). As stated above, there is significant reverberative seismic energy originating within this anomalous lower crustal layer, which may be consistent with the presence of layering within underplated material (e.g. Deemer & Hurich 1994). We therefore suggest that this layer results from magmatic addition to the base of the crust, initially emplaced in the Oligocene, but with a likely recent component derived from the conduit of hot rising material observed in the mantle to the NW of the rift beneath Line 1 (Bastow *et al.* 2005). The emplacement of this underplate, whether Oligocene or recent, at the base of the crust, would have resulted in melting and assimilation of pre-existing lower crust, potentially explaining slight thinning of the lower crust above the anomalous layer. We interpret the upper surface of the anomalous lower crust as the relict Moho prior to plume impingement beneath Ethiopia, being at an equivalent level to the Moho SE of the rift, and as a passive strain marker for subsequent extension across the rift (see below).

## Discussion

### *Magmatic segments*

There is a direct spatial correlation between the surface magmatic segments along the axis of the rift and the 20 km-wide, 50 km-long high-velocity zones identified within the upper crust from the controlled source study (Fig. 6a). This correlation, the high-velocity bodies and the magmatic segments being arranged en echelon along the rift axis, but slightly offset in different senses to the north and south of Boset, strongly implies a causative link. Keranen *et al.* (2004) interpret the high-velocity bodies in terms of cooled gabbroic intrusions accommodating extension at mid-crustal levels. They state there is no marked increase in velocity at shallow depths beneath the surface magmatic segments, suggesting that dykes can only contribute to about 30% of the equivalent extension in the brittle upper crust, the remainder being accommodated by recent dense faulting. NMER seismicity shows a concentration of low-magnitude earthquake swarms at the base of the 6–9 km deep seismogenic layer beneath the magmatic segments (Keir *et al.* 2006) coincident with the top of the zone of extensive mafic intrusions imaged at 7–10 km depth (Fig. 6) (Keranen *et al.* 2004) and also observed below Line 1 (Mackenzie *et al.* 2005). This is consistent with extension above and below the mid-crustal mafic intrusions being accommodated seismically and aseismically respectively (Keir *et al.* 2006; Bendick *et al.* 2006). The small offset of the surface magmatic segments with respect to the high-velocity bodies at depth may result from the basaltic lavas reaching near-surface magma chambers having to 'by-pass' the cooled, rigid gabbroic intrusions at mid-crustal levels, the path of least resistance to the surface being up the side of the gabbroic bodies.

The EAGLE controlled-source project has imaged the main border fault on the southeastern side of the rift (Mackenzie *et al.* 2005; Keranen *et al.* 2004) at 235 km from SP11 (Fig. 5a). Keranen *et al.* (2004) state that the low-velocity near-surface section seen in the 3D model is offset by a subsurface continuation of the Asela (Sire) border fault (Fig. 2), which overlies a high-velocity mid-crustal magmatic intrusion (Fig. 6a). They also show a section across the rift between magmatic segments (Fig. 6b) in which the border fault is also imaged, although a magmatic intrusion is not, which demonstrates support for previous observations that magmatic and structural segmentation of along-axis rift basins are not well correlated (Ebinger & Casey 2001).

The strong reverberatory nature of the Moho reflection from beneath the Boset magmatic segment on Line 1 (Fig. 7b) and the lack of an equivalent signature on the reflection from beneath the southeastern flank of the rift provides a similar relationship to those seen on seismic record sections obtained across the eastern (Kenya) rift. There, reflectivity modelling suggested that the most likely cause of such reverberations resulted from intrusions within the lowermost crust beneath the rift (Thybo *et al.* 2000). Assuming it results from a similar cause beneath Line 1, this reverberatory signature provides evidence for magmatic alteration of the lower crust beneath the Boset magmatic segment. The possible correlation between the high-velocity upper-crustal bodies along the axis of the rift and the along-axis variation in velocities in the lower crust would imply that the magmatic segmentation that is observed on the surface and linked to that identified in the upper crust, also exists in the lower crust. Bastow *et al.* (2005) propose that along the rift axis the segmentation of the low-velocity mantle upwelling beneath the rift is controlled by the mechanically segmented early continental rift structure and that this focuses melt along the rift axis. Keir *et al.* (2006) discuss the possible linkage between magma source regions and surface magmatic segments, commenting on SKS shear-wave splitting studies of lithospheric anisotropy which is likely due to vertically-oriented dykes including partial melt that cross-cut the lithosphere (Kendall *et al.* 2005). Exactly how the melt might then be distributed into the pattern of magmatic segmentation (that is not well-correlated with the early rift structural segmentation) is yet to be resolved, although our results show that this distribution has definitely occurred by the time the melt has reached mid-crustal, and probably lower-crustal levels.

### *Crustal extension*

It would be expected that extension across the rift has been accommodated by faulting and dyking in the brittle upper crust, and ductile flow and magmatic intrusion within the lower crust. Estimates of extension across the rift are complicated by the presence of the Oligocene underplate that has almost certainly eroded or added to the crust beneath the northwestern plateau but by unknown amounts. If we assume that crustal thickness beneath the basement surface was uniform across the rift and equivalent to that identified beneath the southeastern margin (35 km), the present crustal thickness beneath the basement surface below the rift itself (27 km) results in a stretching factor of about 1.3 on Line 1. The intersection point of the two EAGLE profiles lies over the marked decrease in crustal thickness from south to north along the axis of the rift. The crustal thicknesses beneath the basement surface at the south (33 km) and north (21 km) ends of the along-axis profile, Line 2, result in stretching factors of 1.1 and *c.* 1.7 within the Main Ethiopian Rift and in southwestern Afar to the south of Gewane respectively (Fig. 2). These figures correspond with those derived from a recent gravity analysis of 1.2 and 2.0 in the central MER and within the Afar depression to the north of Gewane respectively (Tiberi *et al.* 2005). Eagles *et al.* (2002), from a study of the kinematics of the Danakil microplate, suggest recent extension rates in central and southern Afar are low, being 2–3.5 mm $a^{-1}$ respectively, and Wolfenden *et al.* (2004) arrived at a similar value for the northern MER. Eagles *et al.* (2002) estimate that the crustal stretching factor is >3 in central and southern Afar. The high velocities obtained within the upper crust beneath the rift identified from all the controlled source studies have been interpreted in terms of mafic intrusions beneath the surface magmatic segments. The presence of these intrusions, together with likely associated underplating/intrusion at the base of the crust, will result in our extension factors being underestimations.

While the variation in cross-rift crustal velocities would suggest that modification of the crust has primarily occurred beneath the rift itself, the variation in depth to the upper–lower crust boundary (22–27 km) beneath Line 1 indicates that the upper crust has been thinned over a wider (*c.* 100 km) zone extending to the NW beneath the plateau.

Following primary rift development in the NMER after 11 Ma, there was a hiatus in volcanic activity between 6.5 and 3.2 Ma after which deformation migrated to a narrow zone in the rift centre (Wolfenden *et al.* 2004). Keranen *et al.* (2004), using results from the 2D array data that are consistent with those from Line 1 (Mackenzie *et al.* 2005), state that the *c.* 20 km-wide mafic intrusions in the mid-crust beneath the rift are compatible with the 27 km of extension predicted in the last 3.2 Ma. from plate motions (Horner-Johnston *et al.* 2003). They also suggest that above these intrusions, the maximum estimate of mafic material in the upper crust is about 30% of the 20 km-wide dyke-bearing segments, the remaining extension being taken up by faulting localized to the narrow 20 km-wide segments within the rift. Assuming the width of the extended zone is roughly 100 km as derived from the thinned upper crust across the rift, an overall stretching factor of 1.3 would imply that the amount of extension prior to the migration of deformation at around 3.2 Ma towards the centre of the NMER was about 10 km.

One of the most significant results from our controlled-source survey is the decrease in crustal thickness along Line 2 from south to north into Afar. The degree of extension increases markedly towards Afar. Hypotheses based on the difference in amount of thinning between the upper and lower crustal layers beneath the northeastern end of Line 2 are difficult to test. This results from the difficulty of interpreting seismic models in regions affected by significant magmatic intrusion, owing to the tendency of low-velocity crustal material to be overprinted by higher-velocity material as a result of mafic intrusion. It is possible the mid-crustal layer boundary at the northern end of Line 2 identifies the top of these intrusives, which have penetrated the upper crust.

However, analysis of the crustal velocities and layer thicknesses along this profile might suggest that continental extension (with relatively high crustal velocities associated with the magmatic segments at the southern end of the profile) apparently passes through a transition between SP15/25 and SP27 into near pure spreading at the northern end of the profile. The results from our study, together with those of Berckhemer *et al.* (1975), Makris & Ginzburg (1987), Tiberi *et al.* (2005) and Stuart *et al.* (2006), all indicate an increase in the amount of intrusive material in the crust beneath southwestern Afar. Stuart *et al.* (2006) also show that, within the rift, the $V_p/V_s$ ratio increases to greater than 2.0 (Poisson's ratio, $\sigma > 0.33$) northwards towards the Afar depression, being indicative of partial melt in the crust. Bastow *et al.* (2005), using the methods of Gao *et al.* (2004) to relate seismic-velocity variations to temperature, suggest that lateral variations of between roughly 125 to 800 °C are present beneath the study region if the velocity variations are attributed solely to temperature effects. However, from examination of P and S relative arrival time residuals, they show that the values imply the presence of partial melt, especially beneath the rift, consistent with the predictions of Kendall *et al.* (2005) from analysis of SKS shear-wave splitting observations across the EAGLE network. Melt reduces the temperature contrasts necessary to explain the low velocity anomalies.

If the excess mantle temperature were of the order of 400 °C, this coupled with a stretching factor of 2 derived for the northern end of Line 2 would suggest about 20 km of the present crustal thickness could have resulted from melt produced from adiabatic decompression alone (e.g. White & McKenzie 1995). However, with the excess temperature not being defined, exact determination of the melt fraction within the crust beneath southwestern Afar is beyond the scope of this review. It should be noted that Eagles *et al.* (2002) suggest the possibility of presently active sea-floor spreading in central and southern Afar.

The recent focusing of faulting and intrusion of mafic material towards the axis of the rift, the increased amount of mafic intrusion with increased amount of extension from south to north along the axis of the NMER, the link between crustal magmatic processes and the deeper low-velocity mantle zones interpreted in terms of high temperatures and the presence of partial melt are all consistent with the magma-assisted rifting hypothesis of Buck (2004) as discussed in other EAGLE publications (e.g. Bastow *et al.* 2005; Stuart *et al.* 2006; Kendall *et al.* 2005, 2006).

## *Implications of the underplated lower crust*

The presence of thick, high-velocity crust under the Ethiopian plateau NW of the rift raises several questions. The first is simply its origin, and Mackenzie *et al.* (2005) argue that this layer is due to underplating that was initiated during the 31–29 Ma flood basalt event. Underplating can be the result of a number of magmatic processes, but the end result is thickening of the crust, which from a simple Airy isostasy viewpoint would cause uplift. The Ethiopian plateau (Fig. 2) has an excess elevation above the mean continental elevation of up to approximately 2 km (e.g. Nyblade & Sleep 2003).

Thus, the question arises, how much of this uplift is a result of the underplating? Based on its velocity of 7.4–7.7 $km/s^{-1}$, compared to a mantle velocity of 8.05 $km/s^{-1}$ (Fig. 5), the 15 km-thick lens of underplate at the base of the crust beneath Line 1 has a density contrast with the upper mantle of 125 to 230 $kg\ m^{-3}$ (Brocher 2005), yielding a permanent isostatic uplift of 0.6 to 1.0 km. New gravity modelling, based on our Line 1 refraction data, corroborate this expected density contrast (Cornwell *et al.* 2006) and hence the expected uplift due to underplating. This indicates that the elevation of the plateau must be supported in part by the buoyancy of the mantle. Impingement of the plume head beneath Ethiopia would have produced a transient dynamic uplift up to *c.* 500 m but this would have decayed to less than half its initial value in the 30 Ma that have elapsed since (Nyblade & Sleep 2003). If the permanent uplift due to underplating and lingering dynamic effects of the plume total only around 1 km, the remaining excess elevation must be due to regional mantle density contrasts.

The tomographic results of Bastow *et al.* (2005) document that relatively low seismic velocities in the upper mantle extend west from the Main Ethiopian Rift under the plateau and also extend to considerable depth (>100 km). The lowest velocities are of limited lateral extent under the plateau, but even a small density contrast of 50 $kg\ m^{-3}$ extending to 10 km could produce the needed 1 km of uplift. Although the EAGLE mantle tomography (Bastow *et al.* 2005) and even the broader study of Benoit *et al.* (2006) are too spatially limited to properly demonstrate that mantle velocities beneath the plateau are lower than beyond the plateau, the implication of our EAGLE refraction study which finds insufficient underplating to explain the observed uplift is that mantle densities (hence velocities) beneath the plateau must be lower than beneath surrounding areas, as shown by Cornwell *et al.* (2006). This result echoes the recognition of different upper-mantle structure beneath the rifts and shields further south in East Africa (e.g. Nyblade 2002).

## *The mantle reflector*

The mantle reflector identified beneath both Line 1 and Line 2 was also observed beneath the previous controlled-source profiles beneath Afar (Berckhemer *et al.* 1975; Makris & Ginzburg 1987). Interestingly, a similar feature was identified beneath all of the KRISP seismic profiles in Kenya (Keller *et al.* 1994; Maguire *et al.* 1994; Prodehl *et al.* 1994; Byrne *et al.* 1997). The distinctive features of the reflector are that it exists at a depth of approximately 10–25 km beneath the Moho. The northern extent of the KRISP 90 along-axis profile (Keller *et al.* 1994) and the KRISP 94 cross-rift profile through southern Kenya from Lake Victoria towards the Chyulu Hills to the east of the rift, (Byrne *et al.* 1997) also observed a second reflector approximately 15 km deeper than the first. The reflector underlies the crust beneath the rift flanks and beneath the rift itself in both northern Ethiopia and Kenya, and in southern Kenya it underlies both Proterozoic and Archaean terranes. Beneath those rifts involving small amounts of extension, the reflection is apparently shallower by a few kilometres beneath the rift itself than beneath the rift flanks. The reflector identified beneath the northern end of Line 2 in southwestern Afar rises by about 20 km beneath the highly thinned crust at the north-eastern end of this profile.

Makris & Ginzburg (1987) offer evidence that the mantle reflector continues at least 10 km north of our Line 2 (their Profile III), and less convincingly across Afar to the Danakil Horst (their Profile IV, inset to Fig. 2). While the situation beneath the along-axis profile beneath the eastern (Kenya) rift is well documented, it is not impossible for the reflector identified ~25 km beneath the thick southern section of this rift (Keller *et al.* 1994) to be correlated with the one identified at a similar depth beneath the thinned crust at the northern end of the KRISP along-axis profile rather than a deeper mantle interface.

Levin & Park (2000) provide a brief review of the possible origins of mantle discontinuities in the 50–100 km depth range:

- the phase change from spinel peridotite to garnet peridotite (Hales 1969);
- anisotropy from localized shear zones (Fuchs 1983);
- a hybrid model of localized shear developed within oceanic crust as it undergoes the gabbro-to-eclogite phase transition during shallow subduction (Bostock 1998); and
- shallow low-velocity zones (Benz & McCarthy 1994) speculatively caused by pervasive partial melt (Thybo & Perchuc 1997).

To this list we can add the interpretation of Keller *et al.* (1994) and Byrne *et al.* (1997) that the reflectors beneath the Eastern (Kenya) rift may be caused by shearing near the base of the lithosphere, separating layers resulting from subhorizontal mantle flow radiating from a region of hot mantle material rising beneath the Kenya dome in the southern Eastern (Kenya) rift.

It is not possible here to unequivocally distinguish between these hypotheses. However:

- That the mantle reflector rises up beneath extended crust, where the geotherm will be raised, is inconsistent with the pressure–temperature relationship for the spinel–garnet phase change which would result in an expected deepening of the transition as the temperature increases.
- Both the Ethiopian and the eastern (Kenya) rifts lie within the Mozambique orogenic belt, a region involved in Pan-African continental collision and possible oceanic lithospheric subduction, potentially providing conditions suitable for the hybridized model of Bostock (1998). The consistency of the reflector beneath the Ethiopian and eastern (Kenya) rifts, would suggest that there is a regional origin of the reflector rather than it resulting just from 'localized shear zones' within the sub-crustal mantle.
- Rooney *et al.* (2005) used xenoliths in Quaternary basalts in the Bishoftu and Butajira regions approximately 20 km west of the NMER to estimate the depth of melting of the host lavas at 53–88 km, equivalent to the depth of the mantle reflector discussed here. Rooney *et al.* (2005) and Furman *et al.* (2006) point out the correspondence of this depth range with the low-velocity P- and S-wave anomaly in the upper 100 km of the mantle interpreted as partial melt (Bastow *et al.* 2005). Locally, it is possible that a layer of melt may occur at the depth of the mantle reflector arising from interaction between the *P–T* curve and solidus for the particular mantle composition. However, the reflector is at approximately the same depth beneath both the southeastern and northwestern flanks of the NMER (the former lacking, the latter having, the HVLC layer) and also within both Archaean and Proterozoic lithosphere beneath the flanks of the Eastern (Kenya) rift, with probably different *P–T* conditions that should not allow melting at a constant depth.
- Analysis of SKS recordings and surface-wave dispersion over the NMER are not consistent with the hypothesis of the reflectors resulting from horizontal mantle flow at the top of a spreading mantle plume head (Gashawbeza *et al.* 2004). Seismic anisotropy may correlate with pre-existing Pan-African lithospheric alignment away from the rift trend (Gashawbeza *et al.* 2004; Kendall *et al.* 2006). Within the uppermost 75 km beneath the rift, the observed anisotropy is primarily due to melt alignment orientated in a rift-parallel direction (Kendall *et al.* 2005).

Hence, we believe the most likely cause of the mantle reflector is a compositional or structural boundary within pre-existing lithosphere, owing to the correlation of reflector depth with the amount of lithospheric extension, rather than a phase change or melt front.

## Conclusions

The EAGLE controlled-source project provides an intriguing new image of crustal and uppermost mantle structure beneath a 400 × 400 km region centred on the Boset magmatic segment of the transitional northern Main Ethiopian Rift (see Fig. 8).

The sedimentary–volcanic layer delineated on the along-axis and cross-rift profiles shows that the Mesozoic sedimentary rocks and Tertiary to Recent volcanics in both the rift and the flanking plateaus have a combined maximum thickness of >5 km. Basement topography shows that the bounding faults extend into the crystalline basement, as identified for example on the southeastern flank of the Adama basin. Along the axis of the rift, the total thickness of the sediment–volcanic layer is fairly constant at about 5 km, but its two units vary in thickness considerably. This may be a result of the location of the seismic line and its passage through along-axis sub-basins, or may truly reflect simple basement architecture in this transitional rift.

The upper crust exhibits normal continental crustal velocities ($V_p$ *c.* 6.1 km s$^{-1}$) except beneath the magmatic segments within the rift, where higher-than-normal upper-crustal velocities are interpreted to result from mafic intrusions beneath the magmatic centres. The magmatic segmentation is identified to mid-crustal levels and possibly into the lower crust. While there is a possible correlation between anomalous low-velocity mantle segmentation and rift valley structural segmentation, the process by which melt transport is transferred from the mantle to a crustal region dominated by magmatic segmentation is not yet resolved.

A 40 km-thick crust with normal continental velocity structure lies above normal upper mantle ($V_p$ 8.05 km s$^{-1}$) on the plateau southeast of the rift valley. Beneath the plateau NW of the rift, a 48 km-thick crystalline crust includes a 15 km-thick high-velocity ($V_p$ 7.4 km s$^{-1}$) layer that is interpreted as underplate associated with the Oligocene flood basalt and more recent magmatic activity. Along the axis of the rift, the sub-Moho velocities are *c.* 7.5 km s$^{-1}$ interpreted as resulting from raised mantle temperatures and a small percentage of partial melt. The crust thins significantly over a 50 km distance from *c.* 40 km in the SW to 26 km at the NE end of the EAGLE axial seismic profile, and the thinning primarily occurs in the upper crust. The apparently uniform velocity of the lower crust may mask an increase in mafic intrusions in the highly extended crust beneath Afar as a result of a higher geotherm. Our model of the crust along the axis of the NMER can in fact be explained from SW to NE in terms of rifted continental crust transitioning into thick, possibly near total igneous crust above a high-temperature mantle providing the necessary basaltic melt. The results from this and other EAGLE studies strongly support the magma-assisted rifting hypothesis of Buck (2004).

A mantle reflector has been identified at a depth of about 60 km beneath the plateaus that bound the rift valley. A correlative reflector is apparently slightly shallower beneath the southern part of the rift and at *c.* 40 km beneath the highly extended northeastern end of the axial profile in southern Afar. While it is not possible to be definitive as to its cause, because its depth apparently correlates with the amount of extension, we believe that it most likely represents a compositional or structural boundary. The fact that an equivalent reflector has been observed beneath the eastern (Kenya) rift and further north in Afar strongly suggests that this may be an important observation with generic implications for the origin and development of magmatic continental rifts and subsequent volcanic margins.

We would like to acknowledge the support, work and many discussions held variously with C. Ebinger, G. Stuart, J.M. Kendall, M. Fowler, A. Ayele, I. Bastow, A. Brisbourne, D. Cornwell, P. Denton, E. Gashaw-Beza, D. Keir, K. Mickus, T. Mammo & K. Tadesse. In addition to these and the authors, participants in the EAGLE project came from:

*Ethiopia*: Addis Ababa University, Ethiopian Geological Survey, Ethiopian Petroleum Operations Department, Ethiopian Commission for Science and Technology, Oromia Council, Addis Ababa.

*Europe*: University of Leicester, University of London, Royal Holloway, University of Leeds, SEIS-UK, University of Edinburgh, University of Copenhagen, Dublin Institute for Advanced Studies, University of Vienna, University of Stuttgart.

*USA*: Stanford University, University of Texas at El Paso, Pennsylvania State University, Missouri State University, US Geological Survey, University of Colorado at Boulder.

Figures 1, 2, 3, 4, 5 & 7 involved the use GMT software (Wessel & Smith 1995). Helpful reviews were provided by C. Ebinger, D. Lizarralde and W. Mooney. Funding for the EAGLE project was provided by NERC (NER/A/S/2000/00563,01003,01004), the National Science Foundation Continental Dynamics program EAR 0208475, the Texas Higher Education Coordinating Board, the Royal Society and the University of Leicester. Instrumentation was provided by SEIS-UK, IRIS-PASSCAL and the University of Copenhagen.

## References

ABDELSALAM, M.G. & STERN, R.J. 1996. Sutures and shear zones in the Arabian–Nubian shield. *Journal of African Earth Sciences*, **23**, 289–310.

ALLEN, A. & TADESSE, G. 2003. Geological setting and tectonic subdivision of the Neoproterozoic orogenic belt of Tuludimtu, western Ethiopia. *Journal of African Earth Sciences*, **36**, 329–343.

AYALEW, D., BARBEY, P., MARTY, B., REISBERG, L., YIRGU, G. & PIK, R. 2002. Source, genesis and timing of giant ignimbrite deposits associated with Ethiopian continental flood basalts. *Geochimica et Cosmochimica Acta*, **66**, 1429–1448.

AYALEW, D., EBINGER, C., BOURDON, E., WOLFENDEN, E., YIRGU, G. & GRASSINEAU, N. 2006. Temporal compositional variation of syn-rift rhyolites along the southwestern margin of the Red Sea and northern Main Ethiopian Rift. *In*: YIRGU, G.,

EBINGER, C.J. & MAGUIRE, P.K.H. (eds) *The Afar Volcanic Province within the East African Rift System*. Geological Society, London, Special Publications, **259**, 121–130.

AYELE, A., STUART, G. & KENDALL, J.-M. 2004. Insights into rifting from shear wave splitting and receiver functions; an example from Ethiopia. *Geophysical Journal International*, **157**, 354–362.

BASTOW, I.D., STUART, G.W., KENDALL, J.-M. & EBINGER, C.J. 2005. Upper mantle seismic structure in a region of incipient continental break-up: northern Ethiopian rift. *Geophysical Journal International*, **162**, 479–493 doi:10.1111/j.1365-246x.2005.02666.x.

BENDICK, R., MCCLUSKY, S., BILHAM, R., ASFAW, L. & KLEMPERER, S.L. 2006. Distributed Nubia–Somalia relative motion and dike intrusion in the Main Ethiopian Rift. *Geophysical Journal International* (in press).

BENOIT, M.H., NYBLADE, A.A. & VANDECAR, J.C. 2006. Upper mantle P wave speed variations beneath Ethiopia and the origin of the Afar Hotspot. *Geology* (in press).

BENZ, H.M. & MCCARTHY, J. 1994. Evidence for an upper mantle low velocity zone beneath the southern Basin and Range–Colorado plateau transition zone. *Geophysical Research Letters*, **21**, 509–512.

BERCKHEMER, H., BAIER, B., ET AL. 1975. Deep seismic soundings in the Afar region and on the highland of Ethiopia. *In*: PILGER, A. & RÖSLER, A. (eds) *Afar Depression of Ethiopia*. Schweizerbart, Stuttgart, **I**, 89–107.

BILHAM, R., BENDICK, R., LARSON, K., MOHR, P., BRAUN, J., TESFAYE, S. & ASFAW, L. 1999. Secular and tidal strain across the Main Ethiopian Rift. *Geophysical Research Letters*, **26**, 2789–2792.

BIRT, C.S., MAGUIRE, P.K.H., KHAN, M.A., THYBO, H., KELLER, G.R. & PATEL, J. 1997. The influence of pre-existing structures on the evolution of the southern Kenya Rift Valley—evidence from seismic and gravity studies. *Tectonophysics*, **278**, 211–242.

BOCCALETTI, M., MAZZUOLI, R., BONINI, M., TRUA, T. & ABEBE, B. 1999. Plio-Quaternary volcanotectonic activity in the northern sector of the Main Ethiopian Rift: relationships with oblique rifting. *Journal of African Earth Sciences*, **29**, 679–698.

BONINI, M., CORTI, G. INNOCENTI, F., MANETTI, P., MAZZARINI, F., TSEGAYE ABEBE, T. & PECSKAY, Z. 2005. Evolution of the Main Ethiopian Rift in the frame of Afar and Kenya rifts propagation. *Tectonics*, **24**, TC1007, 10.1029/2004TC001680,

BOSELLINI, A., RUSSO, A. & ASSEFA, G. 2001. The Mesozoic succession of Dire Dawa, Harar Province, Ethiopia. *Journal of African Earth Sciences*, **32**, 403–417.

BOSTOCK, M.G. 1998. Mantle stratigraphy and the evolution of the Slave province. *Journal of Geophysical Research*, **103**, 21183–21200.

BROCHER, T.M. 2005. Empirical relations between elastic wavespeeds and density in the Earth's crust. *Bulletin of the Seismological Society of America*, **95**, 2081–2092.

BUCK, W.R. 2004. Consequences of asthenospheric variability on continental rifting. *In*: KARNER, G.D., TAYLOR, B., DRISCOLL, N.W. & KOHLSTEDT, D.L. (eds) *Rheology and Deformation of the Lithosphere at Continental Margins*, Columbia University Press, New York, pp. 1–30.

BYRNE, G.F., JACOB, A.W.B., MECHIE, J. & DINDI, E. 1997. Seismic structure of the upper mantle beneath the southern Kenya Rift from wide-angle data. *Tectonophysics*, **278**, 243–260.

CHERNET, T., HART, W., ARONSON, J. & WALTER, R. 1998. New age constraints on the timing of volcanism and tectonism in the northern Main Ethiopian Rift – southern Afar transition zone (Ethiopia). *Journal of Volcanology and Geothermal Research*, **80**, 267–280.

CORNWELL, D.G., MACKENZIE, G.D., MAGUIRE, P.K.H., ENGLAND, R.W., ASFAW, L. & OLUMA, B. 2006. Northern Main Ethiopian Rift crustal structure from new high-precision gravity data. *In*: YIRGU, G., EBINGER, C.J. & MAGUIRE, P.K.H. (eds) *The Afar Volcanic Province within the East African Rift System*. Geological Society, London, Special Publications, **259**, 307–321.

D'ACREMONT, E., LEROY, S., BESLIER, M.-O., BELLAHSEN, N., FOURNIER, M., ROBIN, C., MAIA, M. & GENTE, P. 2005. Structure and evolution of the eastern Gulf of Aden conjugate margins from seismic reflection data. *Geophysical Journal International*, **160**, 869–890, doi: 10.1111/j.1365-246X.2005.02524.x.

DEEMER, S.J. & HURICH, C.A. 1994. The reflectivity of magmatic underplating using the layered mafic intrusion analog. *Tectonophysics*, **232**, 239–255.

DUGDA, M., NYBLADE, A., JULIA, J., LANGSTON, C., AMMON, C. & SIMIYU, S. 2005. Crustal structure in Ethiopia and Kenya from receiver function analysis: implications for rift development in eastern Africa. *Journal of Geophysical Research*, **110**, B101303, doi:10.1029/2004JB003065.

EAGLES, G., GLOAGUEN, R. & EBINGER, C.J. 2002. Kinematics of the Danakil microplate. *Earth and Planetary Science Letters*, **203**, 607–620.

EBINGER, C.J. & CASEY, M. 2001. Continental breakup in magmatic provinces: An Ethiopian example. *Geology*, **29**, 527–530.

FUCHS, K. 1983. Recently formed elastic anisotropy and petrological models for the continental subcrustal lithosphere in southern Germany. *Physics of the Earth and Planetary Interiors*, **31**, 93–118.

FURMAN, T., BRYCE, J., ROONEY, T., HANAN, B., YIRGU, G. & AYALEW, D. 2006. Heads and tails: 30 million years of the Afar plume. *In*: YIRGU, G., EBINGER, C.J. & MAGUIRE, P.K.H. (eds) *The Afar Volcanic Province within the East African Rift System*. Geological Society, London, Special Publications, **259**, 95–119.

GAO, W., GRAND, S., BALDRIDGE, W., WILSON, D., WEST, M., NI, J. & ASTER, R. 2004. Upper mantle convection beneath the central Rio Grande rift imaged by P and S wave tomography. *Journal of Geophysical Research*, **109**, B03305, doi: 10.1029/2003 JB002743.

GASHAWBEZA, E.M., KLEMPERER, S.L., NYBLADE, A.A., WALKER, K.T. & KERANEN, K.M. 2004. Shear-wave splitting in Ethiopia: Precambrian mantle anisotropy locally modified by Neogene rifting. *Geophysical Research Letters*, **31**, L18602, doi:10.1029/2004GL020471.

HALES, A.L. 1969. A seismic discontinuity in the lithosphere. *Earth and Planetary Science Letters*, **7**, 44–46.

HEBERT, L. & LANGSTON, C. 1984. Crustal thickness estimate at AAE (Addis Ababa, Ethiopia) and NAI (Nairobi, Kenya) using teleseismic P-wave conversions. *Tectonophysics*, **111**, 299–327.

HOLE, J.A. 1992. Nonlinear high-resolution three-dimensional seismic travel time tomography. *Journal of Geophysical Research*, **97**, 6553–6562.

HORNER-JOHNSTON, B.C., GORDON, R.G., COWLES, S.M. & ARGUS, D.F. 2003. The angular velocity of Nubia relative to Somalia and the location of the Nubia–Somalia–Antarctica triple junction. *EOS Transactions American Geophysical Union*, **84**(6), abs T52F-01.

KEIR, D., KENDALL, J.-M., EBINGER, C. & STUART, G. 2005. Variations in late syn-rift melt alignment inferred from shear-wave splitting in crustal earthquakes beneath the Ethiopian rift. *Geophysical Research Letters* **32**, L23308, doi: 10.1029/2005GL024150.

KEIR, D., EBINGER, C., STUART, G., DALY, E. & AYELE, A. 2006. Strain accommodation by magmatism and faulting as rifting proceeds to breakup: Seismicity of the northern Ethiopian Rift. *Journal of Geophysical Research* (in press).

KELEMEN, P.B. & HOLBROOK, W.S. 1995. Origin of thick high-velocity igneous crust along the U.S. East coast margin. *Journal of Geophysical Research*, **100**, 10077–10094.

KELLER, G.R., MECHIE, J., BRAILE, L., MOONEY, W.D. & PRODEHL, C. 1994. Seismic structure of the uppermost mantle beneath the Kenya Rift. *Tectonophysics*, **236**, 201–216.

KENDALL, J.-M., PILIDOU, S., KEIR, D., BASTOW, I.D., STUART, G.W. & AYELE, A. 2006. Mantle upwellings, melt migration and the rifting of Africa: Insights from seismic anisotropy. *In*: YIRGU, G., EBINGER, C.J. & MAGUIRE, P.K.H. (eds) *The Afar Volcanic Province within the East African Rift System.* Geological Society, London, Special Publications, **259**, 55–72.

KENDALL, J.-M., STUART, G., EBINGER, C., BASTOW, I. & KEIR, D. 2005. Magma-assisted rifting in Ethiopia. *Nature*, **433**, 146–148.

KERANEN, K., KLEMPERER, S.L., GLOAGUEN, R. & the EAGLE Working Group. 2004. Imaging a protoridge axis in the Main Ethiopian Rift. *Geology*, **32**, 949–952.

KORME, T., ACOCELLA, V. & ABEBE, B. 2004. The role of pre-existing structures in the origin, propagation and architecture of the faults in the Main Ethiopian Rift. *Gondwana Research*, **7**, 467–479.

LEVIN, V. & PARK, J. 2000. Shear zones in the Proterozoic lithosphere of the Arabian shield and the nature of the Hales discontinuity. *Tectonophysics*, **232**, 131–148.

MACKENZIE, G.D., THYBO, H. & MAGUIRE, P.K.H. 2005. Crustal velocity structure across the Main Ethiopian Rift: Results from 2-dimensional wide-angle seismic modelling. *Geophysical Journal International*, **162**, 994–1006 doi: 10.1111/j.1365-246X.2005. 02710.x.

MAGUIRE, P.K.H., SWAIN, C.J., MASOTTI, R. & KHAN, M.A. 1994. A crustal and uppermost mantle cross-sectional model of the Kenya Rift derived from seismic and gravity-data. *Tectonophysics*, **236**, 217–249.

MAGUIRE, P., EBINGER, C., *ET AL.* 2003. Geophysics project in Ethiopia studies continental breakup. *EOS Transactions American Geophysical Union*, **84**, 342–343.

MAHATSENTE, R., JENTZSCH, G. & JAHR, T. 1999. Crustal structure of the Main Ethiopian Rift from gravity data: 3-dimensional modelling. *Tectonophysics*, **313**, 363–382.

MAKRIS, J. & GINZBURG, A. 1987. The Afar Depression: transition between continental rifting and sea floor spreading, *Tectonophysics*, **141**, 199–214.

MECHIE, J., KELLER, G.R., PRODEHL, C., GACIRI, S., BRAILE, L.W., MOONEY, W.D., GAJEWSKI, D. & SANDMEIER, K.-J. 1994*a*. Crustal structure beneath the Kenya Rift from axial profile data. *Tectonophysics*, **236**, 179–200.

MECHIE, J., FUCHS, K. & ALTHERR, R. 1994*b*. The relationship between seismic velocity, mineral composition and temperature and pressure in the upper-mantle – with an application to the Kenya Rift and its eastern flank. *Tectonophysics*, **236**, 453–464.

MENZIES, M.A., KLEMPERER, S.L., EBINGER, C.J. & BAKER, J. 2002. Characteristics of volcanic rifted margins. *In*: MENZIES, M.A., KLEMPERER, S.L., EBINGER, C. & BAKER, J. (eds) *Magmatic Rifted Margins.* Geological Society of America Special Paper, **362**, pp. 1–14.

MOHR, P.A. & ZANETTIN, B. 1988. The Ethiopian flood basalt province. *In*: (ANON.) *Continental Flood Basalts*. Kluwer Academy, Dordrecht, Netherlands, 63–110.

MONTELLI, R., NOLET, G., DAHLEN, F.A., MASTERS, G., ENGDAHL, E.R. & HUNG, S.-H. 2004. Finite frequency tomography reveals a variety of plumes in the mantle. *Science*, **303**, 338–343.

NYBLADE, A.A. 2002. Crust and upper mantle structure in East Africa; implications for the origin of Cenozoic rifting and volcanism and the formation of magmatic rifted margins. *In*: MENZIES, M.A., KLEMPERER, S.L., EBINGER, C.J. & BAKER, J. (eds) *Volcanic Rifted Margins.* Geological Society of America, Special Paper, **362**, 15–26.

NYBLADE, A.A. & SLEEP, N.H. 2003. Long lasting epeirogenic uplift from mantle plumes and the origin of the Southern African Plateau. *Geochemistry, Geophysics, Geosystems*, **4**, 1105, doi:10.1029/2003GC000573.

PIK, R., MARTY, B., CARIGNAN, J. & LAVÉ, J. 2003. Stability of the Upper Nile drainage network

(Ethiopia) deduced from (U–Th)/He thermochronometry: implications for uplift and erosion of the Afar plume dome. *Earth and Planetary Science Letters*, **215**, 73–88.

PRODEHL, C., JACOB, A.W.B., THYBO, H., DINDI, E. & STANGL, R. 1994. Crustal structure on the northeastern flank of the Kenya rift. *Tectonophysics*, **236**, 271–290.

PRODEHL, C., FUCHS, K. & MECHIE, J. 1997. Seismic-refraction studies of the Afro-Arabian rift system – a brief review. *Tectonophysics*, **278**, 1–13.

ROONEY, T.O., FURMAN, T., YIRGU, G. & AYELEW, D. 2005. Structure of the Ethiopian lithosphere: Evidence from mantle xenoliths. *Geochimica et Cosmochimica Acta*, **69**(15), 3889–3910. doi:10.1016/j/gca.2005.03.043.

SCHULL, T.J. 1988. Rift basins of interior Sudan; petroleum exploration and discovery. *American Association for Petroleum Geologists Bulletin*, **72**, 1128–1142.

STERN, R.J. 1994. Arc assembly and continental collision in the Neoproterozoic East African orogeny implications for the consolidation of Gondwana. *Annual Review of Earth and Planetary Sciences*, **22**, 319–351.

STUART, G.W., BASTOW, I.D. & EBINGER, C.J. 2006. Crustal structure of the northern Ethiopian Rift from receiver function studies. *In*: YIRGU, G., EBINGER, C.J. & MAGUIRE, P.K.H. (eds) *The Afar Volcanic Province within the East African Rift System.* Geological Society, London, Special Publications, **259**, 253–267.

TADESSE, S., MILESI, J.-P. & DESCHAMPS, Y. 2003. Geology and mineral potential of Ethiopia: a note on the geology and mineral map of Ethiopia. *Journal African Earth Sciences*, **36**, 273–313.

TIBERI, C., EBINGER, C., BALLU, V., STUART, G. & OLUMA, B. 2005. Inverse models of gravity data from the Red Sea–Aden–East African rifts triple junction zone. *Geophysical Journal International*, **63**, 775–787.

THYBO, H. & PERCHUC, E. 1997. The seismic 8° discontinuity and partial melting in continental mantle. *Science*, **257**, 1626–1629.

THYBO, H., MAGUIRE, P.K.H., BIRT, C.S. & PERCHUC, E. 2000. Seismic reflectivity and magmatic underplating beneath the Kenya Rift. *Geophysical Research Letters*, **27**, 2745–2749.

TRUA, T., DENIEL, C. & MAZZUOLI, R. 1999. Crustal control in the genesis of Plio-Quaternary bimodal magmatism of the Main Ethiopian Rift (MER): geochemical and isotopic (Sr, Nd, Pb) evidence. *Chemical Geology*, **155**, 201–231.

WESSEL, P. & SMITH, W.H.F. 1995. New version of the Generic Mapping Tools released. *EOS Transactions American Geophysical Union*, **76**, 329.

WHALER, K.A. & HAUTOT, S. 2006. The electrical resistivity structure of the crust beneath the northern Main Ethiopian Rift (EAGLE Phase III). *In*: YIRGU, G., EBINGER, C.J. & MAGUIRE, P.K.H. (eds) *The Afar Volcanic Province within the East African Rift System.* Geological Society, London, Special Publications, **259**, 293–305.

WHITE, R.S. & MCKENZIE, D. 1995. Mantle plumes and flood basalts. *Journal of Geophysical Research*, **100**, 17543–17585.

WOLFENDEN, E., EBINGER, C., YIRGU, G., DEINO, A. & AYALEW, D. 2004. Evolution of the northern Main Ethiopian Rift: birth of a triple junction. *Earth and Planetary Science Letters*, **224**, 213–228.

ZELT, C.A. & FORSYTH, D.A. 1994. Modeling of wide-angle seismic data for crustal structure: South-eastern Grenville Province. *Journal of Geophysical Research*, **99**, 11687–11704.

ZELT, C.A. & SMITH, R.B. 1992. Seismic traveltime inversion for 2-D crustal velocity structure. *Geophysical Journal International*, **108**, 16–34.

# The electrical resistivity structure of the crust beneath the northern Main Ethiopian Rift

K.A. WHALER[1] & S. HAUTOT[1,2]

[1]*Grant Institute of Earth Science, University of Edinburgh, West Mains Road, Edinburgh EH9 3JW, Scotland (e-mail: kathy.whaler@ed.ac.uk)*

[2]*Now at IUEM-UBO, UMR 'Domaines Océaniques', Place Nicolas Copernic, F-29280 Plouzane, France*

**Abstract:** 18 audio-frequency magnetotelluric (MT) sites were occupied along a profile across the northern Main Ethiopian Rift. The profile covered the central portion of the Ethiopia Afar Geoscientific Lithospheric Experiment (EAGLE) line 1 along which also a number of broadband seismic receivers were deployed, a controlled-source seismic survey was shot, and gravity data were collected. Here, a two-dimensional model of the MT data is presented and interpreted, and compared with the results of other methods. Shallow structure correlates well with geologically mapped Quaternary to Jurassic age rocks. Within it, a small, shallow conducting lens, at less than 1 km depth, beneath the Boset volcano may represent a magma body. The 100 Ωm resistivity contour delineates the seismically inferred upper crust beneath the northern plateau. The Boset magmatic segment is characterized by conductive material extending to at least lower crustal depths. It has high velocity and density in the upper to mid-crust and upper mantle. Thus, all three results suggest a mafic intrusion at depth, with the MT model indicating that it contains partial melt. There is a second, slightly deeper, more conductive body in the lower crust beneath the northern plateau, tentatively interpreted as another zone containing partial melt. The crust is much more resistive beneath the southern plateau, and has no resistivity contrast between the upper and lower crust. The inferred geoelectric strike direction on the plateaus is approximately parallel to the trend of the rift border faults, but rotates northwards slightly within the rift, matching the orientation of the en echelon magmatic segments within it. This follows the change in orientation of the shear wave splitting fast direction.

Details of the setting, geology, structure and tectonic history of the northern Main Ethiopian Rift (NMER) are given elsewhere in this volume (e.g. Kidane *et al.* 2006; Casey *et al.* 2006; Pizzi *et al.* 2006). This paper concerns a study of the resistivity structure across the rift. As part of the Ethiopia Afar Geoscientific Lithospheric Experiment (EAGLE) project, audio-frequency magnetotelluric (MT) data were collected at 18 sites along a ~220 km-long profile crossing the NMER in an approximately NW–SE direction, to image subsurface electrical resistivity. The profile coincided with the central portion of the EAGLE seismic wide-angle reflection/refraction profile (Line 1 (Mackenzie *et al.* 2006)), and traversed the Boset magmatic segment within the NMER. As well as recording the controlled-source shots, a number of broadband seismic receivers measured passive seismicity both along the line and in the surrounding area. Figure 1 shows the location of the MT sites, seismometers and shotpoints, superimposed on the topography, border faults and magmatic segments. Where possible, MT sites were located close to broadband seismometer sites to enable a comparison between the MT model and the results of receiver function analysis. Subsequently, a gravity survey was also carried out along line 1, and the MT and density models are compared here. The MT sites were approximately 20 km apart on the plateaus, reducing to 5 km in the rift where the expected higher conductivity would limit penetration, both laterally and with depth. The line extended further on the northwestern plateau to try to understand the mechanism for plateau uplift.

The MT method provides information on a physical parameter characterizing the subsurface, namely its resistivity. The resistivity of the subsurface depends on its temperature, mineralogy and fluid content, both interstitial and partial melt, and is sensitive to even small amounts of melt and residual fluids and its connectivity. One of the aims of the survey was therefore to image any melt residing in the crust, which was expected to be hard to detect seismically. Partial melt lowers rigidity and therefore seismic velocity. Low-velocity layers are invisible to the seismic refraction method, and their top surfaces tend to have high seismic impedance contrasts, limiting the penetration depth of reflection studies. With the EAGLE seismometer geometry and frequency

*From*: YIRGU, G., EBINGER, C.J. & MAGUIRE, P.K.H. (eds) 2006. *The Afar Volcanic Province within the East African Rift System*. Geological Society, London, Special Publications, **259**, 293–305.
0305-8719/06/$15.00 

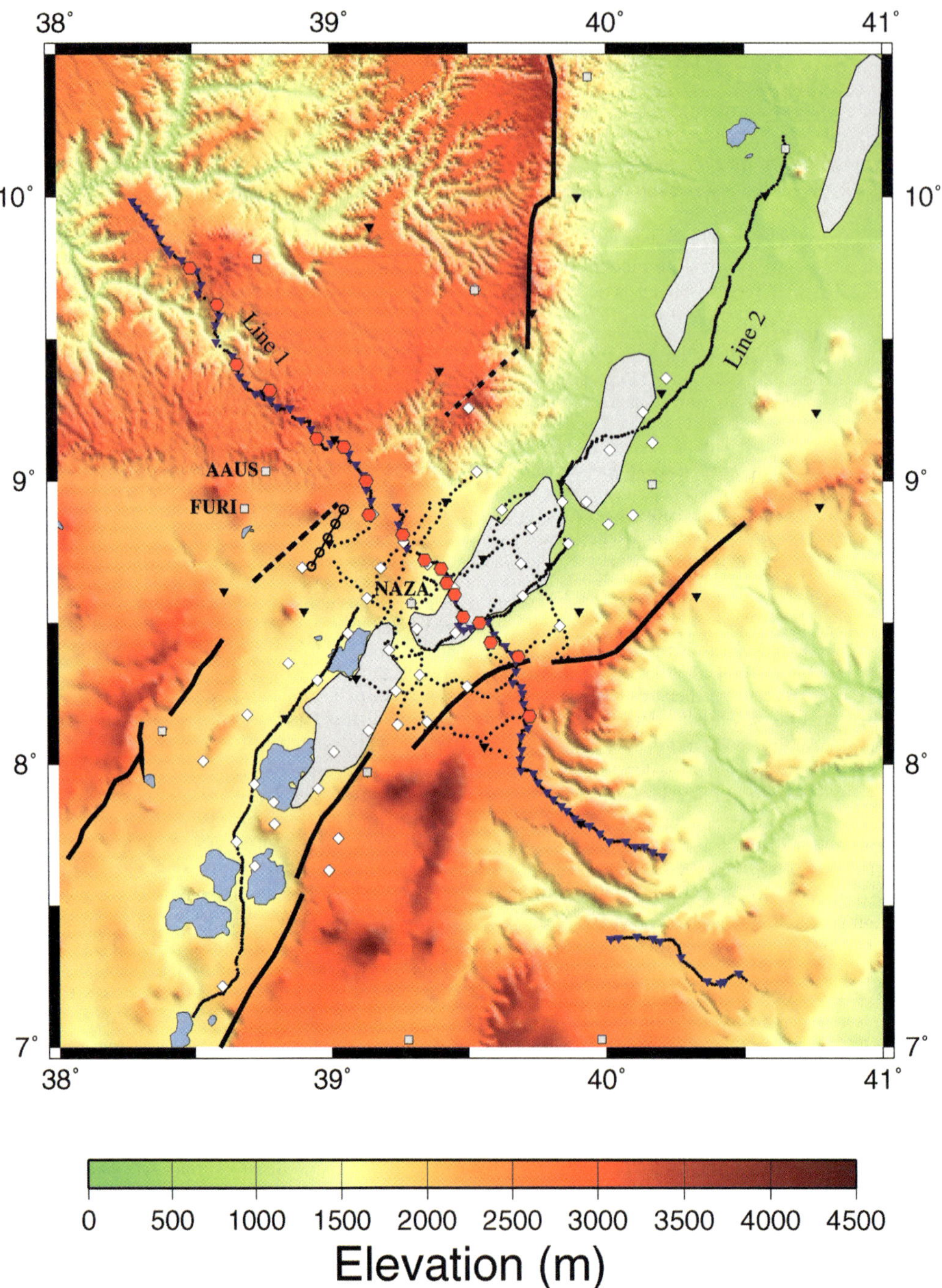


**Fig. 1.** Location of the MT sites (red hexagons) along the central portion of EAGLE line 1 superimposed on the topography, border faults (solid thick black lines), faulted monoclines (dashed thick black lines), Bishoftu Volcanic Chain (linked open circles) and magmatic segments (grey shading) of the northern main Ethiopian rift region. Also shown are the seismic stations, both broadband (black inverted triangles and white diamonds, EAGLE phase I and II respectively; blue inverted triangles, EAGLE line 1; grey squares, Pennsylvania State University (Dugda *et al.* 2005)) and short period (black dots, EAGLE line 2 and central 2D array). Permanent and Pennsylvania State University broadband stations referred to in the text are labelled. The gravity survey was conducted along line 1.

range, receiver function analysis is only able to determine crustal thickness and the average $V_p/V_s$ in the crust, and controlled source tomographic models have resolved just the top *c.* 20 km of crustal structure to date (Keranen *et al.* 2004). Thus, the MT data and their interpretation provide new information on crustal structure, complementing and independent of the results of seismic and gravity surveys.

In the following section, the data acquisition and processing are summarized. The modelling strategy is then described and the preferred two-dimensional (2D) model presented. Results and models from other techniques are compared to aid the interpretation of the MT model. Finally, the main results and conclusions are summarized.

## Data acquisition and processing

SPAM Mark III (Ritter *et al.* 1998) audio-frequency magnetotelluric (MT) instrumentation was used for data acquisition, with Metronix broadband coils and lead–lead chloride porous pot electrodes to measure the horizontal magnetic and electric fields respectively. The highest frequency recorded was 128 Hz; at a number of sites, usable data were collected over almost five decades. Signal strength is weakest in the so-called 'dead band' around 1 Hz. Data quality was poor and restricted to shorter periods at the two sites on the Boset volcano due to strong winds. The data at a nearby site in the rift were too contaminated by mains power and other cultural noise to be used in inversion. At several sites, battery failure restricted the longest period measured. Simultaneous data with GPS timing (accurate to $\pm 2$ μs) were obtained at most sites, allowing remote reference processing (Gamble *et al.* 1979). Direct current (DC) resistivity sounding data with electrode spacings of up to several hundred metres were collected at all sites to determine shallow structure; these models can be used to help resolve static shift arising from near-surface three-dimensional (3D) distortion (e.g. Bahr 1988; Groom & Bailey 1989; Hautot *et al.* 2000). If unaccounted for, static shift can introduce unnecessary structure into the model. Ideally, the profile would have extended further to the SE, but the road quality and network in the area were poor, so there was insufficient time during the fieldwork period to occupy more sites.

Data were post-processed using the robust codes of Chave & Thomson (1989) and Ritter *et al.* (1998) to obtain optimal impedance tensor estimates relating the horizontal magnetic and electric fields in the frequency domain. The profile is close to the magnetic equator and hence the equatorial electrojet (EEJ), which could have polarized the source signals parallel to it during daytime. This can contaminate the data and affect the calculation of the impedance tensor, and complicates the interpretation of the strike direction (V. Haak, pers. comm. 2005). However, no such effect was identified in the period range of these data. The strike is almost constant with period for each site and it is coherent from site to site, whether the data were recorded at night or during the day. The skew values were generally small ($\lesssim 0.1$); this is a necessary but not sufficient condition for the data to be consistent with a 2D structure. In a 2D Earth, the electromagnetic induction equations separate into two modes: the transverse electric (TE) mode or E-polarization (currents parallel to strike); and the transverse magnetic (TM) mode or H-polarization (currents perpendicular to strike). Currents parallel to structure only induce magnetic fields perpendicular to it, and *vice versa* for the TM mode. Thus, the diagonal elements of the impedance tensor vanish. However, this structure only emerges when the electromagnetic field components are measured in, or rotated into, coordinates defined by the strike direction and perpendicular to it. In addition, local 3D distortion needs to be taken into account. This may not be purely galvanic in the conductive environment of the Ethiopian rift, so the general decomposition technique of Counil *et al.* (1986) was used. This determines the strike providing the best 2D approximation to the actual 3D structure, and was determined by rotating the impedance tensor such that its diagonal elements were a minimum, on a site-by-site and period-by-period basis. In a 2D Earth, there is a single rotation angle (with a 90° ambiguity) for which the diagonal elements vanish; shallow 3D galvanic distortion means the diagonal elements will not quite vanish, but a single rotation angle minimizes them.

Counil *et al.* (1986) showed that, in the more general case, there are two angles that minimize the diagonal elements, and they need not be orthogonal. One defines a maximum electric field direction (MED), the other a minimum magnetic field direction (Counil *et al.* 1986). For a 2D geometry, the MED should correspond to the strike of the 2D structure or to the direction perpendicular to it, depending on the nature of the medium underneath. In what follows, it is assumed that the MED indicates the strike direction of the conductivity structure (or perpendicular to it), which is referred to as the geoelectrical strike. 90° rotations of the MED can take place over short distances if there is a large lateral resistivity contrast. The MED was compared to the main tectonic trend in order to determine which of the maximum or minimum components corresponds to the TE (and respectively to the TM) mode. This showed that the

MED is roughly parallel to the rift valley (rotation angles 30–40° from magnetic North), and the maximum impedance is the TE mode, for sites on the plateau, with a rapid switch to the maximum impedance being the TM mode for sites within the rift valley (rotation angles 100–120°). At most sites, the rotation angle was well-defined and constant with period, though others were either scattered or had small diagonal impedance tensor elements for all rotation angles (as would be the case for a 1D Earth). Allowing for the 90° MED rotation, there are two main directions defining the 2D structure along the profile (Fig. 2). Stations on the plateaus tend to have a direction parallel to the rift border faults, while those in the rift valley have a slightly more northerly direction, parallel to the direction of the en echelon magmatic segments (Fig. 1). This rotation is also seen in the fast shear wave splitting direction from seismic anisotropy studies (Maguire *et al.* 2003; Kendall *et al.* 2005, 2006; Keir *et al.* 2005).

The impedance data are most usefully presented as amplitude (in units of resistivity), referred to as the apparent resistivity, and phase of the TE and TM modes, plotted as a function of period (a depth proxy). These quantities are straightforwardly obtained from the rotated real and imaginary impedance tensor components. Examples of the data are given in Figure 3, together with the predictions of the preferred model. Static shifts resulting from 3D near surface distortion are seen as an offset between the two apparent resistivity curves at the shortest periods (e.g. sites 2 and 17). Station numbers can be matched to positions along the profile from the map showing the MED directions in Figure 2 and the 2D model presented in Figure 4.

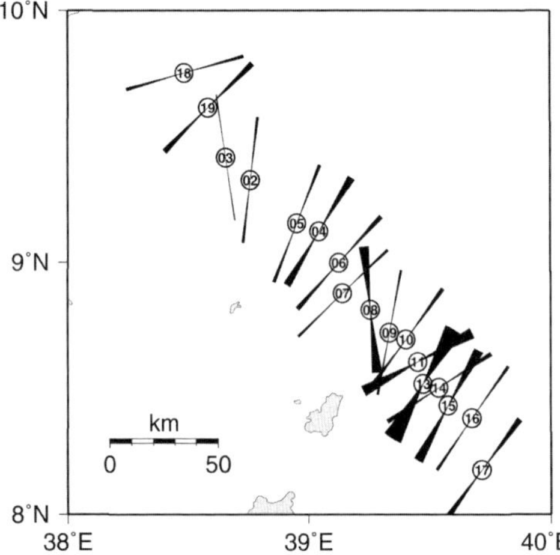


**Fig. 2.** Maximum electrical field direction (MED) at each site at period $T = 0.5$ s. The directions shown are representative of the whole period range. The width of the pie wedge is two standard deviations.

The DC resistivity data were modelled with plane layers overlying an infinite halfspace. The resultant models were then used to predict high-frequency MT data at those sites with static shift. At the two sites (2 and 17) with the largest static shifts, the prediction matched one of the apparent resistivity component curves, so the resistivity of the top layer of the MT model was set to the DC resistivity model half-space value. At other sites, the agreement between the MT data predicted by the DC resistivity model and the actual MT data was less good. This was usually because the near-surface structure was too conductive for the injected DC to reach the shallowest depths sampled by the highest MT frequency, so there was a gap in resistivity structure not resolved by either technique: in this case, the DC resistivity model predictions are not expected to match either component of the actual MT data.

The robustly processed TE and TM mode data were inverted for a 2D model of subsurface resistivity using an algorithm that iteratively minimizes the root-mean-square (rms) difference between the data (weighted by their uncertainties) and those predicted by the model (Hautot *et al.* 2002). The misfit function also contains a regularization term based on the resistivity contrast between adjacent model blocks. The forward problem is solved for both TE and TM modes with a finite difference 2-D algorithm. Data at 17 periods roughly equally spaced in log period and covering the range from 1/128 to 512 s recorded at all sites were selected for modelling; this reduces the numerical effort without significant loss of information.

The top 5 km of the model was well constrained in relatively few iterations by the high quality data at periods less than 1 s. Once a good fit to the short period data was achieved, the structure in the top 3 km was kept constant while some refinement of the horizontal grid was undertaken where deeper structure appeared to change rapidly laterally, in particular between sites 3 and 2, and 16 and 17, allowing a better fit to the data at longer periods, and confirming that some of the key structures discussed below are required by the data. This refinement also involved relaxing the regularization constraint to allow for the more rapid variations in resistivity required by the data. A number of sensitivity tests were run; their results informed the interpretation of the model and its comparison with other data types presented below.

The final model is shown in Figure 4a, with the station numbers indicated at the top, and the upper 5 km in more detail in Figure 4b. Sample data fits are presented in Figure 3. The final weighted rms misfit is 3.1. Although this is larger than the target

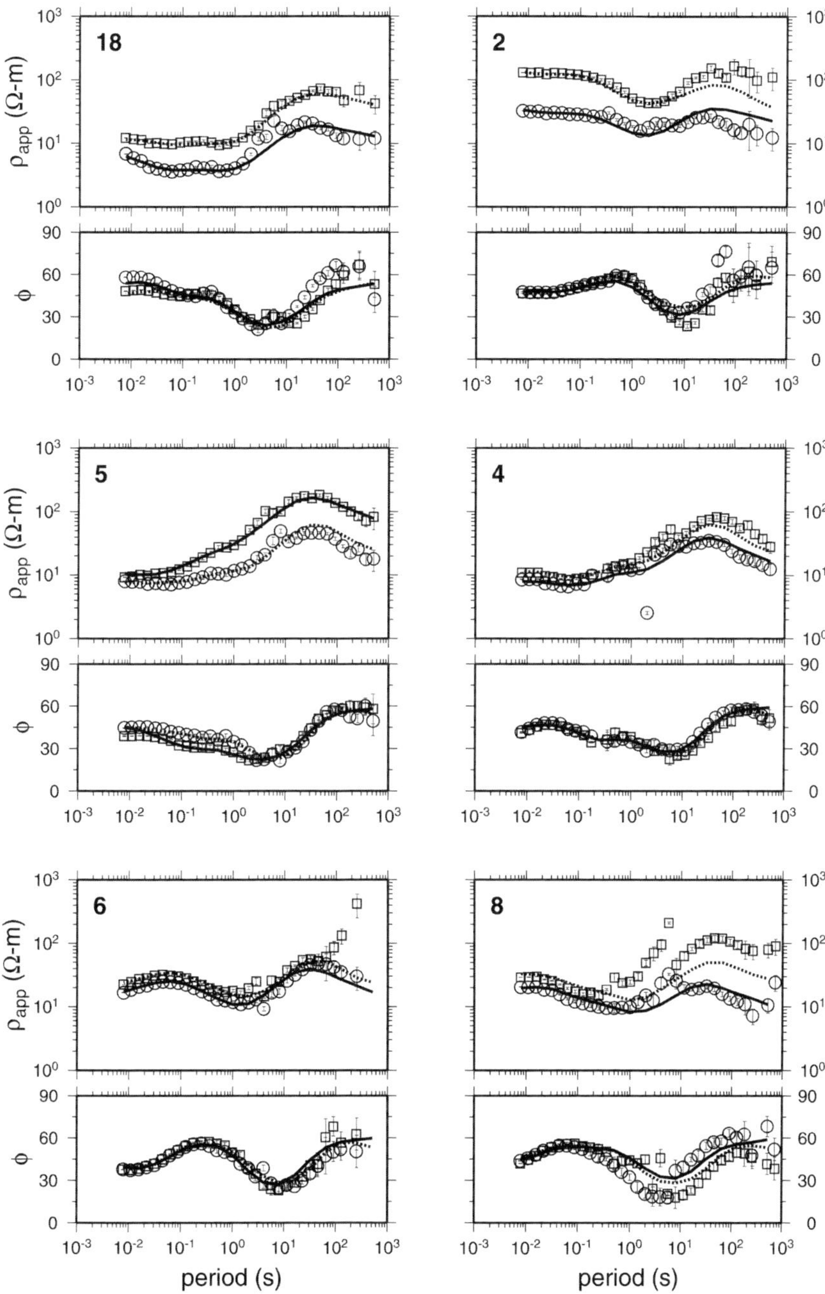


**Fig. 3.** Examples of the MT data and the fit of the 2D model shown in Figure 4, with sites arranged (left to right and top to bottom) from NW to SE along the profile. The squares are the components in the maximum impedance direction, and the circles in the minimum impedance direction. Error bars are one standard deviation. Phases have all been plotted in the first quadrant to show more detail. Dotted line is the TE model response and solid line the TM model response. The maximum impedance is the TE mode for sites 18, 2, 4, 6, 8, 16 and 17, and the TM mode for sites 5, 10 and 11. The DC resistivity data models predicted MT data providing a reasonable match to the short-period TE mode data for sites 2 and 17 so they were used to control static shift at these two sites. The DC resistivity model half-space resistivities were 97 and 123 Ωm, and the models predicted apparent resistivity values of 244 and 270 Ωm at 128 Hz, respectively.

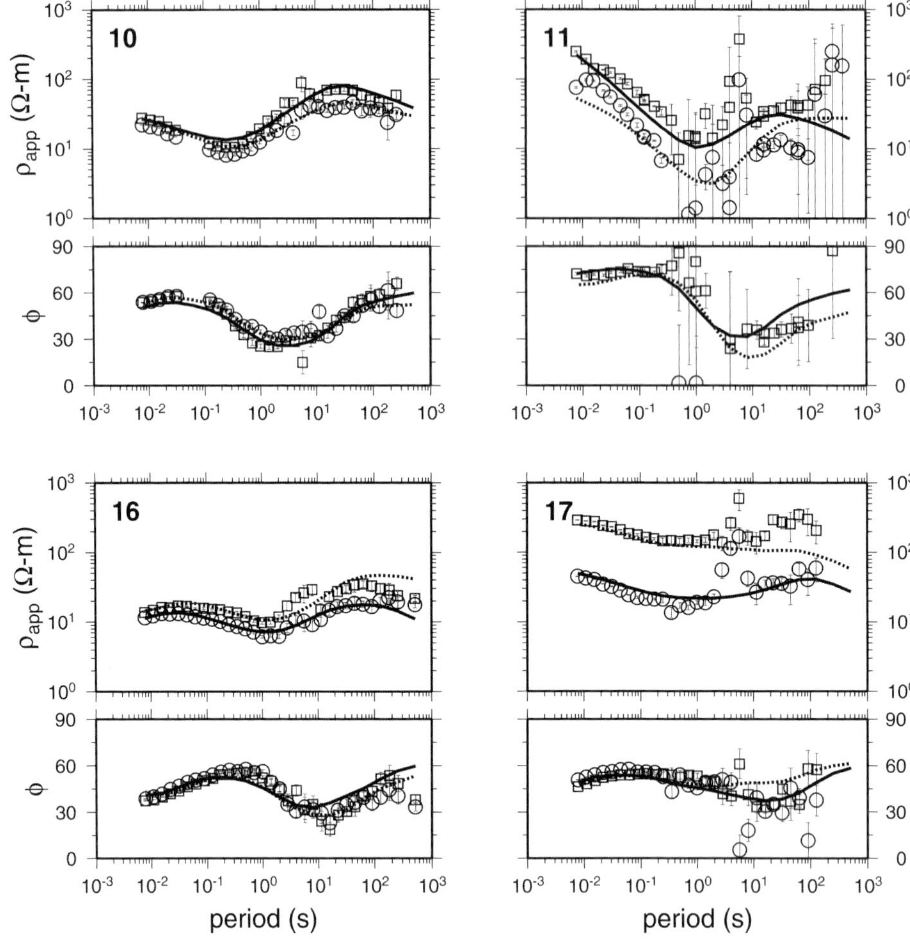


**Fig. 3.** *Continued.*

of 1, it is typical of 2D MT models where errors are not zero-mean Gaussian distributed, and represents a reasonable fit to the data without the model becoming excessively complex.

## Interpretation

The MT model in Figure 4 is a detailed subsurface image of the crust over 40 km depth. The geological field mapping study of Wolfenden *et al.* (2004) is used to interpret the near-surface electrical structure (Fig. 4b). Mackenzie *et al.* (2005) present a model of the seismic wide angle reflection/refraction profile along line 1, referred to hereafter as 'the seismic model', extending to depths of up to 60 km, providing a useful initial comparison with the MT model. Maguire *et al.* (2006) describe the seismic model coordinate system which preserves the curvature of the Earth over the length of the seismic profile. It results in the seismic model zero depth being at the level of the lowest EAGLE shotpoint (at the northeastern end of seismic line 2) which is at an elevation of 555 m above the WGS 84 Geoid used to define zero depth in the MT model. The zero distance point on Figure 4 is approximately 150 km along the seismic model. The resistivity distribution in the top 3 km of the MT model is heterogeneous (Fig. 4b). There is lateral variability such as the resistive material beneath sites 2 and 17, and conductive material beneath sites 3, 7 and 15, which is probably caused by localized variations in composition, fluid content or structure. There is an extremely conductive (resistivity 0.8 Ωm) lens beneath site 11 overlain by a resistor. Sites 11 and

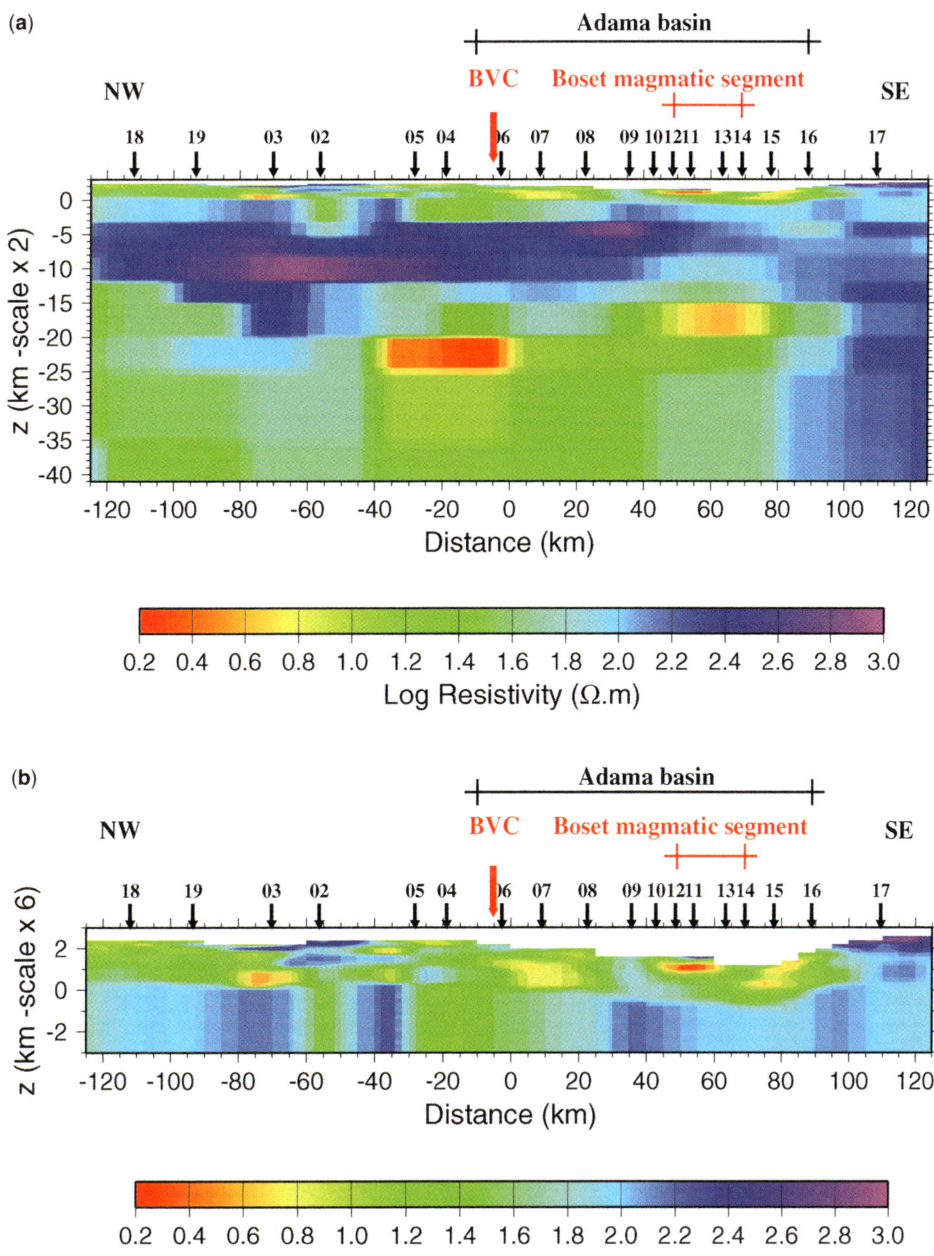


**Fig. 4.** Final 2D model of the resistivity structure across the main Ethiopian rift, from an inversion of both the TE and TM mode data on a finite difference grid. Station numbers, the extent of the Boset magmatic segment and Adama basin, and projection of the Bishoftu Volcanic Chain (BVC) onto the model, are indicated along the top. The Adama basin terminates to the SE against the Arboye rift border fault (Wolfenden *et al.* 2004). Data from site 12 were not included in the inversion because they were too badly contaminated by cultural noise. In addition to the misfit between the data and those predicted by the model, the model minimizes a measure of structural smoothness; the importance assigned to this criterion is variable, to allow for rapid changes in resistivity where the data require them. (**a**) Complete model; (**b**) top 6 km with greater vertical exaggeration.

13 were on the Boset volcano, so it is tempting to attribute the lens to a shallow magma zone, although hot and/or saline fluids or clays, often associated with volcanic activity, would be alternative explanations. The overlying resistor corresponds to fresh volcanic material observable nearby and responsible for extremely high apparent resistivities calculated from the DC resistivity-sounding data. The MT model suggests a maximum thickness for the Adama basin within the rift of around 3 km (based on the rapid increase in resistivity from about 10 Ωm within the basin to about 100 Ωm below), compared to 4 km proposed by Wolfenden *et al.* (2004) and 5 km by Mackenzie *et al.* (2005). The thickness increases towards the SE and reaches a maximum beneath sites 15 and 16, adjacent to the Arboye border fault (site 16), in agreement with these other studies.

Within the rift, Mackenzie *et al.* (2005) deduce a surface layer typically 1 km thick of Late Miocene–Early Pliocene ignimbrites and rhyolites, and Quaternary sediments and volcanics. Beneath that, and cropping out at the surface elsewhere along the line, are a few kilometres thickness of pre-rift Jurassic sediments and Oligocene flood basalts. Together, these loosely match the near-surface relatively conductive material in Figure 4b, both in thickness and geometry of the lower boundary (Mackenzie *et al.* 2005), except on the southern plateau.

There is a good resistivity contrast between the upper and lower crust beneath the northern plateau. The upper crust beneath the northern plateau is delineated approximately by the 100 Ωm resistivity contour. Its lower boundary, generally between 20 and 25 km depth, is elevated by about 5 km beneath the rift. However, the lower crust is markedly more resistive beneath the southern plateau, and there is no obvious resistivity boundary that might distinguish the upper and lower crust. The shallow subsurface beneath the southern plateau is also considerably more resistive than elsewhere along the profile. This major difference in structure can be seen by comparing the data from site 17 with those from other sites (Fig. 3). Note that the DC resistivity data suggest the TE mode curve, with higher apparent resistivity values, represents the data undistorted by static shift. Numerical difficulties accommodating sharp resistivity contrasts, such as at the southern escarpment, and the smoothness constraint imposed on the model, suppress the lateral difference in the model of Figure 4, even after refinement to the horizontal grid. This explains why the long-period TE mode apparent resistivities at site 17 are still underpredicted by the model of Figure 4.

Important features of the model are two conductive bodies at lower crustal depths, one beneath the Boset magmatic segment and the other beneath the plateau to the NW, close to the rift flanks. The body beneath the Boset magmatic segment (below sites 11, 13 and 14, centred at approximately 60 km on the profile) begins at 15 km depth, and becomes even more conductive and wider at depths of 20–25 km. Slightly deeper (25–30 km) and to the NW (beneath sites 5, 4 and 6, centred at −20 km) is a second, extremely conductive body. It terminates to the NW of site 5 to fit the longer-period data at sites 3 and 2. A sensitivity test demonstrates that both conductors are required by the data, with site 5 particularly sensitive to the existence of the more northwesterly conductor. The depth extent of the more northwesterly conductor is better resolved, with sites 4 and 6 constraining the depth of its top, and site 5 its base. Parkinson vectors at periods of 20–180 minutes from an analysis of geomagnetic depth-sounding data collected by H. Porath around 1970 and processed by D. Bennett (both at University of Dallas) indicate elongated conductors in the same positions (trending approximately north-south beneath the northwestern plateau, and NE–SW beneath the rift). An almost contemporaneous long-period MT experiment by V. Haak and colleagues (at University of Munich) also identified spectacular high conductivity structures beneath the northwestern plateau (V. Haak, pers. comm. 2005); however, these studies did not resolve the depths of the conductors.

Gravity data have been collected along line 1 and modelled by Cornwell *et al.* (2006). A starting density model was converted from both Mackenzie *et al.* (2005) and Keranen *et al.*'s (2004) seismic velocity models. This included a high-density body beneath the Boset magmatic segment, about 20 km wide and with its top surface at *c.* 8.5 km below the rift floor extending to at least the lower crust, corresponding to a region of enhanced seismic velocity both on line 1 (Mackenzie *et al.* 2005) and on an adjacent section to the NE derived from the tomography model of Keranen *et al.* (2004). Keranen *et al.* (2004) also noted that the velocity was normal along a parallel section further to the south that passes between the Boset and Koka magmatic segments, and that the Koka and Fantale–Dofan segments to the south and north respectively are also associated with elongate, high-velocity bodies. Thus, the enhanced velocity and density beneath the Boset magmatic segment is thought to be caused by a $\lesssim$1.8 Ma gabbroic mafic intrusion, feeding a series of dykes intruding into the upper crust. Mahatsente *et al.* (1999) and Tadesse *et al.* (2004) model a series of short-wavelength gravity maxima coincident with magmatic segments elsewhere in the main Ethiopian rift as mafic intrusions into the upper crust. The dimensions of the high-velocity body beneath the segment are comparable to those of the conductive body seen in the MT model down to the depth limit of the seismic tomography model (*c.* 15 km; Keranen *et al.* 2004).

The conductive body observed in the MT model beneath the Boset magmatic segment is consistent with a mafic intrusion interpretation, assuming it still contains partial melt at lower crustal depths, since no more than a few percent of interconnected melt can decrease the resistivity significantly (Roberts & Tyburczy 1999). Keranen *et al.* (2004) present the intrusion as cooled, but seismic tomography does not image below depths of about 15 km at which the conductivity starts to increase significantly. Although the seismic model has higher crustal velocities beneath the Boset magmatic segment, the lateral velocity contrast diminishes with depth, vanishing at about the base of the crust. Thus, both the seismic and MT results may indicate that the top part of the intrusion has cooled and solidified, but that it still contains some melt at depth. This is consistent with the gravity model (Cornwell *et al.* 2006) which, although it has a high-density body in the upper crust beneath the Boset magmatic segment, has uniform lower crustal density that could be the result of melt lowering the density of the mafic intrusion. There is extensive seismic evidence for partial melt within the rift. Mackenzie *et al.*'s (2005) low *Pn* velocities indicate its presence immediately beneath the Moho. Kendall *et al.* (2005) attribute observed anisotropy to oriented melt pockets. Receiver function analysis (Dugda *et al.* 2005) gave a Poisson's ratio of 0.34 for station NAZA close to line 1 in the rift (see location on Fig. 1). Values greater than 0.30 are thought to be consistent with the presence of partial melt, although there is a trade-off between crustal thickness and Poisson's ratio. Stuart *et al.* (2006) were unable to perform receiver function analysis of the data from their station (DONE) at site 14 on the edge of the segment, because the receiver function has a poorly developed *Pms* conversion; they suggest this might be caused by attenuation of the converted S-wave. Keir *et al.* (2006) note that seismicity within the rift is confined to the magmatic segment and ceases at a depth of *c.* 10 km. From this they infer that it is caused by dyking into the brittle upper crust, a mechanism which relies on melt still being present.

It is postulated here that the second, deeper (20–25 km), more conductive body to the NW (centred at a distance of −20 km along the MT profile beneath sites 5, 4 and 6) is also caused by partial melt. It is not possible to verify conclusively whether or not the two conductors are connected—sensitivity tests show that the resistivities of the intervening blocks are not well constrained by the data. The seismic and gravity models do not have an associated velocity and density anomaly coincident with the second conductor, and its depth is below the region imaged by the seismic tomography model of Keranen *et al.* (2004). Poisson's ratio values for the crust from receiver function analysis are equivocal. The adjacent permanent station AAUS (Fig. 1) gave a Poisson's ratio of 0.31 (Dugda *et al.* 2005), still high enough to indicate partial melt. Slightly further away, the very late arrivals (compared to the predictions of standard 1D Earth models) of teleseismic phases at the nearby permanent seismic station FURI at all azimuths would also be consistent with partial melt beneath it, as would its Poisson's ratio of 0.33 (Stuart *et al.* 2006). However, at site 4 overlying the conductor, Stuart *et al.* (2006) found a Poisson's ratio of only 0.27 (their station SENE), yet higher values of 0.31 at site 7 (station E70), and 0.30 at site 2 (station CHAE), both constrained by sensitivity tests not to overlie the conductor. Extending from a distance of approximately 20 km orthogonal to the profile to the SW of the conductor, the Bishoftu (sometimes referred to as Debre Zeit) volcanic chain (Fig. 1) is underlain by hot asthenosphere (Bastow *et al.* 2005; Rooney *et al.* 2005). Keir *et al.* (2006) postulate that it is either an unfavourably oriented 'failed' magmatic segment or an incipient zone of strain. Seismic stations overlying it have high (0.31–0.33) Poisson ratios (Stuart *et al.* 2006). Thus, other data indicate partial melt nearby, although none at the location of the conductor.

Beneath the northwestern plateau, the seismic model has an extra *c.* 15 km-thick layer centred on a depth of approximately 40 km, with velocity intermediate between that of the lower crust and upper mantle, interpreted as crustal underplating of Oligocene and/or Recent age (Mackenzie *et al.* 2005). Although melting of the lower crust associated with an Oligocene underplating event would have solidified long ago, the MT model is consistent with a proposed two-stage model (Trua *et al.* 1999) in which underplating is followed by recent moderate degrees of low-pressure fractionation, involving melting of lower crustal rocks. These rocks were probably already close to solidus temperatures owing to the relatively high strain rates associated with the change in rift extensional direction to approximately east–west at the beginning of the Quaternary (Boccaletti *et al.* 1999).

Both lower crustal conductors lie at distances along the seismic profile at which no wide-angle reflections from the base of the underplated layer (*PmP* phase) are received (Mackenzie *et al.* 2005, Fig. 7). This is probably an artefact of the source–receiver geometry (P.K.H. Maguire 2005, pers. comm.) so cannot be used to argue for the presence of partial melt associated with the conductors. In fact, the wide-angle reflections from the base of the lower crust (top of the underplate layer) (the *PxP* phase in Mackenzie *et al.* (2005)) would be expected to be more affected by partial melt in that layer, whereas the *PxP* phase ray coverage is good at the distances of the lower crustal conductors.

There is a region of slightly higher resistivity beneath sites 8 and 9 at 5 km depth. The feature is constrained by the data at site 9, although the fit at both sites 8 and 9 is not as good as elsewhere along the profile. In particular, the apparent resistivities of the TE mode at site 8 and the TM mode at site 9 are underpredicted at periods greater than about 5 s, whereas the apparent resistivities of the other mode at both sites are well reproduced. In fact, it is difficult to posit a 2D structure that will fit both modes at both sites, and 3D structures may be required. Thus, any interpretation of the 2D model must be treated with caution. Cornwell *et al.* (2006) found that a second dense, upper crustal intrusive body was needed to fit their gravity data in the region beneath sites 8 and 9; its top surface is in the same location as the resistive region but it extends further (thickness 10 km) in depth. If the body still contained partial melt, it is possible that the warmer, deeper part could have a similar resistivity to its surroundings and that the MT model images only the coldest, fully solidified upper part, which would be expected to be more resistive than the surrounding rock. If this interpretation is correct, it constrains the timing of the intrusion, though it is difficult to produce a quantitative model predicting resistivity from cooling to compare with the MT model. Cornwell *et al.* (2006) estimated that this and the body beneath the Boset magmatic segment must contain at least 40% gabbro to fit the seismic and gravity data (based upon experimental data of Christensen & Mooney (1995)). In addition, they added a small, relatively high density body to the uppermost 5 km beneath the Boset volcano (with <25% gabbro) where there is no obvious resistivity feature in the MT model, although it might be too small to be resolved. The MT model has another, slightly deeper (*c.* 10 km), resistive zone in the upper crust beneath sites 2 and 3, where the gravity profile has another peak. Again, a dense intrusive body might provide an alternative fit to the gravity data to the model proposed by Cornwell *et al.* (2006). The gravity peak is broader than that beneath sites 8–9, which would imply a deeper body, consistent with the MT model.

The MT model does not image reliably the base of the crust or the upper mantle. However, site 17 has much deeper penetration and constrains the deep heterogeneity (conductive to the NW, resistive to the SE). Long-period MT studies in the 1970s by V. Haak and colleagues (University of Munich) also found a strong resistivity contrast between the very resistive southern plateau and high conductivity northwestern plateau (V. Haak, pers. comm., 2005). This first order difference in the deeper resistivity structure between the plateaus on the two sides of the rift correlates well with variations in the velocity and density structure, with the change occurring in all models beneath the Arboye escarpment (site 16 at *c.* 100 km distance in the MT model). One manifestation of this difference is the layer of underplated material found only beneath the northwestern plateau. Cornwell *et al.* (2006) assign it a density intermediate between that of the lower crust and upper mantle in the gravity model. Also, both the velocity and density of the upper mantle beneath the southern plateau are significantly higher than those beneath the northern plateau. The depth of the velocity and density transitions is about 40 km, whereas it is about 25 km in the MT model.

Dugda *et al.* (2004) and Stuart *et al.*'s (2006) receiver function estimates of crustal thickness beneath the northwestern plateau are up to 10 km less than those of the refraction studies of Mackenzie *et al.* (2005), but compare favourably with those of refraction line I of Berckhemer *et al.* (1975). Stuart *et al.* (2006) state that the discrepancy in crustal thickness estimates can be partly explained if the top of the underplate has been interpreted as the Moho in the receiver function studies. An alternative hypothesis is that the crust contains a low-velocity layer that is below the resolution (at *c.* 30 km depth, *c.* 5 km width, *c.* 40 km length and of small acoustic impedance) of the wide-angle reflection/refraction profiling. This would be expected if the lower crust contains partially molten material, as implied by the MT model, but not by the Poisson's ratio of 0.27 at site 4 (station SENE; Stuart *et al.* 2006). A higher percentage of the lower crustal intrusion beneath site 4 would need to be partially molten than that beneath the Boset magmatic segment to depress the velocity below that of normal lower crust. The extremely conductive lower crust beneath sites 5, 4 and 6 in the MT model supports this hypothesis, but it is inconsistent with the Poisson's ratio.

## Conclusions

The MT model and its interpretation presented here gives a model of crustal structure including a (mainly) conductive surface layer of Quaternary to Jurassic age, a more resistive upper crust, and a lower crust that is conductive beneath the northwestern plateau and much more resistive beneath the southeastern plateau. This fundamental difference in crustal resistivity structure beneath the plateaus on the two sides of the rift follows differences seen in the seismic velocity and density images, as well as earlier electromagnetic induction studies. The enhanced conductivity beneath the Boset magmatic segment, particularly at depths of *c.* 20 km, indicates the presence of partial melt.

A unique feature of the resistivity image is a zone of high conductivity material in the lower crust beneath the northwestern plateau close to the rift. It is proposed that this is a further region of partial melt, although there is no associated feature in the seismic or gravity model. Poisson's ratios for nearby stations are also consistent with the presence of crustal partial melt, although that for the station over the conductor is not. Mantle tomography and xenolith data indicate hot asthenosphere below this region. Seismic anisotropy results beneath the northwestern plateau suggest that melt is still being produced in the top *c.* 75 km (Kendall *et al.* 2006). Teleseismic arrivals are significantly late at the permanent station FURI on the northern rift flank, regardless of azimuth, pointing to extensive (i.e. not confined to the rift) partial melt. There are several lines of Quaternary volcanoes on the plateau (e.g. Bishoftu, and near the Ambo lineament—see e.g. Casey *et al.* 2006, Fig. 2, for location). Late Miocene–Quaternary basaltic magmatism in the Lake Tana area is inferred by Chorowicz *et al.* (1998), which is consistent with an interpretation of a 2D model of MT data collected south and east of Lake Tana (Hautot *et al.* 2006), and Kendall *et al.* (2006) note the existence of Quaternary eruptive centres there. Discrepancies between crustal thickness from the seismic refraction/wide angle reflection study and receiver function analysis might be explained by a thin lower crustal low-velocity layer. Taken together, this implies the existence of lower crustal reservoirs and extensive melt beneath the northwestern plateau, and hence that orientated melt pockets would provide an alternative to the inherited Precambrian fabric explanation of Gashawbeza *et al.* (2004) for the SKS splitting results.

Within the upper crust on the northwestern side of the rift are two regions of slightly higher resistivity. The shallower of the two, which has the smaller resistivity contrast with the surrounding upper crust, has its top surface in the same position as a high-density body inferred from the gravity survey, identified as another mafic intrusion (Cornwell *et al.* 2006). The other coincides with a short wavelength peak on the gravity survey, suggesting it too could be modelled as a mafic intrusion. However, both of these smaller, shallower bodies must have solidified to be imaged as resistivity highs, in contrast to the lower crustal conductors beneath the Boset magmatic segment and further to the NW.

Seismic anisotropy studies show a slight anticlockwise rotation of the fast shear wave direction over magmatic segments; the MT results have a similar rotation of the geoelectric strike direction within the rift. Unfortunately, because the electromagnetic signals recorded in MT satisfy a diffusion equation, rather than the wave equation that governs seismic wave propagation, the data average structure both laterally and vertically and cannot distinguish between rotation confined to sites on the Boset magmatic segment and over the whole of the rift valley. The distinction is important: the rotated direction is parallel to the present-day extension direction inferred geodetically (Bilham *et al.* 1999), from seismicity studies (Keir *et al.* 2006) and field observations (Wolfenden *et al.* 2004), and thus its confinement to magmatic segments supports the magma-assisting rifting hypothesis of Ebinger & Casey (2001) and Buck (2004) (Kendall *et al.* 2005). Whereas shear wave splitting by orientated melt pockets could arise within the crust or upper mantle (down to depths of 75 km; Kendall *et al.* 2006), the MT data demonstrate that geoelectric strike direction rotation takes place at crustal depths, consistent with the melt interpretation for the high conductivity body beneath the Boset segment.

Future work will include joint MT-seismic refraction and MT-gravity inversions (e.g. Gallardo & Meju 2003). These inversions should establish whether the additional mafic intrusions inferred from the gravity and MT data are mutually compatible, and compatible with the seismic data, and may establish whether a lower crustal low-velocity layer could explain the discrepancies in crustal thickness inferred from seismic profiling and receiver function analysis. Li *et al.* (2003) estimated the percentage of partial melt to explain their MT data across the Tibetan plateau. A similar calculation will be undertaken with these MT data, which might constrain the timing of the intrusions.

Thanks to reviewers Volker Haak and Colin Brown and editors Cindy Ebinger and Peter Maguire for useful comments on an earlier version of this manuscript, and to Volker Haak for providing information on previous electromagnetic induction studies in the area. We acknowledge rewarding discussions with EAGLE participants, in particular, Ian Bastow, Dave Cornwell, Cindy Ebinger, Derek Keir, Mike Kendall, Peter Maguire and Graham Stuart, which have aided the interpretation of our results. This work was supported by NERC grant NER/B/S/2001/00863. SPAM equipment was loaned by the NERC Geophysical Equipment Facility, and electrodes by the University of Brest. We are grateful for the assistance and support of all EAGLE participants and field crews, especially Mohammednur Desissa. SH was funded by a Marie Curie Post-Doctoral Fellowship at the University of Edinburgh.

## References


BAHR, K. 1988. Interpretation of the magnetotelluric impedance tensor—regional induction and local telluric distortion. *Journal of Geophysics*, **62**, 119–127.

BASTOW, I.D., STUART, G.W., KENDALL, J.M. & EBINGER, C.J. 2005. Upper mantle seismic structure in a region of incipient continental breakup: northern Ethiopian rift. *Geophysical Journal International*, **162**, 479–493.

BERCKHEMER, H., BAIER, B., *ET AL.* 1975. Deep seismic soundings in the Afar region and on the highland of Ethiopia. *In*: PILGER, A. & ROESLER, A. (eds) *Afar Depression of Ethiopia*, Schweizerbart, Stuttgart, 89–107.

BILHAM, R., BENDICK, R., LARSON, K., MOHR, P., BRAUN, J., TESFAYE, S. & ASFAW, L. 1999. Secular and tidal strain across the Main Ethiopian Rift. *Geophysical Research Letters*, **26**, 2789–2792.

BOCCALETTI, M., MAZZUOLI, R., BONINI, M., TRUA, T. & ABEBE, B. 1999. Plio-Quaternary volcanotectonic activity in the northern sector of the Main Ethiopian Rift: relationships with oblique rifting. *Journal of African Earth Sciences*, **29**, 679–698.

BUCK, W.R. 2004. Consequences of asthenospheric variability on continental rifting. *In*: KARNER, G.D., TAYLOR, B., DRISCOLL, N.W. & KOHLSTEDT, D.L. (eds) *Rheology and Deformation of the Lithosphere at Continental Margins*, Columbia University Press, New York, 1–30.

CASEY, M., EBINGER, C., KEIR, D., GLOAGUEN, R. & MOHAMED, F. 2006. Strain accommodation in transitional rifts: Extension by magma intrusion and faulting in Ethiopian rift magmatic segments. *In*: YIRGU, G., EBINGER, C.J. & MAGUIRE, P.K.H. (eds) *The Afar Volcanic Province within the East African Rift System*. Geological Society, London, Special Publication, **259**, 143–163.

CHAVE, A.D. & THOMPSON, D.J. 1989. Some comments on magnetotelluric response function estimation. *Journal of Geophysical Research*, **94**, 14 202–14 215.

CHOROWICZ, J., COLLET, B., BONAVIA, F.F., MOHR, P., PARROT, J.F. & KORME, T. 1998. The Tana basin, Ethiopia: intra-plateau uplift, rifting and subsidence. *Tectonophysics*, **295**, 351–367.

CHRISTENSEN, M.I. & MOONEY, W.D. 1995. Seismic velocity structure and composition of the continental crust—a global view. *Journal of Geophysical Research*, **100**, 9761–9788.

CORNWELL, D.G., MACKENZIE, G.D., MAGUIRE, P.K.H., ENGLAND, R.W., ASFAW, L.M. & OLUMA, B. 2006. Northern Main Ethiopian Rift crustal structure from new high-precision gravity data. *In*: YIRGU, G., EBINGER, C.J. & MAGUIRE, P.K.H. (eds) *The Afar Volcanic Province within the East African Rift System*. Geological Society, London, Special Publications, **259**, 269–291.

COUNIL, J.-L., LE MOUËL, J.-L. & MENVIELLE, M. 1986. Associate and conjugate direction concepts in magnetotellurics. *Annale Geophysicae*, **4**, 115–130.

DUGDA, M.T., NYBLADE, A.A., JULIA, J., LANGSTON, C.A., AMMON, C.J. & SIMIYU, S. 2005. Crustal structure in Ethiopia and Kenya from receiver function analysis: Implications for rift development in eastern Africa. *Journal of Geophysical Research*, **110**, doi 10.1029/2004JB003065.

EBINGER, C. & CASEY, M. 2001. Continental break-up in magmatic provinces: an Ethiopian example. *Geology*, **29**, 527–530.

GALLARDO, L.A. & MEJU, M.A. 2003. Characterisation of heterogeneous near-surface materials by joint 2D inversion of dc resistivity and seismic data. *Geophysical Research Letters*, **30**, 1658–1661.

GAMBLE, T.D., CLARKE, J. & GOUBAU, W.M. 1979. Magnetotellurics with a remote magnetic reference. *Geophysics*, **44**, 53–68.

GASHAWBEZA, E.M., KLEMPERER, S.L., NYBLADE, A.A., WALKER, K.T. & KERANEN, K.M. 2004. Shear-wave splitting in Ethiopia: Precambrian mantle anisotropy locally modified by Neogene rifting. *Geophysical Research Letters*, **31**, doi:10.1029/2004GL020471.

GROOM, R.W. & BAILEY, R.C. 1989. Decomposition of magnetotelluric impedance tensors in the presence of local 3-dimensional galvanic distortion. *Journal of Geophysical Research*, **94**, 1913–1925.

HAUTOT, S., TARITS, P., WHALER, K., LE GALL, B., TIERCELIN, J.-J. & LE TURDU, C. 2000. The deep structure of the Baringo rift Basin (central Kenya) from 3-D magnetotelluric imaging: Implications for rift evolution. *Journal of Geophysical Research*, **105**, 23 493–23 518.

HAUTOT, S., TARITS, P., PERRIER, F., TARITS, C. & TRIQUE, M. 2002. Groundwater electromagnetic imaging in complex geological and topographical regions: A case study of a tectonic boundary in the French Alps. *Geophysics*, **67**, 1048–1060.

HAUTOT, S., WHALER, K., GEBRU, W. & DESSISA, M. 2006. The structure of the Mesozoic basin beneath the Lake Tana area revealed by magnetotelluric imaging. *Journal of African Earth Sciences* (in press).

KEIR, D., EBINGER, C.J., KENDALL, J.-M. & STUART, G.W. 2005. Variations in late syn-rift melt alignment inferred from shear-wave splitting in crustal earthquakes beneath the Ethiopian Rift. *Geophysical Research Letters*, **32**, L23308.

KEIR, D., EBINGER, C.J., STUART, G.W., DALY, E. & AYELE, A. 2006. Strain accommodation by magmatism and faulting as rifting proceeds to breakup: Seismicity of the northern Ethiopian rift. *Journal of Geophysical Research* (in press).

KENDALL, J.-M., STUART, G.W., EBINGER, C.J., BASTOW, I.D. & KEIR, D. 2005. Magma-assisted rifting in Ethiopia. *Nature*, **433**, 146–148.

KENDALL, J.-M., PILIDOU, S., KEIR, D., BASTOW, I.D., STUART, G.W. & AYELE, A. 2006. Mantle upwellings, melt migration and the rifting of Africa: Insights from seismic anisotropy. *In*: YIRGU, G., EBINGER, C.J. & MAGUIRE, P.K.H. (eds) *The Afar Volcanic Province within the East African Rift System*. Geological Society, London, Special Publications, **259**, 55–72.

KERANEN, K., KLEMPERER, S.L., GLOAGUEN, R. & The EAGLE Working Group. 2004. Three-dimensional seismic imaging of a protoridge axis in the Main Ethiopian rift. *Geology*, **32**, 949–952.

KIDANE, T., EBINGER, C., ABEBE, B., PLATZMAN, E., KEIR, D., LAHITTE, P. & ROCHETTE, P. 2006. Paleomagnetic constraints on continental break-up processes: Observations from the Main Ethiopian rift. *In*: YIRGU, G., EBINGER, C.J. & MAGUIRE, P.K.H. (eds) *The Afar Volcanic Province within the East African Rift System*. Geological Society, London, Special Publications, **259**, 165–183.

LI, S.H., UNSWORTH, M.J., BOOKER, J.R., WEI, W.B., TAN, H.D. & JONES, A.G. 2003. Partial melt or aqueous fluid in the mid-crust of Southern Tibet? Constraints from INDEPTH magnetotelluric data. *Geophysical Journal International*, **153**, 289–304.

MACKENZIE, G.D., THYBO, H. & MAGUIRE, P.K.H. 2005. Crustal velocity structure across the Main Ethiopian Rift: Results from two-dimensional wide-angle seismic modelling. *Geophysical Journal International*, **162**, 994–1006.

MAGUIRE, P.K.H., EBINGER, C.J. ET AL. 2003. Geophysical project in Ethiopia studies continental breakup. *EOS Transactions of the AGU*, **84**, 337, 342–343.

MAGUIRE, P.K.H., KELLER, G.R. ET AL. 2006. Crustal structure of the Northern Main Ethiopian Rift from the EAGLE controlled-sources survey; A snapshot of incipient lithospheric break-up. *In*: YIRGU, G., EBINGER, C.J. & MAGUIRE, P.K.H. (eds) *The Afar Volcanic Province within the East African Rift System*. Geological Society, London, Special Publications, **259**, 269–291.

MAHATSENTE, R., JENTZSCH, G. & JAHR, T. 1999. Crustal structure of the Main Ethiopian Rift from gravity data: 3-dimensional modelling. *Tectonophysics*, **313**, 363–382.

PIZZI, A., COLTORTI, M., ABEBE, B., DISPERATI, L., SACCHI, G. & SALVINI, R. 2006. The Wonji fault belt (Main Ethiopian rift): Structural and geomorphological constraints and GPS monitoring. *In*: YIRGU, G., EBINGER, C.J. & MAGUIRE, P.K.M. (eds) *The Afar Volcanic Province within the East African Rift System*. Geological Society, London, Special Publications, **259**, 191–207.

RITTER, O., JUNGE, A. & DAWES, G.J.K. 1998. New equipment and processing for magnetotelluric remote reference observations. *Geophysical Journal International*, **132**, 535–548.

ROBERTS, J.J. & TYBURCZY, J.A. 1999. Partial-melt electrical conductivity: Influence of melt composition. *Journal of Geophysical Research*, **104**, 7055–7065.

ROONEY, T.O., FURMAN, T., YIRGU, G. & AYELEW, D. 2005. Structure of the Ethiopian lithosphere: Evidence from mantle xenoliths. *Geochimica et Cosmochimica Acta*, **69**, 3889–3910. doi:10.1016/j/gca.2005.03.0.

STUART, G.W., BASTOW, I.D. & EBINGER, C.J. 2006. Crustal structure of the Northern Main Ethiopian rift from receiver function studies. *In*: YIRGU, G., EBINGER, C.J. & MAGUIRE, P.K.H. (eds) *The Afar Volcanic Province within the East African Rift System*. Geological Society, London, Special Publications, **259**, 253–267.

TADESSE, K., MICKUS, K. & KELLER, G.R. 2004. Gravity Field of the Central Portion of the Main Ethiopian Rift. *Eos Transactions of the AGU*, **85**(47), Fall Meeting Supplement, Abstract T41E-1263.

TRUA, T., DENIEL, C. & MAZZUOLI, R. 1999. Crustal control in the genesis of Plio-Quaternary bimodal magmatism of the Main Ethiopian Rift (MER): Geochemical and isotopic (Sr, Nd, Pb) evidence. *Chemical Geology*, **155**, 201–231.

WOLFENDEN, E., EBINGER, C., YIRGU, G., DEINO, A. & AYALEW, D. 2004. Evolution of the northern Main Ethiopian Rift: Birth of a triple junction. *Earth and Planetary Science Letters*, **224**, 213–228.

# Northern Main Ethiopian Rift crustal structure from new high-precision gravity data

D.G. CORNWELL[1], G.D. MACKENZIE[1,2], R.W. ENGLAND[1], P.K.H. MAGUIRE[1], L.M. ASFAW[3] & B. OLUMA[4]

[1]*University of Leicester, University Road, Leicester, LE1 7RH, UK (e-mail: dgc2@le.ac.uk)*

[2]*Present address: Compagnie Générale de Géophysique, Vantage West, Great West Road, Brentford, Middlesex, TW8 9GG, UK*

[3]*Geophysical Observatory, Addis Ababa University, PO Box 1176, Addis Ababa, Ethiopia*

[4]*Geological Survey of Ethiopia, Geophysics Department, PO Box 2302, Addis Ababa, Ethiopia*

**Abstract:** We present analysis of new gravity data to produce a 2D crustal and upper mantle density model across the northern Main Ethiopian Rift (NMER). The magmatic NMER is believed to represent the transitional stage between continental and oceanic rifting. We conclude that beneath our profile, magma emplacement into the upper crust occurs in the form of a 20 km-wide body beneath the axis of the rift, and a 12 km-wide off-axis body beneath the NW margin of the rift. These are coincident with Quaternary volcanic chains, anomalies in seismic velocity and conductivity identified by the Ethiopia Afar Geoscientific Lithospheric Experiment (EAGLE) along the same profile. We also identify a shallow, high-density body beneath the axial Boset volcano interpreted as either a dyke zone or a magma reservoir that may have fed Quaternary felsic volcanism. Our results provide supporting evidence for a *c.* 15 km-thick mafic underplate layer beneath the northwestern rift flank, imaged by the EAGLE controlled- and passive-source seismic data. A relatively low-density upper mantle is required beneath the underplate and the rift to produce the long wavelength features of the gravity anomaly. The resulting model suggests that the lithosphere to the SE of the rift is unaffected by rifting processes. Our results combined with those from other EAGLE studies show that magmatic processes dominate rifting in the NMER.

The magmatic Main Ethiopian Rift (MER), part of the eastern branch of the East African rift system (EARS), constitutes the third arm of a triple junction with the Gulf of Aden and Red Sea rifts (e.g. Wolfenden *et al.* 2004) that has formed within the Afar flood volcanic province. The MER incorporates the transitional stage between continental rifting as observed in south and central Ethiopia (Ebinger *et al.* 1993) and incipient seafloor spreading evident to the north in the Asal rift, the onshore westward extension of the Gulf of Aden spreading ridge (Ruegg & Kasser 1987; Stein *et al.* 1991; De Chabalier & Avouac 1994). Examination of crust and upper mantle structure within the northern Main Ethiopian Rift (NMER) allows detailed study of this transitional stage in which magma injection is believed to be replacing mechanical failure as the main strain accommodation mechanism (Ebinger & Casey 2001). EAGLE, the Ethiopia Afar Geoscientific Lithospheric Experiment, is a passive- and active-source seismology and MT project investigating lithospheric and underlying mantle structure and processes beneath the NMER (e.g. Maguire *et al.* 2003). It included the acquisition of a *c.* 325 km gravity profile across the NMER coincident with the cross-rift seismic profile (Fig. 1). Since there are likely to be strong density contrasts between the existing crustal rocks and igneous material injected as part of the rifting process, modelling the gravity data should reveal the presence of such bodies. The gravity data may also be used to verify the crustal section determined by modelling of the seismic data (Fig. 3). This paper describes the modelling of a 2D profile which demonstrates the accuracy of the seismic velocity model for the crust and reveals the presence of dense bodies of rock that are below the resolution of the seismic data.

*From*: YIRGU, G., EBINGER, C.J. & MAGUIRE, P.K.H. (eds) 2006. *The Afar Volcanic Province within the East African Rift System*. Geological Society, London, Special Publications, **259**, 307–321.
0305-8719/06/$15.00 

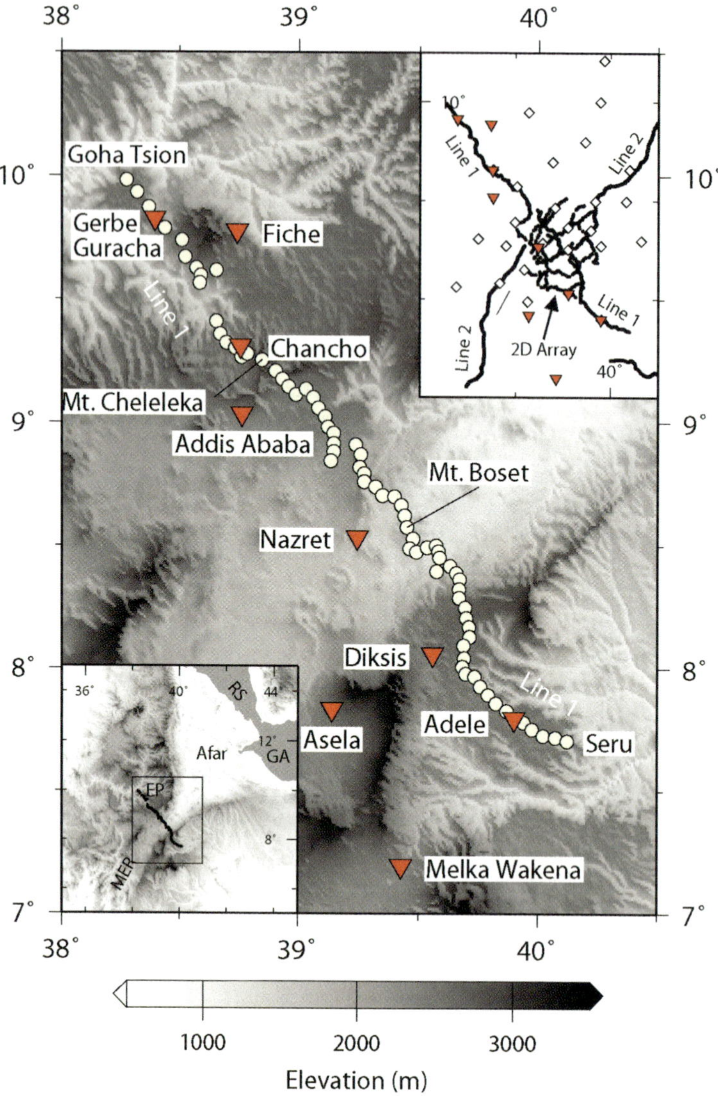


**Fig. 1.** Topography of northern Main Ethiopian Rift with gravity base station (inverted triangles) and roving station (circles) locations. Upper right inset shows location of EAGLE controlled-source seismic stations (black squares on Line 1, Line 2 and 2D array) and EAGLE passive array broadband seismometers (open diamonds) (Maguire *et al.* 2003). Lower left inset shows topography of the northern EARS and location of the Red Sea rift (RS), Gulf of Aden rift (GA), the Main Ethiopian Rift (MER) and the Ethiopian plateau (EP). Rectangle shows extent of main location map.

## Tectonic setting and geological history

The MER, Red Sea and Gulf of Aden (Asal rift) triple junction is situated at *c.* 11.5° N within the central Afar depression (e.g. Acton *et al.* 2000; Kidane *et al.* 2003) and separates the Arabia, Nubia and Somalia plates and the Danakil microplate (e.g. Eagles *et al.* 2002) (Fig. 2). Continental separation has occurred in the Red Sea and Gulf of Aden while seafloor spreading has yet to begin in the MER. NE-directed seafloor spreading in the Gulf of Aden has propagated westward into the Afar depression since 16 Ma (d'Acremont *et al.* 2005) or 25 Ma (Manighetti *et al.* 1998) and NE-directed seafloor spreading commenced at *c.* 4 Ma in the Red Sea (Cochran & Martinez 1988).

The MER formed within the Precambrian metamorphic crustal basement of the Pan-African

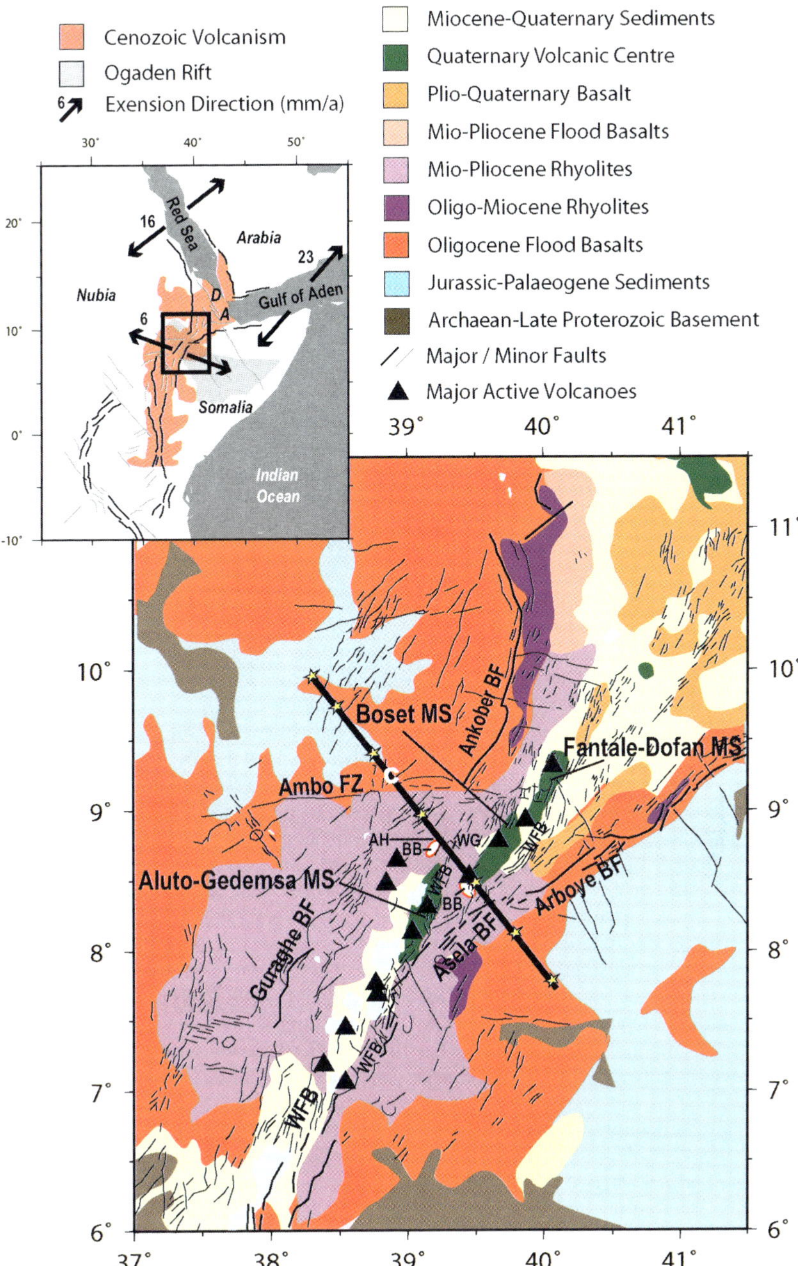


**Fig. 2.** Simplified geological and structural map of northern Main Ethiopian Rift with magmatic segments (MS), border faults (BF) and fault zones (FZ) (after Boccaletti *et al.* 1999; Ebinger & Casey 2001; Acocella & Korme 2002; Korme *et al.* 2004; Wolfenden *et al.* 2004). Locations of the Wonji Fault Belt (WFB), Welenchiti Graben (WG), Cheleleka volcano (C) and near-profile surface extent of Late Pliocene Bofa flood basalts (BB) and Adama Horst (AH) are shown. Bold line indicates best-fit profile for the gravity models with relative location of seismic shotpoints along the profile marked with stars (see Figs 3 & 5). Inset shows tectonic setting of the northern East African rift system with major fault lineation, volcanism and extension rates and directions between tectonic plates (Bilham *et al.* 1999; Chu & Gordon 1999; McClusky *et al.* 2003). Location of Danakil (D) and Asal rift (A) are shown with the extent of the failed Mesozoic Ogaden rift shown in light grey (Korme *et al.* 2004).

shield (Kazmin *et al.* 1978) which exhibits north–south to NNE–SSW suture zones (Vail 1983; Berhe 1990) and NW–SE orientated strike–slip faults (Brown 1970; Purcell 1976). These faults are thought to have been re-activated as normal faults during Mesozoic lithospheric extension to form basins (Korme *et al.* 2004) that were filled with Jurassic marine carbonates and sediments (Bosellini *et al.* 2001) (Fig. 2).

A mantle plume head impacting onto the base of the lithosphere is believed to have initiated volcanism in southwestern Ethiopia (*c.* 45 to 40 Ma) and in the MER (*c.* 31 to 29 Ma). Flood basalts and rhyolites with a total thickness of 500 to 2000 m were emplaced within 1 to 2 million years (Hofmann *et al.* 1997) along the southern Red Sea and to a lesser extent in the MER around 26 to 22 Ma (Keiffer *et al.* 2004; Wolfenden *et al.* 2004). Uplift of the Ethiopian plateau occurred between 29 and 20 Ma (Pik *et al.* 2003). Rifting within the southern and central MER is believed to have occurred between 18 and 15 Ma (WoldeGabriel *et al.* 1990) with extension in the NMER commencing after *c.* 11 Ma (Wolfenden *et al.* 2004).

The present-day rift floor of the NMER is characterized by elongate axial zones of eruptive centres arranged in an en echelon right-stepping pattern that developed after 1.8 Ma (Boccaletti *et al.* 1999; Ebinger & Casey 2001). GPS measurements have been used to show that these 'magmatic segments' accommodate 80% of the extensional strain across the NMER (Bilham *et al.* 1999). The NMER basins are floored by Oligo-Miocene flood basalts, which in turn are overlain by Mid-Miocene–Recent basalt–ignimbrite packages. Some fluvo-lacustrine strata are locally interspersed (WoldeGabriel *et al.* 1990; Wolfenden *et al.* 2004). The major active fault zones identified within the rift are predominantly orientated north–south to NNE–SSW (e.g. the Wonji fault belt) and NE–SW (e.g. the Arboye border fault) (e.g. Korme *et al.* 2004) (Fig. 2).

## Previous geophysical studies

An early gravity map and 2D density model of the MER (Makris *et al.* 1975) suggested that the rift flanks within the Ethiopian plateau are formed from shield-type continental blocks partly underlain by excess low-density material with the crustal thickness varying proportionally with elevation from 30 to 42 km. The western part of the plateau was shown to be isostatically undercompensated at the Moho owing to low-density material at the base of the crust. Low-velocity (and density) upper mantle was identified beneath a thinned (14–22 km) high-velocity crust beneath the northeastern continuation of the rift in Afar. This was interpreted in terms of high temperature mantle underlying a heavily intruded 'partly oceanized' crust. It was also suggested that the model showed asthenosphere penetrating to the base of the crust (Berckhemer *et al.* 1975; Makris & Ginzburg 1987). Interpretation of initial seismic refraction data indicated a *c.* 38 km (Berckhemer *et al.* 1975) to *c.* 44 km (Makris & Ginzburg 1987) thick crust beneath the plateau underlain by normal mantle with a velocity of 8.1 km $s^{-1}$.

Analysis of the new seismic refraction and wide-angle reflection data along the EAGLE (Line 1) cross-rift profile (Fig. 1) (Mackenzie *et al.* 2005) has resulted in a model with a *c.* 48 km-thick crust beneath the western plateau and a *c.* 40 km-thick crust beneath the eastern plateau (Fig. 3). Typical P-wave seismic velocities for continental crust are observed beneath both the eastern and western plateaus with the exception of a *c.* 15 km-thick high-velocity (*c.* 7.4 km $s^{-1}$) lower crustal layer that is imaged only beneath the western plateau. Also, anomalously high upper crustal velocities (greater than 6.5 km $s^{-1}$) occur below the Boset magmatic segment and are interpreted to result from the presence of gabbroic or diabase intrusions (Keranen *et al.* 2004; Mackenzie *et al.* 2005). 3D P-wave velocity tomography (Keranen *et al.* 2004) shows that these high velocities form distinct elongate 50 km by 20 km bodies that mimic the right-stepping en echelon pattern of the magmatic segments along the rift axis. Beneath the centre of the rift, the crust thins and low mantle velocities suggest the presence of hot mantle rocks containing a small percentage (3–5%) of partial melt (Mackenzie *et al.* 2005). The deeper-velocity structure beneath the NMER imaged using teleseismic earthquake arrival times includes low-velocity mantle zones that are often displaced towards the rift flank where topography is greatest and the lithosphere–asthenosphere boundary is steepest. This suggests that pre-existing as well as rift-induced base of lithosphere topography may play an important role in melt transport in the upper mantle beneath the rift system (Bastow *et al.* 2005). These low-velocity anomalies of $\delta V_p \approx -1.5\%$ and $\delta V_s \approx -4.0\%$ also include a limb that extends more than 100 km west from the Boset magmatic segment to the Ambo fault zone (Fig. 2) and are interpreted to be caused by high temperatures and the presence of partial melt (Bastow *et al.* 2005).

Regional gravity data collected by the Ethiopian Geological Survey and by Ebinger (1991) has been analysed to produce 3D density models beneath the MER (Mahatsente *et al.* 1999; Tiberi *et al.* 2005). Mahatsente *et al.* (1999) derived crustal structure beneath the rift between 6.5° and 8.5° N showing a thinned crust beneath the rift axis ($\leq$ 31 km)

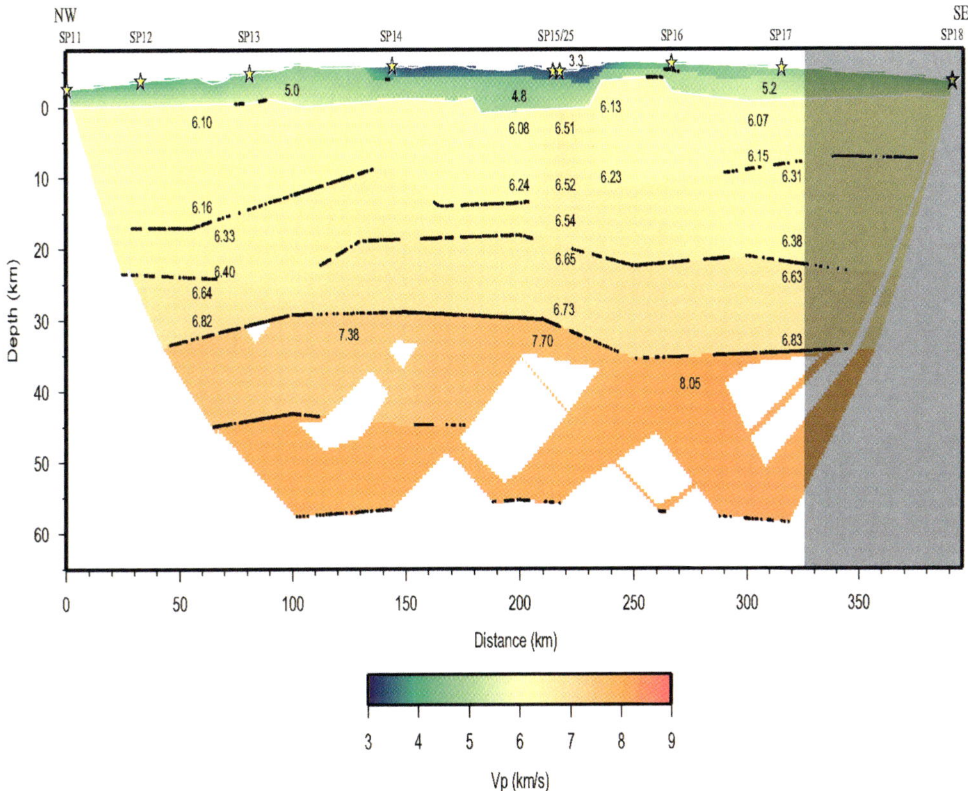


**Fig. 3.** 2D P-wave velocity model derived from Line 1 (see Fig. 1) wide-angle reflection/refraction profile across the Ethiopian rift. Black dots indicate wide-angle reflection points on seismic first-order discontinuities and model outline identifies subsurface ray coverage. Depth scale is relative to mean sea-level. Shaded area indicates region of seismic model not used in this work. Adapted from Figure 6 of Mackenzie *et al.* (2005).

with a segmented high-density upper and lower crustal intrusion beneath the surface magmatic segments. This intrusive material is modelled with a density of 3000 to 3100 kg $m^{-3}$ and is attributed to melt resulting from partial melting of the lithospheric mantle. The crustal thickness beneath the rift flanks varies between 38 and 51 km. Tiberi *et al.* (2005) estimated the MER crustal thickness and 3D density distribution from gravity inversion and forward modelling. Their results suggest a shallower Moho exists beneath the rift (32 to 33 km) that thins (to *c.* 24 km) northwards to Afar. They state that a *c.* 40 km-thick crust is present beneath the broad uplifted Oligocene flood basalt province suggesting that crustal underplating isostatically compensates most of the plateau uplift.

## Gravity data

When used in isolation, potential field data suffer from non-uniqueness, but in conjunction with controlled source seismic data they become a powerful tool for refining the seismic model and providing additional constraints on the physical properties of the lithosphere. The new data presented here consist of 72 stations located between Goha Tsion and Seru along the majority of the EAGLE Line 1 cross-rift seismic profile (Mackenzie *et al.* 2005) at a nominal spacing of 5 km (Fig. 1). The southernmost 75 km of the seismic profile were not occupied for logistic reasons (see Fig. 3). A differential GPS technique was used to locate each gravity station with a maximum error of 0.2 m in elevation, latitude and longitude relative to the WGS84 reference ellipsoid. Station heights were corrected to the EGM96 reference geoid (Lemoine *et al.* 1998) and used to calculate the Bouguer anomaly, following the method described by Swain & Khan (1977) and assuming a constant Bouguer density of 2670 kg $m^{-3}$. Terrain corrections were estimated using conversion tables for near-station (zones A–D) topography

**Table 1.** *Summary of data reduction errors to complete Bouguer anomaly*

| Measurement/process | Error (±mGal) | Cumulative error (±mGal) |
|---|---|---|
| G16 meter reading | 0.03 | 0.03 |
| Tide correction | 0.01 | 0.04 |
| Drift correction | 0.05 | 0.09 |
| Theoretical gravity | 0.02 | 0.11 |
| Geoid correction | 0.05 | 0.16 |
| Free air correction | 0.06 | 0.22 |
| Inner terrain correction (0–170 m) | 0.05 | 0.27 |
| Outer terrain correction (170 m–20 km) | 0.03 | 0.30 |

measurements (Hammer 1939) and calculated to 20 km using the Shuttle-Radar Topography Mission 90 m Digital Elevation Models using the Inner TC program (Cogbill 1990), again assuming a constant density of 2670 kg m$^{-3}$. The maximum Bouguer anomaly error is less than ±0.3 mGal. Experimental errors are summarized in Table 1.

## Constructing the density model

### *Modelling method*

Forward and inverse modelling of the gravity anomaly has been undertaken using GRAVMAG, an interactive 2.5D potential field modelling package (Busby 1987; Pedley *et al.* 1993). The gravity stations were projected onto the best-fit line (Fig. 2) and the modelling was effectively performed in 2D with the model polygons of constant density extending to 1000 km perpendicular to the profile and 1000 km off each end of the profile. Since the profile is only 325 km in length, the base of the model is fixed at 60 km. No attempt has been made to estimate and subtract a regional anomaly from the new data. If the subtracted regional anomaly were incorrect, making the density beneath the Moho in the model constant (to reflect removal of the regional) introduces the possibility that density variations which do occur within the mantle could be mapped into the crust. Instead, the crust and uppermost mantle is modelled using the constraints provided by the controlled source seismic model (see below), assuming that the causes of the observed anomaly lie above 60 km. A minimum-structure modelling approach was employed to ensure that model complexities were only introduced at sub-seismic resolution as additional constant-density polygons.

### *The seismic model*

The initial density model was produced by converting P-wave velocity to density in the EAGLE Line 1 cross-rift seismic refraction / wide-angle reflection velocity model (Fig. 3) (Mackenzie *et al.* 2005). Additional velocity information was obtained from the 3D P-wave velocity tomography model of Keranen *et al.* (2004). The seismic model resolution was used to constrain the gravity modelling so that adjustments to the top or base of polygons corresponding to interfaces in the seismic model in regions of 'good' resolution were no more than the assessed errors defined by Mackenzie *et al.* (2005). These errors are ±1 km for the intracrustal layers and ±2 km for the Moho and no more than ±2 km for the intracrustal layers and ±4 km for the Moho in regions of 'poor' resolution (particularly at the model edges).

### *Initial density model*

The initial density model was constructed by calculating a non-weighted average of the velocity model node values in each seismic layer and then converting each layer to a constant density polygon using a polynomial approximation of the best-fit Nafe–Drake velocity–density relationship (Nafe & Drake 1957):

$$\rho = -168.03 + 1765\,V_{\mathrm{p}} - 481.72\,V_{\mathrm{p}}^2 + 60.973\,V_{\mathrm{p}}^3 - 2.6861\,V_{\mathrm{p}}^4$$

where $\rho$ = density in kg m$^{-3}$ and $V_{\mathrm{p}}$ = P-wave velocity in km s$^{-1}$.

### *Forward modelling and inversion*

Adjusting the seismically defined interfaces within the seismic model error bounds and then inserting constant density polygons into the model at sub-seismic resolution improved the initial model (Figs 3 & 5) and produced a good fit in most places to the observed Bouguer anomaly. Additional polygons were introduced into the model where a poor fit between the initial model and the Bouguer anomaly existed. These new polygons were first assigned an appropriate velocity according to their position within the seismic model and then converted to density using the above relationship. Inversion (optimization) of the model was then undertaken, allowing polygon densities to vary whilst keeping the polygon node positions fixed. The additional polygon node positions were then optimized whilst keeping all polygon densities fixed. These two inversion steps were repeated to find the optimum fit, within geological reasonability. The optimization method fits the long wavelengths in the data with the first iterations and then progresses to successively shorter wavelengths to find the best fit (Pedley *et al.* 1993). Using this trial-and-error method, the RMS error was improved from 0.35 mGal for the initial model to 0.18 mGal for the final model (Fig. 5a,d). The initial and final model densities and related velocities are shown in Figure 4 and summarized in Table 2.

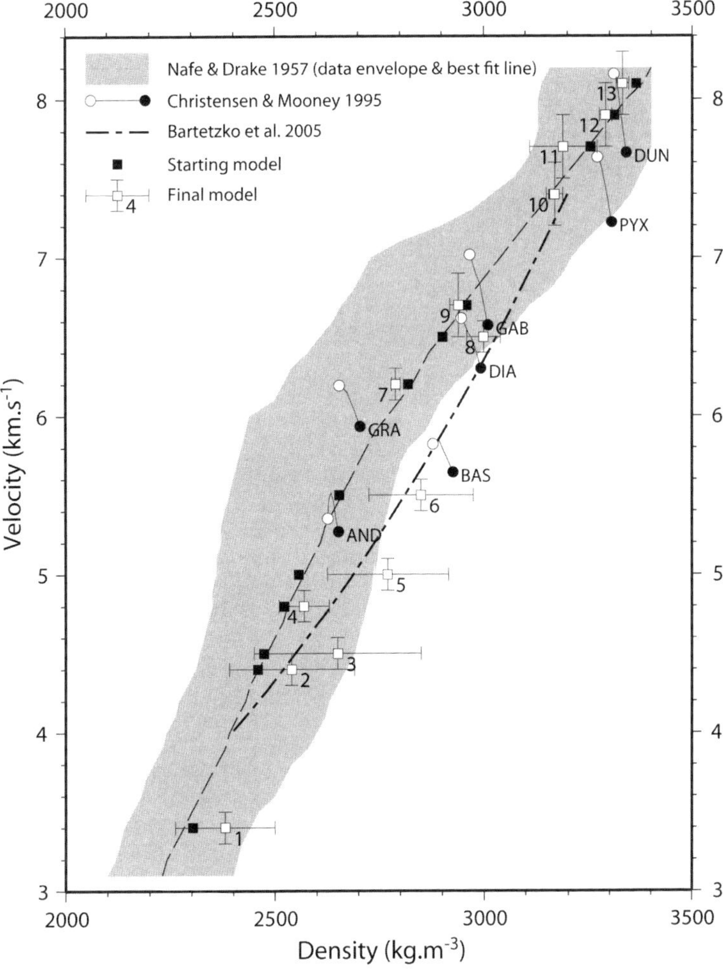


**Fig. 4.** Velocity vs. density plot of the starting (filled squares) and final (open squares) model layers. Velocity error bars for final model values are taken from Mackenzie *et al.* (2005) and density error bars represent the range in densities for each model polygon that invokes a $\pm 10$ mGal offset from the observed Bouguer anomaly (see text for detail). Layer numbers correspond to Table 2. Selected velocity–density relationships are also shown for all crustal rocks (Nafe & Drake 1957; Ludwig *et al.* 1970), oceanic rocks (Bartetzko *et al.* 2005) and igneous rocks between 5 km (filled circles) and 50 km (open circles) depth, assuming high temperatures (i.e. 138 °C at 5 km and 1344 °C at 50 km) (AND, andesite; BAS, basalt; DIA, diabase; DUN, dunite; GAB, gabbro; GRA, granite; PYX, pyroxenite) (Christensen & Mooney 1995).

**Table 2.** *Summary of the model layer velocity and density properties*

| No. | Geological description | Mean velocity ($km\ s^{-1}$) | Starting density ($kg\ m^{-3}$) | Final density ($kg\ m^{-3}$) |
|---|---|---|---|---|
| 1 | Rift sediments | 3.30 | 2302 | 2380 |
| 2 | Flank sediments | 4.20 | 2459 | 2560 |
| 3 | Horst block | 4.50 | 2474 | 2650 |
| 4 | Supra-basement | 4.83 | 2522 | 2570 |
| 5 | Mt. Cheleleka | 5.00 | 2557 | 2770 |
| 6 | Boset dyking | 5.50 | 2654 | 2850 |
| 7 | Upper crust | 6.25 | 2820 | 2790 |
| 8 | Mafic intrusions | 6.50 | 3000 | 3000 |
| 9 | Lower crust | 6.70 | 2959 | 2940 |
| 10 | Underplate | 7.40 | 3167 | 3170 |
| 11 | Axial upper mantle | 7.70 | 3255 | 3190 |
| 12 | NW flank mantle | 7.90 | 3311 | 3290 |
| 13 | SE flank mantle | 8.10 | 3364 | 3330 |

### *Final crustal density model*

The final density model is shown in Fig. 5d. Significant changes to the initial model defined by the seismic model (Fig. 3) and initial densities (Table 2) during the forward and inverse modelling are summarized below. (Note that the approximate model distances ($x$) are measured from the northwestern end of the profile and depths ($z$) are measured from the EGM96 reference geoid (Lemoine *et al.* 1998) (≈ mean sea level)):

1. A *c.* 13 km extension in the 2380 kg $m^{-3}$ layer towards the northwest is required between $x = 122-135$ km.
2. A *c.* 13 km-wide body of a higher density (2650 kg $m^{-3}$) is required at $x = 155$ km.
3. Between $x = 260-290$ km, the 2560 kg $m^{-3}$ layer beneath the SE rift flank has been thickened by *c.* 1 km and the 2570 kg $m^{-3}$ layer thickened by *c.* 2 km resulting in a more pronounced step in basement topography at $x = 265$ km, $z = 5$ km.
4. A near-surface higher-density (2770 kg $m^{-3}$) body is required to fit the short-wavelength anomaly observed at $x = 102$ km, $z = -2$ km. This corresponds to the location of Mount Cheleleka; a volcano of possible Oligocene age.
5. The 2790 kg $m^{-3}$ upper crustal layer is thinned by <1 km between $x = 20-50$ km and by *c.* 2 km between $x = 300-320$ km.
6. A second high-density (3000 kg $m^{-3}$) body is required ($x = 175$ km, $z = 10$ km) *c.* 30 km to the NW of the 3000 kg $m^{-3}$ body included in the initial model ($x = 212$ km, $z = 14$ km). This body coincides with a *c.* 2.5 km step in the upper boundary of the 2790 kg $m^{-3}$ layer (at $x = 180$ km, $z = 5$ km) and is modelled to be *c.* 12 km wide by *c.* 10 km deep.
7. A <10 km-wide body is introduced at $x = 202$ km, $z = 6$ km, modelled with a density of 2850 kg $m^{-3}$. It lies directly beneath the topographical high of the Mount Boset volcano, partially within the 2570 and 2790 kg $m^{-3}$ layers.
8. The 2940 kg $m^{-3}$ lower crustal layer is thinned by *c.* 2 km between $x = 0-20$ km.
9. An axial low-density upper mantle or high-density lower crustal zone is modelled as a *c.* 50 km-wide body with a density of 3190 kg $m^{-3}$, centred at $x = 215$ km, $z = 43$ km.
10. The upper mantle density beneath the NW rift flank has to be reduced to 3290 kg $m^{-3}$ to fit the observed anomaly.

The robustness of the velocity–density relationship and final model was explored by varying each polygon density by an amount that displaced the calculated model anomaly values from the observed Bouguer anomaly values by an average of 10 mGal (in the region of the anomaly affected by that polygon). The results are shown as density error bars in Fig. 4 and generally show that the density of the smaller polygons located at or near the surface are less well constrained than the larger polygons derived directly from the seismic model. Variations in the density (up to 300 kg $m^{-3}$) of some, particularly smaller, polygons produce relatively small changes in the calculated anomaly. This result also shows that the initial velocity–density relationship significantly underestimates the final density of the polygons that constitute layers 1, 2, 3, 5, 6 and 8 (Fig. 4; Table 2). This probably reflects the introduction of mafic intrusive rocks into the continental crust. Polygons 5, 6 and 8 (Table 2), which are related to the magmatic segments at the centre of the rift, lie closest to the oceanic velocity–density relationship as defined

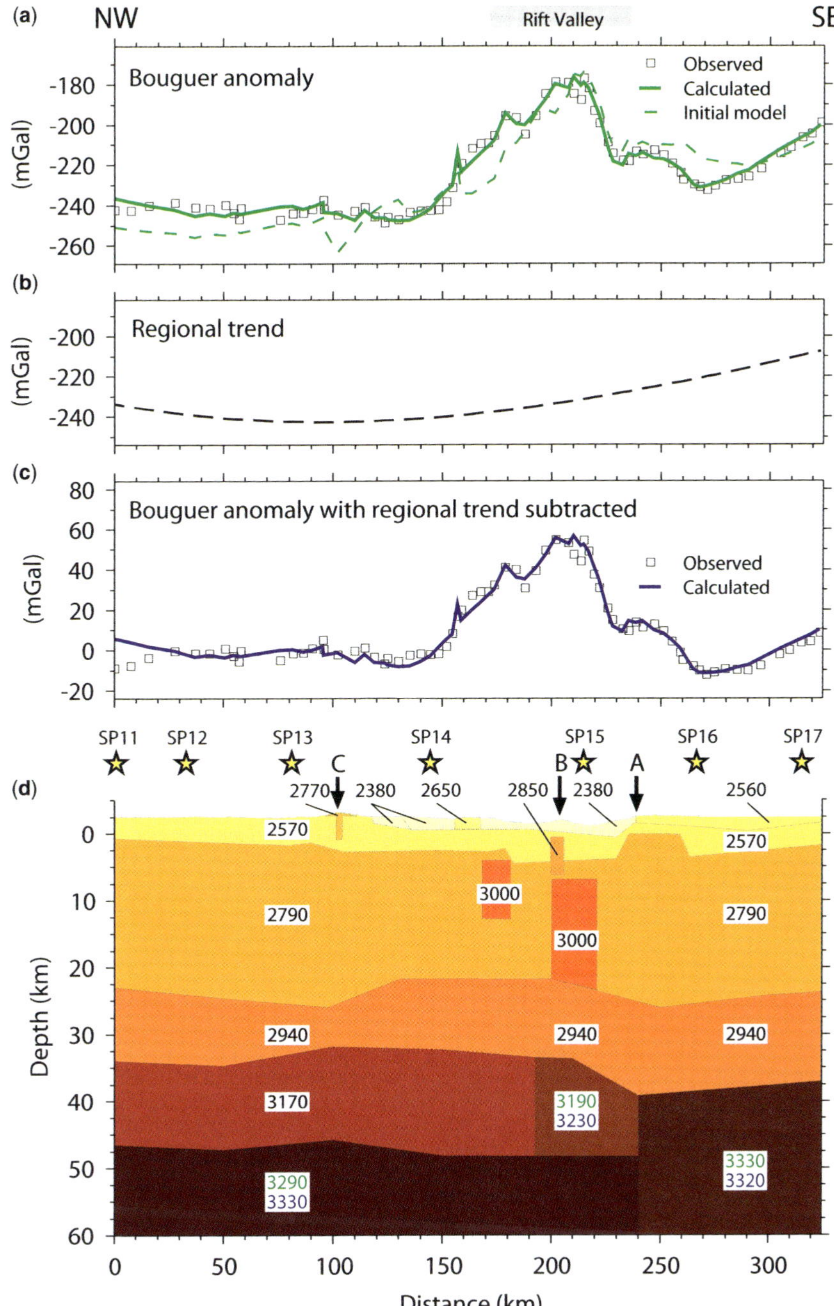


**Fig. 5.** **(a)** Calculated final (solid line) and initial (dashed line) Bouguer anomaly compared with observed Bouguer anomaly (squares) along the EAGLE cross-rift profile. **(b)** Regional long-wavelength trend from low-pass filtering (i.e. wavelengths more than *c.* 80 km) existing Ethiopian gravity data (e.g. Mahatsente *et al.* 1999; Tiberi *et al.* 2005). **(c)** Calculated Bouguer anomaly (solid line) and observed Bouguer anomaly with regional trend subtracted (squares). **(d)** Final density model (values in kg $m^{-3}$). Where two density values are labelled for a particular layer, the upper corresponds to calculated anomaly in (a) and the lower corresponds to calculated anomaly in (c)—see text for details. Locations of the seismic shotpoints (stars) (Maguire *et al.* 2003), Arboye border fault (A), Boset volcano (B), Mount Cheleleka volcano (C) and topographical extent of the rift valley are also shown.

by Bartetzko *et al.* (2005) (Fig. 4). Layer 8 has similar velocity and density properties to those observed for gabbro and diabase (Christensen & Mooney 1995) whilst layer 6 is more comparable to the properties observed for basalt (Christensen & Mooney 1995) (Fig. 4).

### *Mantle densities*

The seismic studies indicate low velocities in the upper mantle beneath the NW part of the profile (Bastow *et al.* 2005). This is supported by modelling of the gravity anomaly. Between 0 km and 190 km distance, the mantle density is lower than would be expected for typical upper mantle. In order to verify that this low density is a real feature and not due to an error in estimating crustal densities, the gravity anomaly along a *c.* 2400 km-long profile (including the EAGLE cross-rift profile) was extracted from existing data across Ethiopia (e.g. Mahatsente *et al.* 1999; Tiberi *et al.* 2005) and low-pass filtered to eliminate wavelengths of less than 80 km (Fig. 5b). This anomaly was subtracted from the observed gravity. This adjusted anomaly (Fig. 5c) should be the result of variable-density crust overlying uniform-density mantle if the long wavelength anomaly is entirely due to density variations in the mantle. Consequently, by adjusting the mantle densities toward a normal mantle density of 3330 kg $m^{-3}$, the calculated anomaly should result in a match between the calculated and the adjusted observed anomaly.

The optimum fit is obtained by raising the density of the axial upper mantle body from 3190 kg $m^{-3}$ to 3230 kg $m^{-3}$, lowering the density of the upper mantle body beneath the southeastern end of the profile from 3330 kg $m^{-3}$ to 3320 kg $m^{-3}$ and raising the density of the body beneath the NW end of the profile to 3330 kg $m^{-3}$ (Fig. 5d). The remaining variations in mantle density (10 kg $m^{-3}$ across the model) can be attributed to uncertainty in the long wavelength anomaly subtracted from the data. This test of the gravity model indicates that, within the limitations of the 2D gravity modelling, the decrease in mantle density to the NW of the rift is a real feature. The magnitude of the decrease is misleading because the model only extends to a depth of 60 km. If the model were extended deeper, the density decrease would still be present, as suggested by the seismic data, but its magnitude would be significantly less.

## Interpretation and discussion

### *Upper sedimentary and volcanic structure*

The uppermost low-density (2380 kg $m^{-3}$) and low-velocity (Mackenzie *et al.* 2005) layer within and to the NW of the rift valley is interpreted to represent interspersed Quaternary lacustrine sediments, Late Pliocene to Early Pleistocene age pyroclastic deposits and Late Miocene to Pliocene age pyroclastic rocks (Wolfenden *et al.* 2004; Abebe *et al.* 2005). The termination of the northwestward extension of this layer in the density model near the Mount Cheleleka volcano (Figs 2 & 5d) is interpreted as the boundary between the pre-rift flood basalts (Fig. 2) and the volcano–sedimentary rift infill. Within this volcano–sedimentary layer, a high-density region (2650 kg $m^{-3}$) corresponds to a surface geology dominated by welded pyroclastic flows with occasional Plio-Pleistocene basalts, underlain by Late Miocene basalts and Oligocene flood basalts (Wolfenden *et al.* 2004; Abebe *et al.* 2005) (Fig. 2). It most likely represents a faulted horst block, locally named the Adama Horst (e.g. Kazmin 1975; Berhe & Kazmin 1978) (Fig. 2) and is marked by a topographical high within the rift. Its location is coincident with the offset between the Aluto–Gedemsa and Boset magmatic segments (Fig. 2). It is also feasible that this high-density feature is a result of deeper, narrow intrusive bodies (similar to that beneath Boset volcano—see below). However, to resolve this feature further is beyond the resolution of the gravity data.

To the SE of the rift, a higher density (2560 kg $m^{-3}$) uppermost layer is justified by Miocene basaltic rocks occurring closer to the surface (Kazmin 1975; Berhe & Kazmin 1978). The most likely composition of the laterally extensive lower supra-basement layer across the model (density 2570 kg $m^{-3}$) is pre-rift Jurassic to Palaeogene sedimentary rocks, combined with Oligocene flood basalts on the northwestern rift flank. The significant basement topography observed on the eastern side of the rift correlates with the uplifted rift flank and is attributed to the Arboye border fault (Fig. 5d) (Mackenzie *et al.* 2005). We suggest that the deep (up to 5 km) basin beneath the southeastern rift flank is a relic Mesozoic extensional basin, following the arguments of Korme *et al.* (2004), who interpret graben structures (e.g. the Welenchiti Graben, Fig. 2) in terms of Mesozoic faulting reactivated by rift extension.

### *Upper crust and volcanic intrusions*

Similar upper crustal velocity (*c.* 6.25 km $s^{-1}$) (Keranen *et al.* 2004; Mackenzie *et al.* 2005) and density (2790 kg $m^{-3}$) values are modelled beneath both rift flanks. These are consistent with values derived for Proterozoic crust beneath the Kenya rift (Maguire *et al.* 1994; Birt *et al.* 1997) and are higher than those associated with Archaean crust. Equivalent velocity and density characteristics beneath both flanks of the NMER suggest that the rift dissects crust of a similar origin (Mackenzie *et al.* 2005).

There is a high-density ($3000\,kg\,m^{-3}$) region between depths of 8–24 km directly beneath the Boset volcano (at *c*. 200 km) (Fig. 5d) with a coincident 5–10% increase in velocity (Fig. 3). These properties would suggest a composition of at least 40% gabbro (e.g. Christensen & Mooney 1995). This body is of similar width (*c*. 20 km) to the surface expression of the Quaternary Boset magmatic segment and the Late Pliocene Bofa flood basalts (Kazmin *et al*. 1980*b*; Abebe *et al*. 2005) (Fig. 2). The width of this body is equivalent to that of the coincident high-velocity upper crustal zone identified by Keranen *et al*. (2004) and to that of a near coincident high conductivity anomaly identified by Whaler & Hautot (2006). The top surface of this body (*c*. 8 km) beneath the rift floor is equivalent to the depth at which Keranen *et al*. (2004) identify the top surface of the high-velocity upper crustal zone, but shallower than the depth (*c*. 15 km) of the conductive body reported by Whaler & Hautot (2006). While Keranen *et al*. (2004) suggested that the body was likely to be a frozen gabbroic intrusion, the deeper high conductivity would suggest that it is still likely to contain some partial melt at depth (Whaler & Hautot 2006). This would be consistent with the crustal shear-wave splitting observations of Keir *et al*. (2005), interpreted in terms of a small degree of partial melt within the crust.

Ebinger & Casey (2001) estimated 35% extension in the rift with as much as 4 km of extension in the upper crust accommodated by dyke intrusion in the past *c*. 1 Ma in the *c*. 20 km-wide magmatic segments. Keranen *et al*. (2004), while not imaging the dykes, state that synthetic seismic modelling indicates that <4 km width of dyking would not be resolved in their data. We model a high-density ($2850\,kg\,m^{-3}$) *c*. 5 km-wide body at a depth between 2 and 7.5 km directly beneath the Boset volcano and above the northwestern edge of the gabbroic intrusion. The density of this body is consistent with a composition of less than 25% gabbroic material (e.g. Christensen & Mooney 1995). There are two possible explanations for this body. It could represent the distribution of dykes proposed by Ebinger & Casey (2001) above the deeper gabbroic, high-velocity intrusion. It is of interest that both our body and the surface magmatic segment are displaced to the NW side of the deep intrusion below *c*. 8 km (Maguire *et al*. 2006). This would suggest that the dykes feeding the magmatic centres result from basaltic lavas at depth having to 'by-pass' cooled, rigid and more-resistive upper levels of the gabbroic intrusion(s) barring vertical ascent to the surface. An alternative explanation for the $2850\,kg\,m^{-3}$ body beneath the Boset volcano is that it represents a body of magma derived by partial fractionation at depth or melting of the existing crust and sediments due to the increased temperature caused by the mafic intrusive material below. This is consistent with the composition of the Pleistocene and Holocene volcanic rocks of the Boset magmatic segment which are mostly (*c*. 80%) felsic eruptive material (Boccaletti *et al*. 1999) derived by partial fractionation from basalts in shallow crustal reservoirs (Hart *et al*. 1989). The very high conductivity anomaly identified by Whaler & Hautot (2006) at very shallow depth beneath the Boset volcano would lend support to the presence of such a magma chamber.

The origin of the small, high-density ($2770\,kg\,m^{-3}$) body beneath the Cheleleka volcano is unknown, but may be due to near-surface dense intrusive material, possibly linked to the Oligocene flood basalt episode.

The second high-density ($3000\,kg\,m^{-3}$) body located beneath the northwestern rift flank does not exhibit a significantly fast velocity anomaly ($5.50–6.00\,km\,s^{-1}$) in the cross-rift seismic model (Mackenzie *et al*. 2005) but a slice through the high-resolution 3D tomography model at 10 km depth reveals an extension of the Koka (a.k.a. Aluto–Gedemsa) magmatic segment (see Fig. 2) high-velocity anomaly to the NE (Keranen *et al*. 2004). This velocity anomaly coincides with the location of the second high-density body and is therefore interpreted as a second, off-axis region of gabbroic intrusive material into the upper crust most likely as a subsurface continuation of the nearby Aluto–Gedemsa magmatic segment. Whaler & Hautot (2006) here identify a resistive upper crustal body. The lack of a high conductivity zone at depth beneath this body would suggest that this intrusion is largely frozen, consistent with the lesser amount of Pleistocene–Holocene surface magmatic activity in this area.

As described above, modelling of the gravity anomaly identifies a number of intrusive bodies beneath the rift in the vicinity of the Boset volcano. The concentration of these intrusions beneath the rift is greater than that which is observed in the seismic data (Keranen *et al*. 2004; Mackenzie *et al*. 2005). This lends support to models for crustal extension and break-up in which there is a clear transition in time between an early tectonic phase, dominated by faulting, to a late stage where extension occurs by the injection of dykes in a narrow region at the centre of the rift (Ebinger & Casey 2001) and that of Buck (2004) in which magma injection weakens the lithosphere to enable rifting to progress.

### *Lower crust*

Mackenzie *et al*. (2005) modelled normal continental lower crust beneath both the eastern and western plateaus with the addition of a *c*. 15 km-thick high-velocity lowest crustal layer beneath the western plateau. They interpreted it to represent Oligocene underplate with possible addition of more recent mafic melt that terminates near the western escarpment of the rift. Such a layer was suggested to exist as a high-density lower crustal body by Berckhemer *et al.* (1975), from joint analysis of gravity and seismic refraction data. This underplate layer, along with a lower-density upper mantle (see below), is required to fit the gravity anomaly along the northwestern rift flank and accounts for the asymmetric Bouguer anomaly values on either side of the rift. The modelled density (3170 kg $m^{-3}$) and velocity (7.40 km $s^{-1}$) for this layer are characteristic of a body with a mafic composition (e.g. Christensen & Mooney 1995).

### *Upper mantle*

The upper mantle is modelled with a typical sub-continental mantle density (3330 kg $m^{-3}$) (e.g. Poudjom Djomani *et al.* 2001) beneath the southeastern rift flank with a reduced density required beneath the northwestern rift flank and the rift axis. These density changes are supported by the observation of average sub-Moho velocities of 8.1 km $s^{-1}$ and less than 8.0 km $s^{-1}$ respectively (Mackenzie *et al.* 2005), indicating a compositional change, a temperature change or a combination of both. A broad low-velocity anomaly imaged by P- and S-wave tomography is concentrated in a region below the northwestern rift flank at *c.* 75 km depth and extends to the southeastern rift wall (Bastow *et al.* 2005). It is interpreted as resulting from anomalously hot upper mantle including local areas of melt or melt ponding and could account for the modelled lower density beneath the northwestern half of the profile.

The final density model indicates that either low-density mantle or high-density underplate is required beneath the rift axis. It is impossible to distinguish between these two interpretations but seismic data strongly suggest the former due to a modelled sub-Moho mantle P-wave velocity of 7.70 km $s^{-1}$ directly beneath the rift (Mackenzie *et al.* 2005). This zone is interpreted as hot mantle material and its density and velocity characteristics are consistent with the presence of 3–5% partial melt, similar to that observed in the Kenya Rift (Mechie *et al.* 1994).

It is possible to identify three lithospheric zones across the model. Between 0 km and 170 km, the upper crust is unaltered, the lower crust is intruded by high-density magmatic underplate and the upper mantle has anomalously low densities. From 170 km to 240 km, the crust is intruded and the upper mantle shows significantly reduced densities. Between 240 km and the end of the profile, the crust is unaltered and the mantle has a typical density of 3330 kg $m^{-3}$ (Poudjom Djomani *et al.* 2001). Both magnetotelluric (MT) (Whaler & Hautot 2006) and earthquake residual travel time studies suggest a difference in properties of the lithosphere across the rift. Such a change in properties may be responsible for constraining the rift, but corresponding lateral changes in density that cannot be attributed to the crustal extension and rifting are not present in the density model.

## Conclusions

We present a new cross-rift 2D crust and upper mantle density model for the northern Main Ethiopian Rift, derived from recent gravity data acquisition. The high quality of these newly acquired data and the regional base station network created for this work will enable the correction of previous data with poorly determined height constraints, as well as facilitating the collection of future gravity data.

A starting model with densities derived from the EAGLE cross-rift wide-angle reflection/refraction seismic velocity model (Mackenzie *et al.* 2005) produces a good first-order fit to the observed Bouguer anomaly profile across the rift (Fig. 5a), supporting the robustness of the velocity model. However, the sensitivity of the gravity method to lateral density changes allows modifications to the starting model to produce a significantly improved fit to the observed data and permits better definition of the rift structure and the distribution of high-density intrusive rocks within the rift. In particular, the gravity modelling shows that the intrusive material is more concentrated than is observed in the seismic models of Keranen *et al.* (2004) and Mackenzie *et al.* (2005).

Beneath the rift axis, low-density and low-velocity uppermost mantle is required by both the gravity anomaly and seismic data. This is interpreted as indicating the presence of partial melt within an upwelling mantle. Low-density upper mantle is also required beneath the northwestern flank of the rift. This coincides with an anomalous low-velocity P- and S-wave region that extends to a depth of at least 300 km (Bastow *et al.* 2005).

The high-density intrusive bodies within the rift are interpreted as mafic ($\geq$40% gabbro) intrusions originating from magma generated by extension and decompression melting of the upper mantle. This is supported by correlations with the conductivity model of Whaler & Hautot (2006). In particular, the results presented here show the existence of a magma body close to the surface beneath the Boset volcano. The spatial relationship of this body to the volcano and other bodies in the mid-crust suggests that the magmatic segments are connected to a magma reservoir deeper in the crust. The subsurface extent of these bodies is coincident with the surface location of the Bofa basalts (Fig. 2), suggesting a possible Late Pliocene age of emplacement. The geometry of this body is consistent with it being a cluster of dykes as suggested by the models of Keranen *et al.* (2004) and Ebinger & Casey (2001). A similar narrow, high-density body required directly beneath the Mount Cheleleka volcano on the edge of the rift is likely to be associated with the earlier Oligocene flood basalt episode. The abrupt change from highly intruded crust and anomalously low-density mantle to typical continental lithosphere near the Arboye rift border fault (Figs 2 & 5d) marks the extent of lithospheric modification by rifting processes. Our density model does not identify pre-existing changes in lithospheric physical properties across the rift that are unattributable to the crustal extension and rifting.

These observations indicate that the magmatism which is accompanying or dominating the rifting process is concentrated into a zone *c.* 50 km wide within the rift beneath the Boset magmatic segment. This magmatic segment is superimposed on a broader zone of tectonic extension. Distributed strain due to faulting is apparently giving way to a narrow zone of extension in which strain is focused into a possibly connected series of magmatic intrusions. This configuration suggests that the profile crosses the rift at the point where extension of the crust by magmatic intrusion is beginning to dominate over tectonic extension, lending support to the models for crustal extension and break-up along magmatic continental margins of Ebinger & Casey (2001) and Buck (2004).

A. Hobbs (NERC Geophysical Equipment Facility), M. Keiding (Danish Lithosphere Centre, University of Copenhagen), M. Zewuge, S. Ashenafi (Ethiopian Geological Survey), A. Ayele (Geophysical Observatory, University of Addis Ababa), P. Denton and A. Brisbourne (SEIS-UK) for their assistance in the data collection. The UK Natural Environment Research Council, the US National Science Foundation Continental Dynamics Program, the Texas Higher Education Coordinating Board, the Royal Society and the University of Leicester provided funding for EAGLE. The NERC Geophysical Equipment Facility and University of Leicester provided instrumentation. D. Cornwell was awarded a Geological Society of London grant from the Timothy Jefferson Field Research Fund to complete the fieldwork. Figures 1, 3, 4 and 5 were plotted using the Generic Mapping Tools software (Wessel & Smith 1995). Reviews by C. Swain and C. Tiberi and editorial comments from C. Ebinger helped to improve the manuscript and are gratefully acknowledged.

## References

ABEBE, T., MANETTI, P., BONINI, M., CORTI, G., INNOCENTI, F., MAZZARINI, F. & PECKSAY, Z. 2005. Geological map (scale 1: 200,000) of the northern Main Ethiopian Rift and its implications for the volcano-tectonic evolution of the rift. *Geological Society of America Map and Chart Series MCH094*, 20pp.

ACOCELLA, V. & KORME, T. 2002. Holocene extension direction along the Main Ethiopian Rift, East Africa. *Terra Nova*, **14**, 191–197.

ACTON, G.D., TESSEMA, A., JACKSON, M. & BILHAM, R. 2000. The tectonic and geomagnetic significance of paleomagnetic observations from volcanic rocks from central Afar, Africa. *Earth and Planetary Science Letters*, **180**, 225–241.

BARTETZKO, A., DELIUS, H. & PECHNIG, R. 2005. Effect of compositional and structural variations on log responses of igneous and metamorphic rocks I: mafic rocks. *In*: HARVEY, P.K., BREWER, T.S., PEZARD, P.A. & PETROV, V.Y. (eds) *Petrophysical Properties of Crystalline Rocks*. Geological Society, London, Special Publications, **240**, 255–278.

BASTOW, I.D., STUART, G.W., KENDALL, J.-M. & EBINGER, C.J. 2005. Upper mantle seismic structure in a region of incipient continental breakup: northern Ethiopian Rift. *Geophysical Journal International*, **162**, 479–493.

BERCKHEMER, H., BAIER, B., ET AL. 1975. Deep seismic soundings in the Afar region and on the highland of Ethiopia. *In*: PILGER, A. & ROESLER, A. (eds) *Afar Depression of Ethiopia*. Schweizerbart, Stuttgart, 89–107.

BERHE, S. & KAZMIN, V. 1978. *Nazret sheet NC37-15 scale 1: 250,000*. Ethiopian Institute of Geological Surveys, Addis Ababa.

BERHE, S.M. 1990. Ophiolites in Northeast and East-Africa—Implications for Proterozoic Crustal Growth. *Journal of the Geological Society*, **147**, 41–57.

BILHAM, R., BENDICK, R., LARSON, K., MOHR, P., BRAUN, J., TESFAYE, S. & ASFAW, L. 1999. Secular and tidal strain across the Main Ethiopian Rift. *Geophysical Research Letters*, **26**, 2789–2792.

BIRT, C.S., MAGUIRE, P.K.H., KHAN, M.A., THYBO, H., KELLER, G.R. & PATEL, J. 1997. The influence of pre-existing structures on the evolution of the southern Kenya Rift Valley—evidence from seismic and gravity studies. *Tectonophysics*, **278**, 211–242.

BOCCALETTI, M., MAZZUOLI, R., BONINI, M., TRUA, T. & ABEBE, B. 1999. Plio-Quaternary volcano-tectonic activity in the northern sector of the Main Ethiopian Rift: relationships with oblique rifting. *Journal of African Earth Sciences*, **29**, 679–698.

BOSELLINI, A., RUSSO, A. & ASSEFA, G. 2001. The Mesozoic succession of Dire Dawa, Harar Province, Ethiopia. *Journal of African Earth Sciences*, **32**, 403–417.

BROWN, G.F. 1970. Eastern margin of Red Sea and coastal structures in Saudi Arabia. *Philosophical Transactions of the Royal Society of London Series*, **267**, 75–87.

BUCK, R. 2004. Consequences of asthenospheric variability on continental rifting. *In*: KARNER, G.D. TAYLOR, B., DRISCOLL, N.W. & KOHLSTEDT, D.L. (eds) *Rheology and Deformation of the Lithosphere at Continental Margins*, Columbia University Press, pp. 1–30.

BUSBY, J.P. 1987. An interactive Fortran 77 program using GKS Graphics for 2.5D modelling of gravity and magnetic data. *Computers and Geosciences*, **13**, 639–644.

CHRISTENSEN, M.I. & MOONEY, W.D. 1995. Seismic velocity structure and composition of the continental- crust—a global view. *Journal of Geophysical Research*, **100**, 9761–9788.

CHU, D.H. & GORDON, R.G. 1999. Evidence for motion between Nubia and Somalia along the southwest Indian Ridge. *Nature*, **398**, 64–67.

COCHRAN, J.R. & MARTINEZ, F. 1988. Evidence from the Northern Red Sea on the transition from continental to oceanic rifting. *Tectonophysics*, **153**, 25–53.

COGBILL, A.H. 1990. Gravity terrain corrections calculated using digital elevation models. *Geophysics*, **55**, 102–106.

D'ACREMONT, E., LEROY, S., BESLIER, M.O., BELLAHSEN, N., FOURNIER, M., ROBIN, C., MAIA, M. & GENTE, P. 2005. Structure and evolution of the eastern Gulf of Aden conjugate margins from seismic reflection data. *Geophysical Journal International*, **160**, 869–890.

DE CHABALIER, J.B. & AVOUAC, J.P. 1994. Kinematics of the Asal Rift (Djibouti) determined from the deformation of Fieale volcano. *Science*, **265**, 1677–1681.

EAGLES, G., GLOAGUEN, R. & EBINGER, C. 2002. Kinematics of the Danakil microplate. *Earth and Planetary Science Letters*, **203**, 607–620.

EBINGER, C.J. 1991. Gravity data acquisition, reduction and error estimation. *Interim Report to the Ethiopian Institute of Geological Surveys*, 9pp.

EBINGER, C.J. & CASEY, M. 2001. Continental breakup in magmatic provinces: An Ethiopian example. *Geology*, **29**, 527–530.

EBINGER, C.J., YEMANE, T., WOLDEGABRIEL, G., ARONSON, J.L. & WALTER, R.C. 1993. Late Eocene–Recent volcanism and Faulting in the Southern Main Ethiopian Rift. *Journal of the Geological Society*, **150**, 99–108.

HAMMER, S. 1939. Terrain corrections for gravimeter stations. *Geophysics*, **4**, 184–194.

HART, W., WOLDEGABRIEL, G., WALKER, R. & MERTZMAN, S. 1989. Basaltic volcanism in Ethiopia: Constraints on continental rifting and mantle interactions. *Journal of Geophysical Research*, **94**, 7731–7748.

HOFMANN, C., COURTILLOT, V., FERAUD, G., ROCHETTE, P., YIRGU, G., KETEFO, E. & PIK, R. 1997. Timing of the Ethiopian flood basalt event and implications for plume birth and global change. *Nature*, **389**, 838–841.

KAZMIN, V. 1975. *Explanatory note to the geology of Ethiopia*. Ethiopian Institute of Geological Surveys, Addis Ababa.

KAZMIN, V., SHIFFERAW, A. & BALCHA, T. 1978. The Ethiopian basement: stratigraphy and possible manner of evolution. *International Journal of Earth Sciences (Geologische Rundschau)*, **67**, 531–546.

KAZMIN, V., BERHE, S.M. & WALSH, J. 1980. *Report on the Geological Map of the Ethiopian Rift Valley*. Ethiopian Institute of Geological Surveys, Addis Ababa.

KEIFFER, B., ARNDT, *ET AL*. 2004. Flood and shield basalts from Ethiopia: Magmas from the African Superswell. *Journal of Petrology*, **45**, 793–834.

KEIR, D., KENDALL, J.-M., EBINGER, C.J. & STUART, G.W. 2005. Variations in late syn-rift melt alignment inferred from shear-wave splitting in crustal earthquakes beneath the Ethiopian rift. *Geophysical Research Letters*, **32**, L23308, doi: 10.1029/2005GL024150.

KERANEN, K.M., KLEMPERER, S.L., GLOAGUEN, R. & The EAGLE Working Group, 2004. Imaging a proto-ridge axis in the Main Ethiopian Rift. *Geology*, **32**, 949–952.

KIDANE, T. *ET AL*. 2003. New paleomagnetic and geochronologic results from Ethiopian Afar: Block rotations linked to rift overlap and propagation and determination of a similar to 2 Ma reference pole for stable Africa. *Journal of Geophysical Research*, **108**, 2102–2122.

KORME, T., ACOCELLA, V. & ABEBE, B. 2004. The role of pre-existing structures in the origin, propagation and architecture of faults in the Main Ethiopian Rift. *Gondwana Research*, **7**, 467–479.

LEMOINE, F.G. *ET AL*. 1998. *The Development of the Joint NASA GSFC and NIMA Geopotential Model EGM96*. NASA Goddard Space Flight Center, Greenbelt, Maryland, 20771 USA.

LUDWIG, J.W., NAFE, J.E. & DRAKE, C.L. 1970. Seismic refraction. *In*: MAXWELL A.E. (ed.) *The Sea*. Wiley, New York, 53–84.

MACKENZIE, G.D., THYBO, H. & MAGUIRE, P.K.H. 2005. Crustal velocity structure across the Main Ethiopian Rift: Results from 2-dimensional wide-angle seismic modelling. *Geophysical Journal International*, **162**, 994–1006.

MAGUIRE, P.K.H., SWAIN, C.J., MASOTTI, R. & KHAN, M.A. 1994. A crustal and uppermost mantle cross-sectional model of the Kenya Rift derived from seismic and gravity data. *Tectonophysics*, **236**, 217–249.

MAGUIRE, P.K.H., EBINGER, C.J., *ET AL*. 2003. Geophysical project in Ethiopia studies continental breakup. *EOS Transactions*, American Geophysical Union, **84**, 342–343.

MAGUIRE, P.K.H., KELLER, G.R., *ET AL*. 2006. Crustal structure of the Northern Main Ethiopian Rift from the EAGLE controlled-source survey; a snapshot of incipient lithospheric break-up. *In*: YIRGU, G., EBINGER, C.J. & MAGUIRE P.K.H. (eds) *The Afar Volcanic Province within the East African Rift System*. Geological Society, London, Special Publications, **259**, 269–291.

MAHATSENTE, R., JENTZSCH, G. & JAHR, T. 1999. Crustal structure of the Main Ethiopian Rift from gravity data: 3-dimensional modeling. *Tectonophysics*, **313**, 363–382.

MAKRIS, J. & GINZBURG, A. 1987. The Afar Depression—Transition between continental rifting and sea-floor spreading. *Tectonophysics*, **141**, 199–214.

MAKRIS, J., MENZEL, H., ZIMMERMANN, J. & GOUIN, P. 1975. Gravity field and crustal structure of north Ethiopia. *In*: PILGER, A. & ROESLER, A. (eds) *Afar Depression of Ethiopia*. Schweizerbart, Stuttgart, 135–144.

MANIGHETTI, I., TAPPONIER, P., GILLOT, P.Y., COURTILLOT, V., JACQUES, E., RUEGG, J.C. & KING, G. 1998. Propagation of rifting along the Arabia-Somalia plate boundary: into Afar. *Journal of Geophysical Research*, **103**, 4947–4974.

MCCLUSKY, S., REILINGER, R., MAHMOUD, S., BEN SARI, D. & TEALEB, A. 2003. GPS constraints on Africa (Nubia) and Arabia plate motions. *Geophysical Journal International*, **155**, 126–138.

MECHIE, J., FUCHS, K. & ALTHERR, R. 1994. The relationship between seismic velocity, mineral-composition and temperature and pressure in the upper mantle—with an application to the Kenya Rift and its eastern flank. *Tectonophysics*, **236**, 453–464.

NAFE, J.E. & DRAKE, C.L. 1957. Variation with depth in shallow and deep water marine sediments of porosity, density and the velocities of compressional and shear waves. *Geophysics*, **22**, 523–552.

PEDLEY, R.C., BUSBY, J.P. & DABEK, Z.K. 1993. GRAVMAG User Manual: Interactive 2.5D gravity and magnetic modelling. *Technical Paper WK/93/26/R*, British Geological Survey.

PIK, R., MARTY, B., CARIGNAN, J. & LAVE, J. 2003. Stability of the Upper Nile drainage network (Ethiopia) deduced from (U–Th)/He thermochronometry: implications for uplift and erosion of the Afar plume dome. *Earth and Planetary Science Letters*, **215**, 73–88.

POUDJOM DJOMANI, Y.H., O'REILLY, S.Y., GRIFFIN, W.L. & MORGAN, P. 2001. The density structure of subcontinental lithosphere through time. *Earth and Planetary Science Letters*, **184**, 605–621.

PURCELL, P.G. 1976. The Marda fault zone, Ethiopia. *Nature*, **261**, 569–571.

RUEGG, J.C. & KASSER, M. 1987. Deformation across the Asal–Ghoubbet Rift, Djibouti, Uplift and Crustal Extension 1979–1986. *Geophysical Research Letters*, **14**, 745–748.

STEIN, R.S., BRIOLE, P., RUEGG, J.C., TAPPONNIER, P. & GASSE, F. 1991. Contemporary, Holocene, and Quaternary deformation of the Asal Rift, Djibouti—Implications for the mechanics of slow spreading ridges. *Journal of Geophysical Research*, **96**, 21789–21806.

SWAIN, C.J. & KHAN, M.A. 1977. A catalogue of gravity measurements in Kenya. Unpublished catalogue, University of Leicester.

TIBERI, C., EBINGER, C.J., BALLU, V., STUART, G.W. & OLUMA, B. 2005. Inverse models of gravity data from the Red Sea–Aden–East African rifts triple junction zone. *Geophysical Journal International*, **163**, 775–787.

VAIL, J.R. 1983. Pan-African crustal accretion in north-east Africa. *Journal of African Earth Sciences*, **1**, 285–294.

WESSEL, P. & SMITH, W.H.F. 1995. New version of the Generic Mapping Tools released. *EOS Transactions*, American Geophysical Union, **76**, 329.

WHALER, K.A. & HAUTOT, S. 2006. Magnetotelluric studies of the northern Ethiopian Rift (EAGLE Phase III). *In*: YIRGU, G., EBINGER, C.J. & MAGUIRE, P.K.H. (eds) *The Afar Volcanic Province within the East African Rift System*. Geological Society, London, Special Publications, **259**, 293–305.

WOLDEGABRIEL, G., ARONSON, J.L. & WALTER, R.C. 1990. Geology, geochronology, and rift basin development in the central sector of the Main Ethiopia Rift. *Geological Society of America Bulletin*, **102**, 439–458.

WOLFENDEN, E., EBINGER, C., YIRGU, G., DEINO, A. & AYALEW, D. 2004. Evolution of the northern Main Ethiopian Rift: birth of a triple junction. *Earth and Planetary Science Letters*, **224**, 213–228.

# Index

Page numbers in *italic* denote figures. Page numbers in **bold** denote tables.


Adama basin 194, *299*, 300
Afar *134*
crustal structure 23, 26, 239–49, 273
depression
basalt 83–4, 87
geology 24–5
plate kinematics 27–38
reconstruction
advanced plate divergence 35-6
continental break-up 25–7
initial rifting 33–5
pre-rifting 24–5, *25*, 31–3
evolutionary 36–7
rhyolites 121–8
silicic volcanism
age 123
petrology 124
tectonics 23, *24*, 123
mantle plume 3–4, 23, 51, 57, 84, 86, 87–9, 95–7, 109, 111–13, 133
triple junction 2, 23, *24*, 25, *25*, 36, 43, 123, 133, 136, 137, 144, 239–40
Afar volcanic province 1, *2*, 3
*see also* Ethiopia–Yemen plateau, flood basalt province
Afro-Arabian rift system 43, *44*
evidence from Gewane earthquake swarm 137, 140
magmatism 43–52, 88–9
Aisha block 23, *24*
reconstruction 25, 31–2, 34, 35–6
Ali Sabieh block *see* Aisha block
Aluto volcano 147
Aluto–Gedemsa magmatic segment *145*, *146*, *150*, 159, 317
faults 147, *149*, 153, 154, 155
*see also* Koka magmatic segment
Amaro basalts 84–6, 88
Angelele magmatic segment *145*, *146*, 147
faults *149*, 151, 153, 155, *158*
reflection/refraction profile 273
anisotropy, seismic 18, *19*, 57–68, 271, 288
LPO vs. OMP 58, *59*, 63–4
Anza rift 217, *218*, 219, *221*, *222*, 223, 225, 226
Arabian Plate 9, *10*
Arabian-Somalia kinematics 9, 28–29, **30**, 34, 36
Nubia–Arabia kinematics 9, 27–9, **30**, 31, 34, 35, 36, 136, 137, 140
Arboye rift border fault 283, *299*, 300
Asal–Ghoubbet rift 159–60
Asela fault *193*, 197, *198*, 199–200, 203, 204
asthenosphere, mantle 82, 310
asthenosphere–lithosphere flow models 18, *20*, 66
Ayelu–Amoissa lineament 136, 137, 139

basalt 77–8
Afar depression 83–4, 86–7
Ethiopian Rift 86, 98, **99–104**, 105–13
flood 1, 2, 25, 31, 32, 35, 43
Ethiopian 1, 83, 84–5, 95–7, 108, 144, 193–4
magmatism and rifting 43, 121
Getra–Kele 86
Kenya rift 81–2, 86–7
ocean island 81, 82, 87
southern Ethiopian 84–6, 96
Turkana 88, 89, 98, *108*
U-series isotope analysis 86–7
Western rift 80–2
Beseka Lake, uplift 185–186, 189
Bishoftu volcanic chain *146*, 147, 263, *294*, *299*, 301
Boset magmatic segment *145*, *146*, 261–4
gabbroic mafic intrusion 285, 300–1, 302, 310, 317
magnetotelluric survey 293, *299*, 300, 301–3
receiver function analysis 255
reflection/refraction profiles 273–4, *279*, 283, 285, 310
Boset volcano *96*, 97, **101–2**, 106, 144, 148, *150*, 156
intrusions 302, 317
Boset–Kone magmatic segment *150*, 159
faults 147–8, *149*, 151–5
vertical deformation 185–6
Bouguer anomaly
Afar depression 26
Main Ethiopian Rift 147
northern Main Ethiopian Rift 311–15
Red Sea 26
Turkana depression 88
break-up, continental
Afar 25–7
models 3
Buluk fault zone 213, *214*, *223*, *224*, 226, 228
Butajira lineament *146*, 147, 263

Chew Bahir graben 213, *214*, *221*, 224, 226
Choke volcanic centre *96*, 98, 106
cinder cones 156
coupling, viscous 18, *20*, 66
crust
extension 23, 30, 287–8
lower, anomalous 284–5
structure
Afar 26, 239–49
northern Main Ethiopian Rift
controlled-source survey 269–89
gravity survey 307–19
receiver function studies 253–65

Danakil block 23, *24*, *134*
reconstruction 25, 26, 29, 31–2, 34, 35–7
Davie ridge *14*, 15, 16–17
Dead Sea transform 29, 31, 43
Debre Zeit lineament *146*, 147, 149
*see also* Bishoftu volcanic chain
deformation
brittle 48
kinematic model 9–20
vertical, Main Ethiopian Rift 185–9
Wonji Fault Belt, GPS monitoring 201–4
demagnetization
alternating field 169–70, *171*, 172
thermal 169, *170*, 172

density model 312–16
divergence, plate 29–30, 33–6
Djibouti
crustal structure 242–9
GEOSCOPE station ADT 239–49, *240*
Dofan volcano *96*, 97, **99–100**, 106, 148, 150, *150*, 156–7
dykes
magmatic rifting 45–8, 51, 240–1, 317
propagation 49–50, 159
Red Sea 34, 44
straight 48–9

EAGLE project 156, 168, *308*
controlled-source experiment 136, 270, *272*, 273–89
gravity survey 307–19
magnetotelluric survey 293–303
passive experiment stations *254*, 256
seismic anisotropy *56*, 57, 63, 65
earthquakes
Gewane swarm 133–40
shear-wave splitting 64–6, 67
slip vectors *10*, 13, 14, 15, 17, 66
East African plateau 77, *78*, *79*
magmatism and rifting 79–82
East African Rift System 1, 2, 55, *56*, 57, *166*, *270*
GPS data 11, **12**
kinematics 9, 14, *14*, 15–20
seismic anisotropy 57–68
Eastern Rift 15, 18, *19*, *20*, 196, 197, 202
*see also* Kenya Rift
Ethiopia Afar Geoscientific Lithospheric Experiment
*see* EAGLE project
Ethiopia–Yemen plateau
flood basalt province 1, 95–7, *96*, 253
*see also* Afar volcanic province
volcanism 1–2, 25
Ethiopian flood basalts 83, 84–5, 107, 193–4
Ethiopian plateau 77, *78*, *79*
magmatism and rifting 78, 83–6
Ethiopian Rift *see* Main Ethiopian Rift
Euler poles
Arabia–Nubia **27**, **28**, 29, **30**, 34
Somalia–Arabia 28, **28**, 29, **30**, 34
Somalia–Nubia 9, *10*, **28**, 29, **30**, 34
Victoria microplate *14*
extension, crustal
dynamics 226–31
initiation 2
NMER 286–7, 316
*see also* rifting

Fantale magmatic segment *see* Fantale–Dofan magmatic segment
Fantale volcano *96*, 97, **99–100**, 106, 144, 148, *150*, *151*, 185
Fantale–Dofan magmatic segment *145*, *146*, 147, 261, 263
faults 148, *149*, 150, 151–5, *157*
reflection/refraction profile 274, 284
faults
detachment 3
Oligocene 26
Quaternary
Main Ethiopian rift 143–60, *146*
down-dip length 155–6
Wonji Fault Belt 136, 146–60, 196–200, *201*
and stress 45, 48
Turkana rift 212–13
fault interaction types 226, *227*, 228
force
magmatic 45–8
tectonic 45

gabbro, mafic intrusion 286, 287, 300–1, 302, 310, 317
Gademotta formation, aeolian sediment 195
Gamo basalts 84–5
garnet 81, 82, 84
Gedemsa caldera 147
Gedemsa magmatic segment 147, 274, 283
geochronology, Main Ethiopian Rift 176, **177**
geomagnetic poles, virtual 179
GEOSCOPE station ADT, Djibouti 239–249, *240*
Gewane earthquake swarm 133–40
Global Positioning System *see* GPS monitoring
GPS monitoring
Nubia–Arabia plate boundary 27, *28*
Somalia–Nubia plate boundary 9, *10*, 11–12, *14*, 15–17
Wonji Fault Belt 191
gravity anomaly 311–19
*see also* Bouguer anomaly
gravity survey, northern Main Ethiopian Rift 307–19
Guguftu volcanic centre *96*, 98, 106
Gulf of Aden
initiation 2, 23, 26, 28, 123
magnetic anomalies 30
reconstruction *25*, 29, 34, 36–7
Gulf of Tadjoura 31, 36, 239, *240*, *241*, 242, 248

Hanang volcano, Tanzania *150*, 157
high-field-strength elements, enrichment 80, 81, 82
HIMU (high $\mu$) signature 84, 88
hotspots 51, 97
*see also* mantle plumes
HTI *see* isotropy, horizontal transverse

isotopes, radiogenic,
analysis
Afar depression 83–4, 124, **126**, 127–8
East African plateau 80–2
rhyolite 124, **126**, 127–8
U-series, Kenya Rift and Afar 86–7
Ethiopian rift basalt **99–104**, 105–8, 109–111
isotropy
horizontal transverse (HTI) 58
vertical transverse (VTI) 58, 63, 64, 65, 67

K–T experiments 172, *174*
Kataboi fault zone 213, *215*, *223*, *224*, 226, 228
Kenya rift 79–80
basaltic magma composition 80–2
U-series isotope analysis 86–7
*see also* Eastern Rift
Kibaran belt 79, 80
kinematics, plate 9–20
Afar region 23, 24, 26–38, 196–7
Wonji Fault Belt 168, 191

Kino Sogo graben 213
Kivu province, basalt *81*, 82
Koka magmatic segment *186*, 317
  *see also* Aluto–Gedemsa magmatic segment
Kone volcano *96*, 97, **99–102**, 106, 148, 261
Kone–Gariboldi volcanic complex 187, 189

lake evolution, Wonji Fault Belt 194–5
Langano Lake 194, 195
lava
  felsic, Main Ethiopian Rift 121, 124, 317
  mafic, Ethiopian rift 98–113
    *see also* gabbro, mafic intrusion
  palaeomagnetic sampling 168–169
layering, periodic thin (PTL) 57, 58
lithophile elements, large ion, enrichment 80
lithosphere
  magmatic rifting 45–8, 50, *51*
  mantle 50, *51*, 82
Lothidok–Kino Sogo province 219, 223, 224, 226, 228, 229
Love waves 58, 63
LPO *see* orientation, lattice-preferred

magma
  alkaline 83, 86
  mafic, Boset magmatic segment 285, 288, 300–301
  silicic, generation 126–128
magma chamber, size and depth 49
magmatism
  Afro-Arabian rift system 43–52
    migration 88–9
  basaltic
    East African Plateau 79–82
    Ethiopian Plateau 83–86
  continental flood basalt 121
  Main Ethiopian rift 146–51
  relationship with rifting 121
  relationship with strain 228–230
  *see also* rifting, magmatic
magnetization
  anomalies, Gulf of Aden 30
  characteristic remanent 172, 175
  natural remanent 169–170, 172
Main Ethiopian Rift 13, *14*, 15, 28, *79*, 165–167
  basalts 86
    composition 97–98, **99–104**, 105–108
  crustal thickness 144
  development 51, 134, 136, 138, 144–145, 271
  fault systems 143–160, *146*
  geochronology 176, **177**
  GPS monitoring 191–204
  initiation 2, 28, 123
  magmatic segments 146–151
  northern
    crustal extension 271, 285–287
    crustal structure
      anomalous lower crust 284–285, 287
      controlled-source seismic survey 269–289
        modelling 276–284
      electrical resistivity 293–303
      gravity survey 307–319
        density model 312–316
      receiver function studies 253–265, 301
    gabbroic mafic intrusion 300–301, 302, 310, 317
    geology *309*
    magmatic segmentation 271, *272*, 283, 285
    mantle 285, 287–288, 310, 316, 318
    partial melt 284, 287, 301, 302, 310–311, 317
    surface-wave anisotropy 63–64, 66
    tectonic setting 310
  palaeomagnetism 168–181
  plate divergence 29–30, 34–35
  rhyolite 121–128
  strain accommodation 143–160, 167, 177–179, 199
  tectonic setting 308, 310
  vertical deformation 185–189
Malawi rift, kinematics *14*, 15–17
mantle
  composition 82, 284
  density 316, 318
  flow 18, *20*, 66
  partial melt 284, 287, 301, 302, 310–311, 317
  reflector 284, 287–288
  uplift 45, 287
  upwelling 18, 55, 57, 66, 77
mantle plumes 77, 78
  Afar 3–4, 23, 51, 57, 84, 85, 86, 87–89, 95–97, 109, 111–113, 133
  head-tail transition 111–112
  HIMU 84, 88
  Kenya 86, 88, 288
  single vs. multiple plume models 3, 66, 87–89, 96, 112–113
  Tanzanian 57, 88–89
Marda lineament 25, 32
Mekele graben 25, 31
melt pockets, oriented (OMP) 57, 58, *59*, 60, 63–64, 67
Moho depth
  Afar *240*, 241–246, 248, 273
  northern Main Ethiopian Rift 254–5, 261, 263, 264, 271, 288
Mozambique belt 60, 79, 81, 273, 288
Mweru rift *14*, 15, 17

Nazret magmatic segment *see* Boset–Kone magmatic segment
N'Doto–Karisia fault zone 213, *216*, 223, 224, 226, 229
Nubian Plate 9, *10*, 60
  GPS data 11, **12**
  Nubia–Arabia kinematics 27, 28, **28**, 29, **30**, 31, 34, 35, 36, 137, 139–140
  Rovuma–Nubia kinematics 16–17
  Somalia–Nubia kinematics 9, *10*, 13–14, 15–16, 28, **28**, 29, **30**, 34, 168
  Victoria–Nubia kinematics 15

olivine, crystal LPO 58, 66
OMP *see* melt pockets, oriented
orientation, lattice-preferred (LPO) 57, 58, *59*, 63, 66, 67

P-waves
  S-wave conversion 255–7, 263
  velocity 58, *59*
    Afar 241–2, *244–5*, 246, *247*, 248

palaeomagnetism
  Main Ethiopian Rift (MER) 168–81
    direction 172, 175, *181*
palaeosecular variation studies 179
Pan-African orogeny 24
  fabric 67–68
peridotite, olivine crystal LPO 58, *59*
Poisson's ratio
  Afar 241, 242, **243**, 246, *247*, 248–9
  Main Ethiopian Rift 29, 255, 257, 262–264, 301
potassium enrichment 80
PTL *see* layering, periodic thin

rare earth elements
  fractionation
    East African plateau 80, 81–2
    Ethiopian flood basalts 84
    Ethiopian rift basalts 86, 98, *106*
    southern Ethiopian basalts 85
Rayleigh waves 58, 60, 63, 242, 246, *247*
receiver function analysis 242–248, 255–259, 262, 263, 264, 301
Red Sea
  constraints 29
  dyke system 34, 44
  initiation 2, 23, 26, 27, **27**, 28, 29, 30–31, 34, 36, 37, 123
  reconstruction *25*
  rhyolite 121–128
remagnetization, viscous 172
resistivity, electrical 293–303
rhyolite, syn-rift, Afar depression 121–8
rifting
  Cenozoic 2–3, 25–27
    Turkana rift 219–226, 230
  driving forces 44–8, 57
  magmatic 43, 45–48, 50, 51, 66–67, 79–86, 240–241, 263, 287
  and seismic anisotropy 57–68
  tectonic 45, *46*, *48*, 51
  transtensional 158–9
rotation, tectonic 177–9
Rovuma block, kinematics *14*, 15–17, *16*, *17*
Ruhuhu graben 15
Rukwa rift *14*, 15
Rungwe province, basalt *81*, 82

S-waves *see* shear-waves
Sabure magmatic segment *see* Fantale–Dofan magmatic segment
sediment
  aeolian, Gademotta formation 194
  lacustrine, Wonji Fault Belt 193, 194–195
segmentation, magmatic 3, 136, 137, 146–151, 159–160, 167–168, 283, 285
seismicity 14, 15, 17, 18
shear-waves
  SKS splitting 18, *19*, 57, 58, *59*, 60–63, 66, 263, 273
  splitting, in earthquakes 64–6, 67
  velocity 58, 246
Shukra el-Sheikh fracture zone 26, 30, 34, 36
Sinai triple junction 27, 29
slip vectors, eathquake *10*, 13, 14, 15, 17, 66
Somalian Plate 9, *10*, 60
  GPS data 11, **12**
  Somalia–Arabia kinematics 9, 28, 29, **30**, 34, 36
  Somalia–Nubia kinematics 9, *10*, 13–14, 15–17, 28, 29, **30**, 34, 168
  Victoria–Somalia kinematics 15, 17
spreading, sea-floor
  Gulf of Aden 26, 30
  Red Sea 37
  Wonji fault belt 167–168
strain, indicated by volcanoes 156–158
strain accommodation 3, 57, 137, 143–160, 167–168, 177–179, 199, 249, 263
stress, extensional 45–47, 48, 50
Suez rift 29, 43
Suguta graben 211, 213, 225
superplume, African 18, *20*, 55, 57, 60, 67, 96, 271, 273
superswell
  African 57, 66, 77, *78*
  Ethiopian 84
surface-waves
  seismic anisotropy 57
    regional anisotropy 60
      Northern Ethiopian Rift 63–64
susceptibility vs. temperature (K–T) experiments 172, *174*

Tanzanian craton 15, 18, *19*, *20*, 43, *44*, *51*, 61–62, 79, 81, 82
  mantle plume 57, 88–89
Tendaho–Goba'ad discontinuity *134*, 136, *145*
thinning, crustal 121, 283–284, 285
tholeiite 83, 98
titanium, Ethiopian flood basalts 84–85
trace element enrichment 80, 81–82
trachyte 124
traps *see* basalt, flood
triple junction *see* Afar triple junction
Turkana basalts 88, 89, 98, *108*
Turkana rift 209–31
  fault interaction types 226, *227*, 228
  geology
    Cenozoic *212*, 213, 219–226
    Cretaceous 217, *218*, 219
    Proterozoic basement 213, 215, 217
  rifting stages 219–26, *227*, 230
    controlling factors 226, 228–30
  strain/magmatism relation 228–30
  transverse fault zones 212–13

U-series isotope analysis 86–7
Ukerewe–Nyanza block *see* Victoria block
underplating 264, 283, 285–286, 287, 301, 302, 318
uplift
  Lake Beseka 185–186, 189
  mantle 45, 287
Usangu graben 15

Victoria microplate, kinematics *14*, 15, 17, *17*, 18, *20*
Virunga province, basalt *81*, 82
volcanism
  Ethiopian Volcanic Province 95–6, 144
  Ethiopian–Yemen Plateau 1–2, 25–26, 31–32, 77, 97–113

felsic 2, 121, 124
mafic 97–113, 211
Miocene 34–35, 83
silicic 121–128
volcanoes, as strain indicators 156–158
VTI *see* isotropy, vertical transverse

Western Rift 15, *19*, 79–80
basaltic magma composition 80–82
extension 18

Wonji Fault Belt 86, 97, *134*, 136, *145*, 146–147, 158, 167, *192*
GPS monitoring 191, 193, 201–204
lake evolution 194–5
mechanism of faulting 196–7, 199–200, *201*
present-day kinematics 195–6

Zambezi rift *14*, 15
Ziway Lake *193*, 194, *195*, 198, 203, 204